electronic devices and circuit theory

electronic devices and circuit theory

second edition

PRENTICE-HALL, INC. ENGLEWOOD CLIFFS, NEW JERSEY 07632

ROBERT BOYLESTAD

Professor
Queensborough College (City University of New York)

LOUIS NASHELSKY

Professor
Queensborough College (City University of New York)

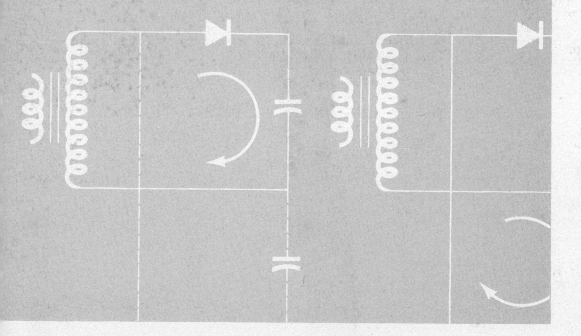

Library of Congress Cataloging in Publication Data

Boylestad, Robert L
 Electronic devices and circuit theory.

 Includes index.
 1. Electronic circuits. 2. Electronic apparatus and appliances. I. Nashelsky, Louis, joint author. II. Title.
TK7867.B66 1978   621.3815   77–7565
ISBN 0–13–250340–9

ELECTRONIC DEVICES AND CIRCUIT THEORY, *second edition*
Robert Boylestad/Louis Nashelsky

© 1978, 1972 BY PRENTICE-HALL, INC., ENGLEWOOD CLIFFS, NEW JERSEY 07632

10 9 8

Printed in the United States of America

PRENTICE-HALL INTERNATIONAL, INC., *London*
PRENTICE-HALL OF AUSTRALIA, PTY. LTD., *Sydney*
PRENTICE-HALL OF CANADA, LTD., *Toronto*
PRENTICE-HALL OF INDIA PRIVATE LIMITED, *New Delhi*
PRENTICE-HALL OF JAPAN, INC., *Tokyo*
PRENTICE-HALL OF SOUTHEAST ASIA PTE. LTD., *Singapore*
WHITEHALL BOOKS LIMITED, *Wellington, New Zealand*

contents

2 diode rectifiers and filters

3 transistors (BJTs) and vacuum tubes

4 dc biasing

5 small-signal analysis

6 field-effect transistors

7 multistage systems, decibels (dB), and frequency considerations

8 large-signal amplifiers

9 pnpn and other devices

10 integrated circuits (ICs)

11 differential and operational amplifiers

12 feedback amplifiers and oscillator circuits

13 pulse and digital circuits

14 regulators and miscellaneous circuit applications

15 cathode ray oscilloscope

appendices

answers to selected odd-numbered problems 691

index 695

preface

This text is designed primarily for use in a two-semester or three-trimester sequence in the basic electronics area. It is expected that the student has taken a course in dc circuit analysis and has either taken or is taking a course in ac circuit analysis. This text requires only a mathematical background similar to that required for the ac circuit analysis course.

In an effort to aid the student, the text contains extensive examples that stress the main points of each chapter. There are also numerous illustrations to guide the student through the new concepts and techniques. Important conclusions are emphasized by boxed equations or boldface answers to make the student aware of the essential points covered.

The text is the result of a two-semester electronics course sequence which both authors were actively involved in teaching over a period of years. However, the fifteen chapters actually contain more material than can be covered in two fifteen-week semesters (or three ten-week trimesters). This preface will show how the authors feel the material can be organized.

This second edition was necessary to update material in a number of areas. The majority of the changes appear in the first seven chapters. The fundamental content of each chapter, however, remains the same. Techniques of analysis have been improved to provide a clearer, more meaningful development. Material on popular new devices has been added or expanded to maintain relevancy.

Essentially, the first six chapters provide the basic background to electronic devices—including construction, biasing, and operation as single stages. The material in these chapters can be included in the first semester with the option left to the teacher of stressing some areas more than others, or some not at all. The course would begin with the theory and operation of two-terminal devices, stressing semiconductor diodes. Since the theory course is usually taught in conjunction with a laboratory course the material has been organized with regard to providing practical circuit examples which can be operated in the lab. New material on LED, LCD, and solar cells have been added.

Chapter 2 (on diode rectifiers and filters) provides some practical examples of diode application to the basic electronic area of power supplies. Other texts generally place this material at the end of the text. Our own experience shows that this practical study serves as an interlude between the chapters on basic device theory and provides some valuable lab experiments. Material on the capacitor filter has been improved, while the emphasis on less popular filter circuits has been reduced.

Chapter 3 covers the BJT transistor device, its construction and theory of operation. As mentioned earlier it is possible to follow Chapter 1 with this chapter. The operation of the transistor is presented both mathematically and graphically; the amplifying action of the transistor is defined and demonstrated. Actual current directions are used in this introductory area, as teaching experience shows that students understand initial concepts best this way.

It is the authors' experience that the student can better comprehend the operation of the BJT transistor device if, initially, the dc bias and ac operation are treated separately. Thus, Chapter 4 deals only with the dc bias of the BJT transistor (and tube). This is done for common-emitter, common-base, and common-collector (emitter follower) configurations for a variety of bias circuit types. Numerous examples help to demonstrate the theory presented. Also, some design problems are included to provide a well-rounded treatment.

Chapter 5 is one of the most important in the basic coverage area and should be given sufficient time in any course. The development of the BJT transistor ac equivalent circuit model is covered in detail, followed by analysis of the ac operation of the full small-signal circuit. The treatment in this chapter (as in Chapter 4) is essentially mathematical. However, the mathematics are kept short and direct, with a generous number of examples provided so that students will be able to follow the ideas presented. The hybrid equivalent circuit of the transistor is presented, and then the usual engineering simplifications are included in ac analysis to provide a more practically meaningful treatment. This is followed by an introduction to a simplified model which has received increased interest in the analysis of BJT circuits.

If possible the material on the field effect transistor (FET) should also be covered in the first semester of electronics. After having presented and developed the concepts of dc bias and ac analysis of the BJT, Chapter 6 then covers a number of practical FET circuits. We had considered including the FET dc biasing in Chapter 4 and ac analysis in Chapter 5. It was our feeling from classroom experience that this would require spending too much time on each topic, and the FET would appear to be a minor device to the student. By covering the FET in a separate chapter, its significance is stressed and its operation can be properly presented. The chapter has been extensively revised to include graphical techniques which permit the student to directly obtain dc levels for any FET device.

Chapter 7 would be the first topic in the second semester and covers the operation of multistage BJT and FET transistor circuits. Stage loading, overall gain calculations, and use of decibels are all covered in this important chapter. A number of examples help emphasize the main points of the chapter. Increased emphasis has been placed on use of the approximate analysis techniques for multistage amplifiers. The material on frequency has been totally revised for increased clarity.

Chapter 8 covers the operation of power transistors in a few basic power amplifier circuits. Most important is the operation of the push-pull circuit. Transistor push-pull circuits containing a transformer as well as transformerless circuits are covered. Additional material on quasi-complementary push-pull amplifiers and on class-B power and efficiency is provided.

Chapter 9 is a "catch-all" of a number of *pnpn* devices—covering their construction, operation, and circuit applications. It can be covered quickly or even passed over, if desired, without loss of continuity. This edition includes an introduction to the modern VFET and its higher power capabilities.

Chapter 10 is a short treatment of the fabrication and construction of integrated circuits (IC) and can be assigned mainly as student reading.

Chapter 11 provides coverage of two very important topics and should be considered essential to the second semester coverage. Due to the popularity of linear IC units, both the differential and operation amplifier are now regarded as basic units. A comprehensive treatment is accordingly given each topic as well as examples and practical applications.

Chapter 12 on feedback amplifiers and oscillators should be covered at least partially in the second semester. The material can also be deferred to a third electronics course on communications if desired. The chapter is extensive and need not be fully covered if time is limited.

Chapter 13 on digital circuits provides a good survey. It is so important in the present electronics field to know this area well. If no course devoted exclusively to computer circuits and logic is taught in your curriculum, the material of this chapter should be closely covered.

Chapter 14 provides coverage of voltage regulators. The miscellaneous circuits provided are for the student's own practical study and can be used to stimulate or motivate his interest.

Chapter 15 can be integrated anywhere in the two semesters. Although this material may not be covered in classroom lectures, it is quite important to the student and can be used to supplement laboratory work on CRO theory and applications. The fundamental operation and use of the CRO is stressed in this chapter. Again, generous examples help emphasize the main points covered.

To improve the use of this text by both student and instructor there are numerous practical examples in most chapters. Problems at the end of these chapters are keyed to the particular section in which the problems are covered.

We wish to thank Professors Aidala and Katz of the Electrical Technology department at Queensborough Community College for their continued help and encouragement over the years. They have provided us with both courses and atmosphere conductive to the best in learning and teaching. We thank Mrs. Doris

Topel and Mrs. Helene Rosenberg, Electrical Technology department secretaries, for their assistance in preparing this revised edition. Finally, we wish to thank each other for a remarkably pleasant and rewarding collaboration.

Robert Boylestad / *Louis Nashelsky*
Hanover, N. H.   *Great Neck, N. Y*

electronic devices and circuit theory

two-terminal devices

1.1 INTRODUCTION

A major portion of the electronic devices in commercial use today have only two terminals. One of the most important is the diode, which is one of the fundamental building blocks of the wide variety of electronic circuits in use today. It is essential to the operation of such representative systems as rectifiers, doublers, limiters, clampers, clippers, modulators and demodulators, waveforming circuits, and frequency converters. The diode is available in many different sizes and shapes with varying modes of operation. The *vacuum* and *semiconductor* diodes will both be considered in detail in this chapter. Later sections will contain brief descriptions of the *Zener, varicap, tunnel, photoelectric, silicon power, and Schottky diodes*. Also to be covered is a temperature-sensitive resistor, the *thermistor*, various types of visual displays, such as the *LED* and *LCD*, and *solar* cells which have enjoyed a renewed interest as a result of energy considerations.

The first diode, called Fleming's valve, was developed by J. Ambrose Fleming in 1902. Its basic construction consisted of two elements, a filament and metallic plate in an evacuated glass envelope, similar in many respects to the modern high-vacuum diode. It was not until the early 1930s that a radically new type of diode became increasingly important: the semiconductor diode. This solid-state device, much smaller than the vacuum diode with characteristics closer to the ideal switching characteristics, led the way to the development of the transistor amplifier (a three-terminal device to be examined in Chapter 3) by J. Bardeen and W. Brattain of Bell Laboratories in 1948.

In recent years emphasis has been almost completely on the development of the semiconductor diode. Except for very high frequencies or high-power applica-

1

tions and special areas of application such as in photoelectric devices, the solid-state diode seems destined to possibly eliminate the vacuum diode from the competitive market.

Before considering the basic theory of operation of the vacuum, and semiconductor diodes, the *ideal* diode is presented to introduce basic diode action and establish a basis for later comparison with actual diode characteristics. The term ideal is used to indicate that the characteristics of the device are those we strive to match with technology.

1.2 IDEAL DIODE

The ideal diode is a *two-terminal* device having the symbol and characteristics shown in Fig. 1.1a and b, respectively.

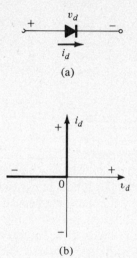

(a)

(b)

Figure 1.1. Ideal diode: (a) symbol; (b) characteristics.

In the description of the elements to follow, it is critical that the various *letter symbols, voltage polarities,* and *current directions* be defined. If the polarity of the applied voltage is consistent with that shown in Fig. 1.1a, the portion of the characteristics to be considered in Fig. 1.1b is to the right of the vertical axis. If a reverse voltage is applied, then the characteristics to the left are pertinent. If the current through the diode has the direction indicated in Fig. 1.1a, the portion of the characteristics to be considered is above the horizontal axis, while a reversal in direction would require the use of the characteristics below the axis. For the majority of the device characteristics to appear in this text the *ordinate* will be the *current axis,* while the *abscissa* will be the *voltage axis.*

One of the important parameters for the diode is the resistance at the point or region of operation. If we consider the region defined by the direction of i_d and polarity of v_d in Fig. 1.1a (upper-right quadrant of Fig. 1.1b), we shall find that the value of the forward resistance, R_f, as defined by Ohm's law is

$$R_f = \frac{V_f}{I_f} = \frac{0}{2, 3 \text{ mA}, \dots, \text{ or any positive value}} = 0\,\Omega$$

where V_f is the forward voltage across the diode and I_f is the forward current through the diode. *The ideal diode, therefore, is a short circuit for the forward region of conduction* ($i_d \neq 0$).

If we now consider the region of negatively applied potential (third quadrant) of Fig. 1.1b,

$$R_r = \frac{V_r}{I_r} = \frac{-5, -20, \text{ or any reverse bias potential}}{0}$$

= very large number, which for our purposes we shall consider to be infinite (∞)

where V_r is the reverse voltage across the diode and I_r is the reverse current in the diode. *The ideal diode, therefore, is an open circuit in the region of nonconduction* $(i_d = 0)$.

In review, the conditions depicted in Fig. 1.2 are true.

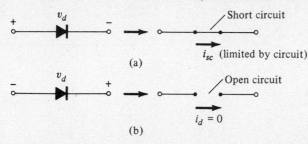

(a)

(b)

Figure 1.2. (a) Conduction and (b) non-conduction states of the ideal diode as determined by the applied bias.

In general, it is relatively simple to determine whether a diode is in the region of conduction or nonconduction by simply noting the direction of the current i_d to be established by an applied emf. For conventional flow (opposite to that of electron flow), if the resultant diode current has the same direction as the arrowhead of the diode symbol, the diode is operating in the conducting region. The above is depicted in Fig. 1.3.

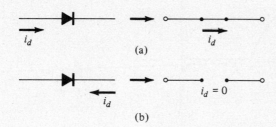

(a)

(b)

Figure 1.3. (a) Conduction and (b) non-conduction states of the ideal diode as determined by current direction of applied network.

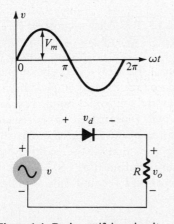

Figure 1.4. Basic rectifying circuit.

As an introductory example of one practical application of the diode let us consider the process of rectification by which an alternating voltage having zero average value is converted to one having a dc or average value greater than zero. The circuit required is shown in Fig. 1.4 with an ideal diode.

For the region defined by $0 \rightarrow \pi$ of the sinusoidal input voltage v, the polarity of the voltage drop across the diode, would be such that the short-circuit representation would result and the circuit would appear as shown in Fig. 1.5a. For the region $\pi \rightarrow 2\pi$ the open-circuit representation would be applicable and the circuit would appear as shown in Fig. 1.5b.

For future reference note the polarities of the input v for each circuit. For sinusoidal inputs the polarity indicated will be for the positive portion of the sinusoidal waveform as shown in Fig. 1.4.

For the situation shown in Fig. 1.5a the output voltage, v_o, will appear exactly

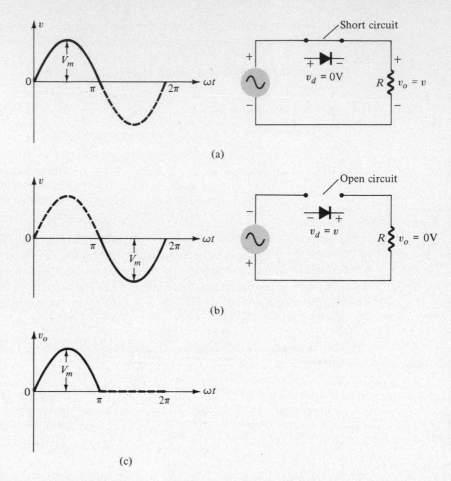

Figure 1.5. Rectifying action of the circuit of Fig. 1.4.

the same as the input voltage, v_i, as long as the diode is forward biased. In Fig. 1.5b, because of the open-circuit representation of the ideal diode, the output voltage v_o equals zero from π to 2π of the impressed voltage v. The complete resultant output waveform is shown in Fig. 1.5c for the entire sinusoidal input. For each cycle of the input voltage v, the waveform of v_o, will repeat itself so that each waveform has the same frequency. A closer examination of the various figures will also reveal that the impressed *emf v* and v_o are *in phase*; that is, the positive pulse of each appears during the same time period. Phase relationships will become increasingly important when we consider semiconductor and vacuum-tube amplifiers.

1.3 VACUUM-TUBE DIODE

The basic vacuum diode consists of a *cathode* and an *anode* (metallic plate) in an *evacuated tube* in the relative positions shown in Fig. 1.6.

A heated cathode will establish a large number of "free" electrons in the region

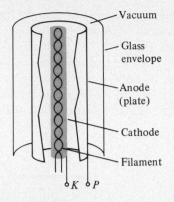

Figure 1.6. Basic construction of the vacuum-tube diode.

between the cathode and the plate. The heating process is accomplished by bringing the cathode to an electron emission temperature by applying a specified heater potential across the heater filament terminals. The applied potential will develop a current I through the filament, resulting in an I^2R heating loss much like the heating element of a toaster. As the temperature of the filament increases, the thermal agitation of the electrons will increase to a point where the electrons have sufficient kinetic energy to leave the surface of the cathode and assume their "free" state. This type of emission is referred to as *thermionic* emission. The tube is evacuated to prevent the cathode from "burning up" due to the oxygen in the air and to increase the mobility of the electrons.

In the directly heated type (the filament *is* the cathode shown in Fig. 1.7a), the electrons are emitted directly by the filament material. In the indirectly heated type shown in Fig. 1.7b, the electrons are emitted from a surface not directly connected to the filament (or heater) that is brought to emission temperature by the radiating heat of the filament. The indirectly heated type is the more commonly used of the two. If a 60-cycle ac potential were applied to the directly heated type, the number of thermionically generated "free" electrons would vary at each instant of time, since the current through the filament is determined by the instantaneous value of the applied signal. This varying emission of electrons may result in a 60-cycle hum at the output of the system. This undesirable effect is negligible in the indirectly heated type. A second advantage of the indirectly heated type is that the entire cathode is at a relatively fixed potential, whereas the potential of the directly heated type will vary from point to point along the filament.

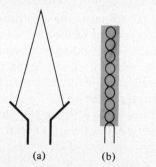

(a)   (b)

Figure 1.7. Cathodes: (a) directly heated; (b) indirectly heated.

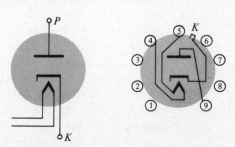

Figure 1.8. Indirectly heated vacuum-tube diode symbol and pin connections.

The graphic symbol for the indirectly heated vacuum-tube diode is shown in Fig. 1.8. There are no special markings on a tube to indicate which pins are connected to the plate, cathode, or filament. A tube manual must be consulted for the pin connections as indicated in Fig. 1.8. The first number of the tube type indicates the rms voltage to be applied to the filament, while the remaining numbers and letters refer to a particular production series.

Let us now examine the basic operation of a vacuum-tube diode using the circuit of Fig. 1.9. Initially the ac heater voltage of 6.3 V rms (typical value) is applied to the filament with the input voltage V set at 0 V. As the temperature of the filament rises, an increasing number of electrons will be emitted by the cathode as shown by Fig. 1.10.

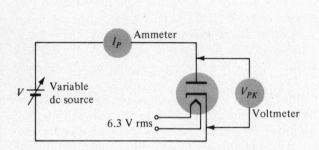

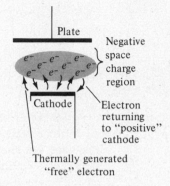

Figure 1.9. Circuit for determining vacuum-tube diode characteristics.

Figure 1.10. Negative space charge region.

Consider, however, that as the negatively charged electrons leave the surface of the cathode, a net positive charge will result on the cathode. This net positive charge will in turn attract the "free" electrons back to the cathode. At the same instant, however, additional electrons are being liberated, resulting in a continual transfer of electrons from the cathode to the region between the plate and cathode. When the full heating effect of the filament has been reached, an equilibrium condition will exist such that the number of electrons leaving the cathode will be equal to the number returning. Once equilibrium is established, there will exist at any instant of time a number of "free" electrons between the plate and cathode. This region or cloud of "free" electrons is referred to as the region of *negative space charge* (Fig. 1.10). The number of electrons in this region can be increased by raising the temperature of the filament by increasing the filament voltage. For the diode of Fig. 1.10, however, increasing the filament voltage above the rated 6.3 V rms may result in permanent damage to the tube.

If the applied voltage V remains set at 0 V, the plate of the tube will remain at zero potential and (ideally) have no effect on the electrons of the space charge region. There are a relatively small number of electrons that are capable of reaching the plate with zero applied potential but this flow of charge or current is never more than a few microamperes. It is due solely to the electrons that are released through heating with sufficient kinetic energy to reach the plate of the diode.

If the potential of the plate is increased by increasing the applied voltage V, a number of "free" electrons of the space charge region will be attracted to the plate as indicated in Fig. 1.11. Further increase in the input voltage V will result in more and more electrons being attracted to the plate, resulting in an increase in the flow

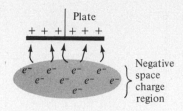

Figure 1.11. The effect of a positive plate on the electrons of the negative space charge region.

of charge or current as shown in Fig. 1.12a. The maximum current is not limited by the applied potential but by the temperature of the cathode material. Once the rate of cathode emission is equal to the rate of absorption by the plate, the current will level off and not increase with further increase in applied voltage. The maximum current can only be increased by increasing the filament temperature. The maximum current for a particular filament temperature is referred to as the *saturation current*.

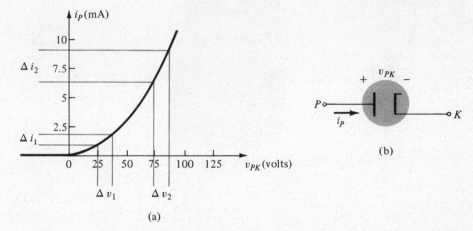

(a)

(b)

Figure 1.12. Vacuum-tube diode: (a) plate characteristics; (b) symbol.

If the polarity of the applied voltage V of Fig. 1.9 is reversed, resulting in a negative potential on the plate, the negatively charged electrons will be repelled and the resulting flow of charge will be zero. This is shown clearly by the horizontal characteristics of Fig. 1.12a for negatively applied plate potentials. It should also be obvious from Fig. 1.12a that the vacuum-tube diode is closely related to the ideal diode in the reverse-bias region ($R_r = \infty\ \Omega$) but is far from ideal in the forward-bias region. In this region, the resistance R_f is not zero or fixed in magnitude but varies from point to point along the curve. The symbol for the vacuum-tube diode is shown in its most common form in Fig. 1.12b with the defined polarities and direction for Fig. 1.12a. Note that the direction of the current (region of conduction) is from plate to cathode as defined by conventional flow. The curve of Fig. 1.12a is called a nonlinear curve since a change in voltage (or current) will not result in an equal change in current (or voltage). This is shown in Fig. 1.12a. The resultant change in i for Δv_2 is almost twice the change for i for Δv_1 although $\Delta v_1 = \Delta v_2$. For a linear curve, such as for a fixed resistance, the per cent of change in one quantity, v or i, will result in a corresponding per cent of change in the other.

The calculation of the resistance of the vacuum-tube diode at various points along its characteristic curve and the analysis of circuits using the vacuum-tube diode will be considered in greater depth in a later section.

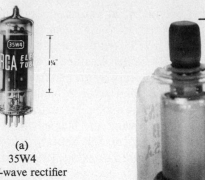

(a)
35W4
Half-wave rectifier
7-pin miniature type
PIV = 330 V
Peak I_P = 600 mA
Heater voltage = 35 V

(Courtesy Radio Corporation of America)

2½"

¼"

(b)
3CN3-B
High voltage rectifier
PIV = 30 KV
PEAK I_P = 110 mA
Heater voltage = 3.65 V

(Courtesy General Electric Company)

(c)
7266
Metal-Ceramic diode
PIV = 600 V
Peak I_P = 11 mA
Heater voltage = 6.3 V

(Courtesy General Electric Company)

Figure 1.13. Various types of vacuum-tube diodes.

In Fig. 1.13 various types of vacuum diodes are shown with pertinent data. Note the ceramic tube, which was designed to withstand extreme shock and high temperatures.

1.4 SEMICONDUCTOR DIODES

As the name implies, the semiconductor diode is constructed of *semiconductor materials*. Semiconductors are neither good conductors, such as copper or aluminum, nor good insulators, such as bakelite or rubber, but belong to a class of materials between the two. There are approximately 10^{28} free electrons/m³ for conductors and 10^7 free electrons/m³ for insulators, at room temperature, with the number of free electrons per unit volume for semiconductors varying within this range. In general, semiconductor materials also have *negative* temperature coefficients, indicating that the resistance of the material decreases as temperature increases. The reverse occurs for most metallic conductors.

To fully appreciate the behavior of the semiconductor diode, a very brief and limited introduction to solid-state physics is necessary. We shall begin by considering the basic structure of the *atom*. The atom is composed of three basic particles; the *electron, proton,* and *neutron*. In the atomic lattice, the neutrons and protons form the *nucleus*, while the electrons revolve around the nucleus in a fixed *orbit*. The atomic structures of the two most commonly used semiconductors, *germanium* and *silicon*, are shown in Fig. 1.14.

As indicated by Fig. 1.14a, the germanium atom has 32 orbiting electrons, while silicon has 14 orbiting electrons. In each case, there are 4 electrons in the outermost (valence) shell. The potential (ionization potential) required to remove any one of these 4 valence electrons is lower than that required for any other electron in the structure. In a pure germanium or silicon crystal these 4 valence electrons are bonded to 4 adjoining atoms, as shown in Fig. 1.15 for germanium.

This type of bonding, formed by *sharing* electrons, is called *covalent bonding*. Although the covalent bond will result in a stronger bond between the valence electrons and their parent atom, it is still possible for the valence electrons to absorb sufficient kinetic energy from natural causes to break the covalent bond and assume the "free" state. These natural causes include effects such as light energy in the form of photons and thermal energy from the surrounding medium. At room temperature there is approximately one "free" electron per 5×10^{10} germanium atoms. These "free" electrons, due only to natural causes, are called *intrinsic* carriers.

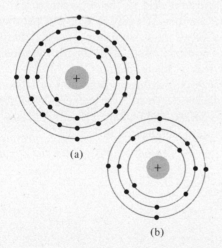

(a)

(b)

Figure 1.14. Atomic structure: (a) germanium; (b) silicon.

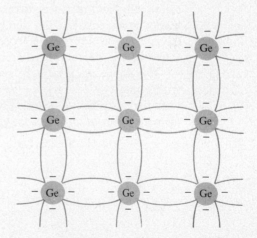

Figure 1.15. Covalent bonding of the germanium atom.

n- and *p*-Type Materials

The semiconductor diode is formed using two types of materials, the *n-type* and the *p-type*. Both are formed by adding a predetermined number of impurity atoms into a germanium or silicon base. The *n*-type is created by adding those

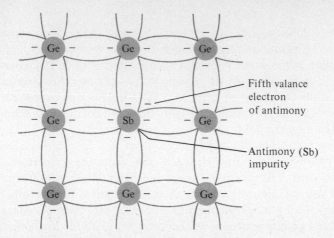

Figure 1.16. Antimony impurity in *n*-type material.

impurity elements that have *five* valence electrons, such as *antimony*, *arsenic*, and *phosphorus*. The effect of such impurity elements is indicated in Fig. 1.16 (using antimony as the impurity in a germanium base). Note that the four covalent bonds are still present. There is, however, an additional fifth electron due to the impurity atom, which is *unassociated* with any particular covalent bond. This remaining electron, loosely bound to its parent (antimony) atom, is relatively free to move within the newly formed *n*-type material. Since the inserted impurity atom has donated a relatively "free" electron to the structure, impurities with five valence electrons are called *donor* atoms. It is important to realize that even though a large number of "free" carriers have been established in the *n*-type material it is still electrically *neutral* since ideally the number of positive charged protons in the nuclei is still equal to the number of "free" and orbiting negatively charged electrons in the structure. The process of adding impurities in the manner described to establish a large number of free carriers is called *doping*.

The *p*-type material is formed by doping a pure germanium or silicon crystal with impurity atoms having *three* valence electrons. The elements most frequently used for this purpose include *boron*, *gallium*, and *indium* The effect of one of these elements, boron, on a base of germanium is indicated in Fig 1.17.

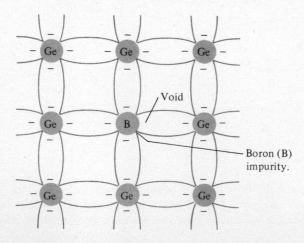

Figure 1.17. Boron impurity in *p*-type material.

Note that there is now an insufficient number of electrons to complete the covalent bonds of the newly formed lattice. The resulting vacancy is called a *hole* and is represented by a small circle or positive sign due to the absence of a negative charge. Since the resulting vacancy will readily accept a "free" electron, the impurities added are called *acceptor* atoms. The resulting *p*-type material is electrically neutral for the same reasons as for the *n*-type material.

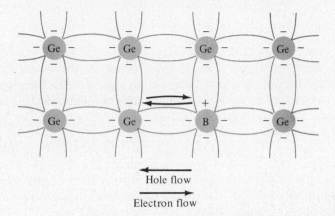

Hole flow

Electron flow

Figure 1.18. Electron vs. hole flow.

The effect of the hole on conduction is shown in Fig. 1.18. If a valence electron gains sufficient kinetic energy to break its covalent bond and fills the void created by a hole, then a vacancy, or hole, will be created in the covalent bond that released the electron. There is therefore a transfer of holes as shown in Fig. 1.18. As pointed out earlier, the direction of flow to be used in this text is that of *conventional* flow, which is indicated by the direction of hole flow.

The graphic representation of the *n*- and *p*-type material is shown in Fig. 1.19. Note the larger number of electrons in the *n*-type material and holes in the *p*-type material. These charge carriers for obvious reasons are called the *majority* carriers. The positive ions of the *n*-type and the negative ions of the *p*-type are the donor and acceptor ions, respectively. The holes and negative ions in the *n*-type material and electrons and positive ions in the *p*-type material are due in part to those

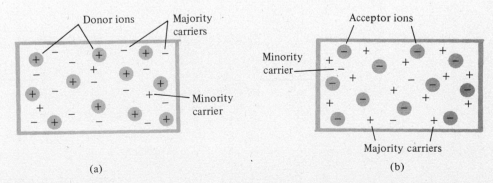

(a)   (b)

Figure 1.19. (a) *n*-type material; (b) *p*-type material.

impurities that exist in every material not through design but because an absolutely *pure* germanium or silicon crystal cannot be obtained. Since carriers of the type just mentioned are very few compared to that of the majority carriers, they are called *minority* carriers. Other sources of minority carriers include thermally generated hole-electron pairs and carriers absorbing sufficient light energy in the form of photons to leave the parent atom.

p-n Junction

The semiconductor diode is formed by "joining" an *n*- and *p*-type material as shown in Fig. 1.20, using techniques to be described below. At the instant the two materials are "joined" the electrons and holes in the region of the junction will combine resulting in a lack of carriers in the region near the junction. This region of uncovered positive and negative ions is called the *depletion* region due to the depletion of carriers in this region.

The minority carriers in the *n*-type material that find themselves within the depletion region will pass directly into the *p*-type material. The closer the minority carrier is to the junction, the greater the attraction for the layer of negative ions and the less the opposition of the positive ions in the depletion region of the *n*-type material. For the purposes of future discussions we shall assume that all the minority carriers of the *n*-type material that find themselves in the depletion region due to their random motion will pass directly into the *p*-type material. Similar discussion can be applied to the minority carriers (electrons) of the *p*-type material. This carrier flow has been indicated in Fig. 1.20 for the minority carriers of each material.

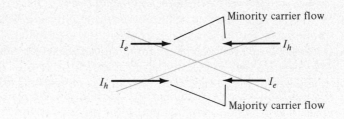

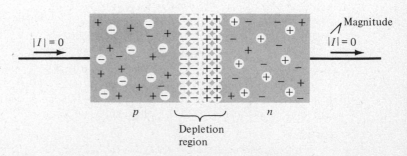

Figure 1.20. *p-n* junction with no external bias.

The majority carriers in the *n*-type material must overcome the attractive forces of the layer of positive ions in the *n*-type material and the shield of negative ions in the *p*-type material in order to migrate into the neutral region of the *p*-type material. The number of majority carriers is so large in the *n*-type material, however, that there will be invariably a small number of majority carriers with sufficient kinetic energy to pass through the depletion region into the *p*-type material. Again, the same type of discussion can be applied to the majority carriers of the *p*-type material. The resulting flow due to the majority carriers is also shown in Fig. 1.20.

A close examination of Fig. 1.20 will reveal that the relative magnitudes of the flow vectors are such that the net flow in either direction is zero. This cancellation of vectors has been indicated by crossed lines. The length of the vector representing hole flow has been drawn longer than that for electron flow to demonstrate that the magnitude of each need not be the same for cancellation and that the doping levels for each material may result in an unequal carrier flow of holes and electrons. In summary, *the net flow of charge in any one direction with no applied emf is zero.*

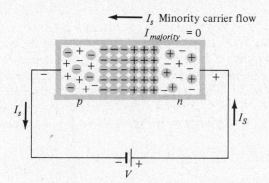

Figure 1.21. Reverse-biased *p-n* junction.

If an external potential of V volts is applied across the *p-n* junction as shown in Fig. 1.21, the number of uncovered negative ions in the depletion region of the *n*-type material will increase due to the large number of "free" electrons drawn to the positive potential of the applied *emf*. For similar reasons, the number of uncovered negative ions will increase in the *p*-type material. The net effect, therefore, is a widening of the depletion region. This widening of the depletion region will establish too great a barrier for the majority carriers to overcome, effectively reducing the majority carrier flow to zero (Fig. 1.21).

The number of minority carriers, however, that find themselves entering the depletion region will not change with reverse bias, resulting in minority carrier flow vectors of the same magnitude indicated in Fig. 1.20 with no applied *emf*. The current that exists under reverse-bias conditions is called the *reverse saturation current* and is represented by the subscript *s*. It is seldom more than a few micro-amperes in magnitude except for high-power devices. The term saturation comes from the fact that it doesn't significantly change with increase in the reverse-bias potential.

The effect of a *forward bias* on the *p-n* junction is indicated in Fig. 1.22. Note that the minority carrier flow has not changed in magnitude, but the reduction in

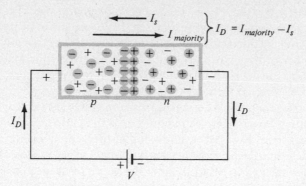

$$I_D = I_{majority} - I_s$$

Figure 1.22. Forward-biased *p-n* junction.

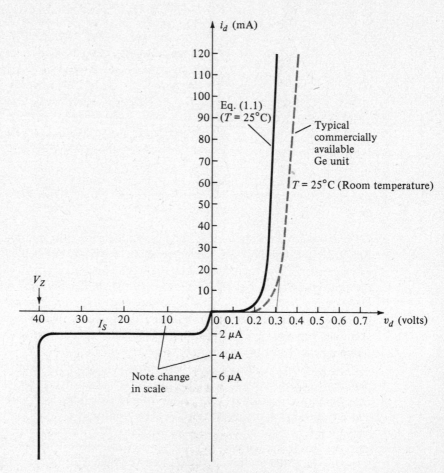

Figure 1.23. Semiconductor diode (Ge) characteristics.

the width of the depletion region has resulted in a heavy majority carrier flow across the junction. The magnitude of the majority carrier flow will increase exponentially with increasing forward bias as indicated in Fig. 1.23. Note that the current scale is in mA above the axis and μA below.

It can be demonstrated through the use of solid-state physics that the diode current can be mathematically related to temperature (T_K) and applied bias (V) in the following manner:

$$I = I_s(e^{kV/T_K} - 1) \tag{1.1}$$

where $\qquad I_s$ = reverse saturation current;

$\qquad$ k = 11,600/η with η = 1 for Ge and 2 for Si;

$\qquad T_K = T_C + 273°$ (degrees Kelvin).

Note the exponential factor that will result in a very sharp increase in I with increasing levels of V. The characteristics of a commercially available germanium (Ge) diode will differ slightly from the characteristics of Fig. 1.23 because of the *body* or *bulk* resistance of the semiconductor material and the *contact* resistance between the semiconductor material and the external metallic conductor. They will cause the curve to shift slightly in the forward-bias region as indicated by the dashed line in Fig. 1.23. As construction techniques improve and these undesired resistance levels are reduced, the commercially available unit will approach the characteristic defined by Eq. (1.1).

In an effort to demonstrate that Eq. (1.1) does in fact represent the curve of Fig. 1.23, let us determine the current I for the forward-bias voltage of 0.22 V at room temperature (25°C).

$$T_K = T_C + 273° = 25° + 273° = 298°$$

$$k(Ge) = \frac{11,600}{1} = 11,600$$

$$\frac{kV}{T_K} = \frac{(11,600)(0.22)}{298} = 8.56$$

and $\qquad I = I_s(e^{8.56} - 1) = (1 \times 10^{-6})(5238 - 1) = 5237 \times 10^{-6}$

so that $\qquad\qquad\qquad I \cong \textbf{5.237 mA}$

as verified by Fig. 1.23.

Temperature can have a marked effect on the diode current. This is clearly demonstrated by the factor T_K in Eq. (1.1). The effect of varying T_K will be determined for the forward-bias condition in the exercises appearing at the end of the chapter. In the reverse-bias region it has been found experimentally that the *reverse saturation current I_s will almost double in magnitude for every 10°C change in temperature*. It is not uncommon for a gemanium diode with an I_s in the order of 1 or 2 μA at 25°C to have a leakage current of 100 μA = 0.1 mA at a temperature of 100°C. Current levels of this magnitude in the reverse-bias region would certainly question our desired open-circuit condition in the reverse-bias region. Fortunately, typical values of I_s for silicon at room temperature range from 1/100 to 1/1000 that of a similar application gemanium diode so that even at higher temperature levels I_s does not usually reach levels of serious concern. In the example above, with

$I_s = 1 \mu A$ for germanium, if I_s were, at the most, 1/100 of 1 μA = 0.01 μA for a silicon diode then at 100°C it would be only 1/100(100 μA) = 1 μA.

Zener Region

Note the sharp change in the characteristics of Fig. 1.24 at the reverse-bias potential V_Z (the subscript Z refers to the name Zener). This constant-voltage effect is induced by a high reverse-bias voltage across the diode. When the applied reverse potential becomes more and more negative, a point is eventually reached where the few free minority carriers have developed sufficient velocity to liberate additional carriers through ionization. That is, they collide with the valence electrons and impart sufficient energy to them to permit them to leave the parent atom. These additional carriers can then aid the ionization process to the point where

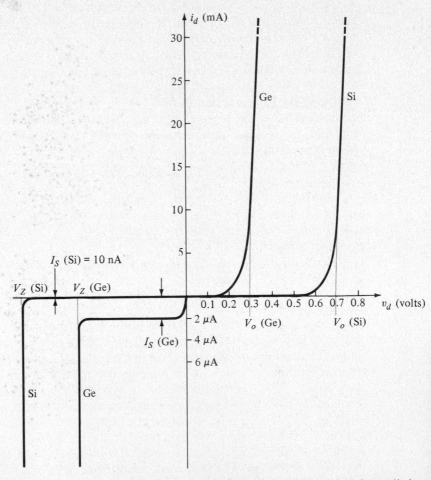

Figure 1.24. Comparison of Si and Ge semiconductor diodes.

a high *avalanche* current is established and the *avalanche breakdown* region determined.

The avalanche region (V_z) can be brought closer to the vertical axis by increasing the doping levels in the *p* and *n*-type materials. However, as V_z decreases to very low levels, such as -5 volts, another mechanism, called *Zener breakdown*, will contribute to the sharp change in the characteristic. It occurs because there is a strong electric field in the region of the junction that can disrupt the bonding forces within the atom and "generate" carriers. Although the Zener breakdown mechanism is only a significant contributor at lower levels of V_z, this sharp change in the characteristic at any level is called the *Zener region* and diodes employing this unique portion of the characteristic of a *p-n* junction are called *Zener diodes*. They will be described in detail in a later section.

The Zener region of the semiconductur diode described must be avoided if the response of a system is not to be completely altered by the sharp change in characteristics in this reverse-voltage region. The maximum reverse-bias potential that can be applied before entering this region is called the *peak inverse voltage* (referred to simply as the PIV rating), or the peak reverse voltage (denoted by PRV rating.).

Silicon and Germanium Characteristics

Silicon diodes have, in general, higher PIV and current ratings and wider temperature ranges than germanium diodes. PIV ratings for silicon can be in the neighborhood of a 1000 V, whereas the maximum value for germanium is closer to 400 V. Silicon can be used for applications in which the temperature may rise to about 200°C (400°F), whereas germanium has a much lower maximum rating (100°F). The disadvantage of silicon, however, as compared to germanium as indicated in Fig. 1.24 is the higher forward-bias voltage required to reach the region of upward swing. It is typically in the order of magnitude of 0.7 V for *commercially* available silicon diodes and 0.3 V for gemanium diodes. The increased off-shoot for silicon is due primarily to the factor η in Eq. (1.1). This factor only plays a part in determining the shape of the curve at very low current levels. Once the curve starts its vertical rise, the factor η drops to 1 (the continuous value for gemanium). This is evidenced by the similarities in the curves once the offshoot potential is reached. The potential at which this rise occurs is very important in the circuit analysis to follow and therefore requires the specific notation V_o as indicated on the figure. In review:

$$\boxed{\begin{aligned} V_o &= 0.7 \text{ Si} \\ V_o &= 0.3 \text{ Ge} \end{aligned}}$$

Obviously, the closer the upward swing is to the vertical axis, the more "ideal" the device. However, the other characteristics of silicon as compared to gemanium make it the choice in the majority of commercially available units.

Semiconductor Diode Ratings

On a specification sheet, or in a manufacturer's data book, the following pieces of information are normally provided:

1. The maximum forward voltage $V_{F(max)}$ (at a specified current and temperature).
2. The maximum forward current $I_{F(max)}$ (at a specified temperature).
3. The maximum reverse current $I_{R(max)}$ (at a specified temperature).
4. The reverse voltage rating (PIV) or PRV or V(BR) where BR comes from the term breakdown (at a specified temperature).
5. The maximum operating (or case) temperature.

Depending on the type of diode being considered, additional data may also be provided, such as frequency range, noise level, capacitance, switching time, peak repetitive values, etc. For the application in mind, the significance of the data will usually be self-apparent if the maximum power or dissipation (D) rating is also provided. It is understood to be the equal to the following product:

$$P_{D_{max}} = V_D I_D \qquad (1.2)$$

where I_D and V_D are the diode current and voltage at a particular point of operation, each variable not to exceed its maximum value. The following information was taken directly from a Texas Instruments Inc. data book. Note that the forward voltage drop does not exceed 1 V, but the current has maximum values of 1 to 200 mA.

General-Purpose Diodes

DEVICE TYPE	FORWARD CURRENT		V_{BR} (V)	MAXIMUM I_R			
				25°C		150°C	
	$I_F(mA)$	$V_F(V)$		(V)	(μA)	(V)	(μA)
1N463	1.0	1.0	200	175	0.5	175	30
1N462	5.0	1.0	70	60	0.5	60	30
1N459A	100.0	1.0	200	175	0.025	175	5
T151	200.0	1.0	20	10	1	—	—

For the 1N463, if we establish maximum forward voltage and current conditions:

$$P_D = V_D \cdot I_D = 1(1) = 1 \text{ mW (a low-power device)}$$

Of course, a device may have a maximum dissipation less than that established by the maximum values. That is, if the voltage is a maximum, the current may have to be less than rated maximum value.

Note the increase in I_R for each device with temperature. For the 1N463, it is $30/0.5 = 60$ times larger.

Semiconductor Diode Notation

The notation most frequently used for semiconductor diodes is provided in Fig. 1.25. Note the carryover of the anode-cathode terminology from the vacuum tube.

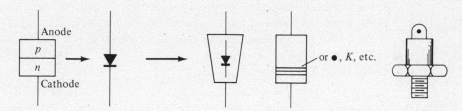

Figure 1.25. Semiconductor diode notation.

Diode Ohmmeter Check

The condition of a semiconductor diode can be quickly determined by using an ohmmeter such as found on the standard VOM. The internal battery (often 1.5 V) of the ohmmeter section will either forward or reverse bias the diode when applied. If the positive (normally the red) lead is connected to the anode and the negative (normally the black) lead to the cathode, the diode is forward biased and the meter should indicate a low resistance. The R × 1000 or R × 10,000 setting should be suitable for this measurement. With the reverse polarity the internal battery will back bias the diode and the resistance should be very large. A small reverse-bias resistance reading indicates a "short" condition while a large forward-bias resistance indicates an "open" situation. The basic connections for the tests appear in Fig. 1.26.

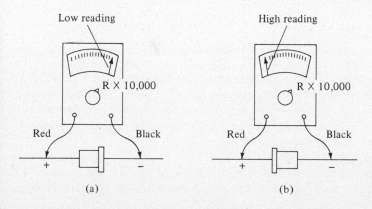

Figure 1.26. Ohmmeter testing of a semiconductor diode: (a) forward bias (b) reverse bias.

A relatively easy way to remember which way to bias a diode to be sure that it is forward biased is to connect the *p*ositive potential to the *p* material (or anode) and the *n*egative potential to the *n*-type material. The reverse is true for the reverse-bias condition.

1.5 SEMICONDUCTOR DIODE FABRICATION

Semiconductor diodes are normally one of the following types: grown junction, alloy, diffused, epitaxial growth, or point-contact.

The first step in the manufacture of any semiconductor device is to obtain semiconductor materials, such as germanium or silicon, of the desired purity level. Impurity levels of *less* than *one* part in *one billion* (1 in 1,000,000,000) are required for most semiconductor fabrications today. The basic processes involved in the production of semiconductor materials with this low level of impurities are indicated in Fig. 1.27.

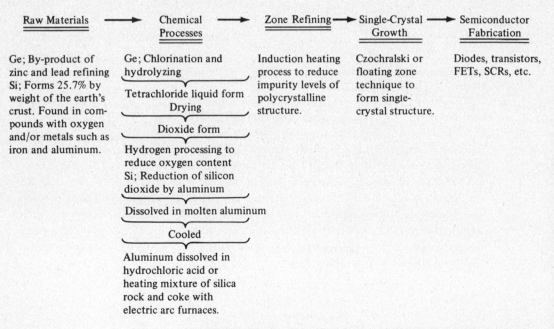

Figure 1.27. Sequence of events leading to semiconductor fabrication.

As indicated in Fig. 1.27, the raw materials are first subjected to a series of chemical reactions and a zone refining process to form a polycrystalline crystal of the desired purity level. The atoms of a polycrystalline crystal are haphazardly arranged, while in the single crystal desired the atoms are arranged in a symmetrical, uniform, geometrical lattice structure.

Zone refining apparatus is shown in Fig. 1.28. It consists of a graphite or quartz boat for minimum contamination, a quartz container, and a set of RF (radio-frequency) induction coils. Either the coils or boat must be movable along

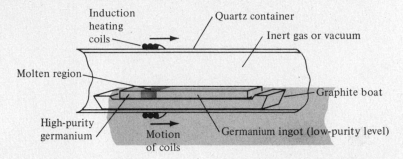

Figure 1.28. Zone refining process.

the length of the quartz container. The same result will be obtained in either case, although moving coils are discussed here since it appears to be the more popular method. The interior of the quartz container is filled with either an inert (little or no chemical reaction) gas, or vacuum, to reduce further the chance of contamination. In the zone refining process, a bar of germanium is placed in the boat with the coils at one end of the bar as shown in Fig. 1.28. The radio-frequency signal is then applied to the coil, which will induce a flow of charge (eddy currents) in the germanium ingot. The magnitude of these currents is increased until sufficient heat is developed to melt that region of the semiconductor material. The impurities in the ingot will enter a more liquid state than the surrounding semiconductor material. If the induction coils of Fig. 1.28 are now slowly moved to the right to induce melting in the neighboring region, the "more fluidic" impurities will "follow" the molten region. The net result is that a large percentage of the impurities will appear at the right end of the ingot when the induction coils have reached this end. This end piece of impurities can then be cut off and the entire process repeated until the desired purity level is reached.

The final operation before semiconductor fabrication can take place is the formation of a single crystal of germanium or silicon. This can be accomplished using either the *Czochralski* or the *floating zone* technique, the latter being the more recently devised. The apparatus employed in the Czochralski technique is shown in Fig. 1.29a. The polycrystalline material is first transformed to the molten state by the RF induction coils. A single crystal "seed" of the desired impurity level is then immersed in the molten germanium and gradually withdrawn while the shaft holding the seed is slowly turning. As the "seed" is withdrawn, a single-crystal germanium lattice structure will grow on the "seed" as shown in Fig. 1.29a. The resulting single-crystal ingot can be as large as 7–10 inches in length and 1–3 inches in diameter (Fig. 1.29b).

The floating zone technique eliminates the need for having both a zone refining and single-crystal forming process. Both can be accomplished at the same time using this technique. A second advantage of this method is the absence of the graphite or quartz boat, which often introduces impurities into the germanium or silicon ingot. Two clamps hold the bar of germanium or silicon in the vertical position within a set of movable RF induction coils as shown in Fig. 1.30. A small single-crystal "seed" of the desired impurity level is deposited at the lower end of the bar

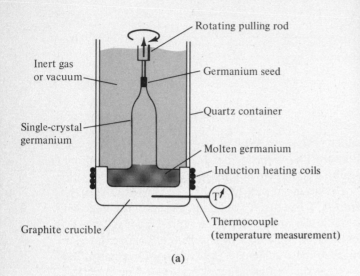

Rotating pulling rod

Inert gas or vacuum

Germanium seed

Quartz container

Single-crystal germanium

Molten germanium

Induction heating coils

Graphite crucible

Thermocouple (temperature measurement)

(a)

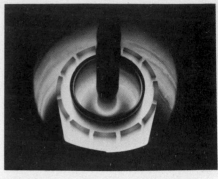

(Courtesy Texas Instruments Incorporated)

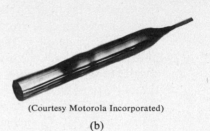

(Courtesy Motorola Incorporated)

(b)

Figure 1.29. Czochralski technique.

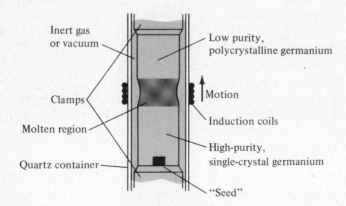

Inert gas or vacuum

Low purity, polycrystalline germanium

Clamps

Motion

Molten region

Induction coils

Quartz container

High-purity, single-crystal germanium

"Seed"

Figure 1.30. Floating zone technique.

and heated with the germanium bar until the molten state is reached. The induction coils are then slowly moved up the germanium or silicon ingot while the bar is slowly rotating. As before, the impurities follow the molten state resulting in an improved impurity level single-crystal germanium lattice below the molten zone. Through proper control of the process, there will always be sufficient surface tension present in the semiconductor material to ensure that the ingot does not rupture in the molten zone.

The single-crystal structure produced can then be cut into wafers sometimes as thin as $\frac{1}{1000}$ (or 0.001) of an inch ($\cong \frac{1}{5}$ the thickness of this paper). This cutting process can be accomplished using the setup of Fig. 1.31a or b. In Fig. 1.31a,

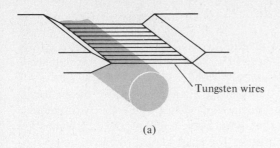

(a)

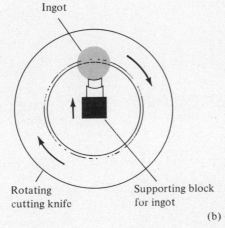

Ingot

Rotating
cutting knife

Supporting block
for ingot

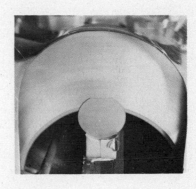

(b)

(Courtesy Texas Instruments Incorporated)

Figure 1.31. Slicing the single-crystal ingot into wafers.

tungsten wires (0.001 in. in diameter) with abrasive deposited surfaces are connected to supporting blocks at the proper spacing and then the entire system is moved back and forth as a saw. The system of Fig. 1.31b is self-explanatory. We shall now consider the four basic processes most commonly used in the manufacture of semiconductor diodes.

Grown Junction

Crystal pulling rod

"Seed"

p–n junction

Slicing process

Melt

Figure 1.32. Grown junction diode.

Diodes of this type are formed during the Czochralski *crystal pulling* process. Impurities of p- and n-type can be alternately added to the molten semiconductor material in the crucible resulting in a p-n junction as indicated in Fig. 1.32 when the crystal is pulled. After slicing, the large-area device can then be cut into a large number (sometimes thousands) of smaller-area semiconductor diodes. The area of grown junction diodes is sufficiently large to handle high currents (and therefore have high power ratings). The large area, however, will introduce undesired junction capacitive effects.

SEC. 1.5 SEMICONDUCTOR DIODE FABRICATION

Alloy

The alloy process will result in a junction-type semiconductor diode that will also have a high current rating and large PIV rating. The junction capacitance is also large, however, due to the large junction area.

The *p-n* junction is formed by first placing a *p*-type impurity on an *n*-type substrate and heating the two until liquefaction occurs where the two materials meet (Fig. 1.33). An alloy will result that, when cooled, will produce a *p-n* junction at the boundary of the alloy and substrate. The roles played by the *n*- and *p*-type materials can be interchanged.

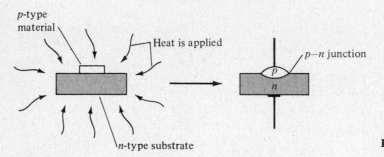

Figure 1.33. Alloy process diode.

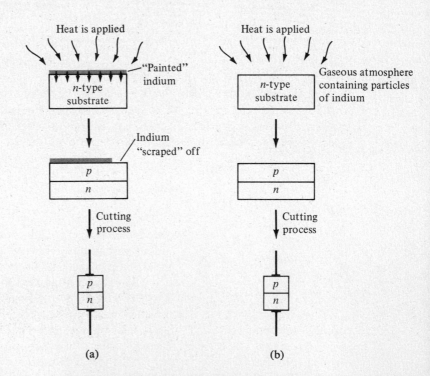

Figure 1.34. Diffusion process diodes: (a) solid diffusion; (b) gaseous diffusion.

Diffusion

The diffusion process of forming semiconductor junction diodes can employ either solid or gaseous diffusion. This process requires more time than the alloy process but it is relatively inexpensive and can be very accurately controlled. Diffusion is a process by which a heavy concentration of particles will "diffuse" into a surrounding region of lesser concentration. The primary difference between the diffusion and alloy process is the fact that liquefaction is not reached in the diffusion process. Heat is applied in the diffusion process only to increase the activity of the elements involved.

The process of solid diffusion commences with the "painting" of an acceptor impurity on an n-type substrate and heating the two until the impurity diffuses into the substrate to form the p-type layer (Fig. 1.34a).

In the process of gaseous diffusion, an n-type material is submerged in a gaseous atmosphere of acceptor impurities and then heated (Fig. 1.34b). The impurity diffuses into the substrate to form the p-type layer of the semiconductor diode. The roles of the p- and n-type materials can also be interchanged in each case. The diffusion process is the most frequently used today in the manufacture of semiconductor diodes.

Epitaxial Growth

The term epitaxial has its derivation from the Greek terms *epi* meaning "upon" and *taxis* meaning "arrangement." A base wafer of n^+ material is connected to a metallic conductor as shown in Fig. 1.35. The n^+ indicates a very high doping level for a reduced resistance characteristic. Its purpose is to act as a semiconductor extension of the conductor and not the n-type material of the p-n junction. The n-type layer is to be deposited on this layer as shown in Fig. 1.35 using a diffusion process. This technique of using an n^+ base gives the manufacturer definite design advantages. The p-type silicon is then applied by using a diffusion technique and the anode metallic connector added as indicated in Fig. 1.35.

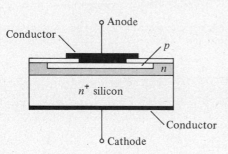

Figure 1.35. Epitaxial growth semiconductor diode.

Point Contact

The point-contact semiconductor diode is constructed by pressing a phosphor bronze spring (called a cat whisker) against an n-type substrate (Fig. 1.36). A high current is then passed through the whisker and substrate for a short period of time, resulting in a number of atoms passing from the wire into the n-type material to create a p-region in the wafer. The small area of the p-n junction results in a very small junction capacitance (typically 1 pF or less). For this reason, the point-contact diode is frequently used in applications where very high frequencies are encountered, such as in microwave mixers and detectors. The disadvantage of the small

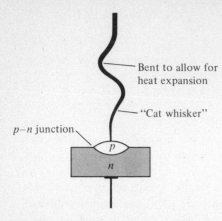

Bent to allow for heat expansion

"Cat whisker"

p–n junction

p

n

Figure 1.36. Point-contact diode.

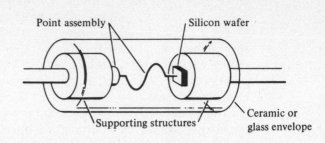

Point assembly

Silicon wafer

Supporting structures

Ceramic or glass envelope

(a)

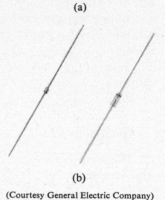

(b)

(Courtesy General Electric Company)

Figure 1.37. Point-contact diodes: (a) basic construction; (b) various types.

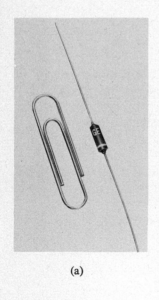

(a)

(b)

Figure 1.38. Various types of junction diodes.

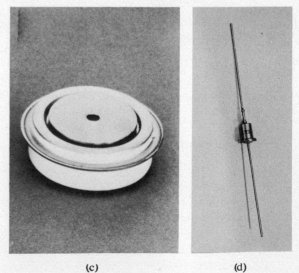

(c)

(d)

(Courtesy International Rectifier Corporation)

contact area is the resulting low current ratings and characteristics less ideal than those obtained from junction-type semiconductor diodes. The basic construction and photographs of point-contact diodes appear in Fig. 1.37. Various types of junction diodes appear in Fig. 1.38.

1.6 LOAD LINE AND QUIESCENT CONDITIONS

In Fig. 1.39 a block symbol has been inserted in the circuit to represent *any* diode that we may choose to use in this particular circuit.

Applying Kirchhoff's voltage law around the indicated loop will result in the following equation:

$$V = V_D + V_R \qquad (1.3)$$

Solving for V_D and substituting $V_R = I_D R$, we have

$$V_D = V - V_R$$

and

$$V_D = V - I_D R \qquad (1.4a)$$

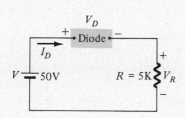

Figure 1.39. Fundamental diode circuit.

Eq. (1.4a) has two dependent variables (V_D and I_D) and two fixed values (V and R). Since a minimum of two equations is required to solve for two unknown dependent variables, Eq. (1.4a) is not sufficient for a complete solution. The second equation necessary to determine the value of V_D and I_D determined by V and R is provided by the characteristics of the diode element in the enclosed container; that is, for the diode employed, we know that the current is a *function of* the voltage across the diode, or mathematically,

$$I_D = f(V_D) \qquad (1.4b)$$

It is necessary, therefore, to find the common solution of the equation determined by the load circuit [Eq. (1.4a)] and the characteristics of the diode. One method of finding this solution is the graphical method, which will now be outlined. It is extremely important that the procedure described in the next few paragraphs be fully understood since similar operations will appear when we consider other devices such as the transistor, vacuum triode, and FET.

Rewriting Eq. (1.4) in a slightly different form, we have

$$I_D = -\frac{1}{R}V_D + \frac{V}{R}$$

(1.5)

$$\begin{array}{ccc} \downarrow & \downarrow\ \downarrow & \downarrow \\ y = & mx\ + & b \end{array} \quad \text{(straight line equation)}$$

Below this newly formed equation, the general equation for a straight line has been included. Note that the slope of the line is negative (I_D decreases in magnitude with increase in V_D) with a magnitude $1/R$, while the y-intercept is V/R and I_D and V_D are the y- and x-variables, respectively. The intercepts of this straight line with

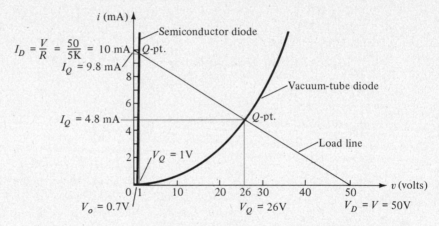

Figure 1.40. Load line and quiescent value determination.

the axes of the graph of Fig. 1.40 can be found rather quickly by applying the following conditions. If we consider first that if $I_D = 0$ mA, we must be somewhere along the horizontal axis of Fig. 1.40 and if we apply this condition to Eq. (1.5), then

$$I_D = 0 = -\frac{V_D}{R} + \frac{V}{R}$$

and solving for V_D

$$V_D = V|_{I_D=0}$$

(1.6a)

The intersection of the straight line with the horizontal axis is the applied voltage V. If we then consider that if $V_D = 0$, we must be somewhere along the vertical axis, and we must apply this condtition to Eq. (1.5); then

$$I_D = -\frac{V_D}{R} + \frac{V}{R} = 0 + \frac{V}{R}$$

and
$$I_D = \frac{V}{R}\bigg|_{V_D=0} \tag{1.6b}$$

The intersection, therefore, of the straight line with the vertical axis is determined by the ratio of the applied voltage and load. Both intersections have been indicated in Fig. 1.40. All that remains is to connect these two points by a straight line to obtain a graphical representation of Eq. (1.4). This resulting line is called the *load line* since it represents the properties of the applied voltage and load and tells us nothing about the diode's characteristics.

In Fig. 1.40 the characteristics of a semiconductor and vacuum-tube diode have also been included. For each, the intersection of the load line and the diode's characteristic curve will determine the point of operation for that diode. This point, due only to the dc input, is called the *quiescent* point—quiescent meaning still, quiet, or inactive. The voltage across and current through the diode can now be found by simply drawing a vertical and horizontal line, respectively, to the voltage and current axis as indicated in Fig. 1.40. Note that each diode has a different potential drop across it, resulting in a different voltage across the load and consequently different current through the load. The subscript Q is used to denote quiescent values of current and voltage as shown in Fig. 1.40. For the semiconductor diode:

$$V_Q = 1\text{ V} \qquad \text{and} \qquad I_Q = 9.8\text{ mA}$$

Substituting into Eq. (1.3), we get

$$V_R = V - V_D = V - V_Q = 50 - 1 = 49\text{ V}$$

or

$$V_R = I_Q R = (9.8 \times 10^{-3})(5 \times 10^3) = 49\text{ V}$$

The power delivered to the load is

$$P_L = I_Q^2 R = (9.8 \times 10^{-3})^2 (5 \times 10^3) = 480.2\text{ mW}$$

or

$$P_L = P_S - P_D$$

where P_S is the power supplied by the source and P_D is the power dissipation of the diode, so that

$$P_L = VI_Q - V_Q I_Q = I_Q(V - V_Q)$$
$$= (9.8 \times 10^{-3})(50 - 1) = 480.2\text{ mW}$$

The calculations for the vacuum-tube diode is left to the reader as an exercise. Take special note of how closely the semiconductor diode approaches that of the ideal diode for the magnitudes of current and voltage indicated.

1.7 STATIC RESISTANCE

A second glance at Fig. 1.40 will reveal that each diode has a fixed voltage and current associated with the diode at the point of operation. Applying Ohm's law to these values for each diode will result in the *static* or *dc* resistance of each diode at the quiescent point.

$$R_{dc} = \frac{V_D}{I_D} \qquad (1.7)$$

For the semiconductor diode:

$$R_{dc} = \frac{V_D}{I_D} = \frac{1}{9.8 \times 10^{-3}} = \mathbf{102 \ \Omega}$$

and the vacuum-tube diode:

$$R_{dc} = \frac{V_D}{I_D} = \frac{26}{4.8 \times 10^{-3}} = \mathbf{5.41 \ K}$$

For the reverse-bias region of a semiconductor diode with $V_D = -20$ V (for example) and $I_S = 1 \ \mu$A:

$$R_{dc} = \frac{V_D}{I_D} = \frac{20}{1 \ \mu A} = \mathbf{20 \ M\Omega}$$

For a vacuum-tube diode with $V_D = -20$ V and $I_D = 0$ A:

$$R_{dc} = \frac{V_D}{I_D} = \frac{20}{0} = \mathbf{\infty \Omega} \text{ (open circuit)}$$

Once the dc resistance has been determined, the diode can be replaced by a resistor of this value. Any change in the applied voltage or load resistor, however, will result in a different Q-point and therefore different dc resistance.

1.8 DYNAMIC RESISTANCE

It is obvious from Fig. 1.40 that the dc resistance of a diode is independent of the shape of the characteristic in the region surrounding the point of interest. If a sinusoidal rather than dc input is applied to the circuit of Fig. 1.40, the situation will change completely. Consider the circuit of Fig. 1.41a, which has as its input

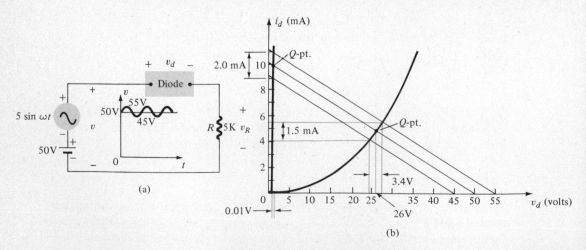

Figure 1.41. ac resistance: (a) circuit; (b) resulting region of operation.

a sinusoidal signal on a dc level. Since the magnitude of the dc level is much greater than that of the sinusoidal signal at any instant of time, the diode will always be forward biased and current will exist continuously in the circuit in the direction shown.

The dc load line resulting from the dc input of 50 V is shown in Fig. 1.41b. The effect of the ac signal is also demonstrated pictorially in the same figure. Note that two additonal load lines have been drawn at the positive and negative peaks of the input signal. At the instant the sinusoidal signal is at its postive peak value the input could be replaced by a dc battery with a magnitude of 55 V and the resultant load line drawn as shown. For the negative peak, $V_{\rm dc} = 45$ V. A moment of thought, however, should reveal the relative simplicity of superimposing the sinusoidal signal on the dc load line and drawing the load lines coinciding with the positive and negative peaks of the sinusoidal signal.

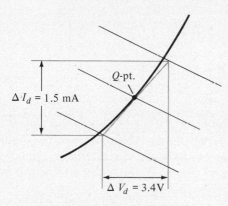

Figure 1.42. Vacuum-tube ac resistance.

Note that we are now interested in a region of the diode characteristics as determined by the sinusoidal signal rather than a single point, as was the case for purely dc inputs. Since the resistance will vary from point to point along this region of interest, which value should we use to represent this portion of the characteristic curve? The value chosen is determined by drawing a straight line tangent to the curve at the quiescent point as shown in Fig. 1.42 for the vacuum-tube diode. The tangent line should "best fit" the characteristics in the region of interest as shown. The resultant resistance, called the *dynamic* or *ac* resistance, is then calculated on an approximate basis, in the following manner:

$$r_d = \frac{\Delta V_d}{\Delta I_d}\bigg|_{\text{tangent line}} \tag{1.8}$$

For the vacuum-tube diode:

$$r_d = \frac{\Delta V_d}{\Delta I_d} \cong \frac{3.4}{1.5 \times 10^{-3}} = \textbf{2.27 K}$$

for the semiconductor diode:

$$r_d = \frac{\Delta V_d}{\Delta I_d} \cong \frac{0.01}{2 \times 10^{-3}} = \textbf{5 }\boldsymbol{\Omega}$$

There is a basic definition in differential calculus that states that *the derivative of a function at a point is equal to the slope of a tangent line drawn at that point.* Eq. (1.8), as defined by Fig. 1.42 is, therefore, essentially finding the derivative of the function at the Q-point of operation. If we find the derivative of the general equation [Eq. (1.1)] for the semiconductor diode with respect to the applied forward bias and then invert the result, we will have an equation for the dynamic or ac resistance in that region. That is, taking the derivative of Eq. (1.1) with respect to the applied bias will result in

$$\frac{dI}{dV} = \frac{k}{T_K}(I + I_s)$$

For values of $I \gg I_s$, $I + I_s \cong I$ and, as indicated earlier, $\eta = 1$ for Ge and Si in the vertical rise section of the characteristics. Therefore,

$$k = \frac{11{,}600}{\eta} = \frac{11{,}600}{1} = 11{,}600$$

with (at room temperature)

$$T_K = T_C + 273° = 25° + 273° = 298°$$

and

$$\frac{dI}{dV} = \frac{11{,}600}{298}I \cong 38.93I$$

or

$$\frac{dV}{dI} = \frac{1}{38.93I} \cong \frac{0.026}{I}$$

and

$$\boxed{r_d' = \frac{dV}{dI} = \frac{0.026 \text{ V}}{I_D} = \frac{26 \text{ mV}}{I_D}}_{\text{Ge, Si}} \qquad (1.9)$$

The significance of Eq. (1.9) must be understood. It implies that the dynamic resistance can be found by simply substituting the quiescent value of the diode current into the equation. There is no need to have the characteristics available or to worry about sketching tangent lines as defined by Eq. (1.8). Its use will be demonstrated below.

We already realize from Eq. (1.8) that the shape of the curve will have an effect on the dynamic resistance. The fact that the silicon and germanium curves in Fig. 1.24 are almost identical after they begin their vertical rise would suggest that the equation for the dynamic resistance of each might be the same as indicated by Eq. (1.9).

It was already noted on Fig. 1.23 that the characteristics of the commercial unit are slightly different from those determined by Eq. (1.1) because of the bulk and contact resistance of the semiconductor device. This additional resistance level must be included in Eq. (1.9) by adding a factor denoted r_B as appearing in Eq. (1.10).

$$\boxed{r_d = \frac{26 \text{ mV}}{I_D \text{ (mA)}} + r_B} \qquad \text{(ohms)} \qquad (1.10)$$

The factor r_B (measured in ohms) can range from typically 0.1 for high-power devices to 2 for some low-power, general-purpose diodes. As construction techniques improve this additional factor will continue to decrease in importance until it can be dropped and Eq. (1.9) applied. For values of I_D in mA, the units of the first term are like those of r_B: ohms. For low levels of current, the first factor of Eq. (1.10) will certainly predominate.

Consider

$$I_D = 1 \text{ mA}$$

with
$$r_B = 2 \, \Omega$$

then,
$$r_d = \frac{26}{1} + 2 = \textbf{28} \, \boldsymbol{\Omega}$$

At higher levels of current the second factor may predominate.
Consider

$$I_D = 52 \text{ mA}$$

with
$$r_B = 2 \, \Omega$$

then,
$$r_d = \frac{26}{52} + 2 = 0.5 + 2 = \textbf{2.5} \, \boldsymbol{\Omega}$$

For the example provided earlier where r_d was graphically determined to be 5 ohms, if we choose $r_B = 2 \, \Omega$, then

$$r_d = \frac{26}{9.8} + 2 = 2.65 + 2 = 4.65 \, \Omega$$

which is very close to the graphically determined value.

The question of how is one to determine which value to choose for r_B will probably arise. For some devices 2 will be an excellent choice while for others the approximate average of 1 will perhaps be more appropriate. Certainly, the value 2 could always be used as a worst-case design approach. However, it would appear that technology is reaching the point where an average value of 1 would, in general, be more appropriate. Of course, the problem of choosing a correct value only arises in the intermediate range of current levels. At low levels of current either choice of r_B would be an insignificant factor. At higher levels the resistance level is so low in comparison to the other series elements that it can probably be ignored. For the purposes of this text the value of r_B chosen for an example will be directly related to the current level; it will extend from a minimum value at high currents of 0.1 to a maximum value of 2 at low levels. Experience will develop a sense for what value to choose, and, indeed, whether it is a factor of significance at all.

In summary, keep in mind that the static or dc resistance of a diode is determined solely by the point of operation, while the dynamic resistance is determined by the shape of the curve in the region of interest.

1.9 AVERAGE AC RESISTANCE

If the input signal is sufficiently large to produce the type of swing indicated in Fig. 1.43, the resistance associated with the device for this region is called the *average ac resistance*. Three values of ac resistance have been calculated and indicated in Fig. 1.43.

The average ac resistance is, by definition, the resistance determined by a straight line drawn between the two intersections determined by the maximum and minimum values of input voltage. In equation form (note Fig. 1.43)

$$r_{av} = \frac{\Delta V_d}{\Delta I_d}\Bigg|_{\text{pt. to pt.}} \tag{1.11}$$

For the situation indicated by Fig. 1.43,

$$r_{av} = \frac{\Delta V_d}{\Delta I_d}\Bigg|_{\text{pt. to pt.}} = \frac{17}{7 \times 10^{-3}} \cong 2.43 \text{ K}$$

The average value of the three ac resistances is

$$\text{Average} = \frac{4.0 \text{ K} + 2.27 \text{ K} + 1.25 \text{ K}}{3} \cong 2.5 \text{ K}$$

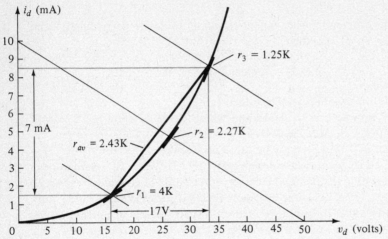

Figure 1.43. Average ac resistance.

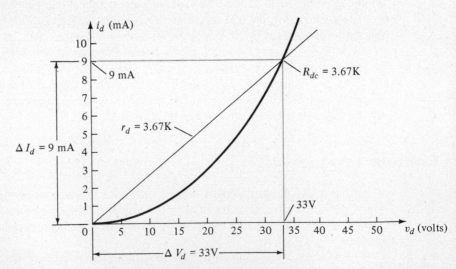

Figure 1.44. Average ac resistance of a region from zero to some positive diode voltage or current.

Note how closely this average value compares to that determined by Eq. (1.11).

If the type of input were such that the region indicated in Fig. 1.44 were employed, the average ac resistance would equal the previously defined dc resistance; that is,

$$r_{av} = \left.\frac{\Delta V_d}{\Delta I_d}\right|_{\text{pt. to pt.}} = \frac{33}{9 \times 10^{-3}} = 3.67 \text{ K}$$

and

$$R_{dc} = \frac{V_D}{I_D} = \frac{33}{9 \times 10^{-3}} = 3.67 \text{ K}$$

It is important to note in this discussion of average ac resistance that the resistance to be associated with the element is determined *only* by the region of interest, *not* by the entire characteristic.

1.10 EQUIVALENT CIRCUITS

An equivalent circuit is a combination of elements properly chosen to best represent the actual terminal characteristics of a device, system, etc. That is, once the equivalent circuit is determined, the device symbol can be removed from a schematic and the equivalent circuit inserted in its place without severely affecting the behavior of the overall system.

One technique for obtaining an equivalent circuit for a diode is to approximate the characteristics of the device by straight line segments such as shown in Fig. 1.45. This type of equivalent circuit is called a *piecewise-linear equivalent circuit*. It should be obvious from each curve that the straight line segments do not result in an exact equivalence between the characteristics and the equivalent circuit. It will, however, at least provide a *first approximation* to its terminal behavior. In each case the resistance chosen is the average ac resistance as defined by Eq. (1.11). The equivalent circuit for each diode appears below the curve in Fig. 1.45. For

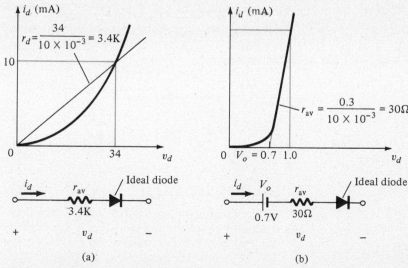

Figure 1.45. Equivalent circuits: (a) vacuum-tube diode; (b) semiconductor diode.

each, an ideal diode was included to indicate that there is only one direction of conduction through the device and that the reverse-bias state is an open-circuit state for each diode. For the vacuum tube, in the forward-bias region, the diode is a short circuit and the diode has a forward resistance of 3.4 K. Although this resistance may be somewhat different from the actual value at a particular point of operation, it does provide an estimate of its value for the diode in use.

Since a silicon semiconductor diode does not reach the conduction state until approximately 0.7 V, an opposing battery V_o of this value must appear in the equivalent circuit. This indicates that the total forward voltage V_D across the diode must be greater than V_o before the ideal diode in the equivalent circuit will be forward biased.

The value of r_{av} for each diode can usually be determined purely from a few numerical values given on a specification sheet. The complete characteristics, therefore, are usually unnecessary for this calculation. For instance, if a tube manual indicates that for a particular vacuum tube that $I_F = 10$ mA at 34 V, then, as above, $r_{av} = 34/10$ mA $= 3.4$ K. For a semiconductor diode, if $I_F = 10$ mA, at 1 V, we know that for silicon a shift of 0.7 V is required before the characteristics rise and

$$r_{av} = \frac{(1 - 0.7)}{10 \text{ mA}} = \frac{0.3}{10 \text{ mA}} = 30 \text{ }\Omega$$

For a germanium diode it would be

$$\frac{(1 - 0.3)}{10 \text{ mA}} = \frac{0.7}{10 \text{ mA}} = 70 \text{ }\Omega$$

The use of the derived equivalent circuit can best be demonstrated by a few examples.

> **EXAMPLE 1.1** For the network of Fig. 1.39, determine the voltage across R, the total diode drop V_D, and the equivalent dc resistance of the diode.
>
> **Solution:** The complete equivalent is substituted in Fig. 1.46.

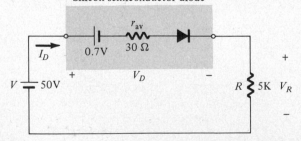

Figure 1.46

Since the applied *emf* of 50 V is much greater than 0.7 V, the ideal diode is forward biased and the short-circuit equivalent can be substituted. Then, using the voltage divider rule, we get

$$V_R = \frac{5 \text{ K}(50 - 0.7)}{5 \text{ K} + 30} = \frac{5 \text{ K}(49.3)}{5030} = 49.006 \cong \mathbf{49} \text{ V}$$

This compares exactly with the value obtained in Section 1.6.

The dc current through the circuit is

$$I_D = \frac{50}{5030} = 9.94 \text{ mA}$$

as compared to 9.8 mA and

$$V_D = 0.7 + I_D(r_{av}) = 0.7 + (9.94 \times 10^{-3})(30) = 0.7 + 0.298 \cong 1 \text{ V}$$

as determined earlier.

Finally,

$$R_{dc} = \frac{V_D}{I_D} = \frac{1}{9.94 \text{ mA}} = 100.6 \text{ }\Omega \text{ versus } 102 \text{ }\Omega$$

The results in this case were excellent. It must be realized, however, that when such an equivalent is used, this type of accuracy cannot always be expected, although a good first approximation is usually provided.

Examining the results above, we see that it should be obvious that the 30 Ω forward resistance of the diode is swamped by the 5 K resistor and could be effectively eliminated from the equivalent circuit and still obtain a good first approximate solution to the circuit. That is,

$$V_R = E - V_0 = 49 \text{ V}$$

$$I_D = \frac{50}{5 \text{ K}} = 10 \text{ mA versus } 9.94 \text{ mA and } V = 0.7 \text{ V}$$

This removal of r_{av} from the equivalent circuit is the same as implying that the characteristics of the diode appear as shown in Fig. 1.47. Indeed, this approximation is frequently employed in semiconductor circuit analysis. The reduced equivalent circuit appears in the same figure. It states that a forward-biased silicon diode in an electronic system under dc conditions has a drop of 0.7 V across it in the conduction state no matter what the diode current (within rated values, of course).

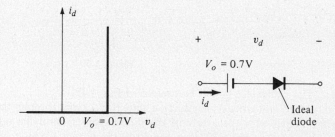

Figure 1.47. Approximate equivalent circuit for the silicon semiconductor diode.

In fact, we can now go a step further and say that the 0.7 V in comparison to the applied 50 V can be ignored leaving only the ideal diode as an equivalent for the semiconductor device. It is for this very reason that a great deal of the applications to follow in later sections use ideal diodes rather than the complete equivalent. Except for small applied voltages or series resistances, it is never too far from the actual response and it doesn't cloud the application with a great deal of mathematical exercises.

EXAMPLE 1.2   For the input shown in Fig. 1.48, determine the output voltage by using the semiconductor diode of Fig. 1.45. Use the complete equivalent circuit.

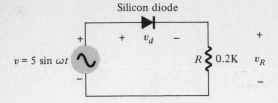

Figure 1.48

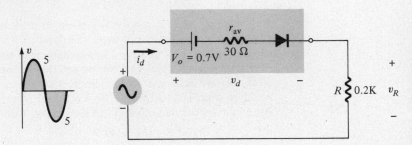

Figure 1.49

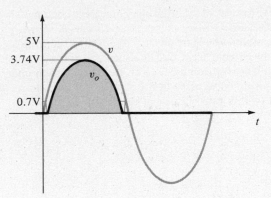

Figure 1.50

Solution: The equivalent circuit is inserted in Fig. 1.49.

The diode will not enter its conduction state (*i* clockwise through the circuit corresponding with the arrow in the diode symbol) until the applied voltage is greater than 0.7 V. This is shown in the output solution of Fig. 1.50. With the input at its maximum value of 5 V, the output is determined by the voltage divider rule.

$$V_o = \frac{(200)(5 - 0.7)}{200 + 30} = \frac{200}{230}(4.3) = 3.74 \text{ V}$$

For an intermediate value such as 3 V:

$$V_o = \frac{(220)(3 - 0.7)}{230} = \frac{200}{230}(2.3) = 2 \text{ V}$$

For e_i less than 0 V, the ideal diode is certainly reverse biased, and $v_0 = 0$ V as shown in Fig. 1.50. In this type of application in which the input swing extends throughout a wide range of the characteristics, the piecewise equivalent circuit that employs r_{av} will give excellent results as a first approximation. However, let us now consider the small-signal situation in which the region of operation is very limited.

At high currents, such as 10 mA, if we use Eq. (1.10), we find the resistance to be

$$r_d = \frac{26}{I_D} + 2 = \frac{26}{10} + 2 = 4.65\,\Omega$$

while at 0.5 mA it is

$$r_d = \frac{26}{I_D} + 2 = \frac{26}{0.5} + 2 = 52 + 2 = 54\,\Omega$$

Depending on the region of operation, therefore, the average value of $30\,\Omega$ may be far from accurate. In this case, since the dc diode current will probably be discernible by first using the approximation of Fig. 1.47, it would be best to substitute this value into Eq. (1.10) and use this value of resistance. For small-signal applications in which the signal rides on a dc level such as in Fig. 1.41, the diode will always be forward biased and the ideal diode can be replaced by its short-circuit equivalent. The dc level of V_o can be removed from the equivalent for the ac response since it will only determine the "riding" dc level and not affect the peak-to-peak (p-p) ac response. We will find in the analysis of electronic systems that the dc and ac response can normally be determined separately. That is, the theorem of superposition can usually be applied. This will be demonstrated in the next example.

EXAMPLE 1.3 Determine the voltage v_d across the diode for the input shown in Fig. 1.51.

Solution: The 2-V dc level will ensure that the diode is always forward biased. The "dc" equivalent appears in Fig. 1.52 using the approximation of Fig. 1.47. The dc diode current is found to be

$$I_D \cong \frac{2 - 0.7}{100} = \frac{1.3}{100} = 13\,\text{mA}$$

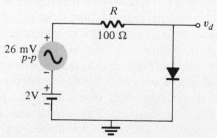

Figure 1.51

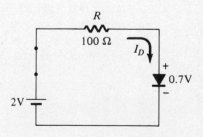

Figure 1.52

Substituting into Eq. (1.10), we get

$$r_d = \frac{26}{13} + 2 = 2 + 2 = 4\,\Omega$$

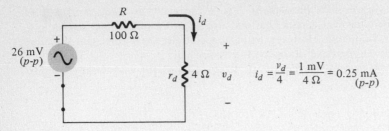

Figure 1.53

$$i_d = \frac{v_d}{4} = \frac{1\,\text{mV}}{4\,\Omega} = 0.25 \begin{array}{c}\text{mA}\\(p\text{-}p)\end{array}$$

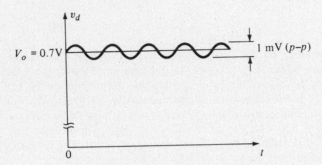

Figure 1.54

The "ac" equivalent is then drawn in Fig. 1.53 and the ac voltage across the diode is determined.

$$v_{d_{ac}} = \frac{4(26\,\text{mV}_{(p-p)})}{4 + 100} = 1\,\text{mV}_{(p-p)}$$

The complete solution (applying the superposition theorem) appears in Fig. 1.54. The technique demonstrated above will be frequently used in the analysis in later chapters.

Electronic devices are inherently sensitive to very high frequencies. Most shunt capacitive effects that can be ignored at lower frequencies ($X_C = 1/2\pi fc$ very large: open-circuit equivalent) cannot be ignored at higher frequencies since X_C will introduce a low-reactance "shorting" path. In the p-n junction semiconductor diode a *diffusion capacitance* will come into play in the forward-bias region. It is a capacitance that appears because of the majority carriers that have diffused into the oppositely polarized semiconductor material during the conduction state. Those electrons that diffuse into the p-type material create a region of charge separated (by a reduced depletion region width) from the holes that have diffused into the n-type material. Recall that capacitive effects are present whenever oppositely charged layers are separated by a finite distance. This capacitance can have a pronounced effect on the diodes' application in high-speed systems since the period of transition from the conduction to nonconduction state is greater. Also recall that the voltage cannot change instantaneously across a capacitor but takes a period of time controlled by the capacitance and series resistive elements. The diffusion capacitance

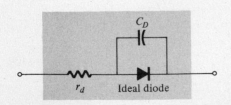

Figure 1.55. ac equivalent for a forward-biased diode.

is typically in the order of magnitude of a few picofarads at low forward-bias potentials, but it does increase almost exponentially with increase in the forward bias. The point-contact diode introduced earlier with its small contact area (low A in $C = \epsilon A/d$) has a high-frequency range of application. Including C_D in the small-signal equivalent model will result in the network appearing in Fig. 1.55.

1.11 CLIPPERS AND CLAMPERS

Clippers and clampers are diode waveshaping circuits. Each performs the function indicated by its name. The output of clipping circuits appears as if a portion of the input signal were clipped off. Clamping circuits simply clamp the waveform to a different dc level.

Clippers

A clipping circuit requires at least two fundamental components, a diode and a resistor. A dc battery, however, is also frequently used. The output waveform can be clipped at different levels simply by interchanging the position of the various elements and changing the magnitude of the dc battery. Only ideal diodes appear in the examples to follow. As indicated in the previous section, however, the response would not be severely altered if semiconductor devices were used.

For networks of this type it is often helpful to consider particular instants of the time varying input signal to determine the state of the diode. Keep in mind that *at any instant of time* a varying signal can simply be replaced by a dc source of the same value. This is clearly shown in Example 1.4.

EXAMPLE 1.4

Clipper: Find the output voltage waveshape (V_o) for the inputs shown in Fig. 1.56.

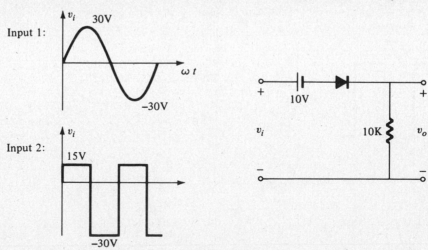

Figure 1.56. Clipping circuit and inputs for Example 1.4.

Solution:

Input 1: For any value of $v_i > 10$ V the ideal diode is forward biased and $v_o = v_i - 10$. For example, at $v_i = 15$ V (Fig. 1.57), the result is $v_o = 15 - 10 = 5$ V.

For any value of $v_i < 10$ V the ideal diode is reverse biased and $v_o = 0$ since the current in the circuit is zero. For example, at $v_i = 5$ V (Fig. 1.58).

The output waveform v_o appears as if the entire input were clipped off except the positive peak (Fig. 1.59).

Input 2: The diode will change state at the same levels indicated for the first input. The output waveform appears as shown in Fig. 1.60.

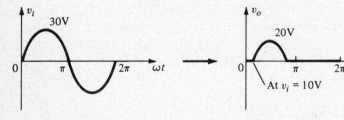

Figure 1.57. Clipping circuit of Fig. 1.56 at instant $V_i = 15$ V of input 1.

Figure 1.58. Clipping circuit of Fig. 1.56 at instant $V_i = 5$ V of input 1.

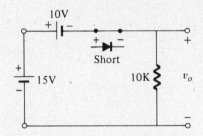

Figure 1.59. Output wave form (v_o) for input 1 to the clipping circuit of Fig. 1.56

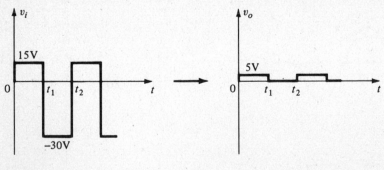

Figure 1.60. Output wave form (v_o) for input 2 to the clipping circuit of Fig. 1.56.

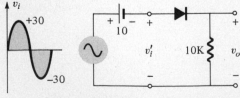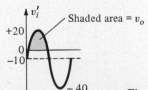

Figure 1.61

SIMPLE SERIES CLIPPERS

POSITIVE   NEGATIVE

BIASED SERIES CLIPPERS

SIMPLE PARALLEL CLIPPERS

BIASED PARALLEL CLIPPERS

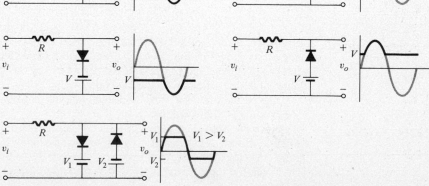

Figure 1.62. Clipping circuits.

Another technique, other than treating instantaneous values of the input as dc levels, is to redraw the applied voltage as shown in Fig. 1.61. Note that the 10-V dc level has only shifted the sinusoidal input down 10 V (Fig. 1.61). In order for the diode to be forward biased, the input v_i' must be positive. This region as clearly shown in the same figure represents the only region that will pass through to the load. A number of clipping circuits and their effect on the applied signal appear in Fig. 1.62.

Clampers

The clamping circuit has a minimum requirement of three elements: a diode, a capacitor, and a resistor. The clamping circuit may also be augmented by a dc battery. The magnitudes of R and C must be chosen such that the time constant $\tau = RC$ is large enough to ensure that the voltage across the capacitor does not change significantly during the interval of time, determined by the input, that both R and C affect the output waveform. The need for this condition will be demonstrated in Example 1.5. Throughout the discussion, we shall assume that for all practical purposes a capacitor will charge to its final value in five time constants.

It is usually advantageous when examining clamping circuits to first consider the conditions that exist when the input is such that the diode is forward biased.

EXAMPLE 1.5

Clamper: Draw the output voltage waveform (v_o) for the input shown (Fig. 1.63).

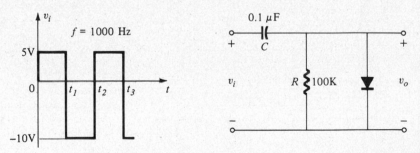

Figure 1.63. Clamping circuit and input for Example 1.5.

Solution: At the instant the input switches to the $+5$-V state the circuit will appear as shown in Fig. 1.64. The input will remain the $+5$-V state for an interval of time equal to one-half the period of the waveform since the time interval $0 \longrightarrow t_1$ is equal to the interval $t_1 \longrightarrow t_2$.

The period of v_i is $T = 1/f = 1/1000 = 1$ ms and the time interval of the $+5$-V state is $T/2 = 0.5$ ms.

Since the output is taken from directly across the diode it is 0 V for this interval of time. The capacitor, however, will rapidly charge to 5 V since the time constant of the network is now $\tau = RC \cong 0C = 0$.

When the input switches to -10 V the circuit of Fig. 1.65 will result.

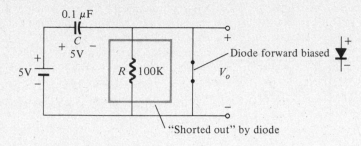

Figure 1.64. Clamping circuit of Fig. 1.63, when $v_i = 5$ V $(0 \longrightarrow t_1)$.

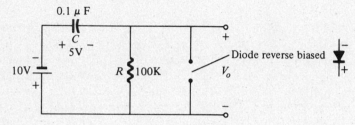

"Shorted out" by diode

Figure 1.65. Clamping circuit of Fig. 1.63, when $v_i = -10$ V $(t_1 \longrightarrow t_2)$.

The time constant for the circuit of Fig. 1.65 is

$$\tau = RC = 100 \times 10^3 \times 0.1 \times 10^{-6} = 10 \text{ ms}$$

Since it takes approximately five time constants or 50 ms for a capacitor to discharge, and the input is only in this state for 0.5 ms, to assume the voltage across the capacitor does not change appreciably during this interval of time is certainly a reasonable approximation. The output is therefore

$$V_o = \underset{\text{supply}\longrightarrow}{-10} - \overset{\text{capacitor}}{5} = -15 \text{ V}$$

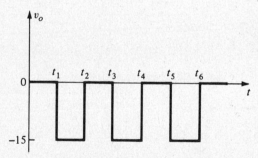

Figure 1.66. Output (v_o) for clamping circuit and input of Fig. 1.63.

The resulting output waveform (v_o) is provided in Fig. 1.66. As indicated, the output is clamped to the negative region and will repeat itself at the same frequency as the input signal. Note that the swing of the input and output voltages is the same: 15 V. *For all clamping circuits the voltage swing of the input and output waveforms will be the same.* This certainly not the case for clipping circuits.

If, for discussion sake, the 100-K resistor were replaced by a 1-K resistor, then the time constant

$$\tau = RC = (10^3)(0.1 \times 10^{-6}) = 0.1 \text{ ms}$$

and $\qquad 5\tau = 0.5$ ms (approximate total discharge time)

The capacitor, therefore, would discharge during the interval in which the voltage is 10 V since the time intervals match. The output waveform would then appear as shown in Fig. 1.67. Clamping networks must have time constants determined by the product RC that will result in $5RC$ being signifi-

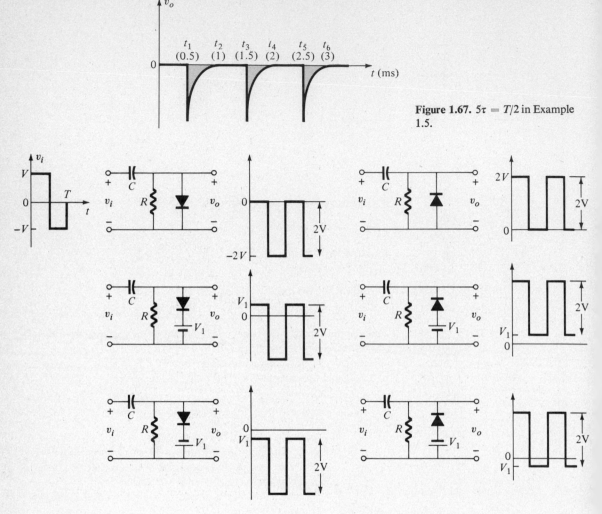

Figure 1.67. $5\tau = T/2$ in Example 1.5.

Figure 1.68. Clamping circuits ($5\tau = 5RC \gg T/2$).

cantly greater than the time interval in which the diode is reverse biased or the waveform will be severely distorted.

A number of clamping circuits and their effect on the input signal appear in Fig. 1.68.

1.12 ZENER DIODES

The Zener and avalanche region of the semiconductor diode were discussed in detail in Section 1.4. It occurs at a reverse-bias potential of V_z for the diode of Fig. 1.69a. For the purposes of introducing notation for the Zener diode and comparing its characteristics to those of the ideal diode, a semiconductor Zener diode characteristic has been drawn as shown in Fig. 1.69b.

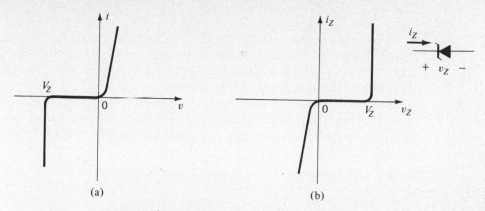

(a)   (b)

Figure 1.69. Zener diodes: (a) Zener potential; (b) characteristics and notation.

The location of the Zener region can be controlled by varying the doping levels. An increase in doping, producing an increase in the number of added impurities, will decrease the Zener potential. Zener diodes are available having Zener potentials. of 2.4 to 200 V with power ratings from $\frac{1}{4}$ to 50 W. Because of its higher temperature and current capability, silicon is usually preferred in the manufacture of Zener diodes.

The complete equivalent circuit of the Zener diode in the Zener region includes a small dynamic resistance and dc battery equal to the Zener potential as shown in Fig. 1.70a. For all applications to follow, however, we shall assume as a first approximation that the external resistors are much larger in magnitude than the Zener-equivalent resistor and the equivalent circuit is simply that indicated in Fig. 1.70b.

A larger drawing of the Zener region is provided in Fig. 1.71 to permit a description of the Zener nameplate data appearing below for a 1N961, Fairchild, 500 mW, 20% diode.

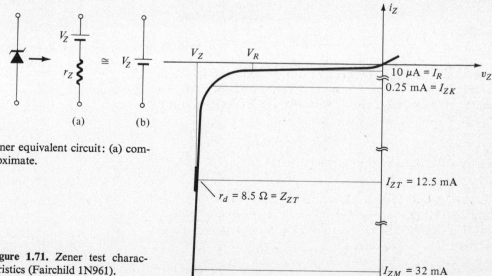

Figure 1.70. Zener equivalent circuit: (a) complete; (b) approximate.

Figure 1.71. Zener test characteristics (Fairchild 1N961).

The term nominal associated with V_Z indicates that it is a typical average value. Since this is a 20% diode, the Zener potential can be expected to vary as 10 V $\pm$ 20% or from 8 to 12 V in its range of application. Also available are 10% and 5% diodes with the same specifications. The test current I_{ZT} is a typical operating level and Z_{ZT} is the dynamic impedance at this current level. The maximum knee impedance occurs at the knee current of I_{ZK}. The reverse saturation current is provided at a particular potential level and I_{ZM} is the maximum current for the 20% unit.

Electrical Characteristics (25°C Ambient Temperature unless otherwise noted)

JEDEC TYPE	ZENER VOLTAGE NOMINAL (V_Z)	TEST CURRENT (I_{ZT})	MAX DYNAMIC IMPEDANCE ($Z_{ZT}@I_{ZT}$)	MAXIMUM KNEE IMPEDANCE ($Z_{ZK}@I_{ZK}$)		MAXIMUM REVERSE CURRENT ($I_R@V_R$)	TEST VOLTAGE (V_R)	MAXIMUM REGULATOR CURRENT (I_{ZM})	TEMPERATURE COEFFICIENT (TYP)
	V	mA	Ω	Ω	mA	μA	V	mA	$\%/°C$
1N961	10	12.5	8.5	700	0.25	10	7.2	32	+.072

The temperature coefficient reflects the change in V_Z with temperature. For this diode it will change 0.072% per degree rise in temperature. A 1-degree rise in temperature will cause an increase in V_Z of (0.072/100 $\times$ 10 V = 0.0072 V, while a 50-degree change would result in a (50)(0.0072) = 0.36 V rise. At I_{ZM}, $V_Z \cong 12$ V for this 20% unit and the power dissipation $P_D = (12)(32 \times 10^{-3}) = 384$ mW which is less than the 500 mW stated above, but we must also consider the power derating curve of Fig. 1.72 provided with the above specifications. At 75°C the power rating has dropped to approximately 390 mW coinciding with the value obtained above. The specification sheet also indicated that for Zeners where the

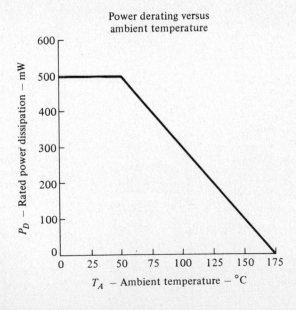

Power derating versus ambient temperature

Figure 1.72. Power derating curve for the 1N961 Fairchild Zener diode.

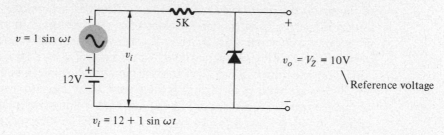

$$v_i = 12 + 1 \sin \omega t$$

Figure 1.73. Reference voltage.

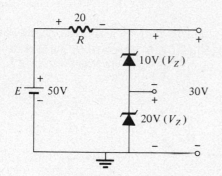

Figure 1.74. Two reference voltages.

actual V_Z is known the I_{ZM} may be increased according to the operating curve. A few applications of the very popular Zener will now be examined.

It is often necessary to have a fixed *reference* voltage in a network for biasing and comparison purposes. This can be accomplished using a Zener diode as shown in Fig. 1.73. The variation in dc supply voltage due to any number of reasons has been included as a small sinusoidal signal.

Since v_i is always greater than 10 V, the Zener diode will always be in the "on" state. The output voltage v_o, therefore, will remain fixed at the Zener potential of 10 V, our reference potential.

Two reference levels can be established by the network of Fig. 1.74. Two back-to-back Zeners can be used as an ac regulator (Fig. 1.75). For the sinusoidal signal

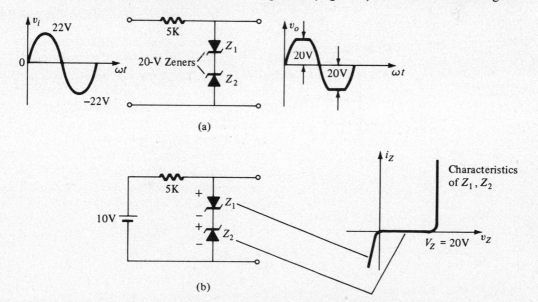

Figure 1.75. Sinusoidal ac regulation: (a) 40-V peak-to-peak sinusoidal ac regulator; (b) circuit operation at $V_i = 10$ V.

v_i the circuit will appear as shown in Fig. 1.75b at the instant $v_i = 10$ V. The region of operation for each diode is indicated in the adjoining figure. Note that the impedance associated with Z_1 is very small, or essentially a short, since it is in series with 5 K, while the impedance of Z_2 is very large corresponding to the open-circuit representation. Since Z_2 is an open circuit, $v_o = v_i = 10$ V. This will continue to be the case until v_i is slightly greater than 20 V. Then Z_2 will enter the low-resistance region (Zener region) and Z_1 will for all practical purposes be a short circuit and Z_2 will be replaced by $V_Z = 20$ V. The resultant output waveform is indicated in the same figure. Note that the waveform is not purely sinusoidal, but its rms (effective) value is closer to the desired 20-V peak sinusoidal waveform than the sinusoidal input having a peak value of 22 V.

The circuit of Fig. 1.75a can be extended to that of a simple square-wave generator (due to its clipping action) if the signal v_i is increased to perhaps 40-V peak with 10-V Zeners. The resultant waveform is indicated in Fig. 1.76.

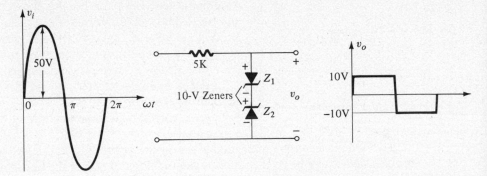

Figure 1.76. Simple square-wave generator.

Lastly, let us use the Zener diode to maintain a fixed voltage across a load for a variation in the load (a voltage regulator). The basic Zener regulator appears in Fig. 1.77. In order to maintain V_Z at 10 V (the Zener potential), the diode must first be in the "on" state. Certainly, for values of v_i less than 10 V the Zener device has not reached its Zener potential. Will it "fire" at 12 V? To determine its state, first remove the Zener diode from the network and find the voltage across its open-circuit terminals using the voltage divider rule (Fig. 1.78).

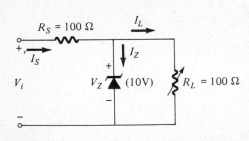

Figure 1.77. Zener regulator.

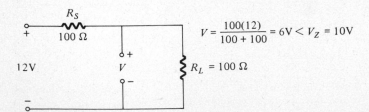

$$V = \frac{100(12)}{100 + 100} = 6\text{V} < V_Z = 10\text{V}$$

Figure 1.78

The resulting voltage is still less than the needed 10 V and the diode remains open. By setting V_Z to 10 V we can calculate the required applied voltage in the following manner:

$$V_Z = 10 = \frac{100(V_i)}{100 + 100} \quad \text{and} \quad V_i = \text{50 V}$$

20 V

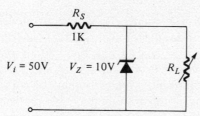

Figure 1.79

Let us now examine the behavior of the network with $V_i = 100$ V (Fig. 1.79). The voltage

$$V_{R_S} = 100 - 10 = 90 \text{ V}$$

$$I_S = \frac{90}{100} = 900 \text{ mA}$$

with

$$I_{R_L} = \frac{10}{100} = 100 \text{ mA}$$

and

$$I_Z = I_S - I_L = 900 - 100 = 800 \text{ mA}$$

EXAMPLE 1.6   For the network of Fig. 1.80, determine the range of I_L that will result in V_{R_L} being maintained at 10 V.

I_z max = 35 ma

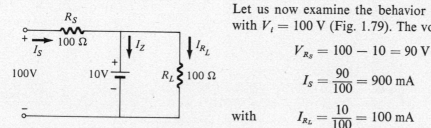

Figure 1.80. Network for Example 1.6.

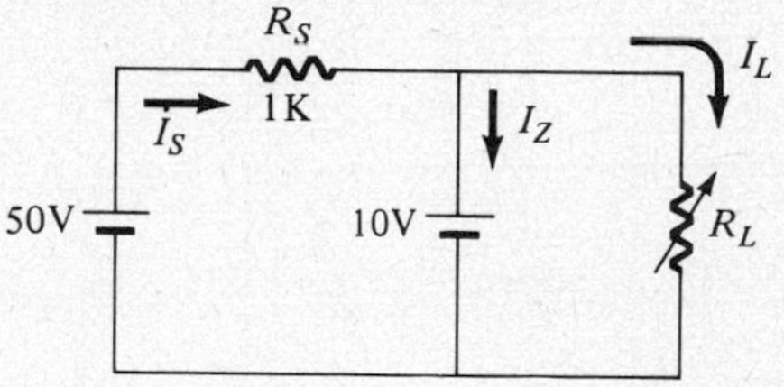

Figure 1.81

Solution: The minimum value of R_L that will ensure that the diode is in the "on" state can be determined through Fig. 1.81.

$$V_{R_L} = V_Z = 10 = \frac{R_{L(\min)}(50 \text{ V})}{R_{L(\min)} + 1000}$$

and

$$10{,}000 + 10R_{L(\min)} = 50R_{L(\min)}$$

or

$$40R_{L(\min)} = 10{,}000$$

and

$$R_{L(\min)} = 250 \text{ }\Omega$$

The minimum load resistance corresponds with the maximum I_L and

$$I_{L(\min)}^{(\max)} = \frac{V_Z}{R_{L(\min)}} = \frac{10}{250} = 40 \text{ mA}$$

$I_{L(\min)}$ is determined by examining the network with maximum I_Z (I_{Z_M}). When the diode is in the "on" state, $V_S = 50 - 10 = 40$ V and I_S is maintained at $I_S = 40/1000 = 40$ mA. Since $I_L = I_S - I_Z$, I_L will be a minimum when I_Z is a maximum, or I_{Z_M}. Therefore,

$$I_{L(\min)} = I_S - I_{Z_M} = 40 - 35 = 5 \text{ mA}$$

and incidently

$$R_{L(\text{max})} = \frac{V_Z}{I_{L(\text{min})}} = \frac{10}{5 \text{ mA}} = 2 \text{ K}$$

A plot of V_L versus I_L appears in Fig. 1.82.

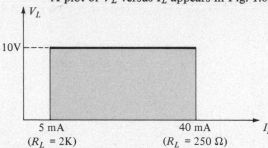

I_L

Figure 1.82. V_{R_L} vs. I_L for the network of Fig. 1.80.

1.13 TUNNEL DIODES

The tunnel diode was first introduced by Dr. Leo Esaki in 1958. Its characteristics, shown in Fig. 1.83, are different from any diode discussed thus far in that it has a negative resistance region. In this region, an increase in terminal voltage results in a reduction in diode current.

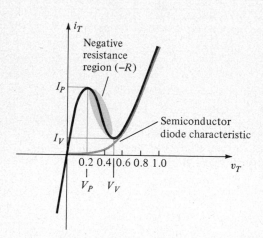

Figure 1.83. Tunnel diode characteristic.

The tunnel diode is fabricated by doping the semiconductor materials that will form the *p-n* junction at a level one hundred to several thousand times that of a typical semiconductor diode. This will result in a greatly reduced depletion region in the order of magnitude of 10^{-6} cm, or typically about $\frac{1}{100}$ the width of this region for a typical semiconductor diode. It is this thin depletion region that many carriers can "tunnel" through, rather than attempt to surmount, at low forward-bias potentials that accounts for the peak in the curve of Fig. 1.83. For comparison purposes, a typical semiconductor diode characteristic has been superimposed on the tunnel diode characteristic of Fig. 1.83.

This reduced depletion region results in carriers "punching through" at velocities that far exceed that available with conventional diodes. The tunnel diode can therefore be used in high-speed applications such as in computers where switching times in the order of nanoseconds or picoseconds are desirable.

You will recall in a previous section on Zener diodes that an increase in the doping level will drop the Zener potential. Note the effect of a very high doping level on this region in Fig. 1.83. The semiconductor materials most frequently used in the manufacture of tunnel diodes are germanium and gallium arsenide. The ratio I_p/I_v is very important for computer applications. For germanium it is typically 10:1 while for gallium arsenide it is closer to 20:1.

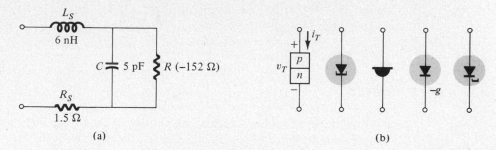

Figure 1.84. Tunnel diode: (a) equivalent circuit; (b) symbols.

The peak current, I_p, of a tunnel diode can vary from a few microamperes to several hundred amperes. The peak voltage, however, is limited to about 600 mV. For this reason, a simple VOM with an internal dc battery potential of 1.5 V can severely damage a tunnel diode if used improperly.

The tunnel diode equivalent circuit in the negative-resistance region is provided in Fig. 1.84, with the symbols most frequently employed for tunnel diodes. The values, for each parameter, are for the 1N2939 GE tunnel diode whose specifications appear below. The inductor L_s is due mainly to the terminal leads. The resistor R_s is due to the leads, ohmic contact at the lead-semiconductor junction, and the semiconductor materials themselves. The capacitance C is the junction diffusion capacitance and the $-R$ is the negative resistance of the region. The negative resistance finds application in oscillators to be described later.

Specifications: Ge 1N2939

ABSOLUTE MAXIMUM RATINGS (25°C)
    current

Forward	(-55 to $+100$°C)	5 mA	
Reverse	(-55 to $+100$°C)	10 mA	

ELECTRICAL CHARACTERISTICS (25°C) (1/8 in. Leads)

	Min.	Typ.	Max.	
I_P	0.9	1.0	1.1	mA
I_V		0.1	0.14	mA
V_P	50	60	65	mV
V_V		350		mV
REVERSE VOLTAGE ($I_R = 1.0$ mA)			30	mV
FORWARD PEAK POINT CURRENT VOLTAGE V_{fp}	450	500	600	mV
I_p/I_v		10		
$-R$		-152		Ω
C		5	15	pF
L_s		6		nH
R_S		1.5	4.0	Ω

Note the lead length of $\frac{1}{8}$ in. included in the specifications. An increase in this length will cause L_s to increase. In fact, it was given for this device that L_s will vary 1 to 12 nH depending on lead length. At high frequencies ($X_{L_s} = 2\pi f L_s$) this factor can take its toll.

The fact that $V_{fp} = 500$ mV (typ.) and $I_{forward}$ (max.) $= 5$ mA indicates that tunnel diodes are low-power devices ($P_D = (0.5 \times 5 \times 10^{-3}) = 2.5$ mW) which is also excellent for computer applications.

1.14 POWER DIODES

There are a number of diodes designed specifically to handle the high-power and high-temperature demands of some applications. The most frequent use of power diodes occurs in the rectification process, in which ac signals (having zero average value) are converted to one having an average or dc level. When used in this capacity, diodes are normally referred to as rectifiers.

The majority of the power diodes are constructed using silicon because of its higher current, temperature, and PIV ratings. The highercurrent demands require that the junction area be larger to ensure that there is a low forward diode resistance. If the forward resistance were too large, the I^2R losses would be excessive. The current capability of power diodes can be increased by placing two or more in parallel and the PIV rating can be increased by stacking the diodes in series.

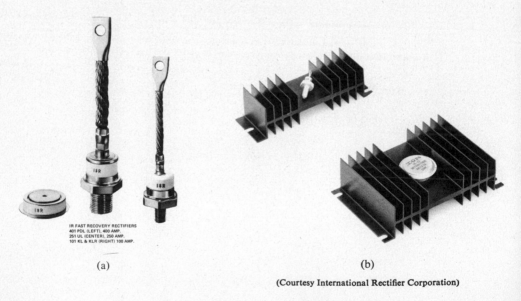

IR FAST RECOVERY RECTIFIERS
401 PDL (LEFT), 400 AMP.
251 UL (CENTER), 250 AMP.
101 KL & KLR (RIGHT) 100 AMP.

(a)   (b)

(Courtesy International Rectifier Corporation)

Figure 1.85. Power diodes and heat sinks.

Various types of power diodes and their current rating have been provided in Fig. 1.85a. The high temperatures resulting from the heavy current flow require, in many cases, that heat sinks be used to draw the heat away from the element. A few of the various types of heat sinks available are shown in Fig. 1.85b. If heat sinks are not employed, stud diodes are designed to be attached directly to the chassis, which in turn will act as the heat sink.

1.15 VARICAP DIODES

Varicap (also called varactor or VVC) diodes are semiconductor, voltage-dependent, variable capacitors. Their mode of operation depends on the capacitance that exists at the *p-n* junction when the element is reverse biased. Under reverse-bias conditions, it was established in Section 1.4 that there is a region of uncovered charge on either side of the junction that in combination make up the depletion region and define the depletion width W_d. The capacitance established by the isolated uncovered charges is determined by

$$C_T = \frac{\epsilon A}{W_d} \tag{1.12}$$

where ϵ is the permittivity of the semiconductor materials, A is the *p-n* junction area, and W_d is the depletion width. Note the similarities between this equation and the basic equation for capacitance: $C = \epsilon A/d$. Recall from the equivalent circuit section that there is a diffusion capacitance in the forward-bias region because carriers have diffused into the opposite region. In the reverse-bias condition the uncovered charges result in a capacitance commonly called the *depletion region* or *transition* (hence the subscript T) *region* capacitance.

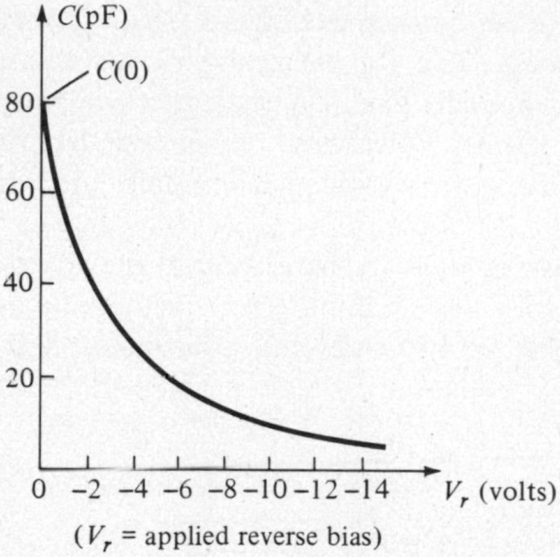

As the reverse-bias potential increases, the width of the depletion region increases which in turn reduces the transition capacitance. The characteristics of a typical commercially available varicap diode appear in Fig. 1.86. Note the initial sharp decline in C_T with increase in reverse bias. The normal range of V_r for VVC diodes is limited to about 20 V. In terms of the applied reverse bias, the transition capacitance is given by

Figure 1.86. Varicap characteristics: C(pF) vs. V_r.

$$C_T = \frac{K}{(V_o + V_r)^n} \tag{1.13}$$

where K = constant determined by the semiconductor material and construction technique;

V_o = knee potential as defined in Section 1.4;

V_r = applied reverse-bias potential:

$n = \frac{1}{2}$ for alloy junctions and $\frac{1}{3}$ for diffused junctions.

In terms of the capacitance at the zero-bias condition $C(0)$, the capacitance as a function of V_r is given by

$$C_T(V_r) = \frac{C(0)}{\left(1 + \frac{V_r}{V_o}\right)^n} \tag{1.14}$$

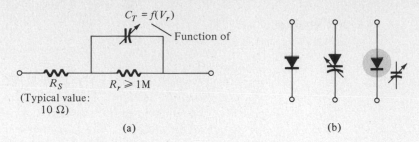

Figure 1.87. Varicap diode: (a) equivalent circuit in the reverse bias region; (b) symbols.

The symbols most commonly used for the varicap diode and a first approximation for its equivalent circuit in the reversebias region are shown in Fig. 1.87. Since we are in the reversebias region, the resistor in the equivalent circuit is very large in magnitude—typically 1 M or larger—while R_s, the geometric resistance of the diode, is, as indicated in Fig. 1.87, very small. The magnitude of C will vary from about 2 to 100 pF depending on the varicap considered. To ensure that R_r is as large (for minimum leakage current) as possible, silicon is normally used in varicap diodes.

The fact that the magnitude of the capacitance is in the picofarad range suggests that this device is used in high-frequency applications. Some of these areas include FM modulators, automatic frequency control devices, adjustable band-pass filters, and parametric amplifiers.

1.16 SCHOTTKY BARRIER (HOT-CARRIER) DIODE

In recent years there has been increasing interest in the Schottky barrier diode which through its unique design has resulted in a number of improved characteristics over the point-contact diode in high-frequency applications.

Its construction is very different from the conventional *p-n* junction in that a metal semiconductor junction is created such as shown in Fig. 1.88. The semiconductor is normally *n*-type silicon (although *p*-type silicon is sometimes used),

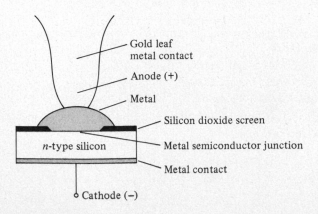

Figure 1.88. Passivated hot-carrier diode.

but a host of different metals such as molybdenum, platinum, chrome, or tungsten are also used. Different construction techniques will result in a different set of characteristics for the device, such as increased frequency range, lower forward bias, etc. Priorities do not permit an examination of each technique here but information will usually be provided by the manufacturer. In general, however, this type of construction results in a more uniform junction region and increased ruggedness compared to the point-contact diode (its prime competitor).

In both materials, the electron is the majority carrier. In the metal, the level of minority carriers (holes) is insignificant. Under forward-bias conditions (indicated by the polarities in the figure), the electrons in the n-type silicon semiconductor material can easily flow into the adjoining metal establishing a heavy flow of majority carriers. Since the injected carriers have a very high kinetic energy level compared to the electrons of the metal, they are commonly called "hot carriers." In the conventional p-n junction it was the injection of minority carriers into the adjoining region. Here the electrons are injected into a region of the same electron plurality resulting in Schottky diodes being unique in that conduction is entirely by majority carriers. In one Hewlett-Packard application note the following analogy was made: "The mechanism of electron flow between the two materials is analogous to the theremionic emission from a hot cathode into a vacuum. The semiconductor in this case acting as the cathode, and the metal as the vacuum."

The exponential rise in current with forward bias is described by Eq. (1.1) but with η dependent on the construction technique (1.05 for the metal whisker type of construction that is somewhat similar to the point-contact diode). In the reverse-bias region the current I_s is due primarily to those electrons in the metal passing into the semiconductor material. One of the areas of continuing research on the Schottky diode centers on reducing the high leakage currents that result with temperatures over 100°C. Through design improvement units are now becoming available that have a temperature range from -65°C to $+150$°C. At room temperature, I_s is typically in the microampere range for low-power units and milliampere range for high-power devices, although it is typically larger than that encountered using conventional p-n junction devices with the same current limits. In addition, even though Schottky diodes exhibit better characteristics than the point-contact diode in the reverse-bias region as shown in Fig. 1.89, the PIV of these diodes is usually significantly less than that of a comparable p-n junction unit. Typically, for a 50-A unit, the PIV of the Schottky diode may be 50 V while it will probably be more like 150 V for the p-n junction variety. It is obvious from the characteristics of Fig. 1.89 that the Schottky diode is closer to the ideal set of characteristics than the point-contact and has levels of V_o less than the typical silicon semiconductor p-n junction. The level of V_o for the "hot-carrier" diode is controlled to a large measure by the metal employed. There exists a required trade-off between temperature range and level of V_o. An increase in one appears to correspond with a resulting increase in the other.

The maximum current rating of the device is presently limited to about 60 A although 100-A units appear to be on the horizon. One of the primary areas of application of this diode is in *switching power supplies* that operate in the frequency range of 20 kHz or more. A typical unit at 25°C may be rated at 50 A at a forward

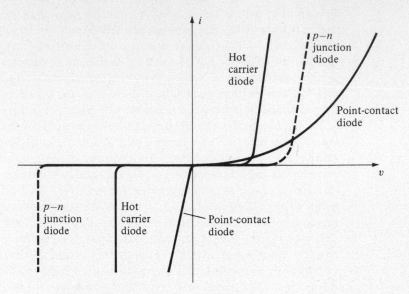

Figure 1.89. Comparison of characteristics of hot-carrier, point-contact and *p-n* junction diodes.

voltage of 0.6 V with a recovery time of 10 ns for use in one of these supplies. A *p-n* junction device with the same current limit of 50 A may have a forward voltage drop of 1.1 V and a recovery time of 30 to 50 ns. The difference in forward voltage may not appear significant, but consider the power dissipation difference: $P_{(\text{hot carrier})} = (0.6)(50) = 30$ W compared to $P_{(p-n)} = (1.1)(50) = 55$ W which is a measureable difference when efficiency criteria must be met. Another important advantage of the Schottky diode is the very low noise figure which is extremely important in communication receivers, radar units, etc. Depending on the construc-

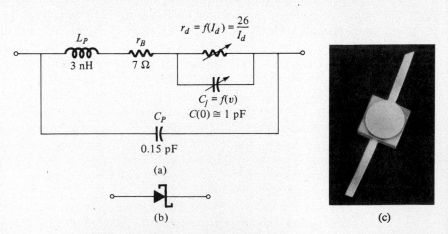

Figure 1.90. Schottky (hot-carrier) diode: (a) equivalent circuit; (b) symbol; (c) photograph of microwave unit (Courtesy of Hewlett-Packard).

tion technique, Schottky diodes can operate up to a limit of about 20 GHz. For higher frequencies the point-contact diode witn its small junction area is still employed.

The equivalent circuit for the device (with typical values) and a commonly used symbol appcar in Fig. 1.90. The inductance L_p and capacitance C_p are package values. R_s is the series resistance which includes the contact and bulk resistance. The resistance r_d and capacitance C_j are values defined by equations introduced in earlier sections. Other areas of application include clipping and clamping networks, computer gating, mixing and detecting networks (in communication systems).

1.17 PHOTOTUBES

The phototube is a two-terminal, light-sensitive device having the basic construction and symbol appearing in Fig. 1.91. The cathode is designed to pick up the maximum incident light possible. It is either constructed of, or coated with, a

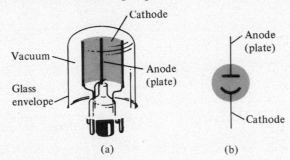

Figure 1.91. Vacuum phototube: (a) basic construction; (b) symbol.

metallic material having photoemissive properties, such as sodium, potassium, and cesium. The material used will be determined by the wavelength of the incident light waves. The wavelength, being the distance between successive peaks of the traveling light wave, usually is measured in angstrom units (Å) or microns (μ), where

$$1 \text{ Å} = 10^{-10} \text{ m} \quad \text{and} \quad 1 \ \mu = 10^{-6} \text{ m}$$

It is related to the frequency of the traveling wave by

$$\lambda = \frac{v}{f} \tag{1.15}$$

where λ = wavelength in meters;
v = velocity of light, 3×10^8 m/s; and
f = frequency in hertz of the traveling wave.

The spectral response of a frequently used phototube appears in Fig. 1.92. Note the effect of wavelength on the color of the visible light. The energy associated with the incident light is directly related to the *frequency* of the traveling wave by

$$\boxed{W_{\text{(joules)}} = \hbar f}$$ (1.16)

where $\hbar$ is called Planck's constant and is equal to 6.624×10^{-34} Joule-seconds.

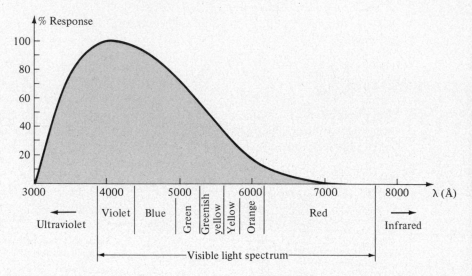

Figure 1.92. Phototube spectral response.

It clearly states that since $\hbar$ is a constant, the increase in energy associated with incident light waves is directly related to the frequency of the traveling wave. It has been further theorized that this light energy exists in the form of discrete packages of energy called *photons* rather than in a continuous distribution.

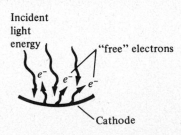

Figure 1.93. Photoemission.

The process of photoemission is clearly depicted in Fig. 1.93. The photons of energy in the incident light are absorbed by the relatively "free" electrons near and on the surface of the photoemissive material. A number of these electrons will then possess sufficient energy to leave the surface of the material as indicated in the figure. The energy associated with these free electrons will be directly proportional to the *frequency* of the incident light as determined by Eq. (1.16). The number of free electrons, however, is proportional to the *intensity* of the incident light. Light intensity is a measure of the amount of luminous flux falling on a particular surface area. Luminous flux is normally measured in lumens (lm) or watts. The two units are related by

$$1 \text{ lm} = 1.496 \times 10^{-3} \text{ W}$$

The light intensity is normally measured in lm/ft², footcandles (fc), or W/m², where

$$1 \text{ lm/ft}^2 = 1 \text{ fc} = 1.609 \times 10^{-12} \text{ W/m}^2$$

Let us now examine how this photoemissive effect will affect the behavior of the relatively simple photoelectric circuit of Fig. 1.94. Note the necessity for a separate dc supply. Consider also that the plate of the tube is, before emission takes

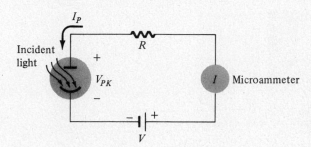

Figure 1.94. Basic phototube circuit.

place, positive with respect to the cathode by V volts (the battery potenital). When a light source of the proper wavelength is applied, photoemission will result and the negatively charged free electrons will travel directly to the positive plate of the tube. The microammeter (I) will indicate the strength of the resulting current flow. The effect of a change in light intensity is clearly indicated by the typical set of characteristics of Fig. 1.95. A luminous flux of 0.1 lm at 200 V (V_{PK}) will result in approximately 4 times the anode current at a luminous flux of 0.02 lm (maintaining $V_{PK} = 200$ V).

A direct application of the circuit of Fig. 1.94 is in a photographic exposure meter. A variable resistor in parallel with the movement would permit the measurement of higher light intensities.

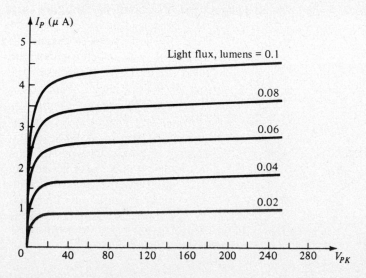

Figure 1.95. Typical set of phototube characteristics.

1.18 SEMICONDUCTOR PHOTOCONDUCTIVE CELL AND PHOTODIODE

The photoconductive cell is a two-terminal semiconductor device whose terminal resistance will vary (linearly) with the intensity of the incident light. For obvious reasons, it is frequently called a photoresistive device. A typical photoconductive cell and the most widely used graphic symbol for the device appear in Fig. 1.96.

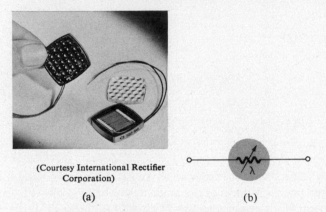

(Courtesy International Rectifier Corporation)

(a)   (b)

Figure 1.96. Photoconductive cell: (a) appearance; (b) symbol.

The photoconductive materials most frequently used include cadmium sulfide (Cds) and cadmium selenide (CdSe). The peak spectral response of CdS occurs at approximately 5100Å and for CdSe at 6150 Å. The response time of CdS units is about 100 ms and 10 ms for CdSe cells.

As the illumination on the device increases in intensity, the energy state of a larger number of electrons in the structure will also increase due to the increased availability of the photon packages of energy. The result is an increasing number of relatively "free" electrons in the structure and a decrease in the terminal resistance. The sensitivity curve for a typical photoconductive device appears in Fig. 1.97. Note the linearity (when plotted using a log-log scale) of the resulting curve

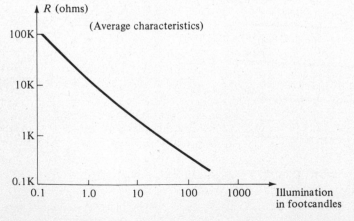

Figure 1.97. Photoconductive cell—terminal characteristics (GE type B425).

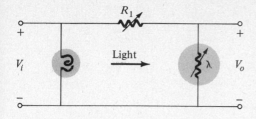

Figure 1.98. Voltage regulator employing a photoconductive cell.

and the large change in resistance ($100 \, \text{K} \rightarrow 100 \, \Omega$) for the indicated change in illumination.

One rather simple, but interesting, application of the device appears in Fig. 1.98. The purpose of the system is to maintain V_o at a fixed level even though V_i may fluctuate from its rated value. As indicated in the figure, the photoconductive cell, bulb, and resistor, all form part of this voltage regulator system. If V_i should drop in magnitude for any number of reasons, the brightness of the bulb would also decrease. The decrease in illumination would result in an increase in the resistance (R_λ) of the photoconductive cell to maintain V_o at its rated level as determined by the voltage divider rule; that is,

$$V_o = \frac{R_\lambda V_i}{R_\lambda + R_i} \qquad (1.17)$$

The photodiode is a semiconductor *p-n* junction device whose region of operation is the reverse-bias region of the junction diode discussed earlier in this chapter. This region is employed to take advantage of the fact that the reverse current increases almost linearly with the increase in the incident light. The basic biasing arrangement, construction, and symbol for the device appear in Fig. 1.99. Compare the defined direction of I_λ to that employed for the reverse saturation current I_s in Section 1.4.

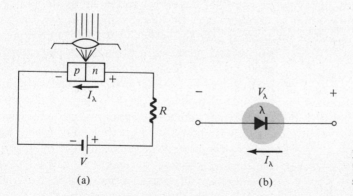

(a)

(b)

Figure 1.99. Photodiode: (a) basic biasing arrangement and construction; (b) symbol.

The characteristics of Fig. 1.100 clearly indicate that the reverse current I_λ will increase with increase in light intensity for the same applied potential. The dark current refers to that current which flows with no incident light. It is the reverse saturation current discussed in Section 1.4. As indicated in Fig. 1.99, a lens is normally employed in the cap of the unit to focus the light on the reverse-biased junction.

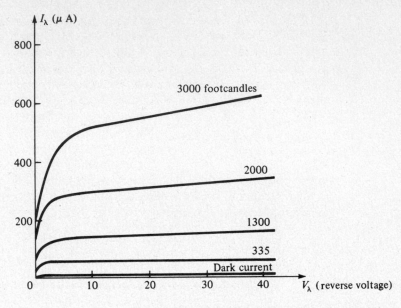

Figure 1.100. Typical set of photodiode characteristics.

1.19 LIGHT EMITTING DIODES (LEDs)

The light emitting diode is, as the name implies, a diode, that will give off visible light when it is energized. In any forward-biased *p-n* junction there is, within the structure, and primarily close to the junction, a recombination of holes and electrons. This recombination requires that the energy possessed by the unbound free electron be transferred to another state. In all semiconductor *p-n* junctions some of this energy will be given off as heat and some in the form of photons. In silicon and germanium the greater percentage is given up in the form of heat and the emitted light is insignificant. In other materials, such as gallium arsenide phosphide (GaAsP) or gallium phosphide (GaP), the number of photons of light energy emitted is sufficient to create a very visible light source. The process of given off light by applying an electrical source of energy is called *electroluminescence*. Of course, no matter what material is used a *p*-type and *n*-type region must be formed and joined as shown in Fig. 1.101. The conducting surface connected to the *p*-material is much smaller to permit the emergence of the maximum number of photons of light energy. Note in the figure that the recombination of the injected carriers due to the forward-biased junction is resulting in emitted light at the site of recombination.

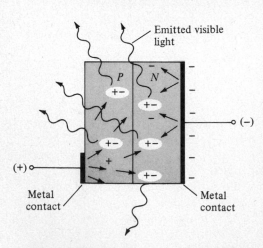

Figure 1.101. The process of electroluminescence in the LED.

The forward-biased diode current will have a direct effect on the intensity of the visible light as shown in Fig. 1.102a. This particular LED, shown in Fig. 1.102b, has a maximum power dissipation of 100 mW and a maximum forward dc current of 50 mA. The type of semiconductor material used will determine the wavelength of the emitted light. In Fig. 1.92 it was noted that the light red and the red region of the visible spectrum extended from about 0.62 to 0.76 μ (6200 to 7600 Å).

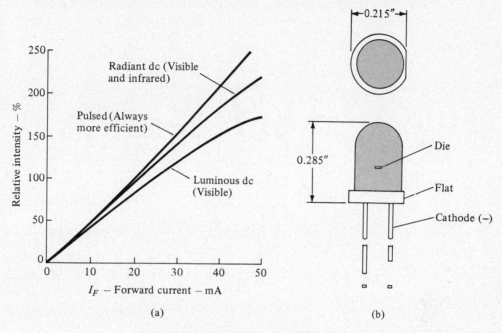

Figure 1.102. Fairchild GaAsP LED: (a) Intensity vs. forward current; (b) construction and terminal identification.

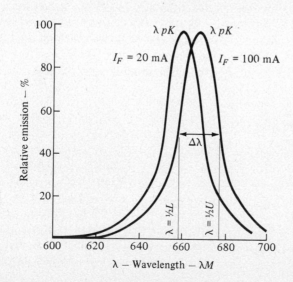

Figure 1.103. Emission spectrum for the Fairchild GaAsP LED.

In Fig. 1.103 we find for this LED that the entire range of the wavelengths is within this region with a peak at 0.66 μ, indicating that it will radiate a deep red light. LED's are commercially available in red, orange, yellow, and green. For the latter three colors gallium phosphicle is frequently used. A green LED will simply require that the wavelength pattern of Fig. 1.103 appear in the green region of Fig. 1.92.

LED displays are available today in many different sizes and shapes. The light emitting region is available in lengths from 0.1 to 1 in. Numbers can be created by segments such as shown in Fig. 1.104. By applying a forward bias to the proper

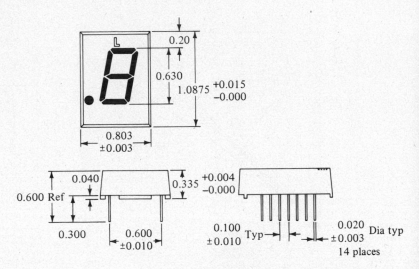

Figure 1.104. Litronix segment display.

p-type material segment, any number from 0 to 9 can be displayed. There are other types of displays such as the LCD to be described in the next section, but LED's predominate in the handheld calculators and in general application today. They operate at voltage levels from 1.7 to 3.3 V which makes them completely compatible with solid-state circuits. They have a fast response time and offer good contrast ratios for visibility. The power requirement is typically from 10 to 150 mW with a lifetime of 100,000+ hours. Its semiconductor construction adds a significant ruggedness factor.

1.20 LIQUID CRYSTAL DISPLAYS

The liquid crystal display (LCD) has the distinct advantage of having a very low power requirement. It is typically in the order of microwatts for the display as compared to the same order of milliwatts for LED's. It does, however, require an external or internal light source and is limited to a temperature range of about 0° to 60°C. The LCD's receiving the major portion of the interest today are either the field-effect or dynamic scattering units. Each will be covered in some detail.

A liquid crystal is a material (normally organic for LCD's) that will flow like a liquid but whose molecular structure has some properties normally associated with solids. For the light scattering units the greatest interest is in the *nematic liquid crystal* having the crystal structure shown in Fig. 1.105. The individual molecules have a rod-like appearance as shown in the figure. The indium oxide conducting surface is transparent and, under the condition shown in the figure, the incident light will simply pass through and the liquid crystal structure will appear clear. If a voltage (for commercial units the threshold level is usually between 6 and 20 V) is applied across the conducting surfaces as shown in Fig. 1.106, the molecular arrangement is disturbed with the result that regions will be established with different indices of refraction. The incident light is, therefore, refracted (referred to as *dynamic scattering*—first studied at RCA labs in 1968) and the result of the scattered light is a frosted glass appearance. Note in Fig. 1.106, however, that the frosted look only occurs where the conducting surfaces are opposite each other and that the remaining areas remain translucent.

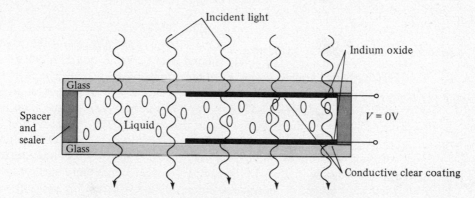

Figure 1.105. Nematic liquid crystal with no applied bias.

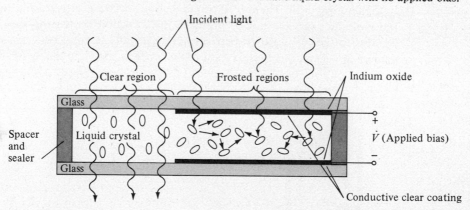

Figure 1.106. Nematic liquid crystal with applied bias.

A digit on an LCD display may have the segment appearance shown in Fig. 1.107. The black area is actually a clear conducting surface connected to the terminals below for external control. Two similar masks are placed on opposite sides of a sealed thin layer of liquid crystal material. If the number 2 were required,

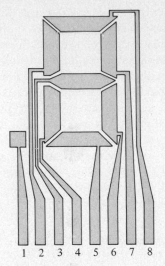

Figure 1.107. LCD 8 segment digit display.

the terminals 8, 7, 3, 4 and 5 would be energized and only those regions would be frosted while the other areas would remain clear.

As indicated earlier, the LCD does not generate its own light but depends on an external or internal source. Under dark conditions it would be necessary for the unit to have its own internal light source either behind or to the side of the LCD. During the day, or in lighted areas, a reflector can be put behind the LCD to reflect the light back through the display for maximum intensity. For optimum operation, current watch manufacturers are using a combination of the transmissive (own light source) and reflective modes called *transflective*.

The *field-effect* or *twisted nematic* LCD has the same segment appearance and thin layer of encapsulated liquid crystal, but its mode of operation is very different. Similar to the dynamic scattering LCD, the field effect can be operated in the reflective mode or transmissive mode with an internal source. The transmissive display appears in Fig. 1.108. The internal light source is on the right and the viewer is on the left. This figure is most noticeably different from Fig. 1.105 in that there is an addition of a *light polarizer*. Only the vertical component of the entering light on the right can pass through the vertical light polarizer on the right. In the field-effect LCD, either the clear conducting surface to the right is chemically etched or an organic film is applied to orient the molecules in the liquid crystal in the vertical plane, parallel to the cell wall. Note the rods to the far right in the liquid crystal. The opposite conducting surface is also treated to ensure that the molecules are 90° out of phase in the direction shown (horizontal) but still parallel to the cell wall. In between the two walls of the liquid crystal there is a general drift from one polarization to the other as shown in the figure. The lefthand

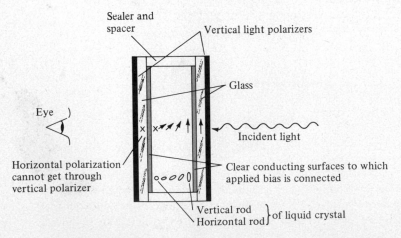

Figure 1.108. Transmissive field effect LCD with no applied bias.

light polarizer is also such that it permits the passage of only the vertically polarized incident light. If there is no applied *emf* to the conducting surfaces, the vertically polarized light enters the liquid crystal region and follows the 90° bending of the molecular structure. Its horizontal polarization at the left-hand vertical light polarizer does not allow it to pass through, and the viewer sees a uniformly dark pattern across the entire display. When a threshold voltage is applied (for commercial units from 2 to 8 V), the rod-like molecules align themselves with the field (perpendicular to the wall) and the light passes directly through without the 90°-shift. The vertically incident light can then pass directly through the second vertically polarized screen and a light area is seen by the viewer. Through proper excitation of the segments of each digit the pattern will appear as shown in Fig. 1.109. The reflective type field effect is shown in Fig. 1.110. In this case the horizontally polarized light at the far left encounters a horizontally polarized filter and passes through to the reflector where it is reflected back into the liquid crystal, bent back to the other vertical polarization, and returned to the observer. If there is no *emf*, there is a uniformly lit display. The application of a voltage results in a vertically incident light encountering a horizontally polarized filter at the left which will not be able to pass through and be reflected. A dark area results on the crystal and the pattern as shown in Fig. 1.111 appears.

Figure 1.109. Reflective-Type LCD (Courtesy of RCA Solid State Division).

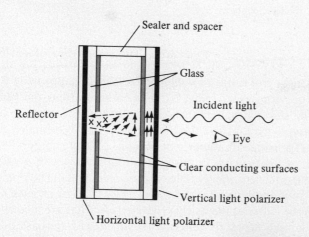

Figure 1.110. Reflective field effect LCD with no applied bias.

Figure 1.111. Transmissive Type LCD (Courtesy of RCA Solid State Division).

Field-effect LCD's are normally used in when a source of energy is a prime factor, for example, in watches, portable instrumentation, etc., since they absorb considerably less power than the light scattering types microwatt range compared to low milliwatt range. The cost is typically higher for field-effect units, and their height is limted to about 2 in. while light scattering units are available up to 8 in. in height.

A further consideration in displays is turn-on and turn-off time. LCD's are characteristically slower than LED's with response times of 100 to 300 μs. The lifetime of LCD units is steadily increasing beyond the 10,000+ hours limit. The color generated by LCD units is dependent on the source of illumination.

1.21 SOLAR CELLS

In recent years there has been increasing interest in the solar cell as an alternate source of energy. When we consider that the power density received from the sun at sea level is about 100 mW/cm² (SI units: 1 kW/m²), it is certainly an energy source that requires further research and development to maximize the conversion efficiency from solar to electrical energy.

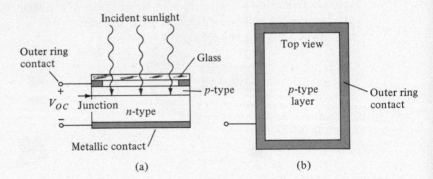

Figure 1.112. Solar cell: (a) cross-section; (b) top view.

The basic construction of a silicon *p-n* junction solar cell appears in Fig. 1.112. As shown in the top view, every effort is made to ensure that the surface area perpendicular to the sun is a maximum. Also, note that the metallic conductor connected to the *p*-type material and the thickness of the *p*-type material are such that they ensure that a maximum number of photons of light energy will reach the junction. A photon of light energy in this region may collide with a valence electron and impart to it sufficient energy to leave the parent atom. The result is a generation of free electrons and holes. This phenomenon will occur on each side of the junction. In the *p*-type material the newly generated electrons are minority carriers and will move rather freely across the junction as explained for the basic *p-n* junction with no applied bias. A similar discussion is true for the holes generated in the *n*-type material. The result is an increase in the minority carrier flow which is opposite in direction to the conventional forward current of a *p-n* junction. This increase in reverse current is shown in Fig. 1.113. Since $V = 0$ anywhere on the vertical axis and represents a short-circuit condition, the current at this interesection is called the *short-circuit current* and is represented by the notation I_{SC}. Under open-circuit conditions ($i_d = 0$) the *photovoltaic* voltage V_{oc} will result. Thus is a logarithmic

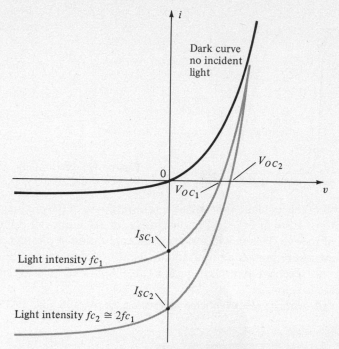

Figure 1.113. Short-circuit current and open-circuit voltage vs. light intensity for a solar cell.

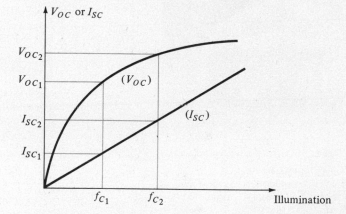

Figure 1.114. V_{oc} and I_{SC} versus illumination for a solar cell.

function of the illumination as shown in Fig. 1.114. V_{oc} is the terminal voltage of a battery under no-load (open-circuit) conditions. Note, however, in the same figure that the shortcircuit current is a linear function of the illumination. That is, it will double for the same increase in illumination (f_{c_1} and $2_{f_{c_1}}$ in Fig. 1.114) while the change in V_{oc} is less for this region. The major increase in V_{oc} occurs for lower-level increases in illumination. Eventually, a further increase in illumination will have very little effect on V_{oc}, although I_{sc} will increase, causing the power capabilities to increase.

Selenium and silicon are the most widely used materials for solar cells, although gallium arsenide, indium arsenide, and cadmium sulfide, among others, are also

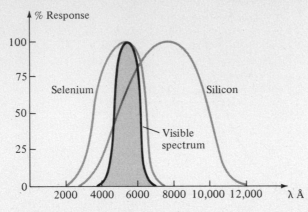

Figure 1.115. Spectral response of Se, Si and the naked eye.

used. The wavelength of the incident light will affect the response of the *p-n* junction to the incident photons. Note in Fig. 1.115 how closely the selenium cell response curve matches that of the eye. This fact has widespread application in photographic equipment such as exposure meters and automatic exposure diaphragms. Silicon also overlaps the visible spectrum but has its peak at the 0.8 μ (8000 Å) wavelength which is in the infrared region.

A typical solar cell, with its electrical characteristics, appears in Fig. 1.116.

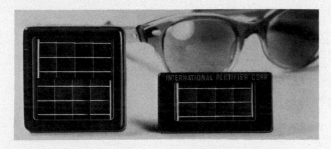

Electrical Characteristics*

IR NUMBER	LOAD VOLTAGE (VOLTS) (MIN.)	LOAD CURRENT (MILLIAMPS) (MIN.)	POWER (MILLIWATTS) (MIN.)
SP2A40B	1.6	36	58
SP2B48B	1.6	40	64
SP4C40B	3.2	36	115
SP2C80B	1.6	72	115
SP4D48B	3.2	40	128
SP2D96B	1.6	80	129
S2900E5M	.4	60	24
S2900E7M	.4	90	36
S2900E9.5M	.4	120	48

*Current Voltage characteristics are based on an illuminational level of 100 mW/cm² (bright average sunlight).

Figure 1.116. Typical solar cell and its electrical characteristics (Courtesy of International Rectifier Corp.).

Efficiency of operation is determined by the electrical power output divided by the power input light energy. That is,

$$\eta\% = \frac{P_{o\text{(electrical)}}}{P_{i\text{(light energy)}}} \times 100\% = \frac{P_{\max\text{(device)}}}{(\text{Area in cm}^2)(100\,\text{mW/cm}^2)} \times 100\% \qquad (1.18)$$

1.22 THERMISTORS

The thermistor is, as the name implies, a temperature-sensitive resistor; that is, its terminal resistance is somehow related to its body temperature. It has a negative temperature coefficient, indicating that its resistance will decrease with an increase in its body temperature. It is not a junctions device but a mixture of oxides of cobalt, nickel, strontium, or manganese.

The characteristics of a fairly representative thermistor are provided in Fig. 1.117, with a commonly used symbol for the diode. Note, in particular, that at room temperature (20°C) the resistance of the thermistor is approximately 5000 Ω, while at 100°C (212°F) the resistance has decreased to 100 Ω. A temperature span of 80°C has therefore resulted in a 50:1 change in resistance. There are, fundamentally, two ways to change the temperature of the device: internally and externally.

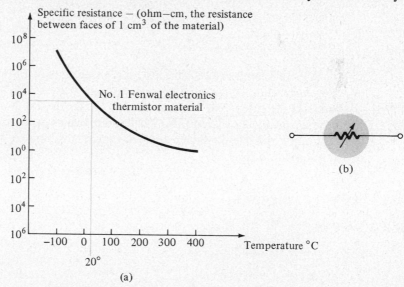

Figure 1.117. Thermistor: (a) typical set of characteristics; (b) symbol.

A simple change in current through the device will result in an internal change in temperature. Externally would require changing the temperature of the surrounding medium or immersing the device in a hot or cold solution.

A photograph of a number of commercially available thermistors is provided in Fig. 1.118.

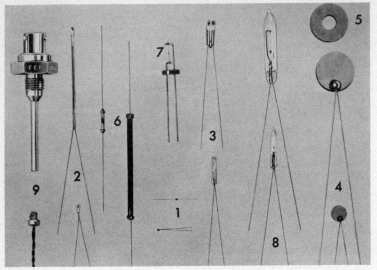

Figure 1.118. Various types of thermistors: (1) beads; (2) glass probes; (3) iso-curve interchangeable probes and beads; (4) discs; (5) washers; (6) rods; (7) specially-mounted beads; (8) vacuum and gas filled probes; (9) special probe assemblies.

(Courtesy Fenwal Electronics, Inc.)

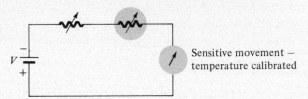

Sensitive movement — temperature calibrated

Figure 1.119. Temperature-indicating circuit.

A simple temperature-indicating circuit appears in Fig. 1.119. Any increase in the temperature of the surrounding medium will result in a decrease in the resistance of the thermistor and an increase in the current I_T. An increase in I_T will produce an increased movement deflection, which when properly calibrated will accurately indicate the higher temperature. The variable resistance was added for calibration purposes.

PROBLEMS

§1.2

1. In your own words describe the characteristics of an ideal diode.

2. (a) Draw the waveform across the resistor of Fig. 1.4 for an input sinusoidal signal of 120 V rms. Indicate the peak voltage magnitude on the waveform.
   (b) Repeat for the case in which the diode is connected opposite to that shown in Fig. 1.4.

3. For the network of Fig. 1.120, determine the voltage across the resistor R_L for the inputs shown in the same figure.

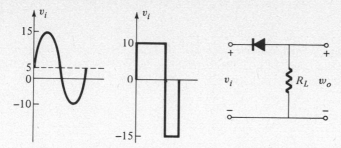

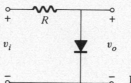

Figure 1.120

4. Determine the voltage across the diode in Fig. 1.121 for the inputs of Fig. 1.120.

Figure 1.121

§ 1.3

5. In your own words describe the operation of a vacuum-tube diode in the conductive and nonconductive states.

6. How do the characteristics of a vacuum-tube diode compare with those of the ideal diode?

§ 1.4

7. Describe the difference between *n*-type and *p*-type semiconductor materials.

8. Describe the difference between donor and acceptor inpurities.

9. Describe the difference between majority and minority carries.

10. (a) What is meant by the condition "forward bias" as applied to a diode.
(b) Repeat part (a) for the "reverse-bias" state.

11. Using Eq. (1.1), determine the diode current at 20°C for a silicon diode with $I_S = 5 \ \mu A$ and an applied forward bias of 0.4 V.

12. Repeat Problem 1.11 for $T = 100°C$ (boiling point of water).

13. In the reverse-bias region the saturation current of a silicon diode is about 3 μA $(T = 20°C)$. Determine its approximate value if the temperature is increased 40°C.

14. Compare the characteristics of a silicon and germanium diode and determine which you would prefer to use for most practical applications. Give some detail. Refer to a manufacturer's listing and compare the characteristics of a germanium and silicon diode of similar maximum ratings.

15. Determine the maximum power dissipation for the T151 diode. What is the maximum reverse-bias dissipation at $V = -10$ V $(T = 25°C)$. What is an approximate value of the reverse saturation current at 150°C for the T151 diode if we assume that I_S will double for every increase in temperature of 10°C.

§ 1.5

16. List the types of diode fabrication listed in Section (1.5) and discuss the disadvantages and advantages of practical applications.

17. Sketch a load line on the characteristics of Fig. 1.122. The supply voltage is 20 V and $R = 1$ K. Determine the Q-point for each diode. What is the dc dissipation of each diode at this quiescent point of operation?

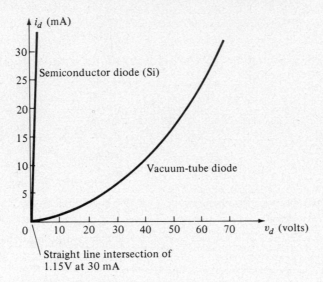

Straight line intersection of
1.15V at 30 mA

Figure 1.122

18. Repeat Problem 1.17 if the supply voltage is increased to 60 V and $R = 2$ K.

§ 1.7

19. Determine the dc or static resistance for each diode at the Q-point determined by Problem 1.17.

§ 1.8

20. Repeat Problem 1.19 for the dynamic resistance. How do the static and dynamic resistances at the Q-point compare? (Make the necessary approximations, due to the size of the characteristic, for the semiconductor diode).

21. Using Fig. 1.24, determine the dynamic resistance of the silicon diode at $I_D = 20$ mA. Use Eq. (1.9) and compare to that determined by Eq. (1.8).

22. Repeat Problem 1.21 for $I_D = 5$ mA.

§ 1.9

23. Determine the average resistance for the diodes of Fig. 1.122 for the range extending from the origin to a maximum current of $I_D = 30$ mA.

24. Repeat Problem 1.23 for the range of I_D extending from 10 to 30 mA.

§ 1.10

25. Find the piecewise linear equivalent circuits for the diodes of Fig. 1.22. Assume that the straight line segment for the semiconductor diode intersects the horizontal axis at 0.7 V (Si).

26. Determine V_R, I_D, V_D, and R_{DC} for the network of Fig. 1.46. Use the equivalent circuits determined in Problem 1.25. Compare the results between the semiconductor and tube diodes.

27. Using the approximate characteristics of Fig. 1.47 for a silicon semiconductor diode, determine the level and appearance of the output voltage for the network of Fig. 1.48 if the supply voltage is 40 sin ωt and $R_L = 2$ K.

28. Determine the peak-to-peak value of v_d in Fig. 1.51 if the dc level is increased to 4 V. Sketch the waveform of the ac sinusoidal voltage across the 100-Ω resistor.

29. Repeat Problem 1.28 for a dc level of 4 V and the 100-Ω resistor replaced by one of 1 K.

§ 1.11

30. Assuming an ideal diode in the circuit of Fig. 1.123, determine the output waveform for each of the input signals of Fig. 1.124.

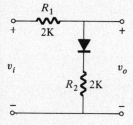

Figure 1.123

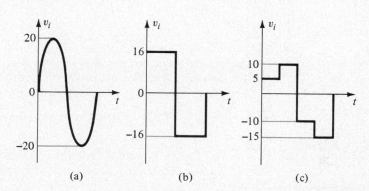

(a)   (b)   (c)

Figure 1.124

31. Repeat Problem 1.30 with the diode reversed.

32. Repeat Problem 1.30 with the diode and resistor R_1 interchanged.

33. Draw the output waveform for the circuit of Fig. 1.125a for each of the input signals of Fig. 1.124.

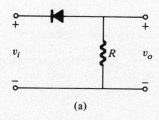

(a)

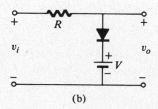

(b)   Figure 1.125

34. Draw the output waveform for the circuit of Fig. 1.25b for each of the input signals fo Fig. 1.124. Use $V = 5$ V.

35. Repeat Problem 1.34 for $V = -10$ V.

36. Sketch the output waveform for the network of Fig. 1.126a for the input of Fig. 1.124b.

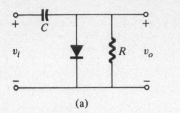 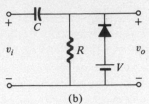

(a)   (b)   **Figure 1.126**

37. Sketch the output waveform for the network of Fig. 1.126b for the input of Fig. 1.124b. Use $V = 5$ V.

38. Repeat Problem 1.37 for $V = -10$ V.

39. Design a network that will only permit the $+5$- to $+10$-V swing of the input of Fig. 1.124c to pass through (riding on a $+5$-V level).

40. Design a network that will shift the input of Fig. 1.124b to a -10 V to -42 V swing.

§ 1.12

41. For the Zener diode network of Fig. 1.77, if $V_i = 40$ V, $R_S = 200$ Ω, and $V_Z = 22$ V:
  (a) Find the minimum value of R_L to ensure that the Zener diode has fired.
  (b) From part (a) determine the maximum load current under Zener regulation.
  (c) Find the minimum I_L if the maximum $R_L = 5$ K.
  (d) Sketch the terminal voltage V_L vs. I_L for the range of I_L.
  (e) At $R_L = 1$ K determine I_Z and the current through the 40-V source.
  (f) From the results of part (e) determine the power dissipated by the Zener diode.

42. Using a 10-V Zener, R_L with a range from 50 Ω to 2 K, $V_i = 30$ V, find a suitable value of R_S to ensure that the Zener diode maintains an "on" condition.

§ 1.13

43. What are the essential differences between a semiconductor junction diode and a tunnel diode?

§ 1.15

44. (a) Determine the transition capacitance of a diffused junction varicap diode at a reverse potential of 4.2 V if $C(0) = 80$ pF and $V_o = 0.7$ V.
  (b) From the information of part (a) determine the constant K in Eq. (1.13).

45. (a) For a varicap diode having the characteristics of Fig. 1.86, determine the difference in capacitance between reverse-bias potentials of -3 and -12 V.
  (b) Determine the incremental rate of change ($\Delta C/\Delta V_r$) at $V = -8$ V. How does this value compare with the incremental change determined at -2 V?

§ 1.16

46. (a) Consult Fig. 1.89. How would you compare the dynamic resistances of the diodes in the forward-bias regions?

(b) Compare the dynamic resistance levels at a reverse current of -0.2 mA.

(c) Sketch the piecewise linear equivalent circuit for each in the forward-bias region.

§ 1.17

47. (a) For the network of Fig. 1.94, if $V = 200$ V and $R = 40$ M, determine the static diode current I_{P_Q} and voltage V_{PK_Q} if the incident luminous flux is 0.06 lm.

(b) If the light intensity drops to 0.02 lm, determine the new static values.

§ 1.18

48. What is the approximate rate of change of resistance with illumination for a photoconductive cell with the characteristics of Fig. 1.97 for the ranges:

(a) 0.1 K $\longrightarrow$ 1 K   (b) 1 K $\longrightarrow$ 10 K   (c) 10 K $\longrightarrow$ 100 K

(Note that this is a log scale.) Which region has the greatest rate of change in resistance with illumination.

49. What is the "dark current" of a photodiode?

§ 1.19

50. Using Fig. 1.102, determine whether or not the relative intensity of the emitted light doubled by an increase in forward current from 20 to 40 mA. If not, what is the approximate rate of increase in intensity for this current range?

§ 1.20

51. Discuss the relative differences between an LED and LCD display. What are the relative disadvantages and advantages of each?

§ 1.21

52. A 1 cm $\times$ 2 cm solar cell has a conversion efficiency of 9%. Determine the maximum power rating of the device.

53. If the power rating of a solar cell is determined on a very rough scale by the product $V_{OC} \cdot I_{SC}$, is the greatest *rate* of increase obtained at lower or higher levels of illumination. Explain your reasoning.

§ 1.22

54. For the thermistor of Fig. 1.117, determine the dynamic rate of change in specific resistance with temperature at $T = 20°C$. How does this compare to the value determined at $T = 300°C$? From the results, determine whether the greatest change in resistance per unit change in temperature occurs at lower or higher levels of temperature.

diode rectifiers and filters

2

2.1 DIODE RECTIFICATION

A rectifying circuit converts ac voltage into pulsating dc voltage. For example, a rectifying circuit will convert the 60-Hz ac voltage obtained from the power line from one having an average voltage over a full cycle of zero to one which has an average value other than zero.

Figure 2.1 shows half-wave rectifier circuits using an ideal diode and the resulting half-wave rectified output voltages developed. When the input ac voltage is positive (positive to negative is measured from top to bottom of the voltage generator) for the circuit connection of Fig. 2.1a, the polarity of voltage across the diode will cause the diode to conduct; that is, the voltage across the diode is positive to negative from anode to cathode, and, in the case of the ideal diode, the forward resistance is zero. The positive half-cycle of the input signal then appears across the resistor as shown in Fig. 2.1a. When the input voltage is negative (measured from top to bottom of the generator) in Fig. 2.1a, the diode is reverse biased, having then infinite resistance and appearing as an open circuit. Since there can be no current flow during the complete time that the voltage at the input causes the diode to be reverse biased, the voltage across the resistor is zero.

The resulting output signal across the resistor due to the half-cycle of diode conduction and the lack of signal during the half-cycle of diode nonconduction is shown in Fig. 2.1a. Notice that although this signal is not steady dc (it is pulsating dc), it nevertheless has an average positive value. If the sinusoidal voltage from the power line were applied to a dc voltmeter the reading obtained would be zero. With the pulsating dc applied to a dc meter there will be a reading representing the average of the applied signal. For the diode connection of Fig. 2.1a, the diode

80

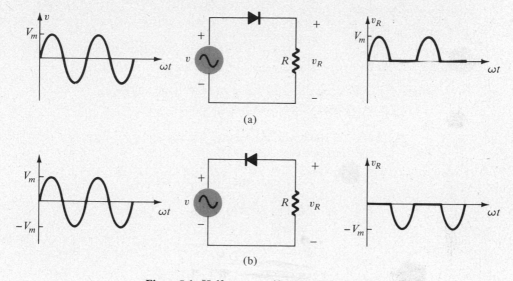

Figure 2.1. Half-wave rectifier circuits: (a) positive half-cycles; (b) negative half-cycles.

allows only the positive half-cycles to appear at the output. Reversing the diode as shown in Fig. 2.1b results in only the negative half-cycles.

Half-wave Average (dc) Voltage

To determine the average value of the rectified signal we can calculate the area under the curve of Fig. 2.2 and divide this value by the period of the rectified waveform. To calculate the area under the half-cycle curve of the rectified signal we must integrate the rectified signal.[1] Doing this integration procedure (and dividing

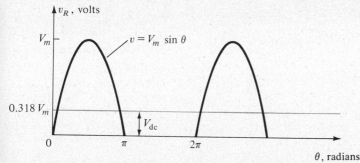

Figure 2.2. Half-wave rectified voltage showing dc value.

---

[1]The dc signal can be expressed as $v = V_m \sin \theta$ for $0 \leq \theta \leq \pi$ radians (see Fig. 2.2). For θ from 0 to 2π radians the average value is calculated to be

$$V_{dc} = V_{av} = \frac{1}{T} \int v \, dt = \frac{1}{2\pi} \int_0^{\pi} (V_m \sin \theta) \, d\theta$$

$$V_{dc} = \frac{V_m}{2\pi} [-\cos \theta]_0^{\pi} = \frac{V_m}{2\pi} [-1(-1) - (-1)] = \frac{V_m}{\pi} = 0.318 \, V_m$$

SEC. 2.1 DIODE RECTIFICATION

by the period) results in

$$V_{dc} = 0.318V_m \qquad \text{(half-wave)} \qquad (2.1)$$

where V_m = maximum (peak) value of ac voltage, and
V_{dc} = average value of rectified voltage.

EXAMPLE 2.1 Calculate the average voltage of the rectified signal obtained from the circuit of Fig. 2.3.

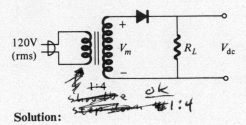

Figure 2.3. Half-wave rectifier circuit for Example 2.1.

Solution:

$$V_m = 4 \ (1.414 \times 120) = 678.8 \text{ V}$$

$$V_{dc} = 0.318V_m = (0.318)(678.8 \text{ V}) = \textbf{215.9 V}$$

Half-wave Peak Inverse Voltage (PIV)

An important diode rating is the peak inverse voltage, PIV of the diode (the maximum voltage across the diode in the direction to block current flow). For the half-wave rectifier circuit of Fig. 2.4 the peak voltage across the diode when the diode is reverse biased is shown to be V_m in value.

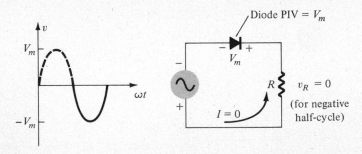

Figure 2.4. Half-wave rectifier circuit showing peak inverse voltage (PIV) across diode.

In Example 2.1 the peak inverse voltage was $V_m = 42.4$ V. The dc voltage obtained, however, was only $V_{dc} = 13.5$ V. This clearly points out one poor feature of the half-wave circuit, namely, the diode PIV rating must be considerably larger than the dc voltage obtained using the circuit. The forward current rating of the diode must equal, at least, the average current through it—V_{dc}/R. The peak current through the diode is V_m/R and must be less than the peak current rating for the diode.

2.2 FULL-WAVE RECTIFICATION

Center-Tapped Full-Wave Rectifier

It would be preferable to obtain a larger dc voltage compared to the maximum voltage than that of $0.318\ V_m$ for a half-wave rectified signal. In addition, we note that although an average voltage is obtained using a half-wave rectifier, no voltage is developed for half of the cycle. Using two diodes, as shown in Fig. 2.5, it is possible to rectify a sinusoidal signal to obtain one having the same polarity half-cycle for *each* of the half-cycles of input signal. This *full-wave* rectified signal provides a signal that has twice the dc value of the comparable half-wave rectified signal.

The full-wave rectifier circuit of Fig. 2.5 requires a center-tapped transformer and two diodes to develop a full-wave rectified output voltage. To understand how the output waveform is developed we shall consider the detailed circuit operation for each half-cycle of secondary voltage. Figure 2.6a, shows the circuit operation for the positive half-cycle of secondary voltage. The transformer is center tapped and a peak voltage, V_m, is developed across each half of the transformer during the positive cycle.

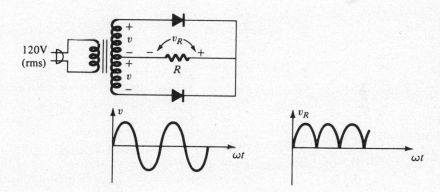

Figure 2.5. Full-wave rectifier circuit.

During the entire positive half-cycle the polarity of the signal across the upper half of the transformer is in a direction to forward bias diode D_1 causing it to conduct. With diode D_1 conducting, a positive half-cycle of voltage is developed across resistor R as shown in Fig. 2.6a. The figure shows the voltages in the circuit at the time of the peak positive voltage and as shown, there is a voltage V_m across the resistor at this time.

The current in the upper transformer half flows through the transformer, diode D_1, and the load resistor. For a perfect diode ($V_D = 0$, when conducting) the voltage across the resistor will equal that of the transformer. At the time the transformer voltage is V_m the voltage across the resistor is also V_m in magnitude as shown in Fig. 2.6a. The voltage developed across the resistor is thus a half-cycle of signal.

The polarity of the voltage developed across the lower half of the transformer results in diode D_2 being back biased. In addition, the reverse-bias voltage across

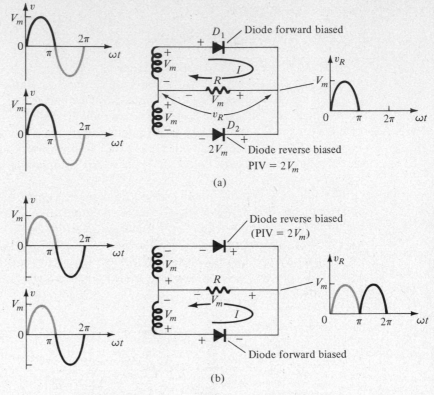

Figure 2.6. Full-wave rectifier, detail circuit operation: (a) positive half-cycle; (b) negative half-cycle.

the diode, which is maximum at the time the maximum voltage V_m is present, is $2V_m$. We can see that this is so by noting that the voltage across reverse-biased diode D_2 is equal to the sum of the voltages across the lower half of the transformer and the load resistor, these voltages being of the same polarity. A diode in this circuit must therefore be capable of handling a reverse-bias voltage equal to twice the value of the peak voltage developed across the output.

During the negative half-cycle diode D_2 in Fig. 2.6b is forward biased, and diode D_1 is reverse biased. Current flows through the lower half of the transformer but in the same direction through resistor R as shown in Fig. 2.6b. The output voltage developed across the resistor for the negative half-cycle of input signal is, then, of the same polarity as for the positive half-cycle of input signal. The peak inverse voltage across diode D_1 is $2V_m$, so that each diode must be capable of withstanding a reverse bias voltage of $2V_m$ sometime during a cycle of operation. The resulting output voltage for a full cycle of input voltage is two positive-going half-cycles.

The average voltage for a full-wave rectified signal is twice that for the half-wave rectified, so that

$$V_{dc} = 2(0.318V_m) = 0.636V_m \qquad \text{(full-wave)} \qquad (2.2)$$

The full-wave rectifier circuit of Fig. 2.5 has the advantage of developing a larger dc voltage for the same peak voltage rating. It has, however, the disadvantage of requiring a diode rating of twice the peak inverse voltage, and a center-tapped transformer having twice the overall voltage rating.

> **EXAMPLE 2.2** Calculate the dc voltage obtained from a center-tapped full-wave rectifier for which the peak rectified voltage is 100 V, and the peak inverse voltage developed across the diode.

Solution:

$$V_{dc} = 0.636 V_m = 0.636(100) = \mathbf{63.6\ V}$$

$$\text{diode PIV} = 2V_m = 2(100) = \mathbf{200\ V}$$

Bridge Rectifier Circuit

Another circuit variation of a full-wave rectifier is the bridge circuit of Fig. 2.7. This circuit requires four diodes for full-wave rectification but the transformer used is not center tapped and develops a maximum voltage of only V_m. In addition, the diode PIV rating will be shown to be only V_m, rather than $2V_m$.

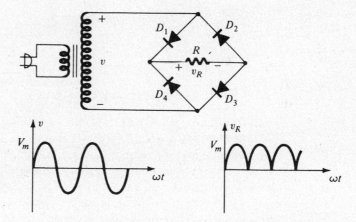

Figure 2.7. Full-wave bridge rectifier circuit.

In considering how the circuit operates we must understand how the conduction and nonconduction paths are formed during each half of the ac cycle. During the positive half-cycle the voltage across the transformer (measured from top to bottom) is positive and the conduction path is shown in Fig. 2.8. Figure 2.8a shows the voltages at the time of the peak positive voltage, V_m. Since the diodes shown are forward biased, the voltage drop across each is 0 V and the peak voltage from the transformer appears across resistor, R, at this time.

At the same time the voltage polarity is such as to reverse bias diodes D_2 and D_4, as shown in Fig. 2.8b. This represents the nonconduction path during the positive half-cycle of the input ac signal. Resistor R has a voltage developed across it by the current flowing through the conducting path of diodes D_1 and D_3. If the voltage

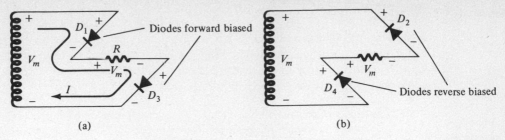

(a)   (b)

Figure 2.8. Bridge circuit, positive half-cycle operation: (a) conduction path; (b) nonconduction path.

drops around the nonconducting loop are summed, the transformer voltage and resistor voltage at the time of the peak voltage add up to $2V_m$. Since there are two diodes in the path, the voltage across each reverse-biased diode is V_m. This is half the developed peak inverse voltage across the diodes in the previous full-wave rectifier circuit (Fig. 2.5).

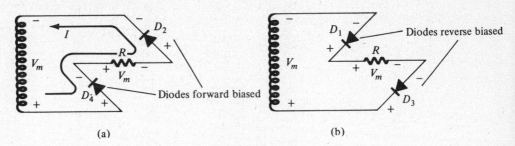

(a)   (b)

Figure 2.9. Bridge circuit, negative half-cycle operation: (a) conduction path; (b) nonconduction path.

During the negative half-cycle the conduction and nonconduction paths are shown in Fig. 2.9. Figure 2.9a shows that diodes D_4 and D_2 are forward biased. Note carefully that the current, I, goes through resistor R in the same direction as did the current on the previous half-cycle. The voltage across resistor R is thus of the same polarity during each half-cycle of the input signal. During the negative-polarity half-cycle the path of diodes D_1 and D_3 is nonconducting as shown in Fig. 2.9b and the peak inverse voltage developed across each of the diodes is V_m.

To summarize the operation of the bridge rectifier circuit, the addition of two extra diodes above the number in the center-tapped full-wave circuit provides improvement in two main factors. One, the transformer used need not be center tapped, requiring a maximum voltage across the transformer of V_m. Two, the peak inverse voltage (PIV) required of each diode is half that for the center-tapped full-wave circuit, only V_m. For low values of secondary maximum voltage the center-tapped full-wave circuit will be acceptable, whereas for high values of maximum secondary voltage the use of the bridge to reduce the maximum transformer rating and diode PIV rating is usually necessary.

EXAMPLE 2.3 The dc voltage developed by a full-wave bridge rectifier circuit is 325 V. Calculate the diode peak inverse voltage rating required for the diodes selected for this circuit.

Solution:

$$V_m = \frac{V_{dc}}{0.636} = \frac{325}{0.636} = 511 \text{ V}$$

For the bridge rectifier the value of diode PIV is V_m so that diode PIV $- V_m$ = **511 V.**

2.3 GENERAL FILTER CONSIDERATIONS

A rectifier circuit is necessary to convert a signal having zero average value to one that has a nonzero average. However, the resulting pulsating dc signal is not pure dc or even a good representation of it. Of course, for a circuit such as a battery charger the pulsating nature of the signal is no great detriment as long as the dc level provided will result in charging of the battery. On the other hand, for voltage supply circuits for a tape recorder or radio the pulsating dc will result in a 60-(or 120-) Hz signal appearing in the output, thereby making the operation of the overall circuit poor. For these applications, as well as for many more, the output dc developed will have to be much "smoother" than that of the pulsating dc obtained directly from half-wave or full-wave rectifier circuits.

A number of different types of filter or smoothing circuits will be considered in this chapter. These will include the popular simple-capacitor filter, the RC filter, the choke filter, the LC filter, and the π-type filter circuits. Voltage regulator circuits using Zener diodes and transistors will be covered in Chapter 14.

Filter Voltage Regulation and Ripple Voltage

Before going into any of the filter circuits it would be appropriate to consider the usual method of rating the circuits so that we are able to compare a circuit's effectiveness as a filter. Figure 2.10 shows a typical filter output voltage, which will be used to define some of the signal factors. The filtered output voltage of Fig. 2.10 has a dc value and some ac variation (*ripple*). Although a battery has essentially a constant or dc output voltage, the dc voltage derived from an ac source signal by rectifying and filtering will have some variation (ripple). The smaller the ac variation *with respect to* the dc level the better the filter circuit operation.

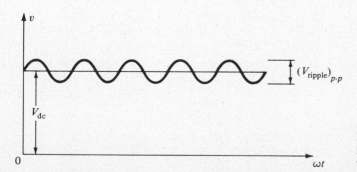

Figure 2.10. Filter voltage waveform showing dc and ripple voltages.

Consider measuring the output voltage of the filter circuit using a dc voltmeter and an ac (rms) voltmeter. The dc voltmeter will read only the average or dc level of the output voltage. The ac (rms) meter will read only the rms value of the ac component of the output voltage (assuming the signal is coupled to the meter through a capacitor to block out the dc level).

Definition: Ripple Factor:

$$r = \text{ripple factor} \equiv \frac{\text{ripple voltage (rms)}}{\text{dc voltage}} = \frac{V_r(\text{rms})}{V_{dc}}$$ (2.3a)

Definition: Percent of Ripple:

$$\% r = \% \text{ ripple} \equiv \frac{V_r(\text{rms})}{V_{dc}} \times 100$$ (2.3b)

EXAMPLE 2.4 Using a dc and ac voltmeter to measure the output signal from a filter circuit, a dc voltage of 25 V and an ac ripple voltage of 2.5 V (rms) are obtained. Calculate the ripple of the filter output.

Solution:

$$r = \text{ripple factor} = \frac{V_r(\text{rms})}{V_{dc}} = \frac{2.5}{25} = \textbf{0.1}$$

$$\% r = \% \text{ ripple} = \frac{V_r(\text{rms})}{V_{dc}}(100) = 0.1(100) = \textbf{10\%}$$

Voltage Regulation

Another factor of importance in a voltage supply is the amount of change in the output dc voltage over the range of the circuit operation. The voltage provided at the output at no-load (no current drawn from the supply) is reduced when load current is drawn from the supply. How much this voltage changes with respect to either the loaded or unloaded voltage value is of considerable interest to anyone using the supply. This voltage change is described by a factor called *voltage regulation.*

Definition: Voltage Regulation:

$$\text{Voltage regulation} \equiv \frac{\text{voltage at no-load} - \text{voltage at full-load}}{\text{voltage at full-load}}$$

$$V.R. = \frac{V_{NL} - V_{FL}}{V_{FL}}$$ (2.4a)

Definition: Percent of Voltage Regulation:

$$\% \, V.R. = \% \text{ voltage regulation} \equiv \frac{V_{NL} - V_{FL}}{V_{FL}} \times 100 \qquad (2.4b)$$

EXAMPLE 2.5 A dc voltage supply provides 60 V when the output is unloaded. When full-load current is drawn from the supply, the output voltage drops to 50 V. Calculate the values of voltage regulation and per cent of voltage regulation.

Solution:

$$V.R. = \frac{V_{NL} - V_{FL}}{V_{FL}} = \frac{60 - 50}{50} = 0.20$$

$$\% \, V.R. = V.R. \times 100 = 0.20 \times 100 = \mathbf{20\%}$$

If the value of full-load voltage is the same as the no-load voltage, the $V.R.$ calculated is 0%, which is the best to expect. This value means that the supply is a true voltage source for which the output voltage is independent of current drawn from the supply. The output voltage from most supplies decreases as the amount of current drawn from the voltage supply is increased. The smaller this voltage decreases the smaller the percent of $V.R.$ and the better the operation of the voltage supply circuit.

Ripple Factor Of Rectified Signal

Although the rectified voltage is not a filtered voltage, it nevertheless contains a dc component and a ripple component. We can calculate these values of dc voltage and ripple voltage (rms) and from them obtain the ripple factor for the half-wave and full-wave rectified voltages. The calculations will show that the full-wave rectified signal has less per cent of ripple and is therefore a better rectified signal than the half-wave rectified signal, if lowest per cent of ripple is desired. The per cent of ripple is not always the most important concern. If circuit complexity or cost considerations are important (and the per cent of ripple is secondary), then a half-wave rectifier may be satisfactory. Also, if the filtered output supplies only a small amount of current to the load and the filtering circuit is not critical, then a half-wave rectified signal may be acceptable. On the other hand, when the supply must have as low a ripple as possible, it is best to start with a full-wave rectified signal since it has a smaller ripple factor, as will now be shown.

For the half-wave rectified signal the output dc voltage is $V_{dc} = 0.318 V_m$. The rms value of the ac component of output signal can be calculated (see Appendix B), and is $V_r \text{ (rms)} = 0.385 V_m$. Calculating the per cent of ripple,

$$\% \text{ ripple} = \frac{V_r \text{ (rms)}}{V_{dc}}(100) = \frac{0.385 V_m}{0.318 V_m}(100)$$

$$= 1.21(100) = 121\% \text{ (half-wave)}$$

For the full-wave rectifier the value of V_{dc} is $V_{dc} = 0.636V_m$. From the results obtained in Appendix B the ripple voltage of a full-wave rectified signal is V_r (rms) $= 0.305V_m$. Calculating the per cent of ripple,

$$\% \text{ ripple} = \frac{V_r(\text{rms})}{V_{dc}}(100) = \frac{0.305V_m}{0.636V_m}(100)$$

$$= 48\% \text{ (full-wave)}$$

The amount of ripple factor of the full-wave rectified signal is about 2.5 times smaller than that of the half-wave rectified signal and provides a better filtered signal. Note that these values of ripple factors are absolute values and do not depend at all on the peak voltage. If the peak voltage is made larger, the dc value of the output increases but then so does the ripple voltage. The two increase in the same proportion so that the ripple factor stays the same.

2.4 SIMPLE-CAPACITOR FILTER

A popular filter circuit is the simple-capacitor filter circuit shown in Fig. 2.11. The capacitor is connected across the rectifier output and the dc output voltage is available across the capacitor. Figure 2.12a shows the rectifier output voltage of a

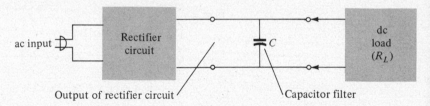

Figure 2.11. Simple-capacitor filter.

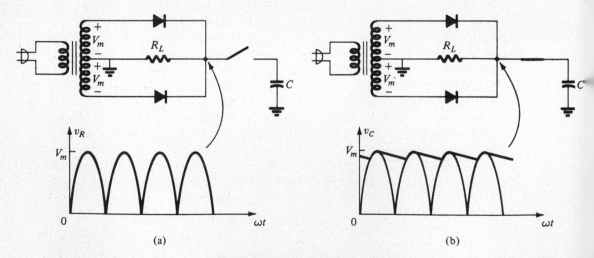

Figure 2.12. Capacitor filter operation: (a) full-wave rectifier voltage; (b) filtered output voltage.

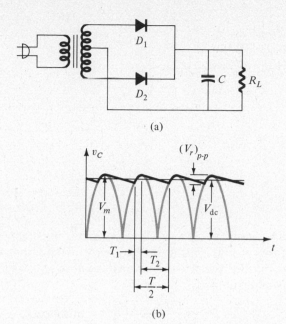

Figure 2.13. Capacitor filter: (a) capacitor filter circuit; (b) output voltage waveform.

full-wave rectifier circuit before the signal is filtered. Figure 2.12b shows the resulting waveform after the capacitor is connected across the rectifier output. As shown this filtered voltage has a dc level with some ripple voltage riding on it.

Figure 2.13a shows a full-wave rectifier and the output waveform obtained from the circuit when connected to an output load. If no load were connected to the filter, the output waveform would ideally be a constant dc level equal in value to the peak voltage (V_m) from the rectifier circuit. However, the purpose of obtaining a dc voltage is to provide this voltage for use by other electronic circuits, which then constitute a load on the voltage supply. Since there will always be some load on the filter, we must consider this practical case in our discussion. For the full-wave rectified signal indicated in Fig. 2.13b there are two intervals of time indicated. T_1 is the time during which a diode of the full-wave rectifier conducts and charges the capacitor up to the peak rectifier output voltage (V_m). T_2 is the time during which the rectifier voltage drops below the peak voltage, and the capacitor discharges through the load.

If the capacitor were to discharge only slightly (due to a light load), the average voltage would be very close to the optimum value of V_m. The amount of ripple voltage would also be small for a light load. This shows that the capacitor filter circuit provides a large dc voltage with little ripple for light loads (and a smaller dc voltage with larger ripple for heavy loads). To appreciate these quantities better we must further examine the output waveform and determine some relations between the input signal to be rectified, the capacitor value, the resistor (load) value, the ripple factor, and the regulation of the circuit.

Figure 2.14 shows the output waveform approximated by straight line charge and discharge. This is reasonable since the analysis with the nonlinear charge and discharge that actually takes place is complex to analyze and because the results

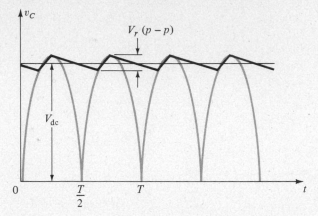

Figure 2.14. Approximate output voltage of capacitor filter circuit.

obtained will yield values that agree well with actual measurements made on circuits. The waveform of Fig. 2.14 shows the approximate output voltage waveform for a full-wave rectified signal. From an analysis of this voltage waveform the following relations can be obtained:

$$V_{dc} = V_m - \frac{V_r(p-p)}{2}$$ (half-wave   (2.5)

   and

$$V_r \text{(rms)} = \frac{V_r(p-p)}{2\sqrt{3}}$$ full wave)   (2.6)

These relations, however, are only in terms of the waveform voltages and we must further relate them to the different components in the circuit. Since the form of the ripple waveform for half-wave is the same as for full-wave, Eqs. (2.5) and (2.6) apply to both rectifier-filter circuits.

Ripple Voltage, V_r (rms)

Appendix B provides the details for determining the value of the ripple voltage in terms of the other circuit parameters. The result obtained for V_r (rms) is the following:

$$V_r \text{(rms)} \cong \frac{I_{dc}}{4\sqrt{3}\,fC} \times \frac{V_{dc}}{V_m} \qquad \text{(full-wave)} \qquad (2.7a)$$

where f is the frequency of the sinusoidal ac power supply voltage (usually 60 Hz), I_{dc} is the average current drawn from the filter by the load, and C is the filter capacitor value.

Another simplifying approximation that can be made is to assume that when used typically for light loads[2] the value of V_{dc} is only slightly less than V_m so that $V_{dc} \cong V_m$, and the equation can be written as

---

[2]Appendix B shows the relation of V_{dc} and V_m based on the amount of ripple. From Fig. B.3 we see that at ripple factors less than 6.5%, V_{dc} is within 10% of V_m. We can therefore define a *light load* as one resulting in a ripple less than 6.5%.

$$V_r \text{ (rms)} \cong \frac{I_{dc}}{4\sqrt{3}\,fC} \qquad \text{(full-wave, light load)} \qquad (2.7b)$$

Finally, we can include the typical value of line frequency ($f = 60$ Hz) and the other constants into the simpler equation

$$\boxed{V_r \text{ (rms)} = \frac{2.4 I_{dc}}{C} = \frac{2.4 V_{dc}}{R_L C}} \qquad \text{(full-wave, light load)} \qquad (2.7c)$$

where I_{dc} is in milliamperes, C is in microfarads, and R_L is in kilohms.

> **EXAMPLE 2.6** Calculate the ripple voltage of a full-wave rectifier with a 100 μF filter capacitor connected to a load of 50 mA.
>
> **Solution:** Using Eq. (2.7c), we get
>
> $$V_r \text{ (rms)} = \frac{2.4(50)}{100} = 1.2 \text{ V}$$

dc Voltage, V_{dc}

Using Eqs. (2.5), (2.6), and (2.7a), we see that the dc voltage of the filter is

$$V_{dc} = V_m - \frac{V_r(p-p)}{2} = V_m - \frac{I_{dc}}{4fC} \times \frac{V_{dc}}{V_m}$$

and

$$V_{dc} = \frac{V_m}{1 + I_{dc}/4fCV_m} \qquad \text{(full-wave)} \qquad (2.8a)$$

Again, using the simplifying assumption that V_{dc} is about the same as V_m for light loads, we get an approximate value of V_{dc} (which is less than V_m), of

$$V_{dc} = V_m - \frac{I_{dc}}{4fC} \qquad \text{(full-wave, light load)} \qquad (2.8b)$$

which can be written (using $f = 60$ Hz):

$$\boxed{V_{dc} = V_m - \frac{4.17 I_{dc}}{C} = \frac{V_m}{1 + \dfrac{4.17}{R_L C}}} \qquad \text{(full-wave, light load)} \qquad (2.8c)$$

where V_m is the peak rectified voltage, in volts, I_{dc} is the load cursent in milliamperes, C is the filter capacitor in microfarads, and R_L is the load resistance in kilohms.

> **EXAMPLE 2.7** If the peak rectified voltage for the filter circuit of Example 2.8 is 30 V, calculate the filter dc voltage.
>
> **Solution:** Using Eq. (2.8c), we get
>
> $$V_{dc} = V_m - \frac{4.17 I_{dc}}{C} = 30 - \frac{4.17(50)}{100} = 27.9 \text{ V}$$

The value of dc voltage is less than the peak rectified voltage. Note, also, from Eq. (2.8c), that the larger the value of average current drawn from the filter the less the value of output dc voltage, and the larger the value of the filter capacitor the closer the output dc voltage approaches the peak value of V_m.

Filter Capacitor Ripple

Using the definition of ripple [Eq. (2.3)] and the equation for ripple voltage [Eq. (2.7c)], we obtain the expression for the ripple factor of a full-wave capacitor filter

$$r = \frac{V_r(\text{rms})}{V_{dc}} \cong \frac{2.4 I_{dc}}{C V_{dc}} \qquad \text{(full-wave, light load)} \qquad (2.9a)$$

Since V_{dc} and I_{dc} relate to the filter load R_L, we can also express the ripple as

$$r = \frac{2.4}{R_L C} \qquad \text{(full-wave, light load)} \qquad (2.9b)$$

where I_{dc} is in milliamperes, C is in microfarads, V_{dc} is in volts, and R_L is in kilohms.

This ripple factor is seen to vary directly with the load current (larger load current, larger ripple factor), and inversely with the capacitor size. This agrees with the previous discussion of the filter circuit operation.

> EXAMPLE 2.8   A load current of 50 mA is drawn from a capacitor filter circuit ($C = 100\ \mu\text{F}$). If the peak rectified voltage is 30 V, calculate the % r.
>
> **Solution:**   Using the results of Examples 2.7 and 2.8 in Eq. (2.9a), we get
>
> $$\% \, r = \frac{2.4 I_{dc}}{C V_{dc}} \times 100 = \mathbf{4.3\%}$$
>
> From the basic definition of % r we could also calculate
>
> $$\% \, r = \frac{V_r(\text{rms})}{V_{dc}} \times 100 = \frac{1.2}{27.9} \times 100 = 4.3\%$$

Relations of V_{dc}/V_m and V_r (rms)/V_{dc} to Ripple

The ratios of V_{dc}/V_m and V_r (rms)/V_m can be related to ripple as shown in Appendix B. Figure B.3 provides a graph showing how the ratio of V_{dc}/V_m varies with % r, while Fig. B.4 provides the information on V_r (rms)/V_m. For a ripple of 5%, for example, we can obtain from Fig. B.3 that $V_{dc}/V_m = 0.92$, or $V_{dc} = 0.92 V_m$. At this same ripple (see Fig. B.4) the ratio of V_r (rms)/V_m is 45×10^{-3}, or, V_r (rms) $= 0.045 V_m$. A few examples will show how this information can help in analyzing or designing filter circuits.

> EXAMPLE 2.9   Determine the dc and ripple voltages across a 500-Ω load connected to a full-wave rectifier and 100-μF filter capacitor, as shown in Fig. 2.15.

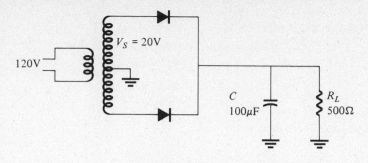

Figure 2.15. Rectifier-filter circuit for Example 2.9.

Solution: The peak voltage is $V_m = \sqrt{2}\,V_s = \sqrt{2}\,(20) = 28.3$ V. Using Eq. (2.9b), we get

$$\% \, r = \frac{2.4}{R_L C} \times 100 = \frac{2.4}{(0.5)(100)} \times 100 = 4.8\%$$

From Figs. B.3 and B.4 we can obtain (for $\% \, r = 4.8\%$)

$$\frac{V_{dc}}{V_m} = 0.925 \quad \text{and} \quad \frac{V_r \,(\text{rms})}{V_m} = 45 \times 10^{-3}$$

from which we calculate

$$V_{dc} = 0.925(28.3) = \mathbf{26.18 \ V}$$

$$V_r \,(\text{rms}) = 0.045(28.3) = \mathbf{1.27 \ V}$$

As a comparison we could calculate V_{dc} and V_r (rms) using Eqs. (2.7c) and (2.8c)

$$V_{dc} = \frac{V_m}{1 + \dfrac{4.17}{R_L C}} = \frac{28.3}{1 + \dfrac{4.17}{(0.5)(100)}} = 26.13 \ V$$

$$V_r \,(\text{rms}) = \frac{2.4 V_{dc}}{R_L C} = \frac{2.4(26.13)}{(0.5)(100)} = 1.25 \ V$$

which compare well with the results obtained from the graph data.

EXAMPLE 2.10   Calculate the per cent of ripple and dc and ripple voltages for the full-wave rectifier and filter circuit of Fig. 2.16.

Solution: Using Eq. (2.9c), we get

$$\% \, r = \frac{2.4}{R_L C} \times 100 = \frac{2.4}{(0.1)(250)} \times 100 = \mathbf{9.6\%} \text{ (not a light load)}$$

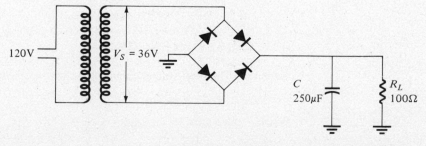

Figure 2.16. Rectifier-filter circuit for Example 2.10.

From Figs. B.3 and B.4 we obtain
$$V_{dc} = 0.86 V_m \quad \text{and} \quad V_r \text{ (rms)} = 0.083 V_m$$
Since
$$V_m = \sqrt{2}\, V_s = \sqrt{2}\,(36) = 50.91 \text{ V}$$
$$V_{dc} = 0.86(50.91) = \textbf{43.78 V}$$
$$V_{r(rms)} = 0.083(50.91) = \textbf{4.23 V}$$

Using these results we find that

$$\% \, r = \frac{V_{dc}}{V_r \text{ (rms)}} \times 100 = \frac{4.23}{43.78} \times 100 = 9.66\%$$

The difference in per cent of ripple is due to the use of Eq. (2.9c) when the load is not light ($\% \, r \geq 6.5\%$). For greater per cent of ripple this difference gets greater. If better accuracy is desired, the equations in Table 2.1 can be used. We should also note at this time that Figs. B.3 and B.4 apply to either half- or full-wave circuits, while Table 2.1 provides the equations to use for either circuit. Since the full-wave rectifier is most common, we shall do examples using mostly full-wave circuits.

A few examples will now show how a capacitor filter circuit can be designed.

EXAMPLE 2.11   Design a filter circuit to provide 18 V dc to a load of 100 mA for less than 5% ripple using a full-wave bridge rectifier.

Solution: For ripple = 5% we obtain (using Fig. B.3)

$$\frac{V_{dc}}{V_m} = 0.92$$

from which we can calculate

$$V_m = \frac{V_{dc}}{0.92} = \frac{18 \text{ V}}{0.92} = 19.6 \text{ V}$$

and
$$V_s = \frac{V_m}{\sqrt{2}} = \frac{19.6}{\sqrt{2}} = 13.8 \text{ V}$$

The load is
$$R_L = \frac{V_{dc}}{I_{dc}} = \frac{18 \text{ V}}{100 \text{ mA}} = 180 \, \Omega$$

Using Eq. (2.9c), we can calculate the value of capacitor, C

$$C = \frac{2.4}{r R_L} \times 100 = \frac{2.4}{5(0.180)} \times 100 = \textbf{266.7 } \boldsymbol{\mu}\textbf{F}$$

Using $C = 300 \, \mu$F will then provide ripple less than 5%.

EXAMPLE 2.12   Using a 12.6-V transformer secondary voltage, design a filter circuit to provide 16 V dc to a 200-Ω load.

Solution: The peak voltage provided by the transformer is

$$V_m = \sqrt{2}\, V_s = \sqrt{2}\,(12.6) = 17.8 \text{ V}$$

We then calculate
$$\frac{V_{dc}}{V_m} = \frac{16}{17.8} = 0.899$$

which occurs at $\% \, r = 6.5\%$ (using Fig. B.3). Since this is a light load, we

TABLE 2.1 Summary of Filter Circuit Operation

FILTER TYPE	NO-LOAD DC VOLTAGE $(V_{dc})NL$	DC VOLTAGE V_{dc}	RMS VALUE OF AC COMPONENT OF RIPPLE VOLTAGE V_r(rms)	RIPPLE FACTOR r	IMPORTANT FACTORS
Capacitor (C)	V_m	$V_{dc} = \dfrac{V_m}{1+(I_{dc}/4fCV_m)}$ $= V_m - \dfrac{4.17I_{dc}}{C} = \dfrac{V_m}{1+\frac{4.17}{R_LC}}$	$V_r\,(\text{rms}) = \dfrac{I_{dc}}{4\sqrt{3}fC} \times \dfrac{V_{dc}}{V_m}$ $\cong \dfrac{2.4I_{dc}}{C} = \dfrac{2.4V_{dc}}{R_LC}$	$r = \dfrac{I_{dc}}{4\sqrt{3}fCV_m}$ $\cong \dfrac{2.4I_{dc}}{CV_{dc}} = \dfrac{2.4}{R_LC}$	full-wave; full-wave, light load
	V_m	$V_{dc} = \dfrac{V_m - (I_{dc}/4fCV_m)}{1+(I_{dc}/4fCV_m)}$ $= V_m - \dfrac{4.17I_{dc}}{C} = \dfrac{V_m}{1+\frac{4.17}{R_LC}}$	$V_r\,(\text{rms}) = \dfrac{I_{dc}}{4\sqrt{3}fC}\left[1+\dfrac{V_{dc}}{V_m}\right]$ $= \dfrac{4.8I_{dc}}{C} = \dfrac{4.8V_{dc}}{R_LC}$	$r = \dfrac{I_{dc}}{4\sqrt{3}fC}\left(\dfrac{1}{V_m}+\dfrac{1}{V_{dc}}\right)$ $= \dfrac{4.8}{R_LC}$	half-wave; half-wave, light load
RC (following C-filter)	V_m	$V_{dc}' = \dfrac{R_L}{R_l+R_L}V_{dc}$	$V_r\,'(\text{rms}) \cong \dfrac{X_c}{R}V_r\,(\text{rms})$	$r' = \dfrac{X_c}{R}r$ $\left(R' = \dfrac{RR_L}{R+R_L}\right)$	$f_r = 120$Hz, full-wave; $f_r = 60$ Hz, half-wave
π-type (C_1-LC_2)	V_m	$V_{dc}' = \dfrac{R_L}{R_l+R_L}V_{dc}$	$V_r\,'(\text{rms}) = \dfrac{1.76}{LC_2}V_r\,(\text{rms})$, full-wave $= \dfrac{7.04}{LC_2}V_r\,(\text{rms})$, half-wave	$r' = \dfrac{3300}{C_1C_2LR_L}$	full-wave
L-type (choke)	$0.636V_m$	$V_{dc}' = 0.636V_m - I_{dc}R_l$	$V_r\,'(\text{rms}) = \dfrac{0.53}{LC}V_m$	$r' = \dfrac{0.83}{LC}$	full-wave; $L_c > \dfrac{R_L}{1000}$
	$0.318V_m$	$V_{dc}' = 0.318V_m - I_{dc}R_l$	$V_r\,'(\text{rms}) = \dfrac{2.49V_m}{LC}$	$r' \cong \dfrac{7.83}{LC}$	half-wave
Half-wave rectifier	$0.318V_m$	$0.318V_m$	$0.385V_m$	1.21	
Full-wave rectifier	$0.636V_m$	$0.636V_m$	$0.305V_m$	0.48	

can calculate C by using Eq. (2.9c)

$$C = \frac{2.4}{rR_L} \times 100 = \frac{2.4}{(6.5)(0.2)} \times 100 = \textbf{184.6 } \boldsymbol{\mu}\textbf{F}$$

A capacitor of value $C = 180\ \mu\text{F}$ will provide the desired output voltage.

EXAMPLE 2.13 Design a filter circuit to provide 160 V at 20 mA with less then 10% ripple.

Solution: At 10% ripple we obtain from Fig. B.3

$$\frac{V_{dc}}{V_m} = 0.85$$

so that
$$V_m = \frac{V_{dc}}{0.85} = \frac{160}{0.85} = 187.7\ \text{V}$$

and
$$V_s = \frac{V_m}{\sqrt{2}} = \frac{187.7}{\sqrt{2}} = 132.7\ \text{V} \quad \text{(transformer secondary voltage, rms)}$$

Since the ripple indicates that the load is not light, we can calculate C by using Eq. (2.7a) and the definition of % r

$$\% \ r = \frac{V_r\ (\text{rms})}{V_{cc}} \times 100 = \frac{I_{dc}}{4\sqrt{3}\ fCV_m} \times 100$$

so that
$$C = \frac{I_{dc}}{4\sqrt{3}\ f\ \%\ rV_m} \times 100 = \frac{20 \times 10}{4\sqrt{3}\ (60)(0.1)(187.7)} \times 100 = \textbf{2.57 } \boldsymbol{\mu}\textbf{F}$$

Using $C = 3\ \mu\text{F}$ will provide ripple less than 10%.

Diode Conduction Period and Peak Diode Current

From the previous discussion it should be clear that larger values of capacitance provide less ripple and higher average voltages, thereby providing better filter action. From this one may conclude that to improve the performance of a capacitor filter it is only necessary to increase the size of the filter capacitor. However, the capacitor also affects the peak current through the rectifying diode and, as will now be shown, the larger the value of capacitance used the larger the peak current through the rectifying diode.

Referring back to the operation of the rectifier and capacitor filter circuit, we see that there are two periods of operation to consider. After the capacitor is charged to the peak rectified voltage (see Fig. 2.13b), a period of diode nonconduction elapses (time T_2) while the output voltage discharges through the load. After T_2 the input rectified voltage becomes greater than the capacitor voltage and for a time, T_1, the capacitor will charge back up to the peak rectified voltage. The average current supplied to the capacitor and load during this charge period must equal the average current drawn from the capacitor during the discharge period. Figure 2.17 shows the diode current waveform for half-wave rectifier operation. Notice that the diode conducts for only a short period of the cycle. In fact, it should be seen that

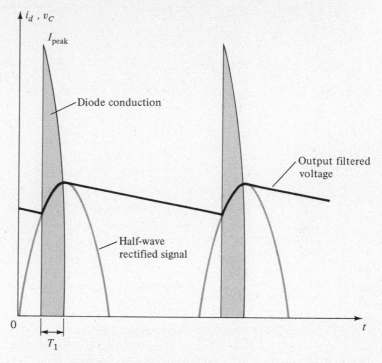

Figure 2.17. Diode conduction during charging part of cycle.

the larger the capacitor the less that amount of voltage decay and the shorter the interval during which charging takes place. In this shorter charging interval the diode will have to pass the same amount of *average current*, and can do so only by passing larger peak current. Figure 2.18 shows the output current and voltage waveforms for small and large capacitor values. The important factor to note is the increase in peak current through the diode for the larger values of capacitance. Since the average current drawn from the supply must equal the average of the current through the diode during the charging period, the following relation can be derived from Fig. 2.18:[3]

$$I_{dc} = \left(\frac{T_1}{T}\right)(I_{peak}) \tag{2.10a}$$

from which we obtain

$$\boxed{I_{peak} = \left(\frac{T}{T_1}\right)I_{dc}} \tag{2.10b}$$

---

[3]Assuming a rectangular-shaped pulse of duration T_1, peak value, I_{peak}, and a period of T, the area under the pulse divided by the period gives the average value, I_{dc}:

$$\frac{1}{T}(I_{peak}T_1) = I_{dc}$$

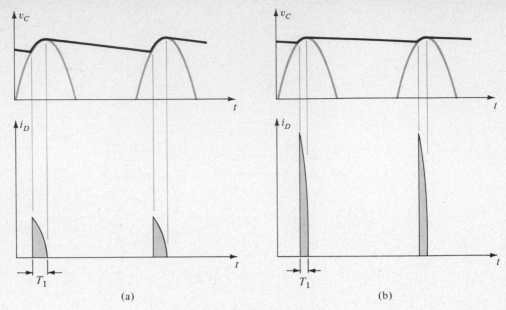

(a) (b)

Figure 2.18. Output voltage and diode current waveforms: (a) small C; (b) large C.

where T_1 = diode conduction time;
$T = 1/f = \frac{1}{60}$ for usual line 60-Hz voltage;
I_{dc} = average current drawn from filter circuit; and
I_{peak} = peak current through the conducting diode.

Relation of I_p/I_{dc} to Per Cent of Ripple

Appendix B.5 shows how the diode conduction angle and therefore the ratio of diode peak current (I_p) to dc current (I_{dc}) varies as a function of ripple. At 10% ripple a full-wave rectifier must pass a peak current that is 3.5 times the dc load current, while at 1% ripple this increases to 11.75 times, and at a low ripple of 0.5% the peak current is about 16.7 times as large as the dc load current. Obviously, the choice of very low ripple using a capacitor filter circuit results in considerable peak current through the rectifier diodes. Figure B.5 provides a graph of I_p/I_{dc} over a range of ripple. Notice that for half-wave operation the ratio of peak to dc current is twice as large as with full-wave rectification—one of the reasons for choosing the latter. Some examples will help show how the information in Fig. B.5 can be used.

EXAMPLE 2.14   A full-wave rectifier and simple capacitor filter circuit provides 40 V, dc with 0.8 V, ripple (rms) to a 200-Ω load. What peak current are the rectifier diodes expected to handle?

Solution: From the given information

$$I_{dc} = \frac{V_{dc}}{R_L} = \frac{40 \text{ V}}{0.2 \text{ K}} = 200 \text{ mA}$$

Also, the % ripple is

$$\%r = \frac{V_r(\text{rms})}{V_{dc}} \times 100 = \frac{0.8 \text{ V}}{40 \text{ V}} \times 100 = 2\%$$

From Fig. B.5 at 2% ripple for full-wave operation we get

$$\frac{I_p}{I_{dc}} = 8.3$$

so that
$$I_p = 8.3 I_{dc} = 8.3(200 \text{ mA}) = \textbf{1.66 A}$$

EXAMPLE 2.15 If a half-wave rectifier, capacitor filter provides 80 V, dc with 2 V, rms of ripple to a 200-Ω load, what peak diode current results?

Solution:

$$I_{dc} = \frac{V_{dc}}{R_L} = \frac{80 \text{ V}}{0.2 \text{ K}} = 400 \text{ mA}$$

$$\%r = \frac{V_r(\text{rms})}{V_{dc}} \times 100 = \frac{2}{80} \times 100 = 2.5\%$$

From Fig. B.5 we then get

$$\frac{I_p}{I_{dc}} = 14.8$$

so that
$$I_p = 14.8(400 \text{ mA}) = \textbf{5.92 A}$$

EXAMPLE 2.16 Design a capacitor filter circuit to provide 40 V, dc at 250 mA with less than 2% ripple.

Solution: From the given information

$$R_L = \frac{V_{dc}}{I_{dc}} = \frac{40 \text{ V}}{250 \text{ mA}} = 0.160 \text{ K}$$

Using Eq. (2.9c), we get

$$C = \frac{2.4}{rR_L} \times 100 = \frac{2.4}{(2)(0.16)} 100 = \textbf{750 } \boldsymbol{\mu}\textbf{F}$$

From Fig. B.5 (for full-wave operation) at $\%r = 2\%$

$$\frac{I_p}{I_{dc}} = 8.3$$

so that
$$I_p = 8.3(250 \text{ mA}) = 2.1 \text{ A}$$

From Fig. B.3 (at $\%r = 2\%$)

$$\frac{V_{dc}}{V_m} = 0.967$$

and
$$V_m = \frac{V_{dc}}{0.967} = \frac{40 \text{ V}}{0.967} = 41.4 \text{ V}$$

The transformer secondary voltage is then

$$V_s = \frac{V_m}{\sqrt{2}} = \frac{41.4 \text{ V}}{\sqrt{2}} = \textbf{29.2 V, rms}$$

A center-tapped rectifier-filter circuit to provide the desired dc voltage is shown in Fig. 2.19.

SEC. 2.4 SIMPLE-CAPACITOR FILTER

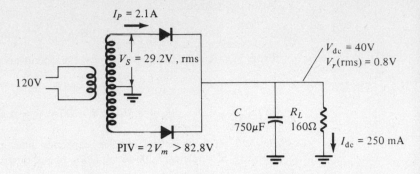

Figure 2.19. Rectifier-filter circuit for Example 2.16.

2.5 RC FILTER

It is possible to further reduce the amount of ripple while reducing the dc voltage by using an additional RC filter section as shown in Fig. 2.20. The purpose of the added network is to pass as much of the dc component of the voltage developed across the first filter capacitor C_1 and to attenuate as much of the ac component of the ripple voltage developed across C_1 as possible. This action would reduce the amount of ripple in relation to the dc level, providing better filter operation than for the simple-capacitor filter. There is a price to pay for this improvement, as will be shown; this includes a lower dc output voltage due to the dc voltage drop across the resistor and the cost of the two additional components in the circuit.

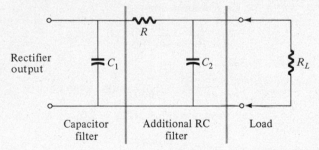

Figure 2.20. RC filter stage.

Figure 2.21 shows the rectifier filter circuit for full-wave operation. Since the rectifier feeds directly into a capacitor, the peak currents through the diodes are many times the average current drawn from the supply [as specified in Fig. B.5.] The voltage developed across capacitor C_1 is then further filtered by the resistor-capacitor section (R, C_2) providing an output voltage having less per cent of ripple than that across C_1. The load, represented by resistor R_L, draws dc current through resistor R with an output dc voltage across the load being somewhat less than that across C_1 due to the voltage drop across R. This filter circuit, like the simple-capacitor filter circuit, provides best operation at light loads, with considerably poorer voltage regulation and higher per cent of ripple at heavy loads.

The analysis of the resulting ac and dc voltages at the output of the filter from that obtained across capacitor C_1 can be carried out by using superposition. We can

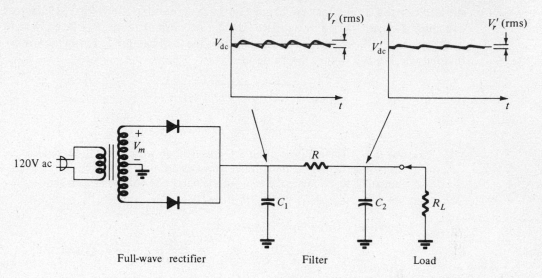

Figure 2.21. Full-wave rectifier and RC filter circuit.

separately consider the RC circuit acting on the dc level of the voltage across C_1 and then the RC circuit action on the ac (ripple) portion of the signal developed across C_1. The resulting values can then be used to calculate the overall circuit voltage regulation and per cent of ripple.

dc Operation of RC Filter Section

Figure 2.22a shows the equivalent circuit to use when considering the dc voltage and currents in the filter and load. The two filter capacitors are open circuit for dc and are thus removed from consideration at this time. Calculation of the dc voltage across filter capacitor C_1 was essentially covered in Section 2.4 and the treatment of the additional RC filter stage will proceed from there. Knowing the dc voltage across the first filter capacitor (C_1), we can calculate the dc voltage at the output of

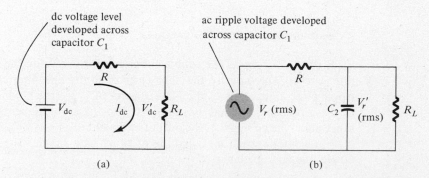

Figure 2.22. dc and ac equivalent circuits of RC filter: (a) dc equivalent circuit; (b) ac equivalent circuit.

the additional RC filter section. From Fig. 2.22a we see that the voltage, V_{dc}, across capacitor C_1 is attenuated by a resistor-divider network of R and R_L (the equivalent load resistance); the resulting dc voltage across the load being V'_{dc}. Using the voltage divider rule, we can obtain the value of V'_{dc}:

$$V'_{dc} = \left(\frac{R_L}{R + R_L} \right)(V_{dc}) \qquad (2.11)$$

EXAMPLE 2.17 The addition of an RC filter section with $R = 120\ \Omega$, reduces the dc voltage across the initial filter capacitor from 60 V (V_{dc}). If the load resistance is 1 K, calculate the value of the output dc voltage (V'_{dc}) from the filter circuit.

Solution: Using Eq. (2.11), we get

$$V'_{dc} = \frac{R_L}{R + R_L} \times V_{dc} = \frac{1000}{120 + 1000} \times 60 = \mathbf{53.6\ V}$$

In addition, we may calculate the drop across the filter resistor and the load current drawn:

$$V_R = V_{dc} - V'_{dc} = 60 - 53.6 = 6.4\ V$$

$$I_{dc} = \left(\frac{V'_{dc}}{R_L} \right) = \left(\frac{53.6}{1 \times 10^3} \right) = 53.6\ mA$$

EXAMPLE 2.18 An RC filter stage connected to a simple-capacitor filter provides 80 V dc to a 500-Ω load. The RC stage is comprised of a 72-Ω resistor and a 40-μF capacitor. Calculate the dc voltage across filter capacitor C_1.

Solution: From the values of load resistance and load voltage the value of dc current is

$$I_{dc} = \frac{V'_{dc}}{R_L} = \frac{80\ V}{0.5\ K} = 160\ mA$$

Reworking Eq. (2.11) we can solve for V_{dc}

$$V_{dc} = \frac{R + R_L}{R_L} \times V'_{dc} = \frac{72 + 500}{500} 80 = \mathbf{91.5\ V}$$

ac Operation of RC Filter Section

Figure 2.22b shows the equivalent circuit for analyzing the ac operation of the filter circuit. The input to the filter stage from the first filter capacitor (C_1) is the ripple or ac signal part of the voltage across C_1, V_r (rms), which is approximated now as a sinusoidal signal. Both the RC filter stage components and the load resistance affect the ac signal at the output of the filter.

For a filter capacitor (C_2) value of 10 μF at a ripple voltage frequency (f) of 60 Hz, the ac impedance of the capacitor is[4]

---

[4] X_C is understood here to represent only the *magnitude* of the capacitor's ac impedance.

$$X_c = \frac{1}{\omega C} = \frac{1}{2\pi fC} = \frac{1}{6.28(60)(10 \times 10^{-6})} = 0.265 \text{ K}$$

Referring to Fig. 2.22b, we see that this capacitive impedance is in parallel with the load resistance. For a load resistance of 2 K, for example, the parallel combination of the two components would yield an impedance of magnitude:

$$Z = \frac{RX_c}{\sqrt{R^2 + X_c^2}} = \frac{2(0.265)}{\sqrt{2^2 + (0.265)^2}} = \frac{2}{2.02}(0.265) = 0.263 \text{ K}$$

This is close to the value of the capacitive impedance alone, as expected, since the capacitive impedance is much less than the load resistance and the parallel combination of the two would be smaller than the value of either. As a rule of thumb we can consider neglecting the loading by the load resistor on the capacitive impedance as long as the load resistance is at least five times as large as the capacitive impedance. Because of the limitation of light loads on the filter circuit the effective value of load resistance is usually large compared to the inpedance of capacitors in the range of microfarads.

In the above discussion it was stated that the frequency of the ripple voltage was 60 Hz. Assuming that the line frequency was 60 Hz, the ripple frequency will also be 60 Hz for the ripple voltage from a half-wave rectifier. The ripple voltage from a full-wave rectifier, however, will be double since there are twice the number of half-cycles and the ripple frequency will then be 120 Hz. Referring to the relation for capacitive impedance $X_c = 1/\omega C$, we have value of $\omega = 377$ for 60 Hz and of $\omega = 754$ for 120 Hz. Using values of capacitance in μF, we can express the relation for capacitive impedance as

$$X_c = \frac{2.65}{C} \quad \text{(half-wave)} \tag{2.12a}$$

$$X_c = \frac{1.33}{C} \quad \text{(full-wave)} \tag{2.12b}$$

where C is in microfarads and X_c is in kilohms.[5]

EXAMPLE 2.19   Calculate the impedance of a 15-μF capacitor used in the filter section of a circuit using full-wave rectification.

Solution:

$$X_c = \frac{1.33}{C} = \frac{1.33}{15} = 0.0886 \text{ K} = \mathbf{88.6\ \Omega}$$

---

[5]Equations (2.12a) and (2.12b) may also be expressed as

$$X_c = \frac{265}{C}\text{(half-wave)}$$

$$X_c = \frac{1326}{C}\text{(full-wave)}$$

where C is in microfarads and X_c is in ohms.

Using the simplified relation that the parallel combination of the load resistor and the capacitive impedance equals, approximately, the capacitive impedance, we can calculate the ac attenuation in the filter stage:

$$V_r'(\text{rms}) = \frac{X_C}{\sqrt{R^2 + X_C^2}} V_r(\text{rms}) \qquad (2.13a)$$

The use of the square root of the sum of the squares in the denominator was necessary since the resistance and capacitive impedance must be added vectorially, not algebraically. If the value of the resistance is larger by a factor of 5 than that of the capacitive impedance, then a simplification of the denominator may be made yielding the following result:

$$\boxed{V_r'(\text{rms}) \cong \frac{X_C}{R} \cdot V_r(\text{rms})} \qquad (2.13b)$$

EXAMPLE 2.20   The output of a full-wave rectifier and capacitor filter is further filtered by an RC filter section (see Fig. 2.23). The component values of the RC section are $R = 500\ \Omega$ and $C = 10\ \mu\text{F}$. If the initial capacitor filter develops 150 V dc with a 15 V ac ripple voltage, calculate the resulting dc and ripple voltage across a 5-K load.

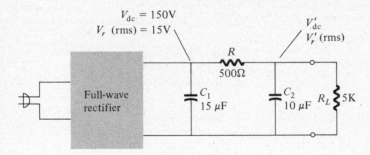

Figure 2.23. RC filter circuit for Example 2.20.

Solution:

dc calculations: Calculating the value of V_{dc}' from Eq. (2.11):

$$V_{dc}' = \frac{R_L}{R + R_L} \times V_{dc} = \frac{5000}{500 + 5000}(150) = \frac{5000}{5500}(150) = \mathbf{136.4\ V}$$

ac calculations: Calculating the value of the capacitive impedance first (for full-wave operation):

$$X_C = \frac{1.33}{C} = \frac{1.33}{10} = 0.133\ \text{K} = 133\ \Omega$$

Since this impedance is not quite 5 times smaller than that of the filter resistor ($R = 500\ \Omega$), we shall use Eq. (2.13a) for the calculation, and then repeat the calculation to show what the difference would have been using Eq. (2.13b) (since the components are almost 5 times different in size). Using Eq. (2.13a), we get

$$V_r'(\text{rms}) = \frac{X_C}{\sqrt{R^2 + X_C^2}}(V_r\,(\text{rms})) = \frac{0.133}{\sqrt{(0.5)^2 + (0.133)^2}}(15)$$

$$= \frac{0.133}{0.518}(15) = \mathbf{3.86\ V}$$

Now using Eq. (2.13b), we get

$$V_r'(\text{rms}) = \frac{X_C}{R} \cdot V_r(\text{rms}) = \frac{0.133}{0.500}(15) = \mathbf{3.99\ V}$$

Comparing the results of 3.86 V and 3.99 V, using Eq. (2.14b), would have yielded an answer within 3.5% of the more exact solution. For values of R much larger than 5 tims X_C it should be clear that the answer obtained using the simplified relation of Eq. (2.13b) would be satisfactory.

Ripple Factor and Per Cent of Ripple with RC Filter Section

One of the more important gains of the added RC filter section is the improvement in ripple factor, which we consider next. The ripple factor at the output of the complete filter circuit (including the RC filter stage) can be expressed as

$$r' = \frac{V_r'(\text{rms})}{V_{dc}'} \tag{2.14}$$

Using Eqs. (2.11) and (2.13) we can express the output ripple factor in terms of the ripple factor at the first filter capacitor and the values of the RC filter stage and load resistance. Doing so we obtain

$$r' = \frac{V_r'(\text{rms})}{V_{dc}'} = \frac{\left(\dfrac{X_C}{R}\right)(V_r(\text{rms}))}{\left(\dfrac{R_L}{R + R_L}\right)(V_{dc})}$$

$$= \left(\frac{X_C}{(R \times R_L)/(R + R_L)}\right)\left(\frac{V_r(\text{rms})}{V_{dc}}\right) \tag{2.15a}$$

$$\boxed{r' = \frac{X_C}{R'} \times r} \tag{2.15b}$$

where

$$R' = \frac{R R_L}{R + R_L}$$

EXAMPLE 2.21 Calculate the ripple at the first filter capacitor and the output of the complete filter circuit for the circuit and component values of Example 2.20.

Solution:

$$r = \frac{V_r(\text{rms})}{V_{dc}} = \frac{15}{150} = 0.10; \qquad \%r = r \times 100 = 10\%$$

$$r' = \frac{V_r'(\text{rms})}{V_{dc}'} = \frac{3.86}{136.4} = 0.0283; \qquad \%r' = r' \times 100 = 2.83\%$$

Using the approximate relation of Eq. (2.15b), we get

$$r' = \frac{X_C}{R'}r = \frac{0.133}{0.455}(0.10) = 0.0292; \qquad \%r' = r' \times 100 = 2.92\%$$

where
$$R' = \frac{RR_L}{R + R_L} = \frac{0.5(5)}{0.5 + 5} = \mathbf{0.455\ K}$$

From the above results we see that the ripple was reduced from 10% to either 2.83 or 2.92% (depending on the simplification used in the calculation). This is improvement by a factor of about 3, as the ratio of X_C to R' indicates.

Voltage Regulation for RC Filter Circuit

The calculation of voltage regulation requires calculating the output voltage at no-load and at full-load. If no-load were connected to the output of the filter circuit, the capacitors would charge up to the peak voltage of the rectifier output (V_m). The relation for voltage regulation is

$$\% \text{ voltage regulation} = \frac{V_{NL} - V_{FL}}{V_{FL}} \times 100$$

For the general RC filter circuit this can be written

$$\% V.R. = \frac{(V'_{dc})_{NL} - (V'_{dc})_{FL}}{(V'_{dc})_{FL}} \times 100 \qquad (2.16a)$$

For the case where the no-load voltage is the peak rectified voltage Eq. (2.16a) can be written:

$$\boxed{\% V.R. = \frac{V_m - V'_{dc}}{V'_{dc}} \times 100} \qquad (2.16b)$$

where
$$V_{NL} = V_m \qquad \text{and} \qquad V_{FL} = V'_{dc}$$

EXAMPLE 2.22   The following filter circuit (Fig. 2.24) is used to supply voltage to a load resistance of 10 K. Calculate the voltage regulation of the circuit at full-load.

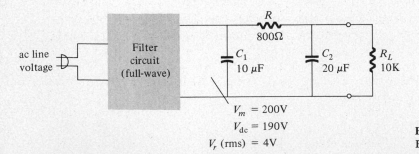

$V_m = 200\text{V}$
$V_{dc} = 190\text{V}$
$V_r \text{ (rms)} = 4\text{V}$

Figure 2.24. RC filter circuit for Example 2.19.

Full-load:

$$(V_{dc}')_{FL} = \frac{R_L}{R + R_L}(V_{dc}) = \frac{10}{10.8}(190) = 175.9 \text{ V}$$

$$\%V.R. = \frac{V_m - V_{dc}'}{V_{dc}'}(100) = \frac{200 - 175.9}{175.9}(100) = 13.7\%$$

2.6 π-TYPE FILTER

The addition of an RC filter section improved the ripple factor by decreasing the ac ripple voltage by a greater amount than it decreased the dc voltage. To decrease the dc output as little as possible the series resistor should be as small as possible. On the other hand, the attenuation of the ripple voltage requires that the RC section capacitor have an impedance that is much smaller than that of the series resistor. Making the series resistor R smaller to pass most of the dc voltage will not provide a large reduction of the ripple voltage. Keeping R large compared to the RC section capacitor will provide attenuation of the ripple voltage but will then result in decrease of the dc voltage.

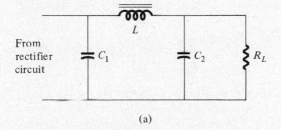

(a)

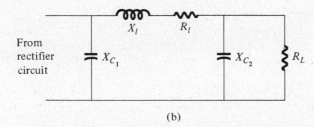

(b)

Figure 2.25. π-type filter circuit: (a) components; (b) impedances.

The ripple factor of the voltage across the first filter capacitor will be reduced most if the filter (attenuation) section following it provides little series dc resistance, while at the same time providing large series ac impedance. A resistor is the same to both ac and dc signals. An inductor, however, can have very low dc resistance, while at the same time having a large ac impedance. Fig. 2.25a shows just such an arrangement, the entire filter circuit of capacitor C_1, inductor L, and capacitor C_2, comprising a π-*type filter* circuit.

The inductor will be considered as a practical component having a dc resistance, R_l (due to wire resistance), and an inductance, L (see Fig. 2.25b). In analyzing the operation of inductor L, and capacitor C_2, in reducing the ripple factor of the initial filtered signal (across C_1) we shall separately consider the dc and ac operation of the circuit.

dc Calculations

The output dc voltage (V'_{dc}) is less than the dc voltage developed across the first filter capacitor (C_1) by the drop in voltage across the resistance of the inductor. This voltage drop depends on the amount of load current being drawn from the filter circuit. If no current is drawn from the filter, the output dc voltage is the same as that across filter capacitor C_1, V_m. When load current is drawn, the output dc voltage can be calculated from

$$V'_{dc} = V_{dc} - I_{dc}R_l \qquad (2.17a)$$

where R_l is the dc resistance of the inductor.

EXAMPLE 2.23 A π-type filter has an input voltage of 150 V dc across the input capacitor. If the filter inductor is a 5-H choke having 300-Ω resistance, calculate the output dc voltage at a load current of 100 mA.

Solution:

$$V'_{dc} = V_{dc} - I_{dc}R_l = 150 - 100(10^{-3}) \times 300$$
$$= 150 - 30 = \textbf{120 V}$$

Another way of obtaining the output dc voltage is to calculate the voltage attenuation of the inductor resistance and the load resistance (see Fig. 2.26a). Considering the dc voltage divider network, the voltage across the load resistor is

$$V'_{dc} = \frac{R_L}{R_L + R_l} V_{dc} \qquad (2.17b)$$

EXAMPLE 2.24 Calculate the output dc voltage of a π-type filter having 150 V dc across the input filter capacitor for a 1.2-K load. The inductor dc resistance is 300 Ω.

Solution:

$$V'_{dc} = \frac{R_L}{R_L + R_l} V_{dc} = \frac{1200}{1200 + 300}(150) = \textbf{120 V}$$

ac Calculations

The output ripple voltage, which is approximated as a sinusoidal signal, can be calculated by considering the action of the ac divider network of Fig. 2.26b. Carrying out the calculation in detail would require using the impedance of X_{C_2} in parallel with R_L, and the impedance of R_l in series with X_l. This detail is usually not necessary since proper selection of filter components should result in X_l being much larger than R_l and in X_{C_2} being much smaller than R_L. An example will help reinforce this consideration.

EXAMPLE 2.25 Calculate the impedance of the series combination of R_l and X_l and the parallel combination of X_{C_2} and R_L, where, $R_l = 300\ \Omega$, $L = 10$ H, $C_2 = 10\ \mu$F, and $R_L = 5$K. The ripple frequency is 60 Hz.

Solution:

$$X_l = 2\pi fL = 6.28(60)(10) = 3768\ \Omega$$

$$Z_l = \sqrt{X_l^2 + R_l^2} = \sqrt{(3768)^2 + (300)^2} = 3780\ \Omega$$

$$X_{C_2} = \frac{1}{2\pi fC_2} = \frac{1}{6.28(60)(10 \times 10^{-6})} = 0.265\ \text{K}$$

$$X_{C_2} \| R_L = \frac{X_{C_2}R_L}{\sqrt{X_{C_2}^2 + R_L^2}} = \frac{(0.265)(5)}{\sqrt{(0.265)^2 + (5)^2}}$$

$$= \frac{5}{\sqrt{25.09}}(0.265)\ \mathbf{0.264\ K}$$

In comparison, the series impedance of the coil inductance and resistance can be approximated by the inductive impedance of 3768 Ω with less than 1% error. The parallel impedance of filter capacitor C_2 and the load resistance is equal to the impedance of the capacitor alone to within 0.1% error.

Figure 2.26c shows the simplified ac circuit representation with the inductor replaced by an impedance X_l and the filter capacitor and load resistor replaced by a capacitive impedance of X_{C_2}. Using the voltage divider rule to calculate the ac voltage across the output due to an input ac voltage V_r (rms), we get

$$V_r'(\text{rms}) = \frac{X_{C_2}}{|X_l - X_{C_2}|} V_r(\text{rms}) \tag{2.18a}$$

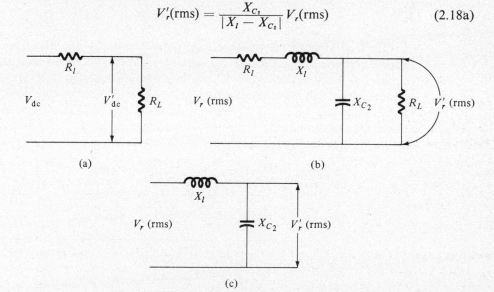

Figure 2.26. Equivalent circuits for π-type filter: (a) dc equivalent circuit; (b) ac equivalent circuit; (c) approximate ac voltage divider circuit.

Again we are able to resort to a simplification replacing the denominator term above by the X_l value alone:

$$V_r'(\text{rms}) \cong \frac{X_{C_2}}{X_l} V_r(\text{rms})$$

(2.18b)

The following example will show this to be a reasonable simplification.

EXAMPLE 2.26 Compare the impedance ratio obtained from Eq. (2.18a) and that from Eq. (2.18b) using the impedance values of Example 2.25.

Solution: Using $X_{C_2} = 0.265$ K and $X_l = 3.77$ K, we get

$$\frac{X_{C_2}}{|X_l - X_{C_2}|} = \frac{0.265}{3.77 - 0.265} \cong 0.0756$$

$$\frac{X_{C_2}}{X_l} = \frac{0.265}{3.77} = 0.0703$$

The results have less than 1% difference, and the simpler expression of Eq. (2.18b) may be used.

EXAMPLE 2.27 A 5-V ripple voltage is developed across the initial capacitor of a π-type filter. If the filter components are $L = 20$ H and $C_2 = 20$ μF, calculate the ripple voltage across the second capacitor filter (C_2). The ripple frequency is 60 Hz.

Solution: Using Eq. (2.18b), we get

$$V_r'(\text{rms}) = \frac{X_{C_2}}{X_l} V_r(\text{rms}) = \frac{1/[6.28(60)(20 \times 10^{-6})]}{6.28(60)(20)}(5)$$

$$= 0.088 \text{ V} = \textbf{88 mV}$$

The frequency of the ripple voltage developed across the first filter capacitor depends on the original sinusoidal signal frequency and on whether the rectifier circuit was half- or full-wave. Using the line ac frequency of 60 Hz, the ripple frequency for half-wave rectification will also be 60 Hz but that with full-wave rectification will be 120 Hz. Referring back to Eq. (2.18b) we can include the frequency and component dimensional units in the equation as a constant to obtain the following equation for the action with a full-wave rectified signal ($f = 120$ Hz):

$$V_r'(\text{rms}) = \frac{X_{C_2}}{X_l} V_r(\text{rms}) = \frac{1/2\pi f C_2}{2\pi f L} = V_r(\text{rms}) = \frac{1.76}{LC_2} V_r(\text{rms})$$

$$V_r'(\text{rms}) = \frac{1.76}{LC_2} V_r(\text{rms}), \qquad \text{full-wave}$$

(2.19a)

and for a half-wave rectified signal ($f = 60$ Hz):

$$V_r'(\text{rms}) = \frac{7.04}{LC_2} V_r(\text{rms}), \qquad \text{half-wave}$$

(2.19b)

In Eqs. (2.19a) and (2.19b) the calculation holds for a 60-Hz sinusoidal supply signal, the inductance being in units of henrys, and the capacitance in units of microfarads.

> **EXAMPLE 2.28** Calculate the ripple voltage out of a π-type filter ($L = 8$ H and $C_2 = 2\ \mu$F) for an input ripple voltage of 4 V (rms) across the initial filter capacitor. The ac signal (supply frequency of 60 Hz) is fed to the filter from a full-wave rectifier.
>
> **Solution:** Using Eq. (2.19a), we get
>
> $$V_r'(\text{rms}) = \frac{1.76}{LC_2}V_r(\text{rms}) = \frac{1.76}{(8)(2)}(4) = \mathbf{0.44\ V}$$

> **EXAMPLE 2.29** Calculate the ripple factor of the output voltage of a π-type filter ($L = 5$ H, $C_2 = 4\ \mu$F, $R_l = 250\ \Omega$) connected to a 4-K load. The voltage across the first filter capacitor (C_1) is 80 V dc with 10 V (rms) ripple at 120 Hz. Compare this to the input voltage ripple factor.
>
> **Solution:**
>
> $$V_{dc}' = \frac{R_L}{R_l + R_L}V_{dc} = \frac{4000}{250 + 4000}(80) = \frac{4}{4.25}(80) = 75.3\ V$$
>
> $$V_r'(\text{rms}) = \frac{1.76}{LC_2}V_r(\text{rms}) = \frac{1.76}{(5)(4)}(10) = 0.88\ V$$
>
> $$r' = \frac{V_r'(\text{rms})}{V_{dc}'} = \frac{0.88}{75.3} = \mathbf{0.0117} \qquad (\%r' = 1.17\%)$$
>
> For the input voltage:
>
> $$r = \frac{V_r(\text{rms})}{V_{dc}} = \frac{10}{80} = 0.125 \qquad (\%r = 12.5\%)$$

2.7 *L*-TYPE FILTER (CHOKE FILTER)

A less popular filter circuit uses a choke or inductor directly from the rectifier circuit thus removing the problem of large diode peak currents. Figure 2.27 shows the *L*-type filter providing a dc voltage to load R_L. The circuit provides a no-load voltage of V_m. Smaller ripple is obtained at larger values of L or C[6].

With the availability of small, inexpensive transistor voltage regulator circuits the need for a filter circuit using large-size inductors has become less popular and is not covered more fully here.

---

[6]It can be shown that

$$r = \frac{0.83}{LC} \text{ for full-wave}$$

and

$$r = \frac{7.83}{LC} \text{ for half-wave}$$

where L is in henrys, C is in microfarads and $f = 60$ Hz.

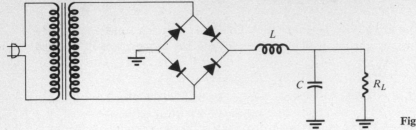

Figure 2.27. *L*-type filter.

2.8 VOLTAGE MULTIPLIER CIRCUITS

Voltage Doubler

A modification of the capacitor filter circuit allows building up a larger voltage than the peak rectified voltage (V_m). The use of this type of circuit allows keeping the transformer peak voltage rating low while stepping up the peak output voltage to two, three, four, or more times the peak rectified voltage.

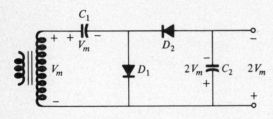

Figure 2.28. Half-wave voltage doubler.

Figure 2.28 shows a half-wave voltage doubler. During the positive-voltage half-cycle across the transformer, secondary diode D_1 conducts (and diode D_2 is cut off), charging capacitor C_1 up to the peak rectified voltage (V_m). Diode D_1 is ideally a short during this half-cycle and the input voltage charges capacitor C_1 to V_m with the polarity shown in Fig. 2.29a. During the negative half-cycle of the secondary voltage, diode D_1 is cut off and diode D_2 conducts charging capacitor C_2. Since diode D_2 acts as a short during the negative half-cycle (and diode D_1 is open), we can sum the voltages around the outside loop (see Fig. 2.29b):

$$-V_{C_2} + V_{C_1} + V_m = 0$$
$$-V_{C_2} + V_m + V_m = 0$$

from which

$$V_{C_2} = 2V_m$$

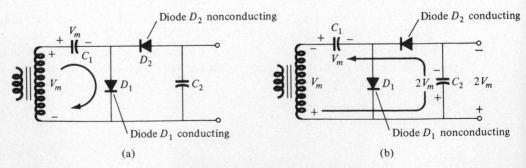

(a)   (b)

Figure 2.29. Double operation, showing each half-cycle of operation: (a) positive half-cycle; (b) negative half-cycle.

CHAP. 2 DIODE RECTIFIERS AND FILTERS   114

On the next positive half-cycle, diode D_2 is nonconducting and capacitor C_2 will discharge through the load. If no load is connected across capacitor C_2, both capacitors stay charged—C_1 to V_m and C_2 to $2V_m$. If, as would be expected, there is a load connected to the output of the voltage doubler, the voltage across capacitor C_2 drops during the positive half-cycle (at the input) and the capacitor is recharged up to $2V_m$ during the negative half-cycle.

The output waveform across capacitor C_2 is that of a half-wave signal filtered by a capacitor filter. The peak inverse voltage across each diode is $2V_m$.

Another doubler circuit is the full-wave doubler of Fig. 2.30. During the positive half-cycle of transformer secondary voltage (see Fig. 2.31a) diode D_1 conducts charging capacitor C_1 to a peak voltage V_m. Diode D_2 is nonconducting at this time.

During the negative half-cycle (see Fig. 2.31b) diode D_2 conducts charging capacitor C_2 while diode D_1 is nonconducting. If no load current is drawn from the circuit, the voltage across capacitors C_1 and C_2 is $2V_m$. If load current is drawn from the circuit, the voltage across capacitors C_1 and C_2 is the same as that across a capacitor fed by a full-wave rectifier circuit. One difference is that the effective capacitance is that of C_1 and C_2 in series, which is less than the capacitance of either C_1 or C_2 alone. The lower capacitor value will provide poorer filtering action than the single-capacitor filter circuit.

The peak inverse voltage across each diode is $2V_m$ as it is for the filter capacitor circuit. In sum-

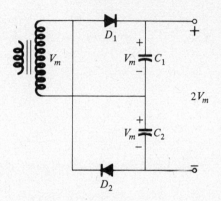

Figure 2.30. Full-wave voltage doubler.

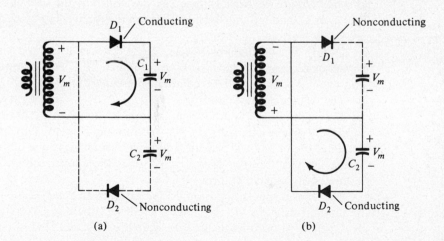

(a)

(b)

Figure 2.31. Alternate half-cycles of operation for full-wave voltage doubler.

mary, the half-wave or full-wave voltage doubler circuits provide twice the peak voltage of the transformer secondary while requiring no centertapped transformer and only $2V_m$ PIV rating for the diodes.

Voltage Tripler and Quadrupler

Figure 2.32 shows an extension of the half-wave voltage doubler, which develops three and four times the peak input voltage. It should be obvious from the pattern of the circuit connection how additional diodes and capacitors may be connected so that the output voltage may also be five, six, seven, etc., times the basic peak voltage (V_m).

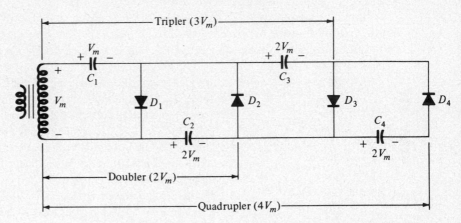

Figure 2.32. Voltage tripler and quadrupler.

In operation capacitor C_1 charges through diode D_1 to a peak voltage, V_m, during the positive half-cycle of the transformer secondary voltage. Capacitor C_2 charges to twice the peak voltage $2V_m$ developed by the sum of the voltages across capacitor C_1 and the transformer, during the negative half-cycle of the transformer secondary voltage.

During the positive half-cycle, diode D_3 conducts and the voltage across capacitor C_2 charges capacitor C_3 to the same $2V_m$ peak voltage. On the negative half-cycle, diodes D_2 and D_3 conduct with capacitor C_3, charging C_4 to $2V_m$.

The voltage across capacitor C_2 is $2V_m$, across C_1 and C_3 it is $3V_m$, and across C_2 and C_4 it is $4V_m$. If additional sections of diode and capacitor are used, each capacitor will be charged to $2V_m$. Measuring from the top of the transformer winding (Fig. 2.32) will provide odd multiples of V_m at the output, whereas measuring from the bottom of the transformer the output voltage will provide even multiples of the peak voltage, V_m.

The transformer rating is only V_m maximum and each diode in the circuit must be rated at $2V_m$ PIV. If the load is small and the capacitors have little leakage, extremely high dc voltages may be developed by this type of circuit, using many sections to step up the dc voltage.

PROBLEMS

§ 2.1

1. Calculate the average (dc) voltage of a 90-V, peak rectified half-wave signal.

2. A half-wave rectifier operates off the 120-V (rms) line voltage through a 3:1 step-down transformer. Calculate the dc voltage of the rectified signal.

3. A half-wave rectifier circuit develops a dc voltage of 180 V. What is the minimum diode peak inverse voltage rating for this circuit?

4. Draw the circuit diagram of a transformer-fed half-wave rectifier providing an output having negative half-cycles.

5. A half-wave rectifier circuit provides 40 mA (dc) to a 2-K load resistance. Calculate the output dc voltage and diode PIV rating for this circuit.

6. A half-wave diode rectifier circuit is transformer-fed from the 120-V line. Calculate the transformer peak voltage rating, turns ratio, and diode PIV rating, if the circuit provides an output of 12 V dc.

§ 2.2

7. Calculate the dc voltage obtained from a full-wave rectifier having a peak rectified voltage of 90 V.

8. Calculate the peak voltage rating of each half of a center-tapped transformer used in a full-wave rectifier circuit whose output dc voltage is 120 V.

9. Calculate the PIV rating for the diodes of a center-tapped full-wave rectifier circuit having an output dc voltage of 80 V.

10. Draw the circuit diagram of a center-tapped full-wave rectifier circuit developing an output rectified voltage of negative-going voltage cycles.

11. Draw the circuit diagram of a bridge full-wave rectifier circuit developing an output rectified voltage of negative-going voltage cycles. Show the conduction path through the circuit for each half-cycle of the input sinusoidal ac voltage.

12. Calculate the diode PIV rating for a bridge rectifier developing 50 V dc.

13. Design rectifier circuits to provide an output of 100 V dc using (a) half-wave; (b) center-tapped full-wave; and (c) bridge full-wave circuits. For each circuit calculate the transformer peak voltage rating, the diode PIV ratings, and the transformer turns ratio, if power is taken from the 120-V line ac supply.

§ 2.3

14. What is the ripple factor of/a filter signal having a peak ripple of 10 V on an average of 150 V?

15. A filter circuit provides an output of 80 V unloaded and 70 V under full-load operation. Calculate the voltage regulation.

16. A half-wave rectifier develops 100 V dc. What is the rms value of the ripple voltage?

17. What is the ripple voltage (rms) of a full-wave rectifier whose output voltage is 80 V dc?

18. A simple-capacitor filter fed by a full-wave rectifier develops 45 V dc at a ripple factor of 8%. What is the output ripple voltage (rms)?

19. A full-wave rectified signal of 80 V peak is fed into a capacitor filter. What is the voltage regulation of the filter circuit, if the output dc voltage is 75 V at full-load?

20. A full-wave rectified voltage of 80 V peak is connected to a 40-μF filter capacitor. What is the dc voltage at 50-mA load?

21. A full-wave rectifier operating from the 60-Hz ac supply line produces a 70-V peak rectified voltage. If a 20-μF filter capacitor is used, calculate the ripple factor and per cent of ripple at 40-mA load.

22. A capacitor filter circuit ($C = 4$ μF) develops a dc voltage of 100 V when connected to a load of 5 K. Using a full-wave rectifier operating from a 60-Hz supply, calculate the per cent of ripple of the output voltage.

23. Calculate the size of the filter capacitor to obtain a filtered voltage with 12% ripple at a load of 150 mA. The full-wave rectified voltage of 80 V peak is obtained from a 60-Hz ac supply.

24. A 100-μF filter capacitor provides 120 mA load current at 5% ripple. Calculate the peak rectified voltage obtained from the 60-Hz line and the dc voltage across the filter capacitor.

25. Calculate the per cent of ripple for the voltage developed across an 80-μF filter capacitor providing 35-mA load current. The full-wave rectifier operating from the 60-Hz line develops a peak rectified voltage of 45 V.

26. Calculate the amount of peak diode current through the rectifier diode of a half-wave rectifier feeding a capacitor filter, if the average current drawn from the filter is 50 mA at 1% ripple.

27. An RC filter stage is added after a capacitor filter to reduce the per cent of ripple to 2%. Calculate the ripple voltage at the output of the RC filter stage providing 80 V dc.

28. An RC filter stage ($R = 320$ Ω, $C = 20$ μF) is used to filter a signal of 120 V dc with 8 V ripple voltage (rms). Calculate the per cent of ripple at the output of the RC section for a load of 10 mA. Also calculate the ripple of the filtered signal applied to the RC stage.

29. A simple-capacitor filter has an input voltage of 40 V dc. If this voltage is fed through an RC filter section ($R = 50$ Ω, $C = 40$ μF), what is the load current for a load resistance of 500 Ω?

30. Calculate the ripple voltage (rms) at the output of an RC filter section feeding a 1-K load, if the filter input is 50 V dc with 2.5 V (rms) ripple from a full-wave rectifier and capacitor filter. The RC section filter components are $R = 100$ Ω and $C = 100$ μF.

31. If the output no-load voltage for the circuit of Problem 2.30 is 60 V, calculate the per cent of voltage regulation with the 1-K load.

§ 2.6

32. A π-type filter of $L = 5$ H ($R_l = 200 \, \Omega$) has an input voltage of 80 V. Calculate the filter output dc voltage at a load current of 30 mA.

33. At a ripple voltage frequency of 120 Hz calculate the ac impedance of an inductor and the effective load impedance ($R_L \| X_{C_2}$) for $L = 8$ H, $R_l = 350 \, \Omega$, $C_2 = 25 \, \mu$F, and $R_L = 4$ K.

34. A π-type filter operating from a full-wave rectifier has components $L = 8$ H, $R_l = 350 \, \Omega$, and $C_2 = 25 \, \mu$F. At a load of $R_L = 4$ K calculate the output ripple voltage, if the ripple voltage input (across first filter capacitor) is 4 V.

35. Calculate the ripple voltage from a π-type filter operating from a half-wave rectifier with 10 V rms across the initial filter capacitor. Filter values are $C_1 = 20 \, \mu$F, $C_2 = 40 \, \mu$F, $L = 2$ H, and $R_L = 2$ K.

36. Design a π-type filter to provide 100 V dc to a 5-K load. Capacitor C_1 (10 μF) has 120 V dc with a ripple voltage of 8 V. A full-wave rectifier is used and the filter output is to have a ripple of less than 1%.

37. A capacitor filter-fed by a half-wave rectifier develops 40 V dc with a ripple of 8%. Determine the values of an additional LC section (making a π-type filter) so that with a 2-K load, the dc voltage developed is at least 36 V with a ripple of less than 1%.

§ 2.8

38. Draw the circuit diagram of a voltage doubler. Indicate the value of the diode PIV rating in terms of the transformer peak voltage, V_m.

39. Draw a voltage tripler circuit. Indicate diode PIV ratings and voltages across each circuit capacitor. Include polarity.

40. Repeat Problem 2.39 for a voltage quadrupler.

transistors (BJTs) and vacuum tubes

3

3.1 INTRODUCTION

During the period 1904–1947 the tube was undoubtedly the electronic device of interest and development. In 1904, as discussed in Chapter 1, the vacuum-tube diode was introduced by J. A. Fleming. Shortly thereafter, in 1906, Lee De Forest added a third element, called the *control grid*, to the vacuum diode, resulting in the first amplifier, the triode. In the following years, radio and television provided great stimulation to the tube industry. Production rose from about 1 million tubes in 1922 to about 100 million in 1937. In the easly 1930s the four-element tetrode and five-element pentode gained prominence in the electron-tube industry. In the years to follow, the electronic industry became one of primary importance and rapid advances were made in design, manufacturing techniques, high-power and high-frequency applications, and miniaturization.

On December 23, 1947, however, the electronic industry was to experience the advent of a completely new direction of interest and development. It was on the afternoon of this day that Walter H. Brattain and John Bardeen demonstrated the amplifying action of the first transistor at the Bell Telephone Laboratories. The original transistor (a point-contact transistor) is shown in Fig. 3.1. The advantages of this three-terminal solid-state device over the tube were immediately obvious: it was smaller and lightweight; had no heater requirement or heater loss; had rugged construction; and was more efficient since less power was absorbed by the device itself; it was instantly available for use, requiring no warm up period; and lower operating voltages were possible. In the early stages of development, transistors were limited to the low-power and low-frequency type. If recent developments

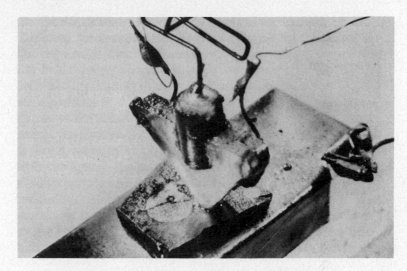

Figure 3.1. The first transistor (Courtesy Bell Telephone Laboratories).

continue, however, the semiconductor element will in all probability equal and eventually surpass the power and frequency ratings of today's tubes.

Just as the triode led the way to four-, five-, and more element tubes, the transistor has paved the way toward four-, five-, and more terminal semiconductor devices, some of which will be discussed in Chapter 9.

3.2 TRANSISTOR CONSTRUCTION

The transistor is a three-layer semiconductor device consisting of either two n- and one p-type layers of material or two p- and one n-type layers of material. The former is called an *npn transistor*, while the latter is called a *pnp transistor*. Both are shown in Fig. 3.2 with the proper dc biasing. The outer layers of the transistor are heavily doped semiconductor materials having widths much greater than that of the

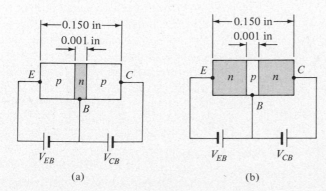

Figure 3.2. Types of transistors: (a) *pnp*; (b) *npn*.

sandwiched *p*- or *n*-type material. For the transistors shown in Fig. 3.2 the ratio of the total width to that of the center layer is 0.150/0.001 = 150 : 1. The doping of the sandwiched layer is also considerably less than that of the outer layers (typically 10 : 1 or less). This lower doping level decreases the conductivity or increases the resistance of this material by limiting the number of "free" carriers.

For the biasing shown in Fig. 3.2 the terminals have been indicated by the capital letters *E* for emitter, *C* for collector, and *B* for base. An appreciation for this choice of notation will develop when we discuss the basic operation of the transistor.

The abbreviation BJT from *bipolar junction transistor* is often applied to this three-terminal device. The term bipolar reflects the fact that holes *and* electrons participate in the injection process into the oppositely polarized material. If only one carrier is employed (electron or hole), it is considered a *unipolar* device.

3.3 TRANSISTOR OPERATION

The basic operation of the transistor will now be described using the *pnp* transistor of Fig. 3.2a. The operation of the *npn* transistor is exactly the same if the roles played by the electron and hole are interchanged.

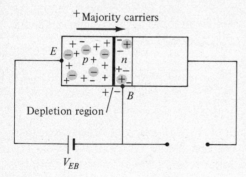

Figure 3.3. Forward-biased junction of a *pnp* transistor.

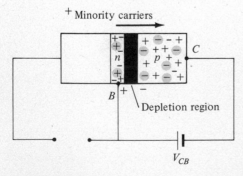

Figure 3.4. Reverse-biased junction of a *pnp* transistor.

In Fig. 3.3 the *pnp* transistor has been redrawn without the base-to-collector bias. Note the similarities between this situation and that of the *forward-biased* diode in Chapter 1. The depletion region has been reduced in width due to the applied bias, resulting in a heavy flow of majority carriers from the *p*-to the *n*-type material.

Let us now remove the base-to-emitter bias of the *pnp* transistor of Fig. 3.2a as shown in Fig. 3.4. Consider the similarities between this situation and that of the *reverse-biased* diode of Section 1.4. Recall that the flow of majority carriers is zero resulting in only a minority carrier flow as indicated in Fig. 3.4. *In summary, therefore, one p-n junction of a transistor is reverse biased, while the other is forward biased.*

In Fig. 3.5 both biasing potentials have been applied to a *pnp* transistor with the resulting majority and minority carrier flow indicated. Note in Fig. 3.5 the widths of the depletion regions, indicating clearly which junction is forward biased and which is reverse biased. As indicated in Fig. 3.5, a large number of majority carriers will diffuse across the forward-biased *p-n* junction into the *n*-type material. The question then is whether these carriers will contribute directly to the base current I_B or pass directly into the *p*-type material. Since the sand-

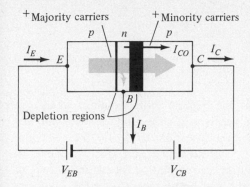

Figure 3.5. Majority and minority carrier flow of a *pnp* transistor.

wiched *n*-type material is very thin and has a low conductivity, a very small number of these carriers will take this path of high resistance to the base terminal. The magnitude of the base current is typically on the order of microamperes as compared to milliamperes for the emitter and collector currents. The larger number of these majority carriers will diffuse across the reverse-biased junction into the *p*-type material connected to the collector terminal as indicated in Fig. 3.5. The reason for the relative ease with which the majority carriers can cross the reverse-biased junction is easily understood if we consider that for the reverse-biased diode the injected majority carriers will appear as minority carriers in the *n*-type material. In other words, there has been an *injection* of minority carriers into the *n*-type base region material. Combining this with the fact that all the minority carriers in the depletion region will cross the reverse-biased junction of a diode accounts for the flow indicated in Fig. 3.5.

Applying Kirchoff's current law to the transistor of Fig. 3.5 as if it were a single node, we get

$$I_E = I_c + I_B \tag{3.1}$$

and we find that the emitter current is the sum of the collector and base currents. The collector current, however, is comprised of two components, due to majority and minority carriers as indicated in Fig. 3.5. The minority current component is called the *leakage current* and is given the symbol I_{CO} (I_C current with emitter terminal open). The collector current, therefore, is determined in total by Eq. (3.2).

$$I_C = I_{C_{\text{majority}}} + I_{CO_{\text{minority}}} \tag{3.2}$$

For general-purpose transistors, I_C is measured in milliamperes, while I_{CO} is measured in microamperes or nanoamperes. I_{CO}, like I_S for a reverse-biased diode, is temperature-sensitive and must be examined carefully when applications of wide temperature ranges are considered. The effects of I_{CO} will be considered in detail when stability is discussed in Chapter 4.

Improvements in construction techniques have resulted in significantly lower levels of I_{CO} to the point where its effect can often be ignored. However, higher-power devices still typically have values of I_{CO} in the microampere range.

The configuration shown in Fig. 3.2 for the *pnp* and *npn* transistors is called the *common-base* configuration since the base is common to both the emitter and collector terminals. For fixed values of V_{CB} in the common-base configuration the ratio of a small change in I_C to a small change in I_E is commonly called the *common-base, short-circuit amplification factor* and is given the symbol α (alpha).

In equation form, the magnitude of α is given by

$$\alpha = \frac{\Delta I_C}{\Delta I_E}\bigg|_{V_{CB}=\text{constant}} \tag{3.3}$$

The term short circuit indicates that the load is short circuited when α is determined. More will be said about the necessity for shorting the load and the operations involved with using equations of the type indicated by Eq. (3.3) when we consider equivalent circuits in Chapter 5. Typical values of α vary from 0.90 to 0.998. For most practical applications, a first approximation for the magnitude of α, usually correct to within a few per cent, can be obtained using the following equation:

$$\alpha \cong \frac{I_C}{I_E} \tag{3.4}$$

where I_C and I_E are the magnitude of the collector and emitter currents, respectively, at a particular point on the transistor characteristics.

Eqs. (3.3) and (3.4) are employed to determine α from the device characteristics or network conditions. However, in the strictest sense, α is only a measure of the percentage of holes (majority carriers) originating in the emitter P-material of Fig. 3.5 that reach the collector terminal. As defined by Eq. (3.2), therefore,

$$I_C = \alpha I_{E_{\text{majority}}} + I_{CO_{\text{minority}}} \tag{3.5}$$

3.4 TRANSISTOR AMPLIFYING ACTION

The basic voltage-amplifying action of the common-base configuration can now be described using the circuit of Fig. 3.6. For the common-base configuration, the input resistance between emitter and base of a transistor will typically vary from 20 to 200 Ω, while the output resistance may vary from 100 K to 1 M. The difference in resistance is due to the forward-biased junction at the input (base to emitter) and the reverse-biased junction at the output (base to collector). Using effective values and an average value of 100 Ω for the input resistance, we get

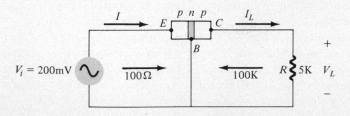

Figure 3.6. Basic voltage amplification action of the common-base configuration.

If we assume for the moment that $\alpha = 1$ $(I_C = I_E)$,

$$I_L = I = 2 \text{ mA}$$

and
$$V_L = I_L R$$
$$= (2 \times 10^{-3})(5 \times 10^{+3})$$
$$V_L = 10 \text{ V}$$

The voltage amplification is

$$A_v = \frac{V_L}{V_i} = \frac{10}{200 \times 10^{-3}} = 50$$

Typical values of voltage amplification for the common-base configuration vary from 20 to 100. The current amplification (I_C/I_E) is always less than one for the common-base configuration. This latter characteristic should be obvious since $I_C = \alpha I_E$ and α is always less than one.

The basic amplifying action was produced by *transferring* a current I from a low- to a high-*resistance* circuit. The combination of the two terms in italics results in the name transistor; that is,

$$transfer + resistor \longrightarrow transistor$$

3.5 COMMON-BASE CONFIGURATION

The notation and symbols used in conjunction with the transistor in the majority of texts and manuals published today are indicated in Fig. 3.7 for the common-base

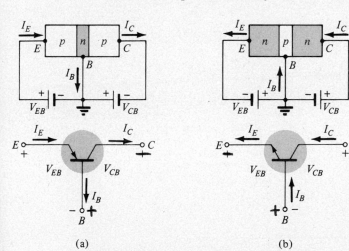

Figure 3.7. Notation and symbols used with the common-base configuration: (a) *pnp* transistor; (b) *npn* transistor.

(a)   (b)

configuration with *pnp* and *npn* transistors. Throughout this text all current directions will refer to the conventional (hole flow) rather than the electron flow. This choice was based partly on the fact that the vast majority of past and present publications in electrical engineering use conventional current flow.

Some texts prefer to show all the currents entering in Fig. 3.7 when they describe the basic operation of the transistor and simply include negative signs when

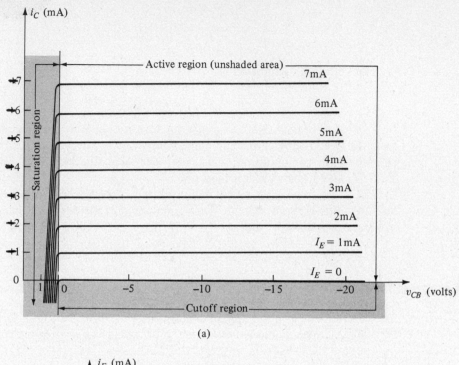

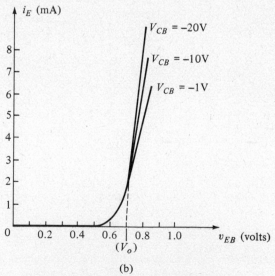

Figure 3.8. Characteristics of a *pnp* transistor in the common-base configuration: (a) collector characteristics; (b) emitter characteristics.

appropriate. In other words, if the actual conventional flow direction is the opposite direction, a negative sign is included along with the magnitude. For clarity, all currents, as indicated in Fig. 3.7, will indicate the actual flow direction for the active region. *Note that the arrow in the symbol is the same as the direction of* I_E *(only true for conventional flow).* On specification sheets negative signs indicate that all currents are entering.

For the common-base configuration the applied potentials are written with respect to the base potential resulting in V_{EB} and V_{CB}. In other words, the second subscript will always indicate the transistor configuration. In all cases the first subscript is defined to be the point of higher potential as shown in Fig. 3.7. For the *pnp* transistor, therefore, V_{EB} is positive and V_{CB} is negative (since the battery V_{CB} sets the collector or the lower potential) as indicated on the characteristics of Fig. 3.8. For the *npn* transistor V_{EB} is negative and V_{CB} is positive.

In addition, note that two sets of characteristics are necessary to represent the behavior of the *pnp* common-base transistor of Fig. 3.7, the driving point (or input), and the output set.

The output or collector characteristics of Fig. 3.8a relate the collector current to the collector-to-base voltage and emitter current. The collector characteristics have three basic regions of interest as indicated in Fig. 3.8a: the *active, cutoff,* and *saturation* regions.

In the active region the collector junction is reverse biased, while the emitter junction is forward biased. These conditions refer to the situation of Fig. 3.5. The active region is the only region employed for the amplification of signals with minimum distortion. When the emitter current (I_E) is zero, the collector current is simply that due to the reverse saturation current I_{CO} as indicated in Fig. 3.8a. The current I_{CO} is so small (microamperes) in magnitude compared to the vertical scale of I_C (milliamperes) that it appears on virtually the same horizontal line as $I_C = 0$.

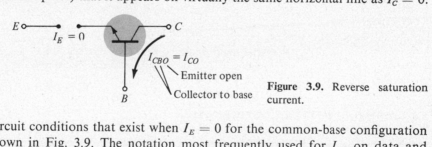

Figure 3.9. Reverse saturation current.

The circuit conditions that exist when $I_E = 0$ for the common-base configuration are shown in Fig. 3.9. The notation most frequently used for I_{CO} on data and specification sheets is, as indicated in Fig. 3.9, I_{CBO}. Due to improved construction techniques the level of I_{CBO} for general-purpose transistors (especially silicon) in the low- and mid-power ranges is usually so low its effect can be ignored. However, for higher power units I_{CBO} will still appear in the microampere range. In addition, keep in mind than I_{CBO} like I_S for the diode (both reverse leakage currents) is temperature-sensitive. At higher temperatures the effect of I_{CBO} for any power level unit may become an important factor since it increases so rapidly with temperature.

Note in Fig. 3.8a that as the emitter current increase above zero, the collector current increases to a magnitude slightly less ($\alpha < 1$) than that of the emitter current as determined by the basic transistor-current relations. Note also the almost negligible effect of V_{CB} on the collector current for the active region. The curves clearly indicate that *a first approximation to the relationship between* I_E *and* I_C *in the active region is given by* $I_E = I_C$.

In the cutoff region the collector and emitter junctions are both reverse biased resulting in negligible collector current as demonstrated in Fig. 3.8a.

The horizontal scale for V_{CB} has been expanded to the left of 0 V to represent clearly the characteristics in this region. *In the region, called the* saturation region, *the collector and emitter junctions are forward biased* resulting in the exponential change in collector current with small changes in collector-to-base potential.

The input or emitter characteristics have only one region of interest as illustrated by Fig. 3.8b. For fixed values of collector voltage (V_{CB}), as the emitter-to-base potential increases the emitter current increases as shown. Increasing levels of V_{CB} result in a reduced level of V_{EB} to establish the same current.

Note the tight grouping of the curves for the wide range of values for V_{CB}. In addition, consider how closely the average value of the curves appears to begin its rise or about $V_O = 0.7$ V for the silicon transistor. As with the semiconductor silicon diode, *a first approximation for the forward-biased base-emitter junction in the dc mode would be that* $V_{EB} \cong 0.7$ V *for all levels of* V_{CB}.

> **EXAMPLE 3.1**   Using the characteristics of Fig. 3.8,
> (a) Find the resulting collector current if $I_E = 3$ mA and $V_{CB} = -10$ V.
> (b) Find the resulting collector current if $V_{EB} = 750$ mV and $V_{CB} = -10$ V.
> (c) Find V_{EB} for the conditions $I_C = 5$ mA and $V_{CB} = -1$ V.
>
> **Solution:**
>
> (a) $\qquad\qquad\qquad\qquad\qquad I_C \cong I_E = 3$ **mA**
>
> (b) On the input characteristics $I_E = 3.5$ mA at the intersection of $V_{EB} = 750$ mV, $V_{CB} = -10$ V, and $I_C \cong I_E = 3.5$ mA.
>
> (c) $\qquad\qquad\qquad\qquad\qquad I_E \cong I_C = 5$ **mA**
>
> On the input characteristics the intersection of $I_E = 5$ mA and $V_{CB} = -1$ V results in $V_{EB} \cong 800$ mV $= 0.8$ V.

3.6 COMMON-EMITTER CONFIGURATION

The most frequently encountered transistor configuration is shown in Fig. 3.10 for the *pnp* and *npn* transistors. It is called the *common-emitter configuration* since the emitter is common to both the base and collector terminals. Two sets of characteristics are again necessary to fully describe the behavior of the common-emitter configuration: one for the input or base circuit and one for the output or collector circuit. Both are shown in Fig. 3.11.

The emitter, collector, and base currents are shown in their actual conventional current flow direction, while the potentials have the capital letter E as the second subscript to indicate the configuration. Even though the transistor configuration has changed, the current relations developed earlier for the common-base configuration are still applicable.

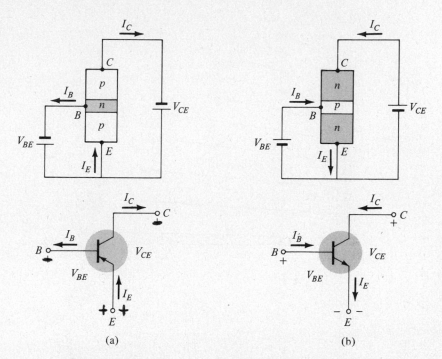

Figure 3.10. Notation and symbols used with the common-emitter configuration: (a) *pnp* transistor; (b) *npn* transistor.

For the common-emitter configuration the output characteristics will be a plot of the input current (I_B) versus the output voltage (V_{CE}) and output current (I_C). The input characteristics are a plot of the output voltage (V_{CE}) versus the input voltage (V_{BE}) and input current (I_B).

Note that on the characteristics of Fig. 3.11 the magnitude of I_B is in microamperes as compared to milliamperes for I_C. Consider also that the curves of I_B are not as horizontal as those obtained for I_E in the common-base configuration indicating that the collector-to-emitter voltage will influence the magnitude of the collector current.

The active region for the common-emitter configuration is that portion of the upper-right quadrant that has the greatest linearity, that is, that region in which the curves for I_B are nearly straight and equally spaced. In Fig. 3.11a this region exists to the right of the vertical dashed line at $V_{CE_{sat}}$ and above the curve for I_B equal to zero. The region to the left of $V_{CE_{sat}}$ is called the saturation region. *In the active region the collector junction is reverse biased, while the base junction is forward biased.* You will recall that these were the same conditions that existed in the active region of the common-base configuration. The active region of the common-emitter configuration can be employed for voltage, current, or power amplification.

The cutoff region for the common-emitter configuration is not as well defined as for the common-base configuration. Note on the collector characteristics of Fig. 3.11 that I_C is not equal to zero when I_B is zero. For the common-base configuration, when the input current I_E was equal to zero, the collector current was equal only to

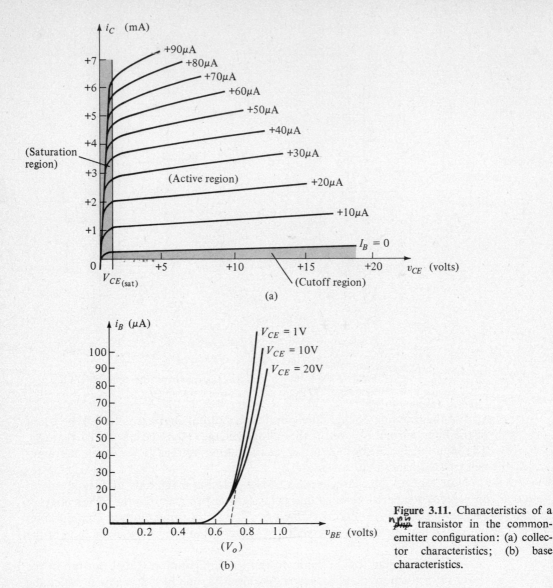

Figure 3.11. Characteristics of a npn transistor in the common-emitter configuration: (a) collector characteristics; (b) base characteristics.

the reverse saturation current I_{co} so that the curve $I_E = 0$ and the voltage axis were, for all practical purposes, one.

The reason for this difference in collector characteristics can be derived through the proper manipulation of Eqs. (3.1) and (3.5). It will appear as an exercise at the end of this chapter. The result:

$$I_C = \frac{I_{CO}}{1 - \alpha} + \frac{\alpha I_B}{1 - \alpha} \qquad (3.6)$$

If we consider the case discussed earlier, where $I_B = 0$, and substitute this value into Eq. (3.6), then

$$I_C = \frac{I_{CO}}{1 - \alpha}\bigg|_{I_B=0} \tag{3.7}$$

For $\alpha = 0.996$

$$I_C = \frac{I_{CO}}{1 - 0.996} = \frac{I_{CO}}{0.004}$$

and

$$I_C = 250 I_{CO}|_{I_B=0}$$

which accounts for the vertical shift in the $I_B = 0$ curve from the horizontal voltage axis.

For future reference, the collector current defined by Eq. (3.7) will be assigned the notation indicated by Eq. (3.8).

$$I_{CEO} = \frac{I_{CO}}{1 - \alpha}\bigg|_{I_B=0} \tag{3.8}$$

In Fig. 3.12 the conditions surrounding this newly defined current are demonstrated with its assigned reference direction.

The magnitude of I_{CEO} is typically much smaller for silicon materials than for germanium materials. For transistors with similar ratings I_{CEO} would typically be a few microamperes for silicon but perhaps a few hundred microamperes for germanium.

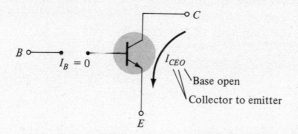

Figure 3.12. Circuit conditions related to I_{CEO}.

For linear (least distortion) amplification purposes, cutoff for the common-emitter configuration will be (for this text) determined by $I_C = I_{CEO}$. In other words, the region below $I_B = 0$ is to be avoided if an undistorted output signal is required.

When employed as a switch in the logic circuitry of a computer, a transistor will have two points of operation of interest: one in the cutoff and one in the saturation region. The cutoff condition should ideally be $I_C = 0$ for the chosen V_{CE} voltage. Since I_{CEO} is typically low in magnitude for silicon materials, *cutoff will exist for switching purposes when $I_B = 0$ or $I_C = I_{CEO}$ for silicon transistors only. For germanium transistors, however, cutoff for switching purposes will be defined as those conditions that exist when $I_C = I_{CBO} = I_{CO}$.* This condition can normally be obtained for germanium transistors by reverse biasing the normally forward-biased base-to-emitter junction a few tenths of a volt.

EXAMPLE 3.2 Using the characteristics of Fig. 3.11,
(a) Find the value of I_C corresponding to $V_{BE} = +800$ mV and $V_{CE} = +10$ V.
(b) Find the value of V_{CE} and V_{BE} corresponding to $I_C = +4$ mA and $I_B = +40$ μA.

Solution: (a) On the input characteristics the intersection of $V_{BE} = +800$ mV and $V_{CE} = +10$ V results in

$$I_B \cong 50 \text{ μA}$$

On the output characteristics the intersection of $I_B = 50$ μA and $V_{CE} = 10$ V results in

$$I_C \cong 5.1 \text{ mA}$$

(b) On the output characteristics the intersection of $I_C = +4$ mA and $I_B = +40$ μA results in

$$V_{CE} = +6.2 \text{ V}$$

On the input characteristics the intersection of $I_B = +40$ μA and $V_{CE} = +6.2$ V results in

$$V_{BE} \cong 770 \text{ mV}$$

In Section 3.3 the symbol alpha (α) was assigned to the forward current transfer ratio of the common-base configuration. For the common-emitter configuration, the ratio of a small change in collector current to the corresponding change in base current at a fixed collector-to-emitter voltage (V_{CE}) is assigned the Greek letter beta (β) and is commonly called the *common-emitter forward current amplification factor*. In equation form, the magnitude of β is given by

$$\beta = \frac{\Delta I_C}{\Delta I_B}\bigg|_{V_{CE}=\text{constant}} \tag{3.9}$$

As a first, but close, approximation, the magnitude of beta (β) can be determined by the following equation:

$$\beta \cong \frac{I_C}{I_B} \tag{3.10}$$

where I_C and I_B are collector and base currents of a particular operating point in the linear region (i.e., where the horizontal base current lines of the common-emitter characteristics are closest to being parallel and equally spaced). Since I_C and I_B in Eq. (3.10) are fixed or dc values, the value obtained for β from Eq. (3.10) is frequently called the *dc beta*, while that obtained by Eq. (3.9) is called the *ac* or *dynamic* value. Typical values of β vary from 20 to 600. Through the proper manipulation of Eqs. (3.1) and (3.4) it can be shown (derived) that the following equations are valid:

$$\beta = \frac{\alpha}{1 - \alpha}$$

(3.11)

or

$$\alpha = \frac{\beta}{\beta + 1}$$

(3.12)

In addition, since

$$I_{CEO} = \frac{I_{CO}}{1 - \alpha} = \frac{I_{CBO}}{1 - \alpha}$$

then

$$I_{CEO} = (\beta + 1)I_{CBO} \cong \beta I_{CBO}$$

(3.13)

EXAMPLE 3.3 (a) Find the dc beta at an operating point of $V_{CE} = +10$ V and $I_C = +3$ mA on the characteristics of Fig. 3.11.
(b) Find the value of α corresponding with this operating point.
(c) At $V_{CE} = +10$ V find the corresponding value of I_{CEO}.
(d) Calculate the approximate value of I_{CBO} using the β_{dc} obtained in part (a).

Solution: (a) At the intersection of $V_{CE} = +10$ V and $I_C = +3$ mA, $I_B = +25\ \mu A$,

so that

$$\beta_{dc} = \frac{I_C}{I_B} = \frac{3 \times 10^{-3}}{25 \times 10^{-6}} = 120$$

(b)

$$\alpha = \frac{\beta}{\beta + 1} = \frac{120}{121} \cong 0.992$$

(c)

$$I_{CEO} = 300\ \mu A$$

(d)

$$I_{CBO} \cong \frac{I_{CEO}}{\beta} = \frac{300\ \mu A}{120} = 2.5\ \mu A$$

The input characteristics for the common-emitter configuration are very similar to those obtained for the common-base configuration (Fig. 3.11). In both cases, the increase in input current is due to an increase in majority carriers crossing the base-to-emitter junction with increasing forward-bias potential. Note also that the variation in output voltage (V_{CE} for the CE configuration and V_{CB} for the CB configuration) does not result in a large relocation of the characteristics. In fact, for the dc voltage levels commonly encountered the variation in base-to-emitter voltage with change in output terminal voltage can, as a first approximation, be ignored. On this basis, if we use an average value, the curve of Fig. 3.13 for the CE configuration will result. Note the similarities with the silicon diode characteristics. Recall also from the description of the semiconductor diode that for dc analysis we approximated the curve of Fig. 3.13 with that indicated in Fig. 3.14. Essentially,

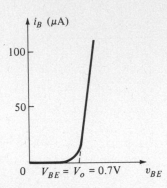

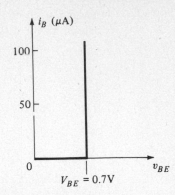

Figure 3.13. Reproduction of Fig. 3.11b ignoring the effects of V_{CE}.

Figure 3.14. Approximate reproduction of Fig. 3.13 for dc analysis.

therefore, for dc analysis a first approximation to the base-to-emitter voltage of a transistor configuration is to assume that $V_{BE} \cong 0.7$ V for silicon and 0.3 V for germanium. If insufficient voltage is present to provide the 0.7 V bias (for silicon transistors) with the proper polarity, the transistor cannot be in the active region. A number of applications of this approximation will appear in Chapter 4. Since the CB characteristics had a similar set of input characteristics [also true for the common-collector (CC) configuration to be discussed], we can conclude *as a first approximation for dc analysis that the base-to-emitter voltage of a BJT is assumed to be* V_0 *when biased in the active region of the characteristics.*

Further, we found for the output characteristics of the CB configuration that $I_C = I_E$. For the CE configuration $I_C = \beta I_B$ where β is determined by the operating conditions.

In manuals, data sheets, and other transistor publications the common-emitter characteristics are the most frequently presented. The common-base characteristics can be obtained directly from the common-emitter characteristics using the basic current relations derived in the past few sections. In other words, for each point on the characteristics of the common-emitter configuration a sufficient number of variables can be obtained to substitute into the equations derived to come up with a point on the common-base characteristics. This process is, of course, time consuming, but it will result in the desired characteristics.

3.7 COMMON-COLLECTOR CONFIGURATION

The third and final transistor configuration is the *common-collector configuration,* shown in Fig. 3.15 with the proper current directions and voltage notation.

The common-collector configuration is used primarily for impedance matching purposes since it has a high input impedance and low output impedance, opposite to that which is true of the common-base and common-emitter configurations.

The common-collector circuit configuration is generally as shown in Fig. 3.16 with the load resistor from emitter to ground. Note that the collector is tied to ground even though the transistor is connected in a manner similar to the common-

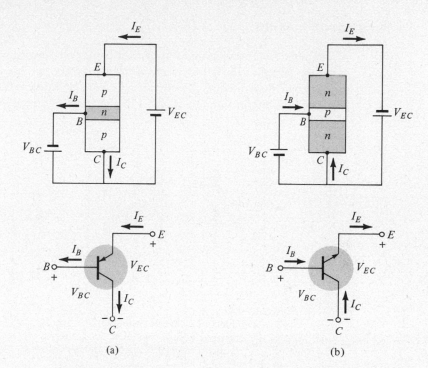

Figure 3.15. Notation and symbols used with the common-collector configuration: (a) *pnp* transistor; (b) *npn* transistor.

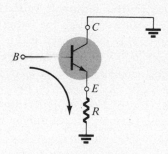

Figure 3.16. Common-collector configuration used for impedance matching purposes.

emitter configuration. From a design viewpoint, there is no need for a set of common-collector characteristics to choose the parameters of the circuit of Fig. 3.16. It can be designed using the common-emitter characteristics of Section 3.6. For all practical purposes, the output characteristics of the common-collector configuration are the same as for the common-emitter configuration. For the common-collector configuration the output characteristics are a plot of the input current I_B versus V_{EC} and I_E. The input current, therefore, is the same for both the common-emitter and common-collector characteristics. The horizontal voltage axis for the common-collector configuration is obtained by simply changing the sign of the collector-to-emitter voltage of the common-emitter characteristics since $V_{EC} = -V_{CE}$. Finally, there is an almost unnoticeable change in the vertical scale of I_C of the common-emitter characteristics if I_C is replaced by I_E for the common-collector characteristics (since $\alpha \cong 1$). For the input circuit of the common-collector configuration the common-emitter base characteristics are sufficient for obtaining any required information by simply writing Kirchhoff's voltage law around the loop indicated in Fig. 3.16 and performing the proper mathematical manipulations.

3.8 TRANSISTOR BIASING

A technique for ensuring that a transistor is properly biased will now be described using the fact that the arrow in the transistor symbol points in the actual emitter conventional flow direction. As an example, consider the *npn* transistor of Fig. 3.17a in the CE configuration. In Fig. 3.17b the current directions for I_B and I_C are

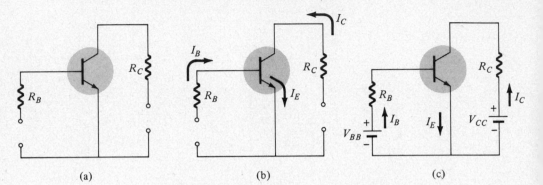

Figure 3.17. Steps leading to the proper biasing of a transistor: (a) unbiased network; (b) conventional current directions inserted; (c) proper biasing inserted as determined by current directions.

included as defined by the fact that the emitter current is the sum of the base and collector currents. That is, if the emitter current leaves the device, both the base and collector currents must enter their respective terminals. Now we have to ensure that the battery inserted between the base and emitter terminals will establish the indicated base current direction (conventional flow through the battery from the negative to positive terminal of the battery). This is indicated in Fig. 3.17c for the base and collector circuits. The transistor junctions are now properly biased. This technique of starting with the necessary current directions as defined by I_E can be applied to any configuration whether it be *pnp* or *npn*.

One easy way to remember whether the arrow of a *pnp* or *npn* transistor symbol points in or out is to associate *pointing in with pnp* and *not pointing in with npn*.

3.9 TRANSISTOR MAXIMUM RATINGS

The standard transistor data sheet will include at least three maximum ratings: *collector dissipation, collector voltage, and collector current.*

For the transistor whose characteristics were presented in Fig. 3.11 the following maximum ratings were indicated.

$$P_{C_{max}} = 30 \text{ mW}$$

$$I_{C_{max}} = 6 \text{ mA}$$

$$V_{CE_{max}} = 20 \text{ V}$$

The power or dissipation rating is the product of the collector voltage and current. For the common-emitter configuration:

$$P_{C_{max}} = V_{CE}I_C \qquad (3.14)$$

The nonlinear curve determined by this equation is indicated in Fig. 3.18. The curve was obtained by choosing various values of V_{CE} or I_C and finding the other variable using Eq. (3.14). For example, at $V_{CE} = 10$ V

$$I_C = \frac{P_{C_{max}}}{V_{CE}} = \frac{30 \times 10^{-3}}{10} = 3 \text{ mA}$$

as indicated in Fig. 3.18. The region above this curve must be avoided in the design of systems using this particular transistor if the maximum power rating is not to be exceeded. The maximum collector voltage, in this case V_{CE}, is indicated as a vertical line in Fig. 3.18. The maximum collector current is also indicated in Fig. 3.18 as a horizontal line.

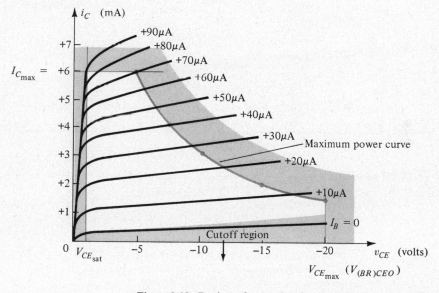

Figure 3.18. Region of operation for amplification purposes.

For the common-base configuration the collector dissipation is determined by the following equation. The maximum collector voltage would refer to V_{CB}.

$$P_{C_{max}} = V_{CB}I_C \qquad (3.15)$$

For amplification purposes the nonlinear characteristics of the saturation and the cutoff regions are also avoided. The saturation region has been indicated by the vertical line at $V_{CE_{sat}}$ and the cutoff region by $I_B = 0$ in Fig. 3.18. The unshaded region remaining is the region employed for amplification purposes. Although it

DESIGNED FOR USE
IN LOW-LEVEL, LOW-NOISE
AMPLIFIERS

- **Guaranteed Low-Noise Characteristics at 10 Hz, 100 Hz, 1 kHz and 10 kHz**
- **Very High Guaranteed h_{FE} at $I_C = 10 \ \mu A$: 400 Minimum**

*mechanical data

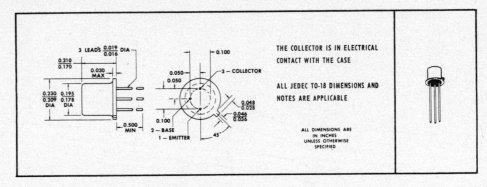

THE COLLECTOR IS IN ELECTRICAL CONTACT WITH THE CASE

ALL JEDEC TO-18 DIMENSIONS AND NOTES ARE APPLICABLE

ALL DIMENSIONS ARE IN INCHES UNLESS OTHERWISE SPECIFIED

*absolute maximum ratings at 25°C free-air temperature (unless otherwise noted)

Collector-Base Voltage	60 V
Collector-Emitter Voltage (See Note 1)	60 V
Emitter-Base Voltage	10 V
Continuous Collector Current	50 mA
Continuous Device Dissipation at (or below) 25°C Free-Air Temperature (See Note 2)	0.3 W
Continuous Device Dissipation at (or below) 25°C Case Temperature (See Note 3)	1.2 W
Storage Temperature Range	−65°C to 200°C
Lead Temperature 1/16 Inch from Case for 10 Seconds	300°C

NOTES: 1. This value applies between 0 and 10 mA when the base-emitter diode is open-circuited.

2. Derate linearly to 175°C free-air temperature at the rate of 2 mw/°C.

3. Derate linearly to 175°C case temperature at the rate of 8 mw/°C.

*JEDEC registered data

USES CHIP N11

TEXAS INSTRUMENTS
INCORPORATED
POST OFFICE BOX 5012 • DALLAS, TEXAS 75222

*electrical characteristics at 25°C free-air temperature (unless otherwise noted)

	PARAMETER	TEST CONDITIONS	MIN	MAX	UNIT
$V_{(BR)CBO}$	Collector-Base Breakdown Voltage	$I_C = 10\ \mu A$, $I_E = 0$	60		V
$V_{(BR)CEO}$	Collector-Emitter Breakdown Voltage	$I_C = 10\ mA$, $I_B = 0$, See Note 4	60		V
$V_{(BR)EBO}$	Emitter-Base Breakdown Voltage	$I_E = 10\ \mu A$, $I_C = 0$	10		V
I_{CBO}	Collector Cutoff Current	$V_{CB} = 45\ V$, $I_E = 0$		10	nA
		$V_{CB} = 45\ V$, $I_E = 0$, $T_A = 150°C$		10	μA
I_{EBO}	Emitter Cutoff Current	$V_{EB} = 5\ V$, $I_C = 0$		10	nA
h_{FE}	Static Forward Current Transfer Ratio	$V_{CE} = 5\ V$, $I_C = 1\ \mu A$	150		
		$V_{CE} = 5\ V$, $I_C = 10\ \mu A$	400	800	
		$V_{CE} = 5\ V$, $I_C = 100\ \mu A$	450		
		$V_{CE} = 5\ V$, $I_C = 1\ mA$	500		
V_{BE}	Base-Emitter Voltage	$V_{CF} = 5\ V$, $I_C = 100\ \mu A$		0.7	V
$V_{CE(sat)}$	Collector-Emitter Saturation Voltage	$I_B = 0.1\ mA$, $I_C = 1\ mA$		0.3	V
h_{ie}	Small-Signal Common-Emitter Input Impedance	$V_{CE} = 5\ V$,	12	42	$k\Omega$
h_{fe}	Small-Signal Common-Emitter Forward Current Transfer Ratio	$I_C = 1\ mA$,	500	1400	
h_{re}	Small-Signal Common-Emitter Reverse Voltage Transfer Ratio	$f = 1\ kHz$		8×10^{-4}	
h_{oe}	Small-Signal Common-Emitter Output Admittance		8	60	μmho
$|h_{fe}|$	Small-Signal Common-Emitter Forward Current Transfer Ratio	$V_{CE} = 5\ V$, $I_C = 0.5\ mA$, $f = 30\ MHz$	3	18	
C_{obo}	Common-Base Open-Circuit Output Capacitance	$V_{CB} = 5\ V$, $I_E = 0$, $f = 1\ MHz$		4.5	pF
C_{ibo}	Common-Base Open-Circuit Input Capacitance	$V_{EB} = 0.5\ V$, $I_C = 0$, $f = 1\ MHz$		6	pF

appears as though the area of operation has been drastically reduced, we must keep in mind that many signals are in the microvolt or millivolt range, while the horizontal axis of the characteristics is measured in volts. In addition to maximum ratings, data and specification sheets on transistors also include other important information about their operation. The discussion of this additional data will not be considered until each parameter is fully defined.

The information on the specification sheet of a TI (Texas Instruments, Inc.) 2N4104 *npn* silicon transistor is provided on pages 138–139. This particular transistor is designed for use in low-level, low-noise amplifiers.

The letter o at the end of the first five parameters listed in the electrical characteristics indicate that the terminal not listed is left open. The terminology h_{FE}, h_{ie}, $|h_{fe}|$, etc., will be described in Chapter 5. The symbols h_{FE} and β_{DC} are synonymous. Even with h_{FE} equal to its maximum value of 800, with $I_{CBO} = 10\ nA$, $I_{CEO} \cong \beta I_{CBO}$ is limited to only $8\ \mu A = 0.008\ mA$ (virtually not noticeable on the vertical scale of $I_C = mA$ resulting in $I_B = 0\ \mu A$ appearing on the horizontal axis). Note at

150°C, however, that I_{CBO} has risen to 10 μA and with a typical value of $h_{FE} = 400$, $I_{CEO} = 400(10\ \mu A) = 4$ mA which is not insignificant. As with the p-n diode, the transistor will also have capacitance levels across its junctions. Since the common-base configuration is used most frequently for high-frequency applications, its capacitance levels are provided.

3.10 TRANSISTOR FABRICATION

The majority of the methods used to fabricate transistors are simply extensions of the methods used to manufacture semiconductor diodes. The methods most frequently employed today include *point-contact, alloy junction, grown junction, and diffusion.* The following discussion of each method will be brief, but the fundamental steps included in each will be presented. A detailed discussion of each method would require a text in itself.

Point-Contact

The point-contact transistor is manufactured in a manner very similar to that used for point-contact semiconductor diodes. In this case two wires are placed next to an *n*-type wafer as shown in Fig. 3.19. Electrical pulses are then applied to each wire resulting in a *p-n* junction at the boundary of each wire and the semiconductor wafer. The result is a *pnp* transistor as shown in Fig. 3.19. This method of fabrication is today limited to high-frequency, low-power devices. It was the method used in the fabrication of the first transistor shown in Fig. 3.1.

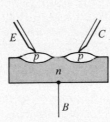

Figure 3.19. Point-contact transistor.

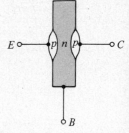

Figure 3.20. Alloy junction transistor.

Alloy Junction

The alloy junction technique is also an extension of the alloy method of manufacturing semiconductor diodes. For a transistor, however, two dots of the same impurity are deposited on each side of a semiconductor wafer having the opposite impurity as shown in Fig. 3.20. The entire structure is then heated until melting occurs and each dot is alloyed to the base wafer resulting in the *p-n* junctions indicated in Fig. 3.20 as described for semiconductor diodes.

The collector dot and resulting junction are larger to withstand the heavy current and power dissipation at the collector-base junction. This method is not employed as much as the diffusion technique to be described shortly, but it is still used extensively in the manufacture of high-power diodes.

Grown Junction

The Czochralski technique (Section 1.4) is used to form the two *p-n* junctions of a grown junction transistor. The process, as depicted in Fig. 3.21, requires that the impurity control and withdrawal rate be such as to ensure the proper base width and doping levels of the *n*- and *p*-type materials. Transistors of this type are, in general, limited to less than $\frac{1}{4}$-W rating.

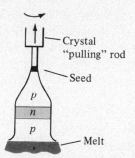

Figure 3.21. Grown junction transistor.

Diffusion

The most frequently employed method of manufacturing transistors today is the diffusion technique. The basic process was introduced in the discussion of semiconductor diode fabrication. The diffusion technique is employed in the production of *mesa* and *planar* transistors, each of which can be of the *diffused* or *epitaxial* type.

In the *pnp*, diffusion-type mesa transistor the first process is an *n*-type diffusion into a *p*-type wafer, as shown in Fig. 3.22, to form the base region. Next, the *p*-type

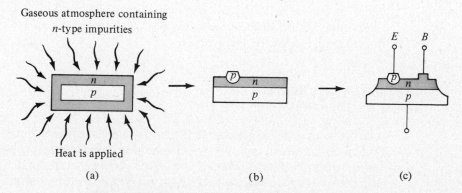

(a)   (b)   (c)

Figure 3.22. Mesa transistor: (a) diffusion process; (b) alloy process; (c) etching process.

emitter is diffused or alloyed to the *n*-type base as shown in the figure. Etching is done to reduce the capacitance of the collector junction. The term mesa is derived from its similarities with the geographical formation. As mentioned earlier in the discussion of diode fabrication, the diffusion technique permits very tight control of the doping levels and thicknesses of the various regions.

The major difference between the epitaxial mesa transistor and the mesa transistor is the addition of an epitaxial layer on the original collector substrate. The term epitaxial is derived from the Greek words *epi*—upon, and *taxi*—arrange, which describe the process involved in forming this additional layer. The original *p*-type substrate (collector of Fig. 3.23) is placed in a closed container having a vapor of the same impurity. Through proper temperature control, the atoms of the vapor will *fall upon* and *arrange* themselves on the original *p*-type substrate resulting in the epitaxial layer indicated in Fig. 3.23. Once this layer is established, the process continues, as above for the mesa transistor, to form the base and emitter regions. The original *p*-type substrate will have a higher doping level and correspondingly less resistance than the epitaxial layer. The result is a low-resistance connection to the collector lead that will reduce the dissipation losses of the transistor.

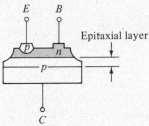

Figure 3.23. Epitaxial mesa transistor.

Figure 3.24. Planar transistor.

The planar and epitaxial planar transistors are fabricated using two diffusion processes to form the base and emitter regions. The planar transistor, as shown in Fig. 3.24, has a flat surface, which accounts for the term planar. An oxide layer is added as shown in Fig. 3.24 to eliminate exposed junctions, which will reduce substantially the surface leakage loss (leakage currents that flow on the surface rather than through the junction).

3.11 TRANSISTOR CASING AND TERMINAL IDENTIFICATION

After the transistor has been manufactured using one of the techniques indicated in Section 3.10, leads of, typically, gold, aluminum, or nickel are then attached and the entire structure is encapsulated in a container such as that shown in Fig. 3.25. Those with the studs and heat sinks are high-power devices, while those with the small can (top hat) or plastic body are low- to medium-power devices.

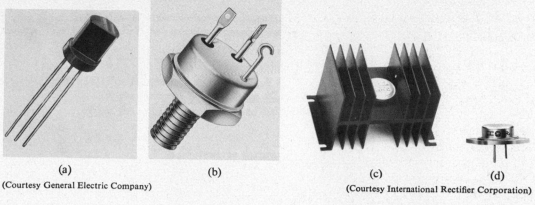

(a)　　　　　　　　　　　　(b)

(Courtesy General Electric Company)

(c)　　　　　　　　　　(d)

(Courtesy International Rectifier Corporation)

Figure 3.25. Various types of transistors.

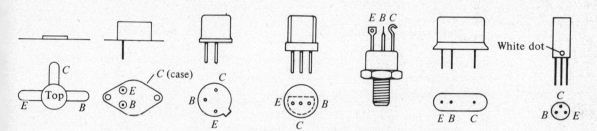

Figure 3.26. Transistor terminal identification.

Whenever possible, the transistor casing will have some marking to indicate which leads are connected to the emitter collector or base of a transistor. A few of the methods commonly used are indicated in Fig. 3.26.

3.12 TRANSISTOR TESTING

Using only the ohmmeter section of a VOM, we can apply a number of tests to a transistor. The first to be described will determine whether a short-circuit or open-circuit state exists within the device.

Short-Circuit or Open-Circuit Test

If the terminal identification and type (*npn* or *pnp*) are known, the tests indicated in Fig. 3.27 will determine whether the device is in the defective short-circuit or open-circuit state. For the *pnp* transistor appearing in Fig. 3.27 the expected readings for the tests applied are indicated. If a low or high reading results between the base and either terminal for each polarity applied, the device is respectively in the short- or open-circuit state. For an *npn* transistor the readings should be the direct opposite.

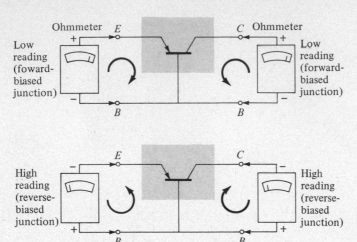

Figure 3.27. Short-circuit or open-circuit test for *pnp* transistor.

Determining the Base Lead

For each set of two terminals (there are three possibilities) measure the resistance between each set with each polarity applied. That set for which the readings remains high for each polarity *does not* include the base terminal.

Determining Whether It Is *pnp* or *npn*

Place the black (or negative) lead on the base terminal and the red (or positive) lead on either of the two remaining terminals. A low reading (as indicated in Fig. 3.27a) indicates a *pnp* type while a high reading indicates an *npn* type.

Determining Which Is the Collector Terminal and Which Is the Emitter Terminal

For the *pnp* transistor the meter is connected as indicated in Fig. 3.27a to the collector and emitter terminals. The measurement that results in the higher reading resistance indicates the emitter terminal. For the *npn* transistor the meter is reversed at each terminal (collector and emitter), but the higher reading again indicates the emitter terminal.

3.13 TRIODE

The *triode* is a three-terminal, high-vacuum-tube device, which, like the transistor, has an amplifying capability. However, there is one distinct difference between the two that must be made clear. The transistor having the input base current controlling the resulting point of operation in the collector characteristics is called a *current-controlled device*. The triode, which has a grid voltage controlling the point of operation in the output or plate characteristics, is called a *voltage-controlled device*. The ac analysis of each, to follow in Chapter 5, will be different due to the

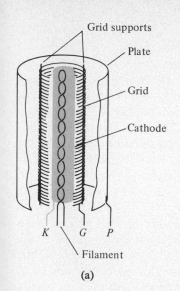

Grid supports

Plate

Grid

Cathode

K   G   P

Filament

(a)

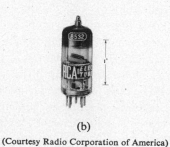

(b)

(Courtesy Radio Corporation of America)

Figure 3.28. Triode: (a) basic construction for octal base type; (b) photograph of high-mn miniature vacuum tube triode.

variation in the mode of operation. The triode construction consists of one more element, called the *control grid*, than was present in the vacuum-tube diode. The relative position of each structure in the vacuum-tube triode is shown in Fig. 3.28 along with a typical triode glass envelope tube. Note that the control grid is placed considerably closer to the cathode than the plate to increase its effectiveness. The control grid structure is very similar to that of a wire mesh fence; that is, the open area is many times greater than that of the wire itself. In this way, the *negative* potential to be applied to the grid can control the flow of charge from cathode to plate without adversely affecting the tube response due to a large number of electrons hitting the grid structure.

The triode symbol and biasing are provided in Fig. 3.29a. Note that a *negative* potential is applied to the grid and a *positive* potential to the plate. The choice of

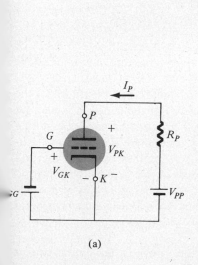

(a)

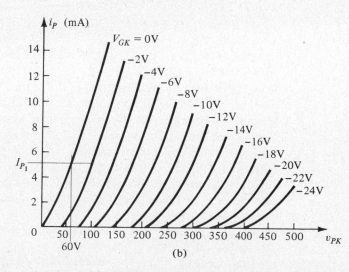

(b)

Figure 3.29. (a) Basic triode biasing circuit; (b) triode plate characteristics.

load resistor and battery potentials will be discussed in detail in Chapter 4. In Fig. 3.29b the *plate* characteristics of the triode indicate clearly the effect that the *grid* potential has on the characteristics. For $V_{GK} = 0$ V the resulting characteristics are essentially the same as those obtained for the vacuum-tube diode. If a negative potential were applied to the grid, the grid would repel a number of the negatively charged electrons emitted by the cathode back in that direction. This would in turn reduce the plate current. For $V_{GK} = -2$ V the curve has shifted somewhat to the right. To demonstrate the effect just discussed consider the vertical dashed line at $V_{PK} = 60$ V. The plate current corresponding with the intersection of this line with $V_{GK} = 0$ is certainly much higher than that for $V_{GK} = -2$ V. To establish the same plate current (I_{P_1} in Fig. 3.29b) for different values of V_{GK} (0, -2) a higher plate-to-cathode potential would be required. The effect of increasing the plate potential is to increase the attraction for the negatively charged electrons being emitted by the cathode. In other words, the higher positive potential is overriding some of the effect of the increased negative bias on the grid resulting in more electrons being able to pass through to the grid and reach the plate.

EXAMPLE 3.4   Using the characteristics of Fig. 3.29:
(a) Find the change in plate current if the grid-to-cathode (V_{GK}) potential is reduced from $V_{GK} = -12$ V to $V_{GK} = -8$ V at $V_{PK} = 240$ V.
(b) Find the increase in plate-to-cathode potential required to maintain the plate current at 8 mA if V_{GK} is changed from -2 to -8 V.

Solution: (a) At $V_{GK} = -12$ V, $I_{P_1} = 0.6$ mA; at $V_{GK} = -8$ V, $I_{P_2} = 6.6$ mA.
The resulting change in plate current is

$$\Delta I_P = I_{P_2} - I_{P_1} = 6.6 - 0.6 = \mathbf{6 \text{ mA}}$$

(b) At $V_{GK} = -2$ V, $V_{PK_1} = 123$ V; at $V_{GK} = -8$ V, $V_{PK_2} = 248$ V.
The change in plate potential is

$$\Delta V_{PK} = V_{PK_2} - V_{PK_1} = 248 - 123 = \mathbf{125 \text{ V}}$$

A second glance at the characteristics of Fig. 3.29 will reveal that the *family* of curves for $V_{GK} = 0, -2, -4, -6$, etc., are fairly linear, parallel, and equidistant from each other in a major portion of the characteristics. It is characteristics of this type that result in the least distorted amplification of signals. For the transistor the input characteristics were of considerable importance. For the triode, however, we assume that there exists an open circuit between the grid and cathode eliminating the necessity for input characteristics. In actual practice, there is a small grid current, even with the negative bias, due to those electrons that bombard the grid structure. In addition, the interelectrode capacitances tie the grid to the plate and cathode with low-impedance values at very high frequencies. The grid of a triode can be made positive with respect to the cathode, but this will result in a heavy grid current and higher internal losses than would result with negative grid voltages for the same amplification.

The extent to which the region described in the previous paragraph can be employed is limited by the ratings of the tube. The *maximum ratings* normally provided on a data sheet include the maximum power or plate dissipation, plate current, and plate-to-cathode potential. For the family of plate characteristics provided in Fig. 3.29b the following maximum values were provided: $P_{D_{max}} = 2.1$ W, $I_{P_{max}} = 12$ mA, $V_{PK_{max}} = 350$ V.

The plate dissipation is equal to the product of the plate potential and corresponding plate current; that is,

$$P_{D_{max}} = V_{PK} I_P \qquad (3.16)$$

The calculations required with Eq. (3.16) to obtain the maximum power curve are exactly the same as those employed with the maximum transistor ratings. The resulting curve is shown in Fig. 3.30, along with the maximum voltage and current ratings. In addition, a horizontal line has been drawn above that nonlinear region to be avoided if distortion in the output waveform is to be kept to a minimum. The resulting unshaded region is that region remaining for the linear (least distorted) amplification of signals within the device limitations.

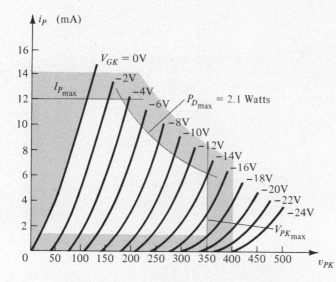

Figure 3.30. Region of operation.

The *pin connections* for the triode amplifier will require, as was true for the vacuum diode, the use of the tube manual. The pin connections and some of the associated data, as they would appear in a tube manual for the 6J5 tube, are provided in Fig. 3.31, along with some of their typical areas of application.

The *interelectrode capacitance* values in the specifications of Fig. 3.31 are the capacitance values that exist between the terminals indicated. At high frequencies these interelectrode capacitances can have a pronounced affect on the triode performance. Consider, for example, the resulting plate-to-cathode capacitive reactance at 1 MHz for the 6J5 triode:

$$X_C = \frac{1}{2\pi f C} = \frac{1}{(6.28)(10^6)(3.6 \times 10^{-12})} \cong 45 \text{ K}$$

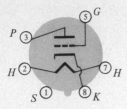

6J5 Triode

Metal and glass octal-type 6J5
used as detectors, amplifiers, or
oscillators in radio equipment

Heater voltage (ac/dc) 6.3V
Heater current 0.3A
Direct interelectrode capacitances (approx.)
    Grid to plate 3.4pF
    Grid to cathode and heater 3.4pF
    Plate to cathode and heater 3.6pF

Figure 3.31. Pin connections and associated data for the 6J5 triode.

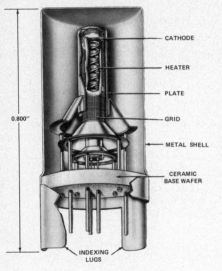

(a) Nuvistor

(Courtesy Radio Corporation of America)

(power amplifier and frequency multiplier):
frequency range—up to 250 MHz
direct interelectrode capacitances:
$$C_{gp} = 2.2 \text{ pf}$$
$$C_{gk} = 4.2 \text{ pf}$$
$$C_{pk} = 0.26 \text{ pf}$$

(b) Ceramic

(Courtesy General Electric Company)

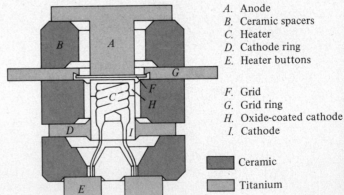

A. Anode
B. Ceramic spacers
C. Heater
D. Cathode ring
E. Heater buttons

F. Grid
G. Grid ring
H. Oxide-coated cathode
I. Cathode

Ceramic

Titanium

frequency range (grounded grid amplifier)—to 450 MHz
direct interelectrode capacitances:
$$C_{gp} = 1.0 \text{ pf}$$
$$C_{gk} = 1.7 \text{ pf}$$
$$C_{pk} = 0.01 \text{ pf}$$

Figure 3.32. General construction and pertinent data of the nuvistor and ceramic triode.

A capacitive reactance of 45 K between plate and cathode will in most applications disallow the use of the open-circuit approximation. In recent years the frequency range has been substantially increased through improved design techniques. Two such triodes with very-high-frequency ranges arc the *Nuvistor* and *Ceramic* triode shown in Fig. 3.32. Note the small size of each triode. Each, by virtue of its construction, can withstand a higher level of shock and vibration. The Nuvistor and 6J5 triodes have a *cylindrical*-type assembly, while the Ceramic triode has a *planar* construction. Note in Fig. 3.32 that the grid, plate, and cathode have a single planar surface and do not completely encompass the heater as shown in the cylindrical structure of Fig. 3.28. The data provided in the same figure indicate clearly the wide frequency range of each device. Note also the low interelectrode capacitance values of each device. The small-size, high-frequency capabilities of each make it suitable for instrumentation equipment, audio and video equipment, and communication systems.

In many cases the same heater is used to operate two or more devices in the same package. This is shown in Fig. 3.33 for a 6GU7 double triode. Another variation is the 6SQ7 twin diode-triode, shown in the same figure.

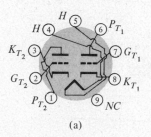

(a)

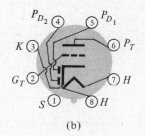

(b)

Figure 3.33. (a) 6GU7 twin triode; (b) 6SQ7 twin diode-triode.

3.14 PENTODE

The *pentode* is a five element vacuum tube having significantly improved characteristics over the triode. The interelectrode capacitance is greatly reduced and its amplifying capabilities can surpass that of a typical triode by 1000 : 1. Two additional grid structures have been added to the basic triode as shown in Fig. 3.34 to form the pentode tube.

The function of the additional *screen* grid is to electrostatically screen the plate from the control grid and cathode. To fully understand the effects of this positive (normally about 100 V positive with respect to the cathode) grid structure in the tube would require a knowledge of electrostatic theory beyond the scope of this text. Nevertheless, some appreciation for its isolating effect can be developed by considering an isolated negatively charged electron that has just passed through the control grid structure. As far as the electron is concerned, the positive potential to which it would now be

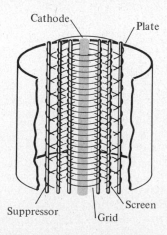

Figure 3.34. Basic pentode construction.

attracted is that of the screen grid, even if the plate were at the same or higher potential. The plate has been effectively *screened* from the control grid and cathode structures. A number of electrons in this region will hit the positive screen and establish a screen current. There will, however, be many other carriers that will simply pass through the grid structure because of their high kinetic energy and hit the plate. The screening effect will significantly reduce the interelectrode capacitance between the plate and control grid (C_{PG}) and between the plate and cathode (C_{PK}) and cause the characteristics to level off at higher values of V_{PK} as shown in the typical set of characteristics of Fig. 3.35. The leveling effect clearly indicates that the plate potential has less effect on the magnitude of the plate current (especially at higher levels). Compare to the characteristics obtained for the vacuum-tube triode. It will be pointed out in a later chapter that the effect of this change in the characteristics is to increase the available gain.

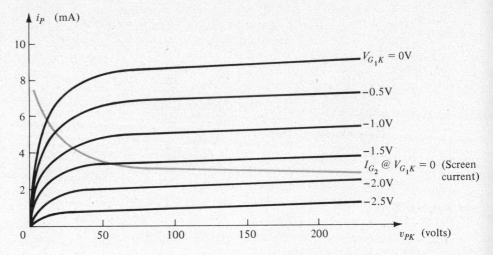

Figure 3.35. Pentode plate characteristics.

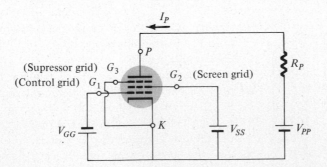

Figure 3.36. Basic pentode biasing circuit.

The last structure to be introduced, called the *suppressor* grid, is placed between the plate and screen grid and is normally connected directly to the cathode as shown in the basic biasing network of Fig. 3.36. The purpose of this additional grid is to *suppress* the secondary emission current from the plate to the screen grid. When

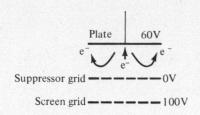

Plate   60V

e^- e^-

e^-

Suppressor grid ———————0V

Screen grid ———————100V

Figure 3.37. Effect of the suppressor grid in the vacuum-tube pentode.

electrons bombard the plate structure at high velocities, additional electrons are often liberated (secondary emission) from the plate surface that may end up on the positive screen grid rather than returning to the plate structure and increasing the plate current. By placing the suppressor grid between the two positive structures as shown in Fig. 3.37 the liberated electrons are left with a choice between going to the suppressor at ground potential (0 V) and returning to the positive plate. Obviously, as shown in the diagram, the vast majority will return to the plate and contribute to the plate current.

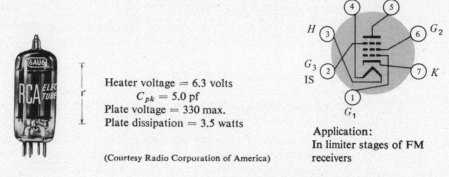

Heater voltage = 6.3 volts
C_{pk} = 5.0 pf
Plate voltage = 330 max.
Plate dissipation = 3.5 watts

(Courtesy Radio Corporation of America)

Application:
In limiter stages of FM receivers

Figure 3.38. Miniature pentode vacuum tube.

The pin connections, pertinent data, and photograph of a miniature pentode-vacuum tube are shown in Fig. 3.38 with an area of application.

PROBLEMS

§ 3.2

1. What are the names applied to the two types of transistors? Sketch each transistor and indicate the type of majority and minority carrier in each layer. Is any of the above altered by changing from a silicon to germanium material?

2. What is the major difference between a bipolar and a unipolar device?

§ 3.3

3. How must the two transistor junctions be biased for proper transistor operation?

4. What is the source of the leakage current in a transistor?

5. Sketch a figure similar to Fig. 3.3 for the forward-biased junction of an *npn* transistor. Describe the resulting carrier motion.

6. Sketch a figure similar to Fig. 3.4 for the reverse-biased junction of an *npn* transistor. Describe the resulting carrier motion.

7. Sketch a figure similar to Fig. 3.5 for the majority and minority carrier flow of an *npn* transistor. Describe the resulting carrier motion.

8. Determine the resulting change in emitter current for a change in collector current of 2 mA and an α (alpha) of 0.98.

9. A transistor has an emitter current of 8 mA and an α of 0.99. How large is the collector current?

§ 3.4

10. Calculate the voltage gain ($A_v = V_o/V_i$) for the circuit of Fig. 3.6 if $V_i = 500$ mV and $R = 1$ K. (The other circuit values remain the same.)

11. Calculate the voltage gain ($A_v = V_o/V_i$) for the circuit of Fig. 3.6 if the source has an internal resistance of 100 Ω in series with V_i.

12. From *memory*, and memory only, sketch the common-base transistor configuration (for *npn* and *pnp*) and indicate the polarity of the applied bias and resulting current directions.

13. Using the characteristics of Fig. 3.8,
    (a) Find the resulting collector current if $I_E = 5$ mA and $V_{CB} = -10$ V.
    (b) Find the resulting collector current if $V_{EB} = 750$ mV and $V_{CB} = -10$ V.
    (c) Find V_{EB} for the conditions $I_C = 4$ mA and $V_{CB} = -15$ V.

14. The characteristics of Fig. 3.8b are for a silicon transistor. How would you expect them to be different for a germanium transistor? What would be a first approximation for the base-to-emitter voltage of the forward-biased junction?

§ 3.6

15. Are I_{CO} and I_{CEO} related? If they are, how and why?

16. Using the characteristics of Fig. 3.11,
    (a) Find the value of I_C corresponding to $V_{BE} = +750$ mV and $V_{CE} = +5$ V.
    (b) Find the value of V_{CE} and V_{BE} corresponding to $I_C = 3$ mA and $I_B = 30$ μA.

17. (a) For the common-emitter characteristics of Fig. 3.11 find the dc beta at an operating point of $V_{CE} = +8$ V and $I_C = 2$ mA.
    (b) Find the value of α corresponding to this operating point.
    (c) At $V_{CE} = +8$ V find the corresponding value of I_{CEO}.
    (d) Calculate the approximate value of I_{CBO} using the dc beta value obtained in part (a).

§ 3.7

18. An input voltage of 2 V rms (measured from base to ground) is applied to the circuit of Fig. 3.16. Assuming that the emitter voltage follows the base voltage exactly and that V_{be} (rms) = 0.1 V, calculate the circuit voltage amplification ($A_v = V_o/V_i$) and emitter current for $R_E = 1$ K.

§ 3.8

19. Using the technique of Section 3.8, sketch a common-emitter *pnp* transistor and determine the proper bias polarities and current directions.

§ 3.9

20. Determine the region of operation for the transistor with the characteristics of Fig. 3.11a if $I_{C_{max}} = 5$ mA, $V_{CE_{max}} = 15$ V, and $P_{C_{max}} = 20$ mW.

§ 3.10

21. Discuss the differences between the four types of transistor construction described in Section 3.10.

§ 3.12

22. Sketch Fig. 3.27 for an *npn* transistor and indicate the expected readings.

§ 3.13

23. Describe in your own words the basic action of a triode vacuum tube.

24. Using the characteristics of Fig. 3.29,
    (a) Find the change in plate current if the grid-to-cathode (V_{GK}) potential is reduced from $V_{GK} = -6$ V to $V_{GK} = -10$ V at $V_{PK} = 200$ V.
    (b) Find the increase in plate-to-cathode potential required to maintain the plate current at 6 mA if V_{GK} is changed from -2 to -12 V.

§ 3.14

25. How does the pentode differ from the triode? Explain in detail the action and effect of the added elements.

26. Determine the change in plate current for a grid-to-cathode voltage change from -1.5 to -2.0 V at a plate-to-cathode voltage of 200 V using the pentode characteristics of Fig. 3.35.

27. Determine the approximate change in plate current using the characteristics of Fig. 3.35 for a plate-to-cathode voltage change from 100 to 200 V at a grid-to-cathode voltage of -1 V.

dc
biasing

4

4.1 GENERAL

Transistors are used in a large variety of applications and in many different ways. It would be difficult if not impossible to learn each area and application. Instead, one studies the more fundamental properties and aspects so that enough is known to carry over this knowledge to slightly different or even completely different applications. This chapter covers the basic concepts in the dc biasing of bipolar transistors (and tube devices).

To use these devices for amplification of voltage or current signals, or as control (ON or OFF) elements, or in any other application, it is necessary first to *bias* the device. The usual reason for this biasing is to turn the device on, and in particular, to place it in operation in the region of its characteristic where the device operates most linearly, providing a constant amount of gain.

Although the purpose of the bias network or biasing circuit is to cause the device to operate in this desired *linear* region of operation (which is defined by the manufacturer for each device type), the bias components are still part of the overall application circuit—amplifier, waveform shaper, logic circuit, etc. We could treat the overall circuit and consider all aspects of the operation at once, but this is more complex and more confusing. Each type of circuit application would have to be studied for all aspects of operation without a more basic understanding of those common features of operation of most other application circuits. This chapter therefore provides basic concepts of dc biasing of the bipolar transistor and tube devices, with the understood aim of getting the device operating in a desired region of the device characteristic. If these concepts are well understood, many different circuits, even new circuit applications, can be studied and analyzed more easily because a basic understanding of the circuit has been established. Amplifier gain and other factors affecting ac operation will be considered in Chapter 5. With this breakdown the basic concepts can be presented and consideration can then be given in Chapters 7, 8, 11, 12, and 14 to the use of these devices in specific areas of application, with the main concern being the problems and operations peculiar to that area of interest.

The dc biasing is a *static* operation since it deals with setting a fixed (steady) level of current flow (through the device) with a desired fixed voltage drop across the device. The necessary information about the device can be obtained from the device's static characteristics, both input and output. As will be shown, the transistor biasing considerations can be obtained from the manufacturer's listing for a particular device, whereas the tube is best described by a set or family of curves of output plate-to-cathode voltage and plate current. Thus, we shall be able, with these two devices, to cover the essential features of the two basic techniques (analytical or mathematical and graphical) used to determine the operation of the circuit and to set the desired operating voltages and currents in the circuit.

4.2 OPERATING POINT

Since the aim of biasing is to achieve a certain condition of current and voltage called the *operating point* (or *quiescent* point), some attention is given to the selection of this point in the device characteristic. Figure 4.1 shows a general device characteristic with four indicated operating points. The biasing circuit may be designed to

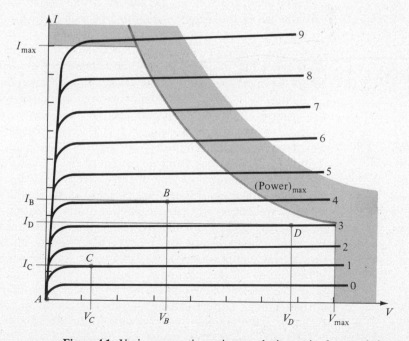

Figure 4.1. Various operating points on device static characteristics.

set the device operation at any of these points or others within the *operating region*. The operating region is the area of current or voltage within the maximum limits for the particular device. These maximum ratings are indicated on the characteristic of Fig. 4.1 by a horizontal line for the maximum current, I_{max}, and a vertical line for the maximum voltage, V_{max}. An additional consideration of maximum power (product of voltage and current) must also be taken into consideration in defining the operating region of a particular device, as shown by the dotted line marked P_{max} on Fig. 4.1.

It should be realized that the device could be biased to operate outside these maximum limit points but that the result of such operation would be either a considerable shortening of the lifetime of the device or destruction of the device. Confining ourselves to the safe operating region we may select many different operating areas or points. The exact point or area often depends on the intended use of the circuit. Still, we can consider some differences between operation at the different points shown in Fig. 4.1 to present some basic ideas about the operating point and, thereby, the bias circuit.

If no bias were used, the device would initially be completely off, which would result in the current of point *A*—namely, zero current through the device (and zero voltage across it). It is necessary to bias the device so that it can respond or change in current and voltage for the entire range of the input signal. While point *A* would not be suitable, point *B* provides this desired operation. If a signal is applied to the circuit, *in addition to the bias level*, the device will vary in current and voltage from operating point *B*, allowing the device to react to (and possibly amplify) both the positive and negative part of the input signal. If, as could be the case, the input signal is small, the voltage and current of the device will vary but not enough to drive the device into *cutoff* or *saturation*. Cutoff is the condition in which the device no longer conducts. Saturation is the condition in which voltage across the device is as small as possible with the current flow in the device path reaching a limiting value depending on the external circuit. The usual amplifier action desired occurs within the operating region of the device, that is, between saturation and cutoff.

Point *C* would also allow some positive and negative variation with the device still operating, but the output could not vary too negatively (left of V_C) because bias point *C* is lower in voltage than point *B*. Point *C* is also in a region of operation in which the current level in the device is smaller and the device gain is *not* linear, that is, the spacing in going from one curve to the next is unequal. This nonlinearity shows that the amount of gain of the device is smaller lower on the characteristic and larger higher up. It is preferable to operate where the gain of the device is most constant (or linear) so that the amount of amplification over the entire swing of input signal is the same. Point *B* is in a region of more linear spacing and, therefore, more linear operation, as shown in Fig. 4.1. Point *D* sets the device operating point near the maximum voltage level. The output voltage swing in the positive direction is thus limited if the maximum voltage is not to be exceeded. Point *B*, therefore, seems the best operating point in terms of linear gain or largest possible voltage and current swing. This is usually the desired condition for small-signal amplifiers (Chapter 5) but not necessarily for power amplifiers and logic circuits, which will be considered in Chapters 8 and 13, respectively. In this discussion, let us concentrate mainly on biasing the device for *small-signal* amplification operation.

One other very important biasing factor must be considered. Having selected and biased for a desired operating point, we see that the effect of temperature must also be taken into account. Temperature causes the device characteristic to change. Higher temperature results in more current flow in the device than at room temperature, thereby upsetting the operating condition set by the bias circuit. Because of this, the bias circuit must also provide a degree of *temperature stability* to the circuit so that temperature changes at the device produce minimum change in its operating point. This maintenance of operating point may be specified by a *stability factor*, S, indicating the amount of change in operating-point current due to temperature. A highly stable circuit is desirable and the stability factors of a few basic bias circuits will be compared.

Tube operation is described by a graphical static characteristic and the selection of a biasing point is obtained from this characteristic. Bipolar transistor operation, however, may be specified sufficiently well by device parameters, and more mathematical (rather than graphical) techniques can be used to determine its biasing. Nevertheless, the transistor characteristic still provides a convenient picture for understanding device operation and will be used on occasion.

4.3 COMMON-BASE (CB) BIAS CIRCUIT

The common-base (CB) configuration provides a relatively straightforward and simple starting point in our dc bias considerations. Figure 4.2a shows a common-base circuit configuration. By CB we mean that the base is the reference point for measurement for both input (emitter, in this case) and output (collector).

The dc supply (battery) terminals are marked with a double-letter designation. V_{EE} is the dc voltage supply associated with the emitter section, and V_{CC} is the dc supply for the collector section of the circuit. Two separate voltage supplies may be required for the CB configuration. Resistor R_E is basically a current-limiting resistor for setting the emitter current I_E. Resistor R_C is the collector (output or load) resistor and the output ac signal is developed across it. It is also one of the components that is used to set a desired operating point.

Figure 4.2b shows the CB collector characteristic, which is the output characteristic for this circuit connection. The abscissa is the collector-to-base voltage V_{CB}, which is a negative voltage for the *pnp* circuit of Fig. 4.2a. The ordinate is the collector current, I_C. The family of curves for the characteristic are for various emitter currents.

The theory developed for either the *pnp* or *npn* transistor device can be equally applied to the other device by merely changing all current directions and all voltage polarities. We shall consider mostly *npn* transistors in developing concepts because it is the more popular unit at present.

It is possible to consider biasing the CB circuit by analyzing separately the input (base-emitter loop) and the output (base-collector loop) sections of the circuit.

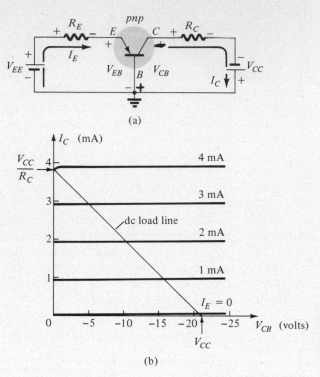

Figure 4.2. Common-base circuit and transistor characteristic: (a) common-base circuit; (b) dc load line on common-base characteristic.

Although in reality there is some interaction between the operation of the base-collector section and that of the base-emitter section, this can be neglected with excellent practical results obtained.

Input Section

The input loop (see Fig. 4.3a) is composed of the battery, V_{EE}, the resistor, R_E, and the emitter-base junction of the transistor, V_{EB}. Writing the voltage loop equation (using Kirchhoff's voltage law) for the input loop

$$+V_{EE} - I_E R_E - V_{EB} = 0 \qquad V_{EE} = I_E R_C + V_{EB}$$
$$+V_{EE} - I_E R_E = V_{EB}$$

$$I_E = \frac{V_{EE} - V_{EB}}{R_E}$$

$$V_{CB} = I_C R_C - V_{CC}$$
$$(I_C \cong I_E)$$

$V_{EB}\ (\text{Ge}) \cong 0.3\text{V}$
$V_{EB}\ (\text{Si}) \cong 0.7\text{V}$

(a)   (b)

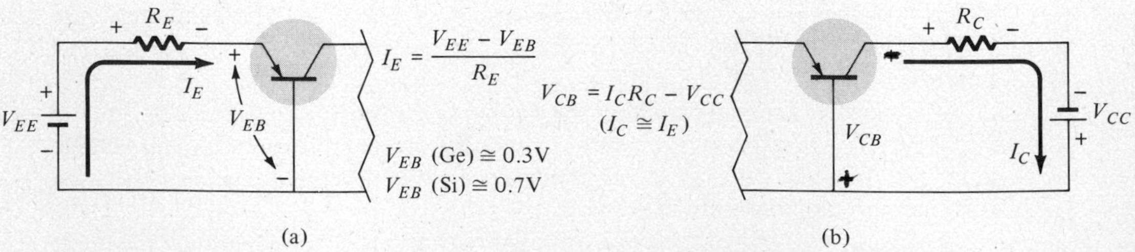

Figure 4.3. Input and output sections for common-base circuit: (a) input (emitter-base) section only; (b) output (collector-base) section only.

from which we get

$$I_E = \frac{V_{EE} - V_{EB}}{R_E}$$

(4.1a)

where all terms has been defined above.

When forward biased, the emitter-base voltage V_{EB} is small—on the order of 0.3 V for germanium transistors and 0.7 V for silicon. Although the actual emitter-base voltage is slightly affected by the collector-base voltage, this effect can be neglected for practical considerations. In fact, if the supply voltage V_{EE} is, say, 10 V or more, the emitter-base voltage could be neglected, giving

$$I_E \cong \frac{V_{EE}}{R_E}$$

(4.1b)

as a good approximation. Observe that the emitter current is set essentially by the emitter supply voltage and the emitter resistor. But the supply voltage is usually fixed since it is required to provide voltage to other parts of the electronic circuit. The emitter current, therefore, is specifically determined by the emitter resistor R_E, whose value is selected to give the desired emitter current.

Output Section

The output loop (see Fig. 4.3b) consists of the battery, V_{CC}, the resistor R_C, and the voltage across the collector-base junction of the transistor, V_{CB}. For operation as an amplifier the *collector-base* junction must be *reverse biased* in addition to the *emitter-base* being *forward biased*. This *reverse bias* is provided by the V_{CC} battery voltage connected in polarity so that the *p*ositive battery terminal connects to the *n*-material and the *n*egative battery terminal to the *p*-material. (Note the letter opposites for *p* and *n* for battery and transistor type.) The result of this consideration is that for *pnp* transistors the battery polarity should be positive terminal to the common-base point and negative terminal to the resistor connected to the collector terminal. (For *npn* transistors we have the opposite—the negative battery terminal connected to the CB terminal and the positive terminal of the battery connected to the resistor, which then connects to the collector terminal.)

Summing the voltage drops around the output or collector-base loop of the circuit of Fig. 4.3b, we get

$$+V_{CC} - I_C R_C - V_{CB} = 0$$

Solving for the collector-base voltage results in

$$V_{CB} = +V_{CC} - I_C R_C$$

(4.2)

The collector current I_C is approximately the same magnitude as the emitter current I_E [obtained from Eqs. (4.1a) or (4.1b)]. This is a very good approximation that will be true for any type of transistor connection used. For the purposes of calculating circuit bias values we may write the relation as

$$\boxed{I_C \cong I_E}$$ (4.3)

Actually, $I_C = \alpha I_E$, where α (alpha) is typically 0.9–0.998 in value.

Complete Solution of Bias Conditions for Common-Base Circuit

Having presented the essential circuit operation of the CB connection we can consider the complete solution of bias currents and voltages. The results obtained will apply to both *npn* and *pnp* CB transistor circuits. To help in the solution of bias conditions for a CB circuit as in Fig. 4.2a and to provide a structured procedure to be followed in later types of circuit connections we shall formalize the solution into a step-by-step calculation. To solve for the bias voltages and currents of a CB bias circuit as in Fig. 4.2a proceed as follows:

1. With emitter-base voltage polarity providing forward bias, assume approximate voltages of

$$V_{EB} \cong 0.3 \text{ V (germanium), or}$$

$$V_{EB} \cong 0.7 \text{ V (silicon)}$$

(Forward bias is provided if the battery voltage connection results in *positive* voltage at *p*-type material and *negative* voltage at *n*-type material of the transistor.)

2. Calculate the emitter current I_E using

$$I_E = \frac{V_{EE} - V_{EB}}{R_E} \cong \frac{V_{EE}}{R_E}$$

3. Collector current is approximately the emitter current calculated in step (2) above:

$$I_C \cong I_E$$

4. Collector-base voltage, is calculated from

$$V_{CB} = V_{CC} - I_C R_C$$

EXAMPLE 4.1 Calculate bias voltages V_{EB} and V_{CB} and currents I_E and I_C for the circuit of Fig. 4.4. The circuit contains a *pnp* silicon transistor with alpha (α) of 0.99.

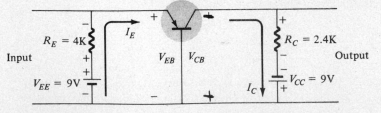

Figure 4.4. Common-base bias circuit for Example 4.1.

Solution: Using the step-by-step procedure outlined previously, we get

(a) $V_{EB} \cong 0.7$ (silicon).

(b) $I_E = (V_{EE} - V_{EB})/R_E = (9 - 0.7 \text{ V})/4 \text{ K} \cong 2.1 \text{ mA}$.

(c) $I_C \cong I_E = 2.1 \text{ mA}$.

(d) $V_{CB} = V_{CC} - I_C R_C = 9 - (2.1 \text{ mA})(2.4 \text{ K}) = \textbf{3.96 V}$

4.4 COMMON-EMITTER (CE) CIRCUIT CONNECTION—GENERAL BIAS CONSIDERATIONS

A more popular amplifier connection applies the input signal to the base of the transistor with the emitter as common terminal. The CE circuit of Fig. 4.5 shows only one supply voltage. Recall that the CB circuit used two supply voltages, one to forward bias the base-emitter and the second to provide reverse bias for the base-collector. Both forward- and reverse-bias conditions are achieved in the CE connection using one voltage supply. In addition, we shall show a number of other important advantages of the CE circuit relating to input and output impedances, current and voltage gain, etc., which apply to the ac operation of the circuit. This chapter deals with the dc biasing of the circuit. Other important circuit factors will be considered in Chapter 5.

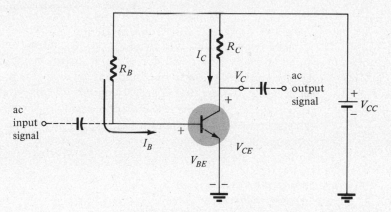

Figure 4.5. Common-emitter fixed-bias circuit.

Although a mathematical (rather than graphical) technique may be used to obtain the bias voltages and currents in the circuit, we shall refer to the CE collector characteristic of the transistor to provide a reason for choosing a particular operating point. Figure 4.6 shows the CE collector characteristic of a transistor indicating a few bias points. The result of connecting a supply voltage V_{CC} and resistors R_B and R_C, as in Fig. 4.5, is to cause both base and collector current flow with a resulting bias voltage V_{CE} measured from collector to emitter (common terminal).

It would seem from the discussion in Section 4.2 about selecting a suitable bias point that point C in Fig. 4.6 would be satisfactory. Bias at point C would allow a range of voltage for the collector-emitter voltage to be either increased or decreased by the input ac signal.

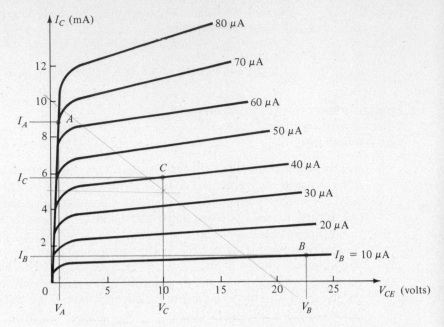

Figure 4.6. Common-emitter collector characteristics showing typical bias points.

The remainder of this chapter will concentrate on (a) determining what bias point actually will result for a given circuit containing given circuit elements (*analysis*) and (b) how to obtain the circuit elements to provide a desired bias point of collector current and collector-emitter voltage (*synthesis*). To present the basic circuit theory and provide some appreciation of the circuit operation, analysis of a given circuit will be considered first. After that, the design (synthesis) of a circuit to obtain a desired bias condition will be covered.

4.5 BIAS CONSIDERATIONS FOR A
FIXED-BIAS CIRCUIT

A fixed-bias circuit is shown in Fig. 4.5 with typical component values. An *npn* transistor is used with current gain (beta or h_{FE}) of 50. Our objective at the moment is to calculate the base current I_B, the collector current I_C in the given circuit, and also the base-emitter (V_{BE}) and collector-emitter (V_{CE}) voltages. This information will be used to determine whether the circuit is properly biased and also to consider how the bias point is affected in case an adjustment is desired.

We shall *not* resort to writing many loop equations but rather we shall use the constraints imposed on the circuit by the transistor to apply a single Kirchhoff voltage or current loop equation to solve for a particular current or voltage quantity. Using, in addition, the voltage and current constraints of the transistor (which are simple) permits a step-by-step solution, which should also provide a more meaningful understanding of the circuit operation and expected current and voltage values. The intent of this approach is that numerical answers are not arrived at mechanically using involved mathematical techniques, which then mean little to the reader. Throughout the procedure the question "Does the value computed make sense?" will and should be asked.

The transistor in Fig. 4.5 can be used as an amplifying device if the base-emitter is forward biased and the base-collector reverse biased. Let us see how this is accomplished in Fig. 4.5 using a single voltage supply. The positive supply voltage V_{CC} connected through resistor R_B provides a positive voltage at the base of the transistor with respect to the common-emitter point (ground). With the base-emitter junction of the transistor forward biased, the voltage drop across the junction is determined mainly by the transistor type and only very slightly by the supply voltage used or even by the base resistor value. What is essential is that the supply voltage polarity be such that the base-emitter is forward biased (and not reverse biased). As a means of remembering the polarity needed for forward bias consider the following: p-material is connected toward the positive-voltage side and n-material is connected toward the negative-voltage side of the supply battery. For the npn transistor in Fig. 4.5 this would require the base (p-material) to be connected toward the positive of the supply voltage and the emitter (n-material) to be connected toward the negative terminal of the battery. Since this is precisely the connection in Fig. 4.5, the base-emitter will be forward biased. For a silicon transistor the forward-bias voltage developed across the transistor base-emitter will be around 0.7 V, typically, with possible values from 0.5 up to 1.0 V, depending on base current, temperature, etc. For present purposes a typical value of 0.7 V will be suitable. For germanium transistors typical emitter-base forward-bias voltage will be 0.3 V.

A reverse-bias voltage must also be provided by the single voltage supply. For the npn transistor this would require a collector voltage more positive than the base voltage; that is, the collector-base voltage must be positive toward n-material and negative toward p-material for reverse bias. Since the supply voltage is connected in the collector-emitter circuit, how does the collector-base become reverse biased? In Fig. 4.5 the base-emitter will be forward biased with a few tenths volt from base to emitter (or from base to ground). As long as the collector voltage is more positive than V_{BE}, the voltage measured from collector to base V_{CB} will be positive to negative (more positive to less positive) and the collector-base junction will be reverse biased. If the collector voltage drops below the few tenths base-emitter voltage, the collector-base is no longer reverse biased and the device no longer acts as an amplifier. If V_{CB} becomes forward biased, the transistor is in saturation—one of the voltage states desired in computer switching circuits, but not in amplifier circuits. This bias condition will be considered in Chapter 13 when switching circuits are covered.

Thus, the single supply V_{CC} provides forward bias for the base-emitter junction and reverse bias for the collector-base junction to bias the transistor into operation as an amplifier device. A pnp transistor would be biased in much the same way as that discussed above. The only difference would be that a single negative battery voltage polarity would be used, all voltages would be negative, and all currents would be opposite those shown in Fig. 4.5. The magnitudes of the calculated values would, however, be the same for either transistor type.

4.6 CALCULATION OF BIAS POINT FOR FIXED-BIAS CIRCUIT

Given the fixed-bias circuit of Fig. 4.5, how can we determine the dc bias currents and voltages in the base and collector of the transistor? This section will develop a step-by-step procedure, which will determine the answers to the above question, thereby providing analysis of a given fixed-bias circuit.

Input Section

To provide for simple step-by-step analysis consider only the base-emitter circuit loop shown in the partial circuit diagram of Fig. 4.7a. Writing the Kirchhoff voltage equation for the given loop, we get

$$+V_{CC} - I_B R_B - V_{BE} = 0$$

$+ V_{cc} - I_B R_B = V_{BE}$

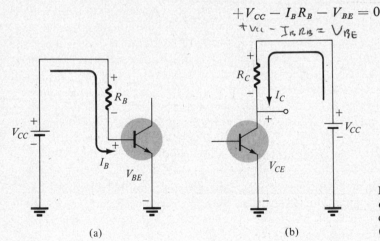

Figure 4.7. Separate input and output loops for fixed-bias circuit: (a) input base-emitter loop; (b) output collector-emitter loop.

We can solve the above equation for the base current I_B

$$I_B = \frac{V_{CC} - V_{BE}}{R_B} \qquad (4.4a)$$

Since the supply voltage V_{CC} and the base-emitter voltage V_{BE} are fixed values of voltage, the selection of a base bias resistor fixes the value of the base current. As a good approximation we may even neglect the few tenths volt drop across the forward-biased base-emitter V_{BE}, obtaining the simplified form for calculating base current,

$$I_B \cong \frac{V_{CC}}{R_B} \qquad (4.4b)$$

Output Section

The output section of the circuit (Fig. 4.7b) consists of the supply battery, the collector (load) resistor, and the transistor collector-emitter junctions. The currents in the collector and emitter are about the same since I_B is small in comparison to either. For linear amplifier operation the collector current is related to the base current by the transistor current gain, beta (β) or h_{FE}. Expressed mathematically,

$$\boxed{I_C = \beta I_B} \qquad \beta = \frac{I_C}{I_\beta} \tag{4.5}$$

The base current is determined from the operation of the base-emitter section of the circuit as provided by Eq. (4.4a) or (4.4b). The collector current as shown by Eq. (4.5) is β times greater than the base current *and* not at all dependent on the resistance in the collector circuit. From the previous consideration of the common-base circuit we know that the collector current is controlled in the base-emitter section of the circuit and not in the collector-base (or collector-emitter, in this case) section of the circuit.

Calculating voltage drops in the output loop, we get

$$V_{CC} - I_C R_C - V_{CE} = 0$$

$$\boxed{V_{CE} = V_{CC} - I_C R_C} \tag{4.6}$$

Equation (4.6) shows that the sum of voltages across the collector-emitter and across the collector resistor is the supply voltage value. This can also be stated as: The supply voltage provides the voltages across the collector resistor and across the collector-emitter—or the voltage across the collector-emitter V_{CE} is the remaining voltage from that of the voltage supply minus that voltage dropped across the collector resistor.

Transistor Saturation

One additional consideration must be included in the above solution steps. The relation between collector and base current, namely, that $I_C = \beta I_B$, is true *only* if the transistor is properly biased in the linear region of the transistor's operation. If the transistor, for example, is biased in the *saturation* region (too large an amount of base bias current), the use of Eqs. (4.5) and (4.6) leads to incorrect results.

For the transistor to be biased in a region of linear amplifier operation (as opposed to regions of cutoff or saturation) the base-emitter junction must be forward biased *and* the base-collector junction reverse biased. Our concern here is with the second bias condition—that the collector-base be properly reverse biased. This is true only as long as the collector-emitter voltage V_{CE} is larger in value than the base-emitter forward-bias voltage V_{BE}. Since the collector-emitter voltage V_{CE} given by Eq. (4.6) is the difference between the supply voltage V_{CC} and the voltage drop across the collector resistor ($I_C R_C$), the latter must be less than V_{CC} or in terms of the collector current, I_C must be less than V_{CC}/R_C. Stated mathematically,

$$I_C < \frac{V_{CC}}{R_C} \tag{4.7}$$

for the transistor to be biased in the active (linear) region of operation. A quick check of Eq. (4.5) would therefore be in order using Eq. (4.7) when performing the calculations of collector-emitter voltage to make sure that the condition just stated is correct in the circuit under consideration. If so, the three solution steps outlined above can be carried out, as representing the operation of the circuit. If, however, the above relation of maximum I_C allowable for operation in the transistor linear region is exceeded, the transistor is operating in the saturation region. In this case the collector current will be the maximum value set by the circuit:

$$I_{C_{\text{sat}}} \cong \frac{V_{CC}}{R_C} \tag{4.8}$$

and $\qquad\qquad V_{CE_{\text{sat}}} \cong 0 \text{ V} \qquad$ (actually a few tenths volt) $\tag{4.9}$

The base current calculated from Eq. (4.4) is correct in any case.

If the circuit to be analyzed is used as an amplifier, we shall not expect it to be biased in the saturation region. If, however, some value used is incorrect, or some wiring error occurs, the resulting operation might possibly bias the transistor into saturation and we must be aware of this condition. (Keep in mind that the saturation condition is undesirable only for amplifier operation. For operation in computer switching circuits the saturation region of operation is important and we shall consider it fully under that topic in Chapter 13.)

EXAMPLE 4.2   Compute the dc bias voltages and currents for the *npn* CE circuit of Fig. 4.8.

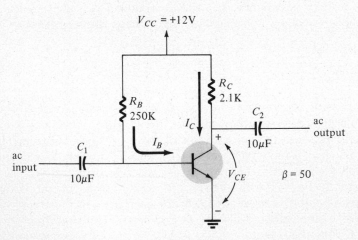

Figure 4.8. dc fixed-bias circuit for Example 4.2.

Solution:
(a) $I_B = (V_{CC} - V_{BE})/R_B \cong V_{CC}/R_B = 12 \text{ V}/250 \text{ K} = \textbf{48 μA}$
(b) $I_C = \beta I_B = 50(48 \text{ μA}) = \textbf{2.4 mA.}$
(c) $V_{CE} = V_{CC} - I_C R_C = 12 - (2.4 \text{ mA})(2.1 \text{ K}) = 12 - 5 = \textbf{7 V.}$

4.7 BIAS STABILIZATION

While the fixed-bias circuit provides suitable gain as an amplifier it has difficulty maintaining bias stability. In any amplifier circuit the collector current, I_C, will vary with change in temperature because of the three following main factors:

1. Reverse saturation current (leakage current), I_{CO}, which doubles for every 10° increase in temperature.
2. Base-emitter voltage, V_{BE}, which decreases by 2.5 mV per degree centigrade.
3. Transistor current gain, β, which increases with temperature.

Any or all of these factors can cause the bias point to shift from the values originally set by the circuit because of a change in temperature. Table 4.1 lists typical parameter values for silicon transistors.

TABLE 4.1 Typical Silicon Transistor Parameters

T (°C)	I_{CO} (nA)	β	V_{BE} (V)
−65	0.2×10^{-3}	20	0.85
25	0.1	50	0.65
100	20	80	0.48
175	3.3×10^{3}	120	0.3

We first demonstrate the effect of leakage current and current gain change on the dc bias point initially set by the circuit. Consider the graphs of Figs. 4.9a and 4.9b,

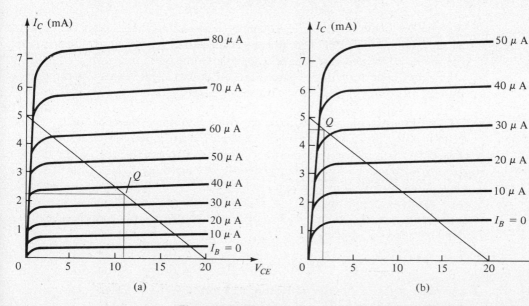

Figure 4.9. Shift in dc bias point (Q-point) due to change in temperature: (a) 25°C; (b) 100°C.

which show a transistor collector characteristic at room temperature (25°C) and the same transistor at some elevated temperature (100°C). Notice that the significant increase of leakage current not only causes the curves to rise but also that an increase in beta occurs as shown by the larger spacing between the curves at the higher temperature.

The operating point may be specified by drawing the circuit dc load line on the graph of the collector characteristic and noting the intersection of the load line and the dc base current set by the input circuit. An arbitrary point is marked as an example in Fig. 4.9a. Since the fixed-bias circuit provides a base current whose value depends approximately on the supply voltage and base resistor, neither of which is affected by temperature or the change in leakage current or beta, the same base current magnitude will exist at high temperatures as indicated on the graph of Fig. 4.9b. As the figure shows, this will result in the dc bias point's shifting to a higher collector current and a lower collector-emitter voltage operating point. In the extremes the transistor could be driven into saturation. In any case the new operating point may not be at all satisfactory and considerable distorion may result because of the bias point shift. A better bias circuit is needed, one that will stabilize or maintain the dc bias initially set, so that the amplifier can be used in a changing-temperature environment.

Stability Factor, S

A stability factor, S, can be defined for each of the parameters affecting bias stability. These are

$$S(I_{co}) = \frac{\Delta I_C}{\Delta I_{co}} \qquad S(V_{BE}) = \frac{\Delta I_C}{\Delta V_{BE}} \qquad S(\beta) = \frac{\Delta I_C}{\Delta \beta}$$

The stability factor is a numerical quantity representing the amount the collector current changes due to changes in each of the parameters because of temperature. The results of detailed mathematical analysis will be used to provide a comparison of how the device and circuit components affect bias stability to allow comparison of the effects on I_C by each of the transistor parameters.

$S(I_{co})$: Figure 4.10a shows a basic transistor circuit and the effect of I_{co}. Figure 4.10b uses the result of analyzing the stability based on change due only to I_{co} (β and V_{BE} considered held constant). Referring to Fig. 4.10b, we see that the stability factor varies from the ideal case (best condition) of $S = 1$ up to a maximum value of $S = \beta + 1$, which occurs for the fixed-bias circuit, or when the ratio R_B/R_E is greater than $\beta + 1$. Essentially, the stability factor is smallest for larger values of R_E so that inclusion of an emitter resistor improves the bias stability (makes S smaller).

$$S(I_{co}) = \frac{(\beta + 1)(1 + R_B/R_E)}{(\beta + 1) + (R_B/R_E)} \qquad (4.10)$$

The value of I_{co} in modern transistors is so low (refer to Table 4.1), however, that the change in bias point in a circuit with even a large value of $S(I_{co})$ would not be considerable, as the following example illustrates.

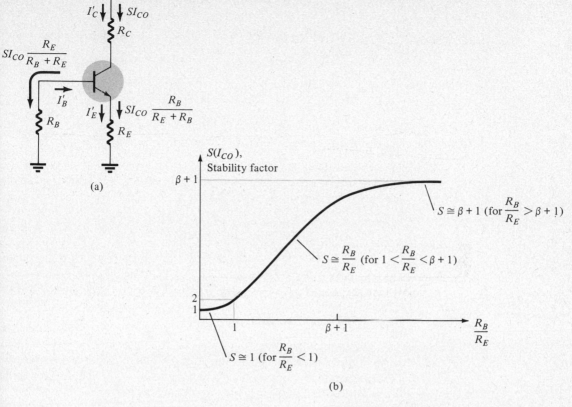

(a)

(b)

Figure 4.10. Effect of I_{CO} on bias point.

EXAMPLE 4.3 In a circuit using a transistor typified by the parameters in Table 4.1 calculate the change in I_C from 25°C to 100°C for (a) fixed bias, (b) $R_B/R_E = 10$, and (c) $R_B/R_E = 1$.

Solution: From 25°C to 100°C the change in I_{CO} is

$$\Delta I_{CO} = (20 - 0.1)\, \text{nA} \cong 20\, \text{nA}$$

(a) For fixed bias: $S = \beta + 1 = 51$.
Using the definition of stability, we get

$$\Delta I_C = S(\Delta I_{CO}) = 51(20\, \text{nA}) = 1\, \mu\text{A}$$

(b) For $S = R_B/R_E = 10$:

$$\Delta I_C = 10 \cdot \Delta I_{CO} = 10(20\, \text{nA}) = 0.2\, \mu\text{A}$$

(c) For $S = 1$:

$$\Delta I_C = 1(20\, \text{nA}) = 20\, \text{nA}$$

While the change in I_C is considerably different in a circuit having ideal stability ($S = 1$) and one having the maximum stability factor ($S = 51$, in this example), the change in I_C is not significant. For example, the amount of change in I_C from a dc bias current set at, say, 2 mA, would be from 2 mA to to 2.001 mA (only 0.05%) in the worst case, which is obviously small enough to be ignored. Some power transistors exhibit larger leakage current, but for most amplifier circuits the effect of I_{CO} change with temperature is slight.

$S(V_{BE})$: Analysis of the stability factor due to change in V_{BE} will result in

$$S(V_{BE}) = \frac{\Delta I_C}{\Delta V_{BE}} = \frac{-\beta}{R_B + R_E(\beta + 1)}$$

$$= \frac{-1}{R_E}, \qquad \text{for } (\beta + 1) \gg \frac{R_B}{R_E} \text{ and } \beta \gg 1 \tag{4.11}$$

Since smaller values of S indicate better stability, the larger the value of R_E the better the circuit stability due to changes in V_{BE} with temperature.

EXAMPLE 4.4 Determine the change in I_C for a transistor having parameters as listed in Table 4.1 over a temperature range from 25°C to 100°C for a circuit having $R_E = 1$ K (and $\beta + 1 \gg R_B/R_E$).

Solution: Using $S(V_{BE}) = -1/R_E = -1/1$ K $= -10^{-3}$ and $V_{BE} = (0.65 - 0.48) = 0.17$ V, from 25°C to 100°C

$$I_C = S \cdot \Delta V_{BE} = -10^{-3}(0.17 \text{ V}) = -170 \text{ } \mu\text{A}$$

which is seen to be a reasonably large current change compared to that resulting from a change in I_{CO}. Using a typical collector current of 2 mA, we see that the collector current would change from 2 mA at 25°C to 1.830 mA at 100°C (a −8.5% change).

The effect of V_{BE} changing with temperature can be somewhat compensated by the use of diode compensation in which the voltage change across a diode compensates for the change in V_{BE} to maintain the bias valus of I_C. Thermistor and transistor compensation techniques are also used when needed.

$S(\beta)$: Analysis of the effect of β changing with temperature on the circuit bias stability results in

$$\frac{\Delta I_C}{I_C(T_1)} = \left(1 + \frac{R_B}{R_E}\right)\frac{\Delta \beta}{\beta(T_1)\beta(T_2)} = \left(1 + \frac{R_B}{R_E}\right)\frac{\frac{\beta(T_2)}{\beta(T_1)} - 1}{\beta(T_2)} \tag{4.12}$$

where $\beta(T_1)$ — beta at temperature T_1;

$\beta(T_2) =$ beta at temperature T_2;

$I_C(T_1) =$ collector current at temperature T_1.

EXAMPLE 4.5 Calculate the change in collector current for the transistor having parameters as given in Table 4.1 from room temperature to 100°C. Assume that $R_B/R_E = 20$ for the circuit used and that I_C at room temperature is 2 mA.

Solution: The change in I_C is calculated to be

$$\Delta I_C = I_C(T_1)\left[\left(1 + \frac{R_B}{R_E}\right) \cdot \frac{\frac{\beta(T_2)}{\beta(T_1)} - 1}{\beta(T_2)}\right]$$

$$= 2 \text{ mA}\left[(1 + 20)\frac{\left(\frac{80}{50} - 1\right)}{80}\right] = 0.315 \text{ mA} = \textbf{315 } \boldsymbol{\mu}\textbf{A}$$

The collector current changing from 2 mA at room temperature to 2.135 mA at 100°C represents a change of about 16%.

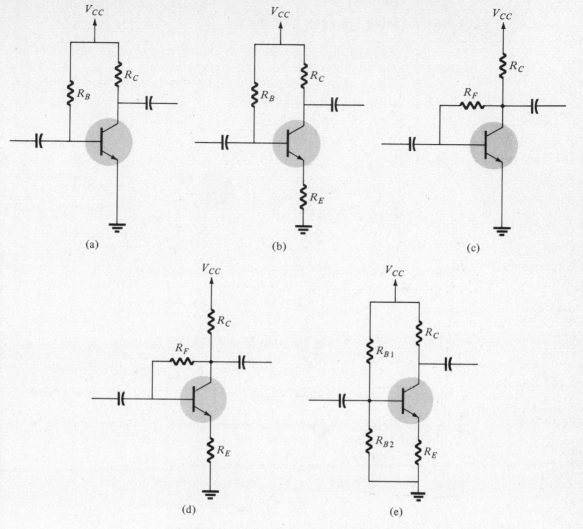

Figure 4.11. Circuits providing voltage and current feedback: (a) fixed bias; (b) current feedback; (c) voltage feedback; (d) current and voltage feedback; (e) voltage-divider biasing.

Comparison of the three examples shows that of the three parameters affecting bias stability the change due to β variation is probably greatest. These changes in parameter value need not be only due to temperature. Although the value of I_{co} at room temperature varies negligibly between transistors, even of the same manufacture type, as does V_{BE}, the value of β varies considerably. For example, the same numbered transistor may have $\beta = 125$ for one device and $\beta = 300$ for another. In addition, the value of β for specific transistors will be different at different values of bias current. For all these reasons the design of a good bias stabilized circuit usually concentrates most on stabilizing the effect of changes in transistor beta.

Figure 4.11 shows a few bias stabilized circuits, most of which are covered in detail in the next few sections.

4.8 DC BIAS CIRCUIT WITH EMITTER RESISTOR

The dc bias circuit of Fig. 4.12 contains an emitter resistor to provide better bias stability than the fixed-bias circuit considered in Section 4.6. For the analysis of the circuit operation we shall deal separately with the base-emitter loop of the circuit and the collector-emitter loop of Fig. 4.12.

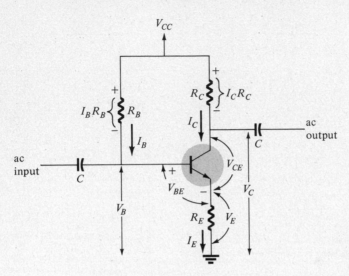

Figure 4.12. dc bias circuit with emitter stabilization resistor.

Input Section (Base-Emitter Loop)

A partial circuit diagram of the base-emitter loop is shown in Fig. 4.13a. Writing Kirchhoff's voltage equation for the loop, we get

$$V_{CC} - I_B R_B - V_{BE} - I_E R_E = 0$$

We can replace I_E with $(\beta + 1)I_B$ so that the above equation can be written as

$$V_{CC} - I_B R_B - V_{BE} - (\beta + 1)I_B R_E = 0$$

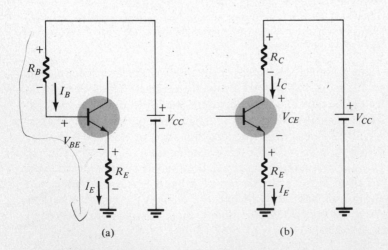

(a)

(b)

Figure 4.13. Input and output loops of the circuit of Fig. 4.12: (a) input loop; (b) output loop.

Solving for the base current, we get

$$\text{ok} \rightarrow ?\qquad \boxed{I_B = \frac{V_{CC} - V_{BE}}{R_B + (\beta + 1)R_E} \cong \frac{V_{CC}}{R_B + \beta R_E}} \qquad \text{(4.13)}$$

Note that the difference between the fixed-bias current calculation [Eq. (4.4)] and Eq. (4.13) is the additional term of $(\beta + 1)R_E$ in the denominator.

Output Section (Collector-Emitter Loop)

The collector-emitter loop is shown in Fig. 4.13b. Writing the Kirchhoff voltage equation for this loop, we get

$$V_{CC} - I_C R_C - V_{CE} - I_E R_E = 0$$

Using the relation

$$I_C \cong I_E$$

we can solve for the voltage across the collector-emitter:

$$\boxed{V_{CE} \cong V_{CC} - I_C(R_C + R_E)} \qquad \text{(4.14)}$$

The voltage measured from emitter to ground is

$$V_E = I_E R_E \cong I_C R_E$$

and the voltage measured from collector to ground is

$$V_C = V_{CC} - I_C R_C$$

The voltage at which the transistor is biased is measured from collector to emitter, V_{CE}, which is given by Eq. (4.14) and may also be calculated as

$$V_{CE} = V_C - V_E \qquad = V_{cc} - I_c R_c - V_e$$

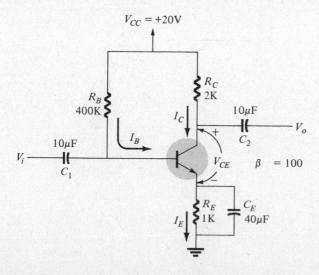

$V_{CC} = +20V$

R_B 400K

R_C 2K

$10\mu F$ C_1 V_i

I_B

I_C

$10\mu F$ C_2 V_o

V_{CE} $\beta = 100$

R_E 1K C_E 40μF

I_E

Figure 4.14. Emitter-stabilized bias circuit for Example 4.6.

EXAMPLE 4.6   Calculate all dc bias voltages and currents in the circuit of Fig.
4.14.

Solution:
(a) $I_B \cong V_{CC}/(R_B + \beta R_E) = 20\text{ V}/[400\text{ K} + 100(1\text{ K})] = 20\text{ V}/500\text{ K}$
    $= \mathbf{40\ \mu A}$.
(b) $I_C = \beta I_B = 100(40\ \mu A) = \mathbf{4\ mA} \cong I_E$.
(c) $V_{CE} = V_{CC} - I_C R_C - I_E R_E = 20 - (4\text{ mA})2\text{ K} - (4\text{ mA})1\text{ K}$
    $= 20 - 8 - 4 = \mathbf{8\ V}$.

4.9 DC BIAS CIRCUIT INDEPENDENT OF BETA

In the previous dc bias circuits the values of the bias current and voltage of the
collector depended on the current gain (β) of the transistor. But the value of beta is
temperature-sensitive, especially for silicon transistors and since, also, the nominal
value of beta is not well defined, it would be desirable for these as well as other
reasons (transistor replacement and stability) to provide a dc bias circuit that is
independent of the transistor beta. The circuit of Fig. 4.15 meeets these conditions
and is thus a very popular bias circuit.

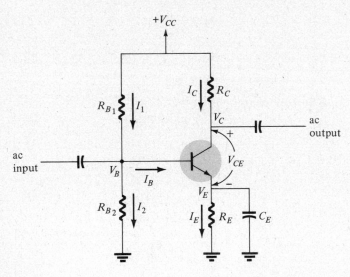

Figure 4.15. Beta-independent dc bias circuit.

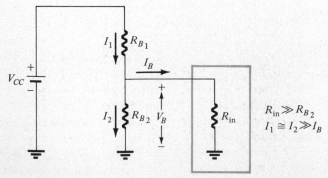

Figure 4.16. Partial bias circuit for calculating base voltage V_B.

Let us first analyze the base-emitter input circuit. A basic assumption (which will later be proved) is that the resistance seen looking into the base (see Fig. 4.16) is much larger than that of resistor R_{B2}. If this is so, then the current through R_{B1} flows almost completely into R_{B2} and the two resistors may be considered effectively in series. The voltage at the junction of the resistors, which is also the voltage of the base of the transistor, is then determined simply by the voltage divider network of R_{B1} and R_{B2} and the supply voltage. Calculating the voltage at the transistor base due to the voltage divider network of resistors R_{B1} and R_{B2}; we get

$$V_B = \frac{R_{B2}}{R_{B1} + R_{B2}} V_{CC} \qquad (4.15)$$

where V_B is the voltage measured from base to ground.

We can calculate the voltage at the emitter from

$$V_E = V_B - V_{BE} \qquad (4.16)$$

The current in the emitter may then be calculated from

$$I_E = \frac{V_E}{R_E} \qquad (4.17a)$$

and the collector current is then

$$I_C \cong I_E \qquad (4.17b)$$

The voltage drop across the collector resistor is

$$V_{R_C} = I_C R_C$$

The voltage at the collector (measured with respect to ground) can then be obtained

$$V_C = V_{CC} - V_{R_C} = V_{CC} - I_C R_C \qquad (4.18)$$

and, finally, the voltage from collector to emitter is calculated from

$$V_{CE} = V_C - V_E$$

$$V_{CE} = V_{CC} - I_C R_C - I_E R_E \qquad (4.19)$$

Look back at the procedure just outlined and notice that the value of beta was never used in Eqs. 4.15–4.19. The base voltage is set by resistors R_{B1} and R_{B2} and the supply voltage. The emitter voltage is fixed at approximately the same voltage value as the base. Resistor R_E then determines emitter and collector currents. Finally, R_C determines the collector voltage and, thereby, the collector-emitter bias voltage.

The base voltage V_B is best adjusted using resistor R_{B2}, the collector current by resistor R_E, and the collector-emitter voltage by resistor R_C. Varying other components will have less effect on the dc bias adjustments. The capacitor components

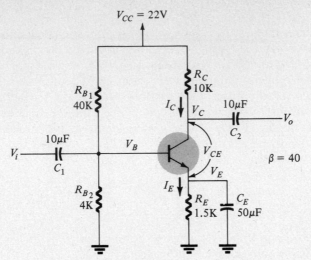

Figure 4.17. Beta-stabilized bias circuit for Example 4.7.

are part of the ac amplifier operation but have no effect on the dc bias and will not be discussed at this time.

EXAMPLE 4.7 Calculate the dc bias voltages and currents for the circuit of Fig. 4.17.

Solution:
(a) $V_B = [R_{B2}/(R_{B1} + R_{B2})](V_{CC}) = [4/(40 + 4)](22) = \mathbf{2\ V}$.
(b) $V_E = V_B - V_{BE} = 2 - 0.7 = \mathbf{1.3\ V}$.
(c) $I_E = V_E/R_E \cong I_C = 1.3\ \text{V}/1.5\ \text{K} = \mathbf{0.87\ mA}$.
(d) $V_C = V_{CC} - I_C R_C = 22 - (0.87\ \text{mA})(10\ \text{K}) = \mathbf{13.3\ V}$.
(e) $V_{CE} = V_C - V_E = 13.3 - 1.3 = \mathbf{12\ V}$.

4.10 DC BIAS CALCULATIONS FOR VOLTAGE FEEDBACK CIRCUITS

Apart from the use of an emitter resistor to provide improved bias stability, voltage feedback also provides improved dc bias stability. Figure 4.18 shows a dc bias

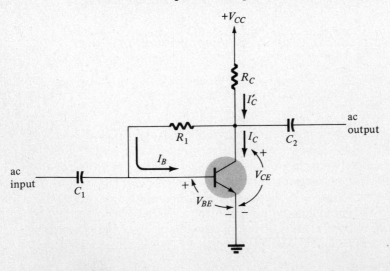

Figure 4.18. dc bias circuit with voltage feedback.

circuit with voltage feedback. This section shows how to calculate the dc currents and voltages of this circuit.

Input Section

Figure 4.19a shows the input section (base-emitter) loop of the voltage feedback circuit. Writing the Kirchhoff voltage equation around the loop gives

$$+V_{CC} - I'_C R_C - I_B R_1 - V_{BE} = 0$$

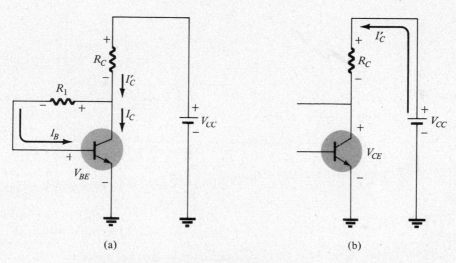

Figure 4.19. Input and output sections of a voltage feedback dc bias circuit: (a) input section; (b) output section.

The current I'_C is the sum of I_C and I_B, but I_B is so much smaller than I_C that we can write the approximation:

$$I'_C \cong I_C \cong \beta I_B$$

which, substituted into the Kirchhoff voltage equation, yields

$$V_{CC} - \beta I_B R_C - I_B R_1 - V_{BE} = 0$$

Solving for the base current I_B, we get,

$$I_B = \frac{V_{CC} - V_{BE}}{R_1 + \beta R_C} \simeq \frac{V_{CC}}{R_1 + \beta R_C} \qquad (4.20)$$

Output Section

From the partial circuit diagram of the output section shown in Fig. 3.19b the Kirchhoff voltage equation is

$$+V_{CC} - I'_C R_C - V_{CE} = 0$$

and using $I'_C = I_C$, we can solve for V_{CE}

$$V_{CE} \cong V_{CC} - I_C R_C \qquad (4.21)$$

EXAMPLE 4.8 Calculate the dc bias currents and voltages for the practical dc bias circuit of Fig. 4.20 using voltage feedback.

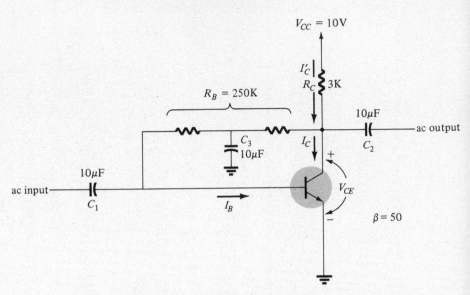

Figure 4.20. Voltage feedback stabilized bias circuit for Example 4.8.

Solution: The feedback resistor R_B is the sum of the resistors between collector and base (the capacitor in the feedback path provides for attenuating or blocking the ac feedback signal and has no effect on the dc bias calculation).

(a) $(V_{CC} - V_{BE})/(R_B + \beta R_C) \cong V_{CC}/(R_B + \beta R_C)$
$$= 10 \text{ V}/[250 \text{ K} + 50(3 \text{ K})] = 10 \text{ V}/400 \text{ K}$$
$$= \textbf{25 } \boldsymbol{\mu}\textbf{A.}$$

(b) $I'_C \cong I_C = \beta I_B = 50(25 \ \mu\text{A}) = \textbf{1.25 mA.}$

(c) $V_{CE} \cong V_{CC} - I_C R_C = 10 - (1.25 \text{ mA})(3 \text{ K}) = 10 - 3.75 = \textbf{6.25 V.}$

The dc bias circuit may include *both* emitter resistor stabilization and voltage feedback stabilization as shown in the practical circuit of Fig. 4.21. An example will help demonstrate calculation of dc bias values.

EXAMPLE 4.9 Calculate the dc bias currents and voltages for the bias circuit of Fig. 4.21.

Solution:

(a) $I_B = \dfrac{V_{CC} - V_{BE}}{R_B + \beta R_C + (\beta + 1)R_E} = \dfrac{18 - 0.7 \text{ V}}{300 \text{ K} + 75(2.5 \text{ K}) + 76(0.5 \text{ K})}$
$$\cong \textbf{33 } \boldsymbol{\mu}\textbf{A.}$$

(b) $I'_C \cong I_C = \beta I_B = 75(33 \ \mu\text{A}) \cong \textbf{2.5 mA.}$

(c) $V_{CE} \cong V_{CC} - I_C(R_C + R_E) = 18 - (2.5 \text{ mA})(3 \text{ K}) = \textbf{10.5 V.}$

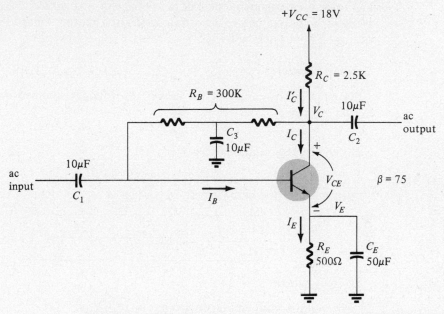

Figure 4.21. dc bias circuit with emitter resistor and voltage feedback stabilization.

4.11 COMMON-COLLECTOR (EMITTER-FOLLOWER) DC BIAS CIRCUIT

A third connection for the transistor provides input to the base circuit and output from the emitter circuit, with the collector common (CC) to the ac input and output signals. A simple CC circuit (usually referred to as emitter-follower) is shown in Fig. 4.22. The collector voltage is fixed at the positive supply voltage value. For

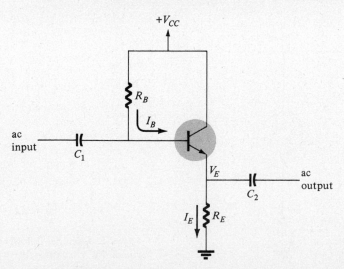

Figure 4.22. Emitter-follower dc bias circuit.

V_{CE} to be approximately one-half the voltage of V_{CC}, allowing the widest voltage swing in the output before distortion occurs, the emitter voltage should be set at a voltage of about one-half V_{CC}.

Input Section

For the input section of the circuit the voltages summed around the base-emitter loop give

$$+V_{CC} - I_B R_B - V_{BE} - I_E R_E = 0$$

Using the current relation, we get

$$I_E = (\beta + 1)I_B \cong \beta I_B$$

we can solve for the base current, obtaining

$$I_B = \frac{V_{CC} - V_{BE}}{R_B + (\beta + 1)R_E} \cong \frac{V_{CC}}{R_B + \beta R_E} \qquad (4.22)$$

Output Section

The voltage from emitter to ground is

$$V_E = I_E R_E \qquad (4.23)$$

and the collector-emitter voltage is

$$V_{CE} = V_{CC} - V_E = V_{CC} - I_E R_E \qquad (4.24)$$

EXAMPLE 4.10   Calculate all dc bias currents and voltages for the circuit of Fig. 4.23.

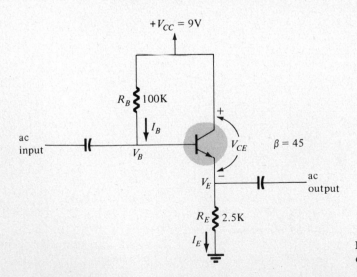

Figure 4.23. Emitter follower bias circuit for Example 4.10.

Solution:

(a) $I_B \cong V_{CC}/(R_B + \beta R_E) = 9 \text{ V}/(100 + 45 \times 2.5) \cong \mathbf{42 \ \mu A}$.

(b) $I_E = (\beta + 1)I_B = 46(42 \ \mu A) \cong \mathbf{1.9 \ mA}$.

(c) $V_{CE} = V_{CC} - I_E R_E = 9 - (1.9 \text{ mA})(2.5 \text{ K}) = \mathbf{4.25 \ V}$.

(d) $V_E = I_E R_E = (1.9 \text{ mA})(2.5 \text{ K}) = \mathbf{4.75 \ V}$.

A second emitter-follower dc bias circuit is shown in Fig. 4.24. Like the similar CE dc bias circuit of Fig. 4.15, this circuit provides a bias operating condition that depends not on the current gain (β) of the transistor, but only on the resistor components and the supply voltage.

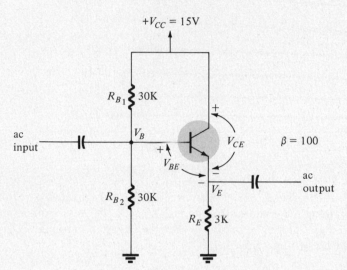

Figure 4.24. Beta-stabilized emitter follower bias circuit for Example 4.11.

EXAMPLE 4.11 Calculate all dc bias currents and voltages for the CC circuit of Fig. 4.24.

Solution:

(a) $V_B = (R_{B2}/[R_{B1} + R_{B2}])(V_{CC}) = (30/[30 + 30])(15) = \mathbf{7.5 \ V}$.

(b) $V_E = V_B - V_{BE} = 7.5 - 0.7 = \mathbf{6.8 \ V}$.

(c) $I_E = (V_E/R_E) = 6.8 \text{ V}/3 \text{ K} \cong \mathbf{2.3 \ mA}$.

(d) $V_{CE} = V_{CC} - I_E R_E = 15 - (2.3 \text{ mA})(3 \text{ K}) = \mathbf{8.1 \ V}$.

4.12 GRAPHICAL DC BIAS ANALYSIS

The previous analysis of the dc bias currents and voltages was carried out mathematically for a number of transistor circuits. The only factors of interest used were the current gain (β) and base-emitter voltage (V_{BE}) when forward biased. This section shows a graphical technique for finding the operating point of a biased transistor circuit. The graphical method demonstrated provides additional insight into the choice of operating point and leads into Section 4.14 on the design (or synthesis) of a dc bias circuit.

The typical CE collector characteristic, shown in Fig. 4.25, only defines the

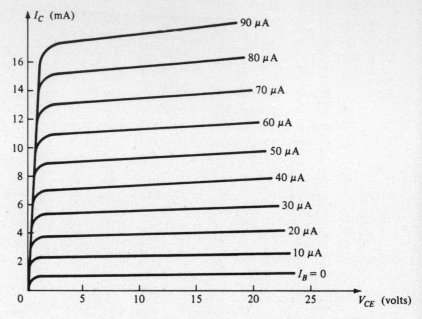

Figure 4.25. Transistor collector characteristics.

overall operation of the transistor device. The circuit constraints must also be taken into account in obtaining the actual operating point (called the *quiescent operating point* or *Q*-point). Eq. (4.25) for the output loop of the circuit is a straight line in a voltage-current plot (as in Fig. 4.26).

$$I_C = -\frac{1}{R_C}V_{CE} + \frac{V_{CC}}{R_C}$$

$$y = \quad m \ x \ + \ b$$

(4.25)

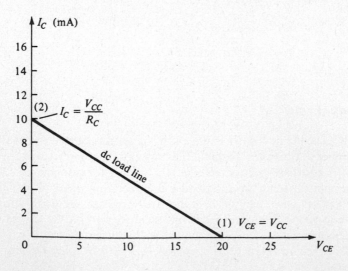

Figure 4.26. Direct-current load line.

The straight line representing Eq. (4.25) can be drawn on the graph of Fig. 4.25 by obtaining the two extreme points of the straight line as follows:

1. For $I_C = 0$, $V_{CE} = V_{CC}$ in Eq. (4.25).
2. For $V_{CE} = 0$, $I_C = V_{CC}/R_C$ in Eq. (4.25).

These points are marked in Fig. 4.26 as (1) and (2), respectively, and the straight line connecting them is called the *dc load line*. Although the same voltage-current axis as that of the transistor collector characteristic is used, no characteristic is shown to reinforce the fact that the dc load line has nothing to do with the device itself. The load line drawn depends only on the supply voltages, V_{CC}, and the value of R_C, the collector resistor.

The slope of the load line depends only on the value of R_C. Figure 4.27a shows the load line slopes for R_C values smaller and larger than that of Fig. 4.26. Figure 4.27b shows that changing only the supply voltage will move the load line parallel to that of Fig. 4.26, and the slope remains the same since R_C has not changed.

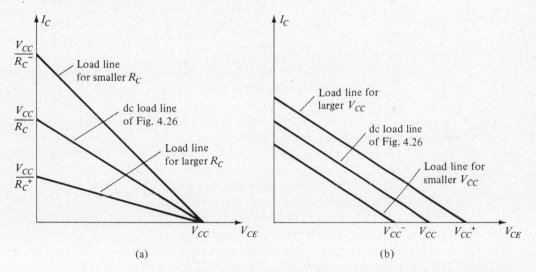

Figure 4.27. Effect of varying R_C or V_{CC} on dc load line: (a) effect of resistor on dc load line; (b) effect of supply voltage on dc load line.

Since circuit operation depends on both the transistor characteristic and the circuit elements, plotting *both* curves (transistor characteristic and dc load line) on *one* graph allows determination of the circuit Q-point. This is shown in Fig. 4.28. The typical dc bias point shown in Fig. 4.28 is somewhat in the center of the voltage range (0 to V_{CC}) and the center of the current range (0 to V_{CC}/R_C). A large-signal amplifier with output voltage swing near the voltage range set by the voltage supply value would require a centered operating point.

For circuits other than amplifiers different bias points may be desired. The dc load line describes all the possible values of voltage and current in the output section of the circuit. Figure 4.28 shows a typical bias point set by the amount of

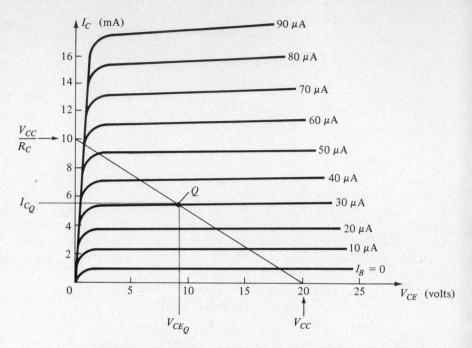

Figure 4.28. Using transistor collector characteristic and dc load line to obtain quiescent operating point (Q-point).

base current and the dc load line. Adjusting the base current to higher values moves the operating point toward saturation along the load line, whereas reducing the base current moves the bias point toward transistor cutoff. For the characteristic and load line of Fig. 4.28 base currents in excess of 60 μA will drive the transistor into saturation. Note that for the load line and operating point indicated, the ac input, which adds to the dc base bias current, can go positive only by about 25 μA (from 30 to 55 μA) before the limiting condition (saturation) occurs. The variation of the ac base current, on the other hand, can go negative by 30 μA (30 to 0 μA) before cutoff is reached so that the particular bias point in Fig. 4.28 is not centered. For small-signal amplifiers with output voltage swings of less than 1 V, the exact centering of the Q-point is not essential—usually a region of largest transistor gain or most linear operation is sought.

To see how the centering of the bias point is important for large-signal amplifiers, Fig. 4.29 shows the resulting distortion for a few different bias points and signal amplitude, which result in the output's distorting because the transistor is driven into saturation, into cutoff, or both (too large an input signal). When distortion occurs only in saturation or only in cutoff as shown in Fig. 4.29a and b, the bias point could be adjusted correspondingly to correct the distortion. With too large an input signal as shown in Fig. 4.29c the resulting distortion cannot be helped by operating-point adjustment.

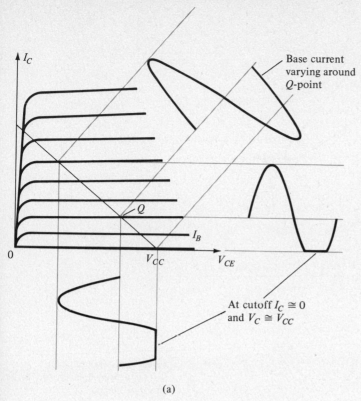

Base current
varying around
Q-point

At cutoff $I_C \cong 0$
and $V_C \cong V_{CC}$

(a)

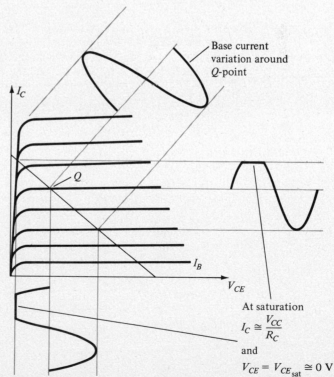

Base current
variation around
Q-point

At saturation

$$I_C \cong \frac{V_{CC}}{R_C}$$

and

$$V_{CE} = V_{CE_{sat}} \cong 0 \text{ V}$$

Figure 4.29. Effect of bias point on output signal distortion: (a) distortion at cutoff condition; (b) distortion at saturation condition.

(b)

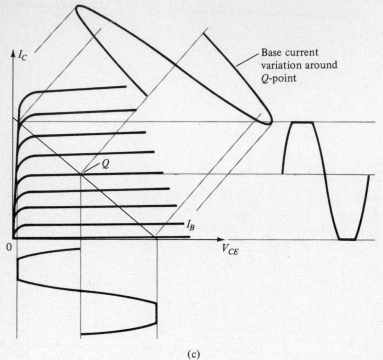

Figure 4.29. (continued): (c) distortion due to too large an input signal.

(c)

EXAMPLE 4.12 Given the circuit of Fig. 4.30(a) and the transistor collector characteristic of Fig. 4.30b:
(a) Plot the dc load line and obtain the Q-point.
(b) Find V_{CE}, I_C, $I_C R_C$, and I_E from the graph.

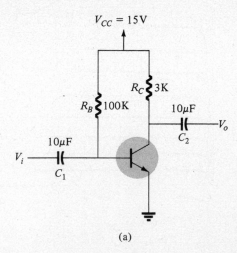

$V_{CC} = 15\text{V}$

$R_B \lessgtr 100\text{K}$

$R_C \lessgtr 3\text{K}$

$10\mu\text{F}$
C_2
$-V_o$

$10\mu\text{F}$
V_i
C_1

(a)

Figure 4.30. Graphical analysis of fixed-bias circuit for Example 4.12: (a) transistor circuit.

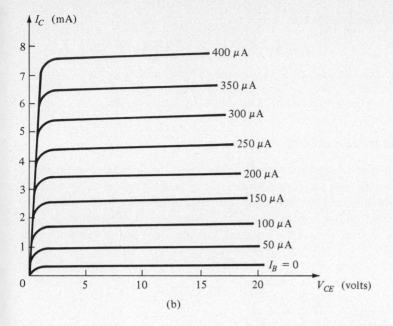

(b)

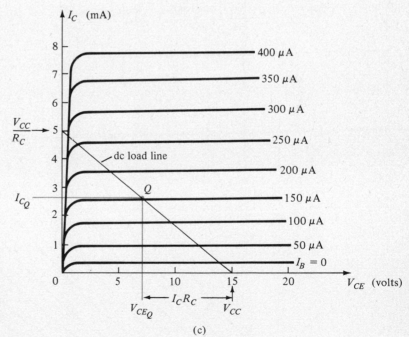

Figure 4.30. (continued):
(b) transistor collector characteristics; (c) dc load line and Q-point.

(c)

Solution:

(a) 1. Draw the dc load line. The two points for the load line are
a. at $I_C = 0$, $V_{CE} = V_{CC} = 15$ V.
b. at $V_{CE} = 0$, $I_C = V_{CC}/R_C = 15$ V/3 K = 5 mA.
Connect a straight line between these points for the dc load line.

2. Calculate the base current using

$$I_B \cong \frac{V_{CC}}{R_B} = \frac{15 \text{ V}}{100 \text{ K}} = 150 \text{ } \mu\text{A}$$

3. The intersection of the load line and the transistor curve for $I_B = 150$ μA defines the quiescent operating point (see Fig. 4.30c).

(b) From curve $V_{CE} = 7$ V, $I_C R_C = 8$ V, $I_C = 2.6$ mA, and $I_E \cong 2.6$ mA.

4.13 DC BIAS OF VACUUM-TUBE CIRCUITS

Mathematical methods using equations for calculating the dc bias point (as in the case of a transistor circuit) are not applied to the bias of a vacuum-tube triode. A graphical procedure is necessary because the relation of input grid voltage and output plate current and voltage is best described by a relatively nonlinear plate characteristic. As will be shown the graphical calculations for the vacuum triode are straightforward.

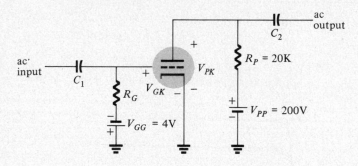

Figure 4.31. Triode amplifier circuit.

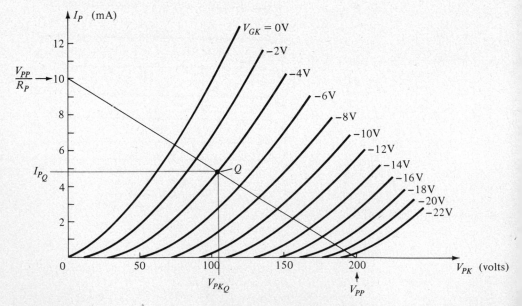

Figure 4.32. Triode plate characteristics showing dc load line.

Figure 4.31 shows a fixed-bias circuit using a vacuum triode as a voltage amplifier. A triode plate characteristic is shown in Fig. 4.32. Writing the voltage equation around the output plate-cathode loop results in

$$V_{PK} = V_{PP} - I_P R_P \qquad (4.26)$$

Equation 4.26 can be plotted on the plate characteristic as the dc load line by interconnecting the following two points by a straight line:

1. For $I_P = 0$, $V_{PK} = V_{PP}$.
2. For $V_{PK} = 0$, $I_P = V_{PP}/R_P$.

Figure 4.32 shows the load line for the circuit of Fig. 4.31 drawn between points 1 and 2 as determined for the values given in the circuit. The grid voltage is set by the grid battery

$$V_{GK} = V_{GG} \qquad (4.27)$$

which is -4 V in Fig. 4.31. For the particular grid-cathode voltage and the dc load line set by the plate resistor and plate supply voltage, the quiescent operating point marked on Fig. 4.32 is obtained. The vacuum triode is thus biased into the amplifier region of operation and will provide an output signal that is an amplified version of the input signal applied through the input coupling capacitor to the grid.

Resistor R_G has no effect on the dc bias since there is no dc grid current for negative grid-cathode voltages. The grid resistor has effect on the ac operation of the circuit as will be shown in Chapter 5.

EXAMPLE 4.13 Calculate the quiescent operating point of the vacuum triode in the circuit of Fig. 4.33a using the plate characteristic of Fig. 4.33b.

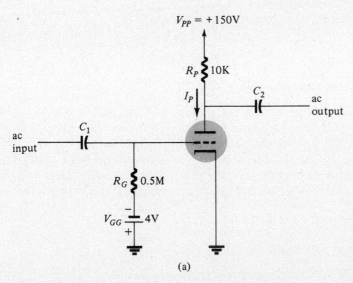

(a)

Figure 4.33. Circuit and characteristics for Example 4.13: (a) triode amplifier circuit.

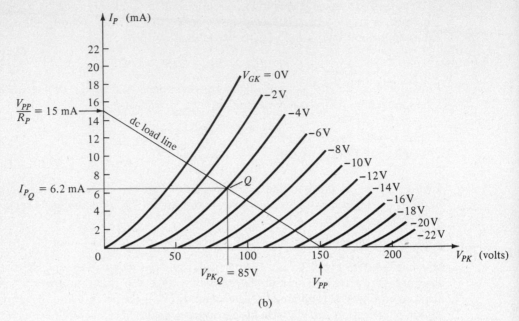

Figure 4.33. (continued): (b) triode characteristics with dc load line and Q-point.

Solution:
(a) Obtain the dc load line from the following points:
    1. At $V_{PK} = V_{PP} = 150$ V along the x-axis.
    2. At $I_P = V_{PP}/R_P = 150$ V/10 K = 15 mA along the y-axis. For the load line drawn in Fig. 4.33b:
(b) At the intersection of $V_{GK} = V_{GG} = -4$ V and the dc load line, obtain quiescent operating point Q.
From the graph the quiescent operating point is

$$V_{PK_Q} = 85 \text{ V}$$

$$I_{P_Q} = 6.2 \text{ mA}$$

Pentode Bias Calculations

A pentode amplifier circuit is shown in Fig. 4.34. As part of the usual bias connection the suppressor grid is connected to ground and the screen grid to a fixed positive dc voltage, V_{SS}, somewhat lower than the plate supply, V_{PP}. The plate supply voltage, V_{PP}, and the plate resistor, R_P, provide much the same operation as in the triode circuit. A dc load line for the circuit operation is obtained using the same method as for the triode circuit. The two points that are used to draw the load line are

1. The point $V_{PK} = V_{PP}$ along the x-axis.
2. The point $I_P = V_{PP}/R_P$ along the y-axis.

Figure 4.35 shows a pentode plate characteristic and a dc load line drawn on the

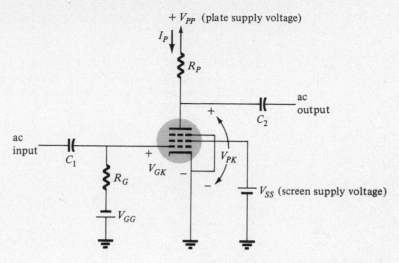

Figure 4.34. Pentode amplifier circuit.

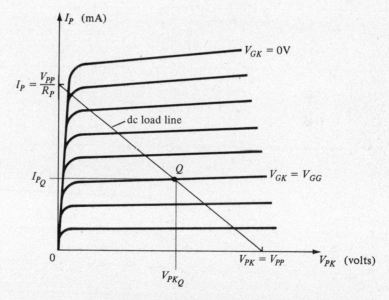

Figure 4.35. Pentode plate characteristics and dc load line.

same graph. The intersection of the load line and the grid-cathode voltage curve defines the operating Q-point of the pentode. The Q-point is thus dependent on the plate supply voltage, the plate resistor, the grid voltage, and, of course, the pentode characteristic. A change in the screen voltage V_{SS} could change the operating point by modifying the pentode characteristic curves. However, the screen voltage is usually kept at a constant dc potential and under this restriction it has no effect on the bias point.

EXAMPLE 4.14 Calculate the quiescent operating point of the pentode in the circuit of Fig. 4.36a using the pentode characteristic of Fig. 4.36b.

Solution:

(a) 1. $V_{PK} = V_{PP} = 300$ V along the x-axis.
   2. $I_P = V_{PP}/R_P = 300$ V/10 K $= 30$ mA along the y-axis.

(b) The intersection of the dc load line and $V_{GK} = V_{GG} = -2$ V gives an operating point of

$$V_{PK_Q} = 150 \text{ V}, \qquad I_{P_Q} = 15 \text{ mA}$$

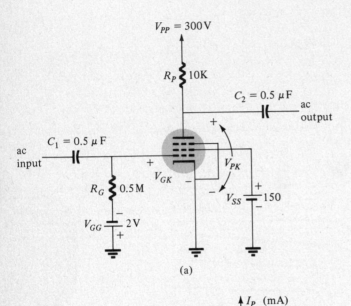

(a)

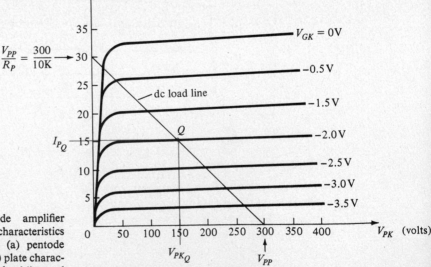

Figure 4.36. Pentode amplifier circuit and plate characteristics for Example 4.14: (a) pentode amplifier circuit; (b) plate characteristics showing dc load line and operating point.

(b)

4.14 DESIGN OF DC BIAS CIRCUITS

Up to now the discussion has been directed to the techniques of analyzing a given transistor or tube circuit to determine the dc operating point. Although it is often necessary to determine the Q-point of a given circuit, it is also important to be able to design a circuit to operate at a desired or specified bias point. Often the manufacturer's specification (spec) sheets provide information stating a suitable operating point (or operating region) for the particular transistor. In addition, other circuit factors connected with the given amplifier stage may also dictate some conditions of current swing, voltage swing, value of common supply voltage, etc., which can be used in determining the Q-point in a design.

The techniques of synthesis (or design) readily follow from the previous discussions of circuit analysis. In almost all cases the calculation of the circuit elements proceeds in the reverse to those in the analysis consideration. Basically, the problem of concern in this section can be briefly stated as follows.

Given a desired point or region of operation for a particular transistor, design the bias circuit (resistor and supply voltage values) to obtain the specified operating point.

In actual practice many other factors may have to be considered and may go into the select on of the desired operating point. For the moment we shall concentrate, however, on determining the component values to obtain a specified operating point. Since the basic relations and operation of a number of bias circuits have already been considered, no new theory has to be developed.

Design of a Fixed-Bias Circuit

Consider designing a fixed-bias amplifier (Fig. 4.37) using a 2N2192 transistor. The manufacturer's spec sheet states that the transistor has a minimum current gain (β) of 75 at a collector current of 10 mA. The maximum collector supply voltage is 40 V, and an operating point at, say, $V_{CE_Q} = 10$ V would be proper. From the above information the circuit of Fig. 4.37 is to be biased for operation at $I_C = 10$ mA, $V_{CE} = 10$ V.

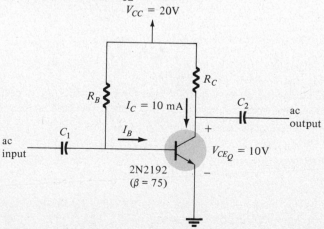

Figure 4.37. Fixed-bias circuit for design consideration.

Solving for the unknown value of resistance using Eq. (4.6), we get

$$R_C = \frac{V_{CC} - V_{CE_Q}}{I_{C_Q}} = \frac{20 - 10 \text{ V}}{10 \text{ mA}} = 1 \text{ K}$$

We need only calculate the value of the base resistor to complete the design. For this calculation we need the value of the base current, which is

$$I_B = \frac{I_C}{\beta} = \frac{10 \text{ mA}}{75} \cong 133 \text{ } \mu A$$

From the equations developed in analyzing the fixed-bias circuit we can solve for the value of base resistance by rewriting Eq. (4.14a) in the following form.

$$R_B = \frac{V_{CC} - V_{BE}}{I_B} = \frac{20 - 0.7 \text{ V}}{133 \text{ } \mu A} \cong 145 \text{ K}$$

This completes the design of the bias section of the fixed-bias amplifier for the given transistor and operating point. Looking back over the design procedure, we should now see that it is obvious that the same equations as in the analysis section were used, but in reverse order. In addition, the equations were rewritten to solve for a different unknown quantity—the collector resistance instead of the collector-emitter voltage or the base resistance instead of the base current.

> EXAMPLE 4.15 Design a fixed-bias circuit using a 2N3402 *npn* transistor (see Fig. 4.37). The manufacturer's spec sheet states that the transistor has a minimum current gain, β of 75 at a collector current of 2 mA. The maximum collector supply voltage is 25 V.
>
> Design Solution: Design the circuit for an operating point $I_{C_Q} = 2 \text{ mA}$ and $V_{CE_Q} = 10 \text{ V}$.
> Calculating the collector resistance, we get
>
> $$R_C = \frac{V_{CC} - V_{CE_Q}}{I_C} = \frac{20 - 10 \text{ V}}{2 \text{ mA}} = 5 \text{ K}$$
>
> Using the value of transistor current gain, we see that the base current is calculated to be
>
> $$I_B = \frac{I_C}{\beta} = \frac{2 \text{ mA}}{75} = 26.7 \text{ } \mu A$$
>
> Calculating the base resistance, we get
>
> $$R_B = \frac{V_{CC} - V_{BE}}{I_B} = \frac{20 - 0.7 \text{ V}}{26.7 \text{ } \mu A} \cong 723 \text{ K} \quad \text{(use 750 K)}$$
>
> Considering only the dc bias restrictions or requirements for a fixed-bias circuit, we see that the resulting values of collector and base resistance are: $R_C = 500 \text{ } \Omega$ and $R_B = 750 \text{ K}$.

Design of Bias Circuit with Emitter Feedback Resistor

We now proceed to the design of the dc bias components of an amplifier circuit having emitter-resistor bias stabilization. (See Fig. 4.38.) The supply voltage and

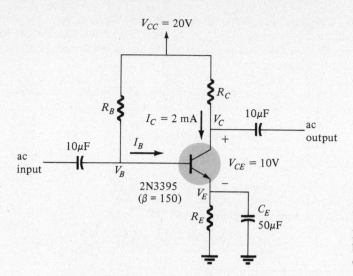

$V_{CC} = 20V$

R_B

R_C

$I_C = 2$ mA V_C

$10\mu F$

I_B

ac
output

$10\mu F$

ac
input

V_B

$V_{CE} = 10V$

2N3395
($\beta = 150$) V_E

R_E

C_E
$50\mu F$

Figure 4.38. Emitter-stabilized bias circuit for design consideration.

operating point will be selected from the manufacturer's information on the transistor used in the amplifier.

The selection of collector and emitter resistors cannot proceed directly from the information just specified. In Eq. (4.14), which relates the voltages around the collector-emitter loop, there are two unknown quantities present—the values of the collector and emitter resistors, R_C and R_E, respectively. To make the solution to the problem easy (and meaningful), some engineering judgment may be used; that is, if some reasonable approximation of, say, the emitter voltage, can be made, the problem will then be straightforward.

Recall that the need for including a resistor from emitter to ground was to provide a means of dc bias stabilization so that the change of collector current due to leakage currents in the transistor would not cause a large (if any) shift in the operating point. The emitter resistor cannot be unreasonably large because the voltage developed across it limits the range of voltage swing of the voltage from collector to emitter. The examples in Section 4.8 show that the voltage from emitter to ground, V_E, is typically around one-fifth to one-tenth of the supply voltage, V_{CC}. Selecting the emitter voltage in this way will permit calculating the emitter resistor, R_E, and then the collector resistor, R_C. Carrying out these calculations we get

$$V_E \cong \frac{1}{10}V_{CC} = \frac{20 \text{ V}}{10} = 2 \text{ V}$$

$$R_E = \frac{V_E}{I_E} \cong \frac{V_E}{I_C} = \frac{2 \text{ V}}{2 \text{ mA}} = 1 \text{ K}$$

$$R_C = \frac{V_{CC} - V_{CE_Q} - V_E}{I_C} = \frac{20 - 10 - 2 \text{ V}}{2 \text{ mA}} = 4 \text{ K}$$

$$I_B = \frac{I_C}{\beta} = \frac{2 \text{ mA}}{150} = 13.3 \text{ } \mu A$$

$$R_B = \frac{V_{CC} - V_{BE} - V_E}{I_B} = \frac{20 - 0.7 - 2 \text{ V}}{13.3 \text{ } \mu A} = 1.3 \text{ M}$$

EXAMPLE 4.16 Calculate the resistor values R_E, R_C, and R_B for a transistor amplifier circuit having emitter-resistor stabilization (Fig. 4.38). The current gain of an *npn* 2N3396 transistor is typically 90 at a collector current of 5 mA. Use a supply voltage of 20 V.

Design Solution: The operating point selected from the information of supply voltage and transistor is $I_{C_Q} = 2$ mA and $V_{CE_Q} = 10$ V.

$$V_E \cong \frac{1}{10}(V_{CC}) = \frac{1}{10}(20) = 2 \text{ V}$$

The emitter resistor is then

$$R_E \cong \frac{V_E}{I_C} = \frac{2 \text{ V}}{5 \text{ mA}} = 400 \text{ } \Omega$$

The collector resistor is calculated to be

$$R_C = \frac{V_{CC} - V_{CE_Q} - V_E}{I_C} = \frac{20 - 10 - 2 \text{ V}}{5 \text{ mA}} = \frac{8 \text{ V}}{5 \text{ mA}} = 1.6 \text{ K}$$

Calculating the base current using

$$I_B = \frac{I_C}{\beta} = \frac{5 \text{ mA}}{90} \cong 55.6 \text{ } \mu A$$

we see that the base resistor is calculated to be

$$R_B = \frac{V_{CC} - V_{BE} - V_E}{I_B} = \frac{(20 - 0.7 - 2) \text{ V}}{55.6 \text{ } \mu A} = \frac{17.3 \text{ V}}{55.6 \text{ } \mu A}$$

$$= 311 \text{ K} \quad (\text{use } R_B = 300 \text{ K})$$

Design of Current Gain Stabilized Circuit

The circuit of Fig. 4.39 provides stabilization both for leakage current and current gain changes. The values of the four resistors shown must be obtained for a

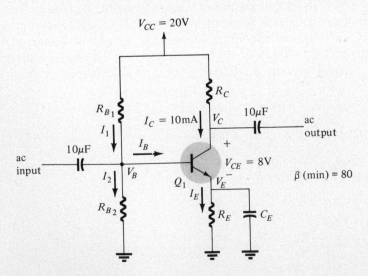

Figure 4.39. Current gain stabilized circuit for design considerations.

specified operating point. Engineering judgment in selecting a value of emitter voltage, V_E, as in the previous design consideration, leads to a simple straight-forward solution for all the resistor values. The design steps are as follows:

The emitter voltage will be selected to be approximately one-tenth of the supply voltage (V_{cc}).

$$V_E \cong \frac{1}{10} V_{cc} = \frac{1}{10}(20 \text{ V}) = 2 \text{ V}$$

Using this value of V_E, we see that the emitter-resistor value is calculated to be

$$R_E \cong \frac{V_E}{I_C} = \frac{2 \text{ V}}{10 \text{ mA}} = 200 \text{ }\Omega$$

The collector resistance is then obtained using

$$R_C = \frac{V_{cc} - V_{CE_Q} - V_E}{I_C} = \frac{20 - 8 - 2 \text{ V}}{10 \text{ mA}} = 1 \text{ K}$$

The base voltage is approximately equal to the emitter voltage or, more exactly,

$$V_B = V_E + V_{BE} = 2 \text{ V} + 0.7 \text{ V} = 2.7 \text{ V}$$

The equation for calculation of the base resistors R_{B1} and R_{B2} will require a little thought. Using the values of base voltage calculated in step (4) and the value of the supply voltage will provide one equation—but there are two unknowns, R_{B1} and R_{B2}. An additional equation can be obtained from an understanding of the operation of these two resistors in providing the base voltage. For the circuit to operate properly the current through the two resistors should be approximately equal and therefore larger than the base current by an order of magnitude (at least 10 times larger). The two equations that will enable calculating the resistors R_{B1} and R_{B2} are

$$V_B = \frac{R_{B2}}{R_{B1} + R_{B2}}(V_{cc})$$

$$R_{B2} \leq \frac{1}{10}(\beta R_E)$$

Solving these equations results in

$$R_{B1} \cong 10 \text{ K}, \qquad R_{B2} \cong 1.6 \text{ K}$$

EXAMPLE 4.17 Design a dc bias circuit for an amplifier circuit as in Fig. 4.39 using a 2N3565 *npn* transistor. The manufacturer's spec states that the transistor has a current gain of 150, typical, at a collector current of 1 mA. The supply voltage for the present circuit is 16 V.

Design Solution: The bias point to be designed is for $I_{C_Q} = 1$ mA, $V_{CE_Q} = 6$ V.
(a) Selecting $V_E = \frac{1}{10}(V_{cc}) = \frac{1}{10}(16) = 1.6$ V
(b) Calculating R_E:

$$R_E \cong \frac{V_E}{I_C} = \frac{1.6 \text{ V}}{1 \text{ mA}} = \textbf{1.6 K}$$

(c) Calculating R_C:

$$R_C = \frac{V_{cc} - V_{CE_Q} - V_E}{I_C} = \frac{16 - 6 - 1.6 \text{ V}}{1 \text{ mA}} = \textbf{8.4 K}$$

(d) Calculating V_B:

$$V_B = V_E + V_{BE} = 1.6 + 0.7 = 2.3 \text{ V}$$

(e) Calculating R_{B1} and R_{B2}:

$$R_{B2} \leq \frac{1}{10}(\beta R_E) = \frac{150(1.6 \text{ K})}{10} = 24 \text{ K}$$

and since

$$\frac{R_{B2}}{R_{B1} + R_{B2}} \cdot V_{CC} = V_B = 2.3 \text{ V}$$

$$R_{B1} \cong 143 \text{ K}$$

4.15 MISCELLANEOUS BIAS CIRCUITS

In practice, one finds that bias circuits do not always conform to the basic forms considered in this chapter. It should not be difficult to analyze the bias operation of a circuit slightly modified from those considered previously. To show various bias circuits a few examples are included in this section to demonstrate the use of the basic bias concepts discussed in this chapter.

EXAMPLE 4.18 Obtain the bias voltages and currents for the circuit of Fig. 4.40.

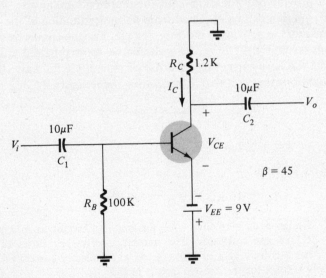

Figure 4.40. Bias circuit for Example 4.18.

Solution: Input loop:

$$-I_B R_B - V_{BE} + V_{EE} = 0$$

$$I_B = \frac{V_{EE} - V_{BE}}{R_B} = \frac{9 - 0.7 \text{ V}}{100 \text{ K}} = 83 \ \mu\text{A}$$

$$I_C = \beta I_B = 45(83 \ \mu\text{A}) = 3.74 \text{ mA}$$

Output loop: $-I_C R_C - V_{CE} + V_{EE} = 0$

$$V_{CE} = V_{EE} - I_C R_C = 9 - (3.74 \text{ mA})(1.2 \text{ K})$$

$$= 9 - 4.5 = 4.5 \text{ V}$$

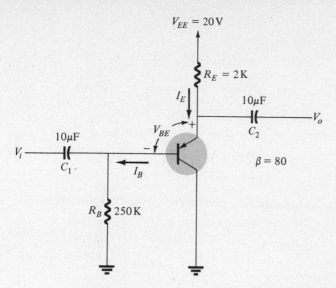

Figure 4.41. Bias circuit for Example 4.19.

EXAMPLE 4.19 Calculate the bias voltages and currents for the circuit of Fig. 4.41.

Solution: Writing the voltage loop equation

$$V_{EE} - I_E R_E - V_{EB} - I_B R_B = 0$$

$$(V_{EE} - V_{EB}) = (\beta + 1)I_B R_E + I_B R_B$$

$$I_B = \frac{V_{EE} - V_{EB}}{R_B + (\beta + 1)R_E} = \frac{V_{EE}}{R_B + \beta R_E} = \frac{20 \text{ V}}{250 \text{ K} + 80(2 \text{ K})}$$

$$= \frac{20 \text{ V}}{410 \text{ K}} = 48.8 \; \mu\text{A}$$

$$I_C = \beta I_B = 80(48.8 \; \mu\text{A}) = \textbf{3.9 mA} \cong I_E$$

$$V_E = V_{EE} - I_E R_E = 20 - (3.9 \text{ mA})(2 \text{ K}) = 20 - 7.8$$

$$= \textbf{12.2 V}$$

EXAMPLE 4.20 Calculate the bias voltages and currents for the circuit of Fig. 4.42.

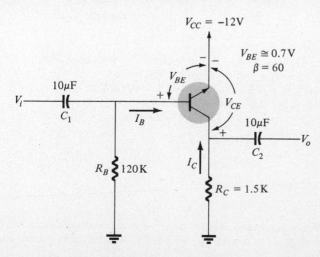

Figure 4.42. Bias circuit for Example 4.20.

SEC. 4.15 MISCELLANEOUS BIAS CIRCUITS

199

Solution: Input loop:

$$-I_B R_B - V_{BE} + V_{CC} = 0$$

$$I_B = \frac{V_{CC} - V_{BE}}{R_B} = \frac{12 - 0.7\ \text{V}}{120\ \text{K}} = 94\ \mu\text{A}$$

$$I_C = \beta I_B = 60(94\ \mu\text{A}) = \mathbf{5.6\ mA}$$

Output loop:

$$-I_C R_C - V_{CE} + V_{CC} = 0$$

$$V_{CE} = V_{CC} - I_C R_C = 12 - (5.6\ \text{mA})(1.5\ \text{K})$$

$$= 12 - 8.5 = \mathbf{3.5\ V}$$

EXAMPLE 4.21 Calculate the bias currents and voltages for the circuit of Fig. 4.43.

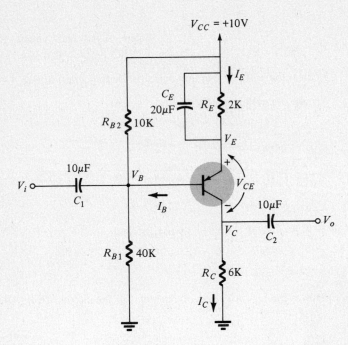

Figure 4.43. Bias circuit for Example 4.20.

Solution:
(a) $V_B = [R_{B1}/(R_{B1} + R_{B2})](V_{CC}) = [40/(10 + 40)](10) = \mathbf{8\ V}$.
(b) $V_E = V_B - V_{BE} = 8 + 0.2 = \mathbf{8.2\ V}$.
(c) $I_E = (V_{CC} - V_E)/R_E = (10 - 8.2)/2 = \mathbf{0.9\ mA} \cong I_C$.
(d) $V_C = I_C R_C = (0.9\ \text{mA})(6\ \text{K}) = \mathbf{5.4\ V}$.
(e) $V_{CE} = V_C - V_E = 5.4 - 8.2 = \mathbf{-2.8\ V}$.

PROBLEMS

§ 4.3

1. Calculate the dc bias voltages and currents for a *pnp* common-base bias circuit. Assume transistor values of $\alpha = 0.985$ and $V_{BE} = -0.2$ V. Circuit components are $R_E =$

720 Ω and $R_C = 3.9$ K and supply voltages are $V_{CC} = 9$ V and $V_{EE} = 1.5$ V. (See Fig. 4.2 for circuit connection.)

2. Calculate the collector-base voltage for *npn* common-base bias circuit (as in Fig. 4.5) *4.4* for the following circuit values: $R_E = 1.8$ K, $R_C = 2.7$ K, $V_{EE} = 9$ V, $V_{CC} = 22$ V, $\alpha = 0.995$, and $V_{BE} = +0.7$ V.

§ 4.6

3. For a fixed-bias common-emitter circuit, as in Fig. 4.6, *4.5* calculate the bias currents and voltages for the following circuit values: $R_B = 150$ K, $R_C = 2.1$ K, $V_{CC} = 9$ V, $V_{BE} = +0.7$ V, and $\beta = 45$.

4. Using an *npn* fixed-bias circuit as in Fig. 4.8, calculate the collector-emitter bias voltage (V_{CE}) for the following circuit values: $R_B = 250$ K, $R_C = 1.8$ K, $V_{CC} = 12$ V, $V_{BE} = 0.7$ V, and $\beta = 70$.

§ 4.8

5. Calculate the dc bias voltages and currents for an emitter-stabilized bias circuit as in Fig. 4.12 for the following circuit values: $R_B = 47$ K, $R_E = 750$ Ω, $R_C = 0.5$ K, $V_{BE} = 0.7$ V, $\beta = 55$, and $V_{CC} = 18$ V.

6. Calculate the collector-emitter voltage (V_{CE}) for an *npn* emitter-stabilized bias circuit as in Fig. 4.12 for the following circuit values: $R_B = 75$ K, $R_C = 0.5$ K, $R_E = 470$ Ω, $V_{BE} = 0.7$ V, $V_{CC} = 10$ V, and $\beta = 80$.

§ 4.9

7. Calculate the bias voltages and currents for a circuit as in Fig. 4.15 for the following circuit values: $R_{B1} = 56$ K, $R_{B2} = 4.7$ K, $R_E = 750$ Ω, $R_C = 6.8$ K, $V_{CC} = 24$ V, $V_{BE} = 0.7$ V, and $\beta = 55$.

8. Calculate the collector voltage V_C for a bias circuit as in Fig. 4.15 for the following values: $R_{B1} = 12$ K, $R_{B2} = 1.5$ K, $R_E = 1$ K, $R_C = 4.7$ K, $V_{CC} = 9$ V, $V_{BE} = +0.7$ V, and $\beta = 75$.

§ 4.10

9. Calculate the bias currents and voltages for a circuit as in Fig. 4.20 using the following circuit values: $R_B = 100$ K, $R_C = 5$ K, $V_{CC} = 10$ V, $V_{BE} = 0.7$ V, and $\beta = 60$.

10. For a bias circuit as in Fig. 4.20 calculate the dc voltage measured from collector to ground (V_C) for the following circuit values: $R_B = 68$ K, $R_C = 2.4$ K, $V_{CC} = 15$ V, $V_{BE} = 0.7$ V, and $\beta = 48$.

11. Calculate the bias voltages and currents for a circuit as in Fig. 4.21 for the following circuit values: $R_B = 200$ K, $R_C = 3.6$ K, $R_E = 270$ Ω, $V_{BE} = 0.7$ V, $\beta = 40$, and $V_{CC} = 16$ V.

§ 4.11

12. For an emitter follower circuit as in Fig. 4.22 calculate the dc bias currents and voltages for the following circuit values: $R_B = 240$ K, $R_E = 1.8$ K, $V_{CC} = 9$ V, $V_{BE} = 0.7$ V, and $\beta = 85$.

13. For an emitter-follower circuit as in Fig. 4.22 calculate the dc voltage across the emitter resistor for the following circuit values: $R_B = 91$ K, $R_E = 1.2$ K, $V_{CC} = 25$ V, $V_{BE} = 0.7$ V, and $\beta = 60$.

14. Calculate the emitter voltage (with respect to ground) for a bias circuit as in Fig. 4.24 for the following values: $R_{B1} = 270$ K, $R_{B2} = 33$ K, $R_E = 4.7$ K, $V_{CC} = 15$ V, $V_{BE} = 0.7$ V, and $\beta = 50$.

§ 4.12

15. For a fixed-bias circuit as in Fig. 4.30a with circuit values of $R_B = 80$ K, $R_C = 4$ K, $V_{CC} = 20$ V, $V_{BE} = 0.7$ V, and transistor collector characteristic as shown in Fig. 4.44, do the following:

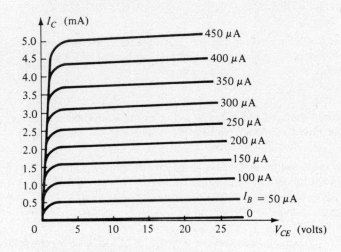

Figure 4.44. Transistor collector characteristics for Problem 4.15.

(a) Draw the dc load line.
(b) Obtain the quiescent operating point (Q-point).
(c) Find the operating point for $R_C = 8$ K.
(d) Find the operating point for $V_{CC} = 15$ V (R_C still is 4 K).

16. Determine graphically the operating point for a fixed-bias circuit using a *pnp* transistor having a collector characteristic as in Fig. 4.45. The circuit values are as follows: $R_B = 150$ K, $R_C = 2$ K, $V_{CC} = 20$ V, and $V_{BE} = -0.3$ V.

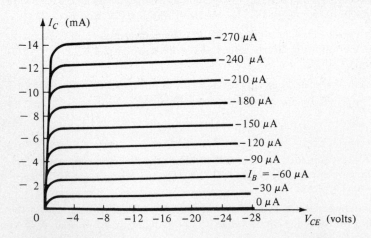

Figure 4.45. Transistor collector characteristics for Problem 4.16.

§ 4.13

17. Calculate the quiescent operating point of the vacuum triode in the circuit of Fig. 4.31 using the characteristic of Fig. 4.32. Circuit values are as follows: $R_G = 1$ M, $R_P = 18$ K, $V_{PP} = 200$ V, and $V_{GG} = 6$ V.

18. What value of grid supply voltage is needed in Problem 4.17 to obtain an operating point at $V_{PK_Q} = 100$ V?

19. Calculate the quiescent operating point of a pentode as in the circuit of Fig. 4.34 for the following circuit values: $R_G = 0.47$ K, $V_{GG} = -3$ V, $V_{PP} = 400$ V, and $R_P = 16$ K. Use the tube characteristic of Fig. 4.36b.

20. What value of load resistor is required for a quiescent plate voltage of $V_{PK_Q} = 200$ V in Problem 4.19?

§ 4.14

21. Design a fixed-bias common-emitter circuit using a 2N2192 transistor. The *npn* transistor has a current gain (β) of 80 and is to be operated at $I_{C_Q} = 2$ mA and $V_{CE_Q} = 10$ V. Use a collector supply of 22 V.

22. Design a fixed-bias common-emitter circuit using a 2N5234 transistor rated to have a current gain of 250° at a collector current of 10 mA and collector-emitter voltage of 10 V. Use a supply of 22 V.

23. Calculate the resistor values for an emitter-stabilized amplifier circuit as in Fig. 4.38. Use a 2N5234 silicon *npn* transistor having a current gain of 250 at 10 mA and 20 V. Use a 30-V supply voltage.

24. Design a dc bias circuit as in Fig. 4.38 using a 2N5235 *npn* transistor having a current gain of 400 at $I_{C_Q} = 10$ mA and $V_{CE_Q} = 10$ V. Use a supply voltage of 22 V.

§ 4.15

25. Obtain the value of dc voltage measured from collector to ground for the circuit of Fig. 4.40 for the following circuit values: $V_{EE} = 15$ V, $R_B = 47$ K, $R_C = 1.2$ K, and $\beta = 30$.

26. Calculate the base voltage (with respect to ground) for the circuit of Fig. 4.41 using the following values: $R_B = 120$ K, $R_E = 8.2$ K, $V_{EE} = 12$ V, $V_{BE} = -0.2$ V, and $\beta = 20$.

27. Calculate the collector current for the circuit of Fig. 4.42 using the following values: $R_B = 80$ K, $R_C = 1.8$ K, $V_{CC} = 9$ V, $V_{BE} = 0.7$, and $\beta = 35$.

28. Calculate the collector-emitter voltage for a circuit as in Fig. 4.43 for the following circuit values: $R_1 = 120$ K, $R_2 = 15$ K, $R_E = 3.9$ K, $R_C = 12$ K, $V_{CC} = 18$ V, $V_{BE} = 0.7$ V, and $\beta = 200$.

small-signal analysis

5.1 INTRODUCTION

Chapter 4 was a detailed discussion of transistor and tube circuits purely from a dc viewpoint. We must now begin to investigate the response of these circuits with a sinusoidal ac signal applied.

Of first concern is the magnitude of the input signal. It will determine whether *small-signal* or *large-signal* techniques must be used. There is no set dividing line between the two, but the application, and the magnitude of the variables of interest (i, v) relative to the scales of the device characteristics, will usually make it clear which is the case in point. The small-signal technique will be discussed in this chapter; large-signal applications will be considered in Chapter 8.

The key to the small-signal approach is the use of equivalent circuits to be derived in this chapter. It is that combination of circuit elements, properly chosen, that will best approximate the actual semiconductor or tube device in a particular operating region. Once the ac equivalent circuit has been determined, the graphic symbol of the device can be replaced in the schematic by this circuit and the basic methods of ac circuit analysis (branch-current analysis, mesh analysis, nodal analysis, and Thévenin's theorem) can be applied to determine the response of the circuit.

There are two schools of thought in prominence today regarding the equivalent circuit to be substituted for the transistor. For many years the industrial and educational institutions relied heavily on the *hybrid parameters* (to be introduced shortly). The hybrid parameter equivalent continues to be very popular, although it must now share the spotlight with an equivalent circuit derived directly from the operating conditions of the transistor. Manufacturers continue to specify the

204

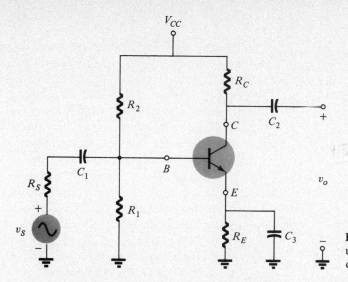

Figure 5.1. Transistor circuit under examination in this introductory discussion.

hybrid parameters for a particular operating region on their specifications sheets. The parameters (or components) of the other equivalent circuit can be derived directly from the hybrid parameters in this region. However, the hybrid equivalent circuit suffers from being limited to a particular set of operating conditions if it is to be considered accurate. The parameters of the other equivalent circuit can be determined for any region of operation within the active region and are not limited by the single set of parameters provided by the specification sheet. For the purposes of this text, if the operating region corresponds with that indicated on the specification sheet, then either equivalent will be used. If not specified, the equivalent circuit derived from the operating conditions will be used. Be encouraged by the fact that both equivalent circuits are very similar in appearance and application. Developed skills with one will result in the same measure of ability with the other.

In an effort to demonstrate the effect that the ac equivalent circuit will have on the analysis to follow, consider the circuit of Fig. 5.1, discussed in detail in Chapter 4. Let us assume for the moment that the small-signal ac equivalent circuit for the transistor has already been determined. Since we are interested only in the ac response of the circuit, all the dc supplies can be replaced by a zero potential equivalent (short circuit) since they determine only the dc or quiescent level of the output voltage and not the magnitude of the swing of the ac output. This is clearly demonstrated by Fig. 5.2. The dc levels were simply important for determining the proper Q-point of operation. Once determined, the dc levels can be ignored for the ac analysis of the network. In addition, the coupling capacitors C_1 and C_2 and bypass capacitor C_3 were chosen to have a very small reactance at the frequency of application. Therefore, they too, may for all practical purposes be replaced by a low-resistance path (short circuit). Note that this will result in the "shorting out" of the dc biasing resistor R_E. Connecting common grounds will result in a parallel combination for resistors R_1 and R_2, and R_C will appear from collector to emitter as shown in Fig. 5.3. Since the components of the transistor equivalent circuit inserted in Fig. 5.3 are those we are already familiar with (resistors, controlled sources, etc.), analysis techniques such as superposition, Thevenin's theorem, etc., can be applied to determine the desired quantities.

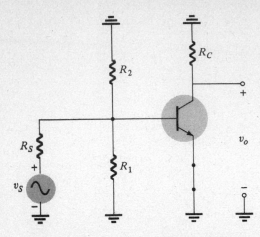

Figure 5.2. The network of Fig. 5.1 following the removal of the dc supply and inserting the short-circuit equivalent for the capacitors.

Let us further examine Fig. 5.3 and identify the important quantities to be determined for the system. Certainly, we would like to know the input and output impedance Z_i and Z_o as shown in Fig. 5.3. Since we know that the transistor is an amplifying device, we would expect some indication of how the output current i_o is related to the input current —the *current gain*. Note in this case that $i_o = i_C$ and $i_i = i_B$. The ratio of these two quantities certainly relates directly to the β of the transistor. In Fig. 3.11b we found that the collector-to-emitter voltage did have some effect (if even slight) on the input relationship between i_B and v_{BE}. We might, therefore, expect some "feedback" from the output to input circuit in the equivalent circuit. The following section, through its brief introduction to *two-port theory*, will develop the hybrid equivalent circuit, which will have parameters that will permit a determination of each of the quantities discussed above.

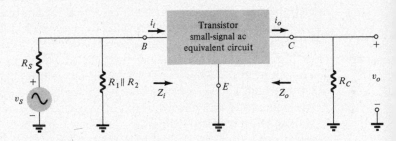

Figure 5.3. Circuit of Fig. 5.2 redrawn for small-signal ac analysis.

5.2 TRANSISTOR HYBRID EQUIVALENT CIRCUIT

The development that follows is an introduction to a subject called *two-port* theory. For the basic three-terminal device (such as the transistor or triode) it is obvious, from Fig. 5.4, that there are two ports (pairs of terminals) of interest. For our purposes, the set at the left will represent the input terminals, and the set at the right, the output terminals. Note that, for each set of terminals, there are two variables of interest.

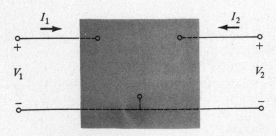

Figure 5.4. Two-port system.

The following set of equations (5.1) is only one of a number of ways in which the four variables can be related. It is the most frequently employed in transistor circuit analysis, however, and, therefore, will be discussed in detail in this chapter.

$$V_1 = h_{11}I_1 + h_{12}V_2$$

(5.1a)

[handwritten: input V, Leakage V, $Z_i I_{in} + \frac{1}{he}V_2$]

$$I_2 = h_{21}I_1 + h_{22}V_2$$

(5.1b)

[handwritten: $\beta I_1 + \frac{1}{R}V_2$]

The parameters relating the four variables are called *h-parameters* from the word hybrid. The term hybrid was chosen because the mixture of variables (*v & i*) in each equation results in a "hybrid" set of units of measurement for the *h*-parameters.

A clearer understanding of what the various *h*-parameters represent and how we can expect to treat them later can be developed by isolating each and examining the resulting relationship.

If we arbitrarily set $V_2 = 0$ (short circuit the output terminals), and solve for h_{11} in Eq. (5.1a), the following will result:

$$h_{11} = \frac{V_1}{I_1}\bigg|_{V_2=0}$$

(ohms) *[handwritten: Z_{in}]*

(5.2)

The ratio indicates that the parameter h_{11} is an impedance parameter to be measured in ohms. Since it is the ratio of the *input* voltage to the *input* current with the output terminals *shorted*, it is called the *short-circuit input impedance parameter*.

If I_1 is set equal to zero by opening the input leads, the following will result for h_{12}:

$$h_{12} = \frac{V_1}{V_2}\bigg|_{I_1=0}$$

[handwritten: $\approx \frac{1}{he}$?]

(5.3)

The parameter h_{12}, therefore, is the ratio of the input voltage to the output voltage with the input current equal to zero. It has no units since it is a ratio of voltage levels. It is called the *open-circuit reverse transfer voltage ratio parameter*. The term reverse is included to indicate that the voltage ratio is an input quantity over an output quantity rather than the reverse, which is usually the ratio of interest.

If in Eq. (5.1b), V_2 is set squal to zero by again shorting the output terminals, the following will result for h_{21}:

$$h_{21} = \frac{I_2}{I_1}\bigg|_{V_2=0}$$

[handwritten: $\approx A_i$]

(5.4)

Note that we now have the ratio of an output quantity to an input quantity. The term *forward* will now be used rather than *reverse* as indicated for h_{12}. The parameter h_{21} is the ratio of the output current divided by the input current with the output terminals shorted. It is, for most applications, the parameter of greatest interest. This parameter, like h_{12}, has no units since it is the ratio of current levels. It is formally called the *short-circuit forward transfer current ratio parameter*.

The last parameter, h_{22}, can be found by again opening the input leads to set $I_1 = 0$ and solving for h_{22} in Eq. (5.1b).

$$h_{22} = \frac{I_2}{V_2}\bigg|_{I_1=0} \qquad \text{(mhos)} \qquad (5.5)$$

Since it is the ratio of the output current to the output voltage, it is the output conductance parameter and is measured in *mhos*. It is called the *open-circuit output conductance parameter*.

Since each term of Eq. (5.1a) has the units of volts, let us apply Kirchhoff's voltage law in reverse to find a circuit that "fits" the equation. Performing this operation will result in the circuit of Fig. 5.5. Since the parameter h_{11} has the units of ohms, it is represented as an impedance which for the transistor becomes a resistor in Fig. 5.5. h_{12} is a dimensionless quantity. Note that it is a "feedback" of the output voltage to the input circuit.

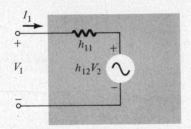

Figure 5.5. Hybrid input equivalent circuit.

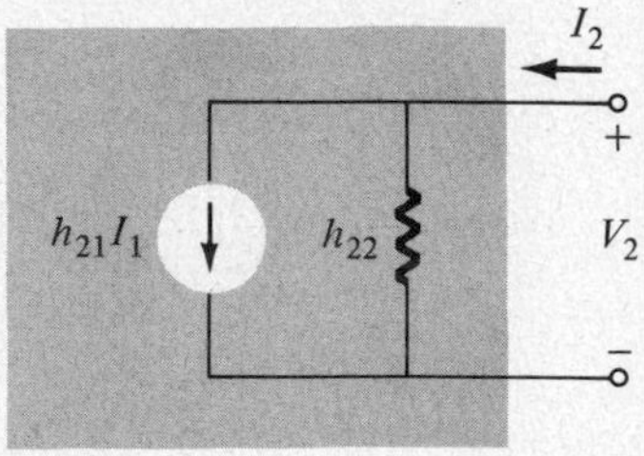

Figure 5.6. Hybrid output equivalent circuit.

Each term of Eq. (5.1b) has the units of current. Let us now apply Kirchhoff's current law in reverse to obtain the circuit of Fig. 5.6.

Since h_{22} has the units of conductance, it is represented by the resistor symbol. Keep in mind, however, that the resistance in ohms of this resistor is equal to the reciprocal of conductance ($1/h_{22}$).

The complete "ac" equivalent circuit for the basic three-terminal linear device is indicated in Fig. 5.7 with a new set of subscripts for the *h*-parameters.

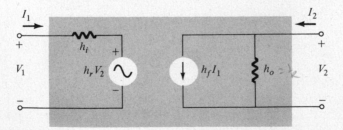

Figure 5.7. Complete hybrid equivalent circuit.

The notation of Fig. 5.7 is of a more practical nature since it relates the *h*-parameters to the resulting ratio obtained in the last few paragraphs. The choice of letters is obvious from the following listing:

$h_{11} \longrightarrow$ *i*nput resistance $\longrightarrow h_i$ Z_{in}

$h_{12} \longrightarrow$ *r*everse transfer voltage ratio $\longrightarrow h_r = \frac{1}{Ae}$

$h_{21} \longrightarrow$ *f*orward transfer current ratio $\longrightarrow h_f = A_i$

$h_{22} \longrightarrow$ *o*utput conductance $\longrightarrow h_o$ $\frac{1}{R_{ow}} = \frac{1}{Z_{ow}}$

The circuit of Fig. 5.7 is applicable to any linear three-terminal device with no internal independent sources. For the transistor, therefore, even though it has three basic configurations, *they are all three-terminal configurations*, so that the resulting equivalent circuit will have the same format as shown in Fig. 5.7. The *h*-parameters, however, will change with each configuration. To distinguish which parameter has been used or which is available, a second subscript has been added to the *h*-parameter notation. For the common-base configuration the lowercase letter *b* was added, while for the common-emitter and common-collector configurations the letters *e* and *c* were added, respectively. The hybrid equivalent circuit for the common-base and common-emitter configurations with the standard notation is presented in Fig. 5.8. The circuits of Fig. 5.8 are applicable for *pnp* or *npn* transistors.

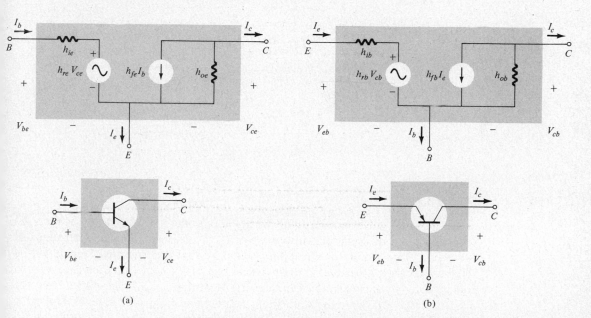

Figure 5.8. Complete hybrid equivalent circuits: (a) common-emitter configuration; (b) common-base configuration.

The hybrid equivalent circuit of Fig. 5.7 is an extremely important one in the area of electronics today. It will appear over and over again in the analysis to follow. It would be time well spent, at this point, for the reader to memorize and draw from memory its basic construction and define the significance of the various parameters [see Eqs. (5.6)–(5.9)]. The fact that both a Thévenin and Norton circuit appear in the circuit of Fig. 5.7 was further impetus for calling the resultant circuit a *hybrid* equivalent circuit. Two additional transistor equivalent circuits, not to be discussed in this text, called the Z-parameter and Y-parameter equivalent circuits, use either the voltage source or the current source but not both in the same equivalent circuit. In Section 5.3 the magnitude of the various parameters will be found from the transistor characteristics in the region of operation resulting in the desired *small-signal equivalent circuit* for the transistor.

5.3 GRAPHICAL DETERMINATION OF THE *h*-PARAMETERS

Using partial derivatives (calculus), we can show that the *h*-parameters for the small-signal transistor equivalent circuit in the region of operation for the common-emitter configuration can be found using the following equations:

$$
h_{ie} = \frac{\partial v_1}{\partial i_1} = \frac{\partial v_{BE}}{\partial i_B} \cong \frac{\Delta v_{BE}}{\Delta i_B}\bigg|_{v_{CE}=\text{constant}}
\tag{5.6}
$$

$$
h_{re} = \frac{\partial v_1}{\partial v_2} = \frac{\partial v_{BE}}{\partial v_C} \cong \frac{\Delta v_{BE}}{\Delta v_{CE}}\bigg|_{i_B=\text{constant}}
\tag{5.7}
$$

$$
h_{fe} = \frac{\partial i_2}{\partial i_1} = \frac{\partial i_C}{\partial i_B} \cong \frac{\Delta i_C}{\Delta i_B}\bigg|_{v_{CE}=\text{constant}}
\tag{5.8}
$$

$$
h_{oe} = \frac{\partial i_2}{\partial v_2} = \frac{\partial i_C}{\partial v_{CE}} \cong \frac{\Delta i_C}{\Delta v_{CE}}\bigg|_{i_B=\text{constant}}
\tag{5.9}
$$

In each case the symbol Δ refers to a small change in that quantity around the quiescent point of operation. In other words, the *h*-parameters are determined in the region of operation for the applied signal, so that the equivalent circuit will be the most accurate available. The constant values of V_{CE} and I_B in each case refer to a condition that must be met when the various parameters are determined from the characteristics of the transistor. For the common-base and common-collector configurations the proper equation can be obtained by simply substituting the proper values of v_1, v_2, i_1, and i_2. *In Appendix A a table has been provided that relates the hybrid parameters of the three basic transistor configurations.* In other words, if the *h*-parameters for the common-emitter configuration are known, the *h*-parameters for the common-base or common-collector configurations can be found using these tables.

The parameters h_{ie} and h_{re} are determined from the input or base characteristics, while the parameters h_{fe} and h_{oe} are obtained from the output or collector characteristics. Since h_{fe} is usually the parameter of greatest interest, we shall discuss the operations involved with equations, such as Eqs. (5.6)–(5.9), for this parameter

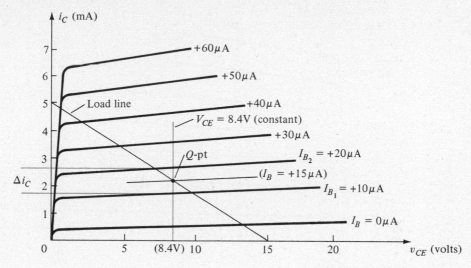

Figure 5.9. h_{fe} determination.

first. The first step in determining any of the four hybrid parameters is to find the quiescent point of operation as indicated in Fig. 5.9. In Eq. (5.8) the condition V_{CE} = constant requires that the changes in base voltage and current be taken along a vertical straight line drawn through the Q-point representing a fixed collector-to-emitter voltage. Equation (5.8) then requires that a small change in collector current be divided by the corresponding change in base current. For the greatest accuracy these changes should be made as small as possible.

In Fig. 5.9 the change in i_B was chosen to extend from I_{B_1} to I_{B_2} along the perpendicular straight line at V_{CE}. The corresponding change in i_C is then found by drawing the horizontal lines from the intersections of I_{B_1} and I_{B_2} with V_{CE} = constant to the vertical axis. All that remains is to substitute the resultant changes of i_B and i_C into Eq. (5.8); that is,

$$|h_{fe}| = \frac{\Delta i_C}{\Delta i_B}\bigg|_{V_{CE}=\text{constant}} = \frac{(2.7 - 1.7) \times 10^{-3}}{(20 - 10) \times 10^{-6}}\bigg|_{V_{CE}=8.4\text{ V}}$$

$$= \frac{10^{-3}}{10 \times 10^{-6}} = \mathbf{100}$$

In Fig. 5.10 a straight line is drawn tangent to the curve I_B through the Q-point to establish a line I_B = constant as required by Eq. (5.9) for h_{oe}. A change in v_{CE} was then chosen and the corresponding change in i_C is determined by drawing the horizontal lines to the vertical axis at the intersections on the I_B = constant line. Substituting into Eq. (5.9), we get

$$|h_{oe}| = \frac{\Delta i_C}{\Delta v_{CE}}\bigg|_{i_B=\text{constant}} = \frac{(2.2 - 2.1) \times 10^{-3}}{10 - 7}\bigg|_{I_E=+15\,\mu A}$$

$$= \frac{0.1 \times 10^{-3}}{3} = \mathbf{33\ \mu A/V}$$

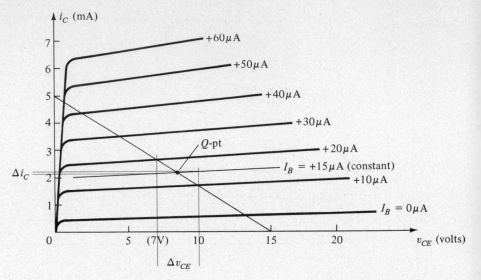

Figure 5.10. h_{oe} determination.

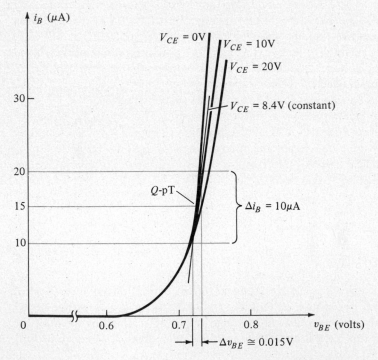

Figure 5.11. h_{ie} determination.

To determine the parameters h_{ie} and h_{re} the Q-point must first be found on the input or base characteristics as indicated in Fig. 5.11. For h_{ie}, a line is drawn tangent to the curve $V_{CE} = 8.4$ V through the Q-point, to establish a line $V_{CE} =$ constant as required by Eq. (5.6). A small change in v_{BE} was then chosen, resulting in a corresponding change in i_B. Substituting into Eq. (5.6), we get

$$|h_{ie}| = \frac{\Delta v_{BE}}{\Delta i_B}\bigg|_{V_{CE}=\text{constant}} = \frac{(733-7.18)\times 10^{-3}}{(20-10)\times 10^{-6}}\bigg|_{V_{CE}=8.4\text{V}}$$

$$= \frac{15\times 10^{-3}}{10\times 10^{-6}} = \mathbf{1.5\,K}$$

The last parameter, h_{re}, can be found by first drawing a horizontal line through the Q-point at $I_B = 15\ \mu$A. The natural choice then is to pick a change in v_{CE} and find the resulting change in v_{BE} as shown in Fig. 5.12.

Substituting into Eq. (5.7), we get

$$|h_{re}| = \frac{\Delta v_{BE}}{\Delta v_{CE}}\bigg|_{I_B=\text{constant}} = \frac{(733-725)\times 10^{-3}}{20-0} = \frac{8\times 10^{-3}}{20} = \mathbf{4\times 10^{-4}}$$

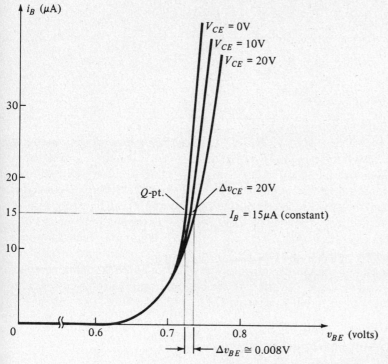

Figure 5.12. h_{re} determination.

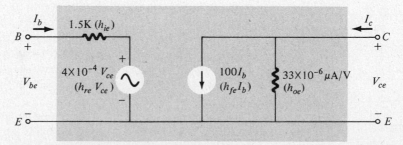

Figure 5.13. Complete hybrid equivalent circuit for a transistor having the characteristics that appear in Figs. 5.9–5.12.

For the transistor whose characteristics have appeared in Figs. 5.9–5.12 the resulting hybrid small-signal equivalent circuit is shown in Fig. 5.13.

As mentioned earlier the hybrid parameters for the common-base and common-collector configurations can be found using the same basic equations with the proper variables and characteristics.

Typical values for each parameter for the broad range of transistors available today in each of its three configurations are provided in Table 5.1. The minus sign indicates that in Eq. (5.8) as one quantity increased in magnitude, within the change chosen, the other decreased in magnitude.

TABLE 5.1 Typical Parameter Values for the CE, CC, and CB Transistor Configurations

PARAMETER	CE	CC	CB
h_i	1 K	1 K	20 Ω
h_r	2.5×10^{-4}	$\cong 1$	3.0×10^{-4}
h_f	50	-50	-0.98
h_o	25 μA/V	25 μA/V	0.5 μA/V
$1/h_o$	40 K	40 K	2 M

Note in retrospect (Section 3.4: Transistor Amplifying Action) that the input resistance of the common-base configuration is low, while the output impedance is high. Consider also that the short-circuit gain is very close to 1. For the common-emitter and common-collector configurations note that the input impedance is much higher than that of the common-base configuration and that the ratio of output to input resistance is about 40 : 1. Consider also for the common-emitter and common-base configuration that h_r is very small in magnitude. Transistors are available today with values of h_{f_e} that vary from 20 to 600. For any transistor the region of operation and conditions under which it is being used will have an effect on the various h-parameters. The effect of temperature and collector current and voltage on the h-parameters will be discussed in Section 5.4.

5.4 VARIATIONS OF TRANSISTOR PARAMETERS

There are a large number of curves that can be drawn to show the variations of the h-parameters with temperature, frequency, voltage, and current. The most interesting and useful at this stage of the development include the h-parameter variations with junction temperature and collector voltage and current.

In Fig. 5.14 the effect of the collector current on the h-parameter has been indicated. Take careful note of the logarithmic scale on the vertical and horizontal axes. The parameters have all been normalized to unity so that the relative change in magnitude with collector current can easily be determined. On every set of curves, such as in Fig. 5.14, the operating point at which the parameters were found is always indicated. For this particular situation the quiescent point is at the

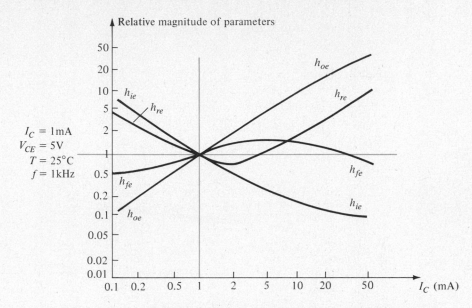

Figure 5.14. Hybrid parameter variations with collector current.

intersection of $V_{CE} = 5.0$ V and $I_C = 1.0$ mA. Since the frequency and temperature of operation will also affect the h-parameters, these quantities are also indicated on the curves. At 0.1 mA, h_{fe} is about 0.5 or 50% of its value at 1.0 mA, while at 3 mA, it is 1.5 or 150% of that value. In other words, h_{fe} has changed from a value of 0.5(50) = 25 to 1.5(50) = 75 with a change of I_C from 0.1 mA to 3 mA. In Section 5.6 we shall find that for the majority of applications it is a fairly good approximation to neglect the effects of h_{re} and h_{oe} in the equivalent circuit. Consider, however, the point of operation at $I_C = 50$ mA. The magnitude of h_{re} is now approximately 11 times that at the defined Q-point, a magnitude that may not permit eliminating this parameter from the equivalent circuit. The parameter h_{oe} is approximately 35 times the normalized value. This increase in h_{oe} will decrease the magnitude of the output resistance of the transistor to a point where it may approach the magnitude of the load resistor. There would then be no justification in eliminating h_{oe} from the equivalent circuit on an approximate basis.

In Fig. 5.15 the variation in magnitude of the h-parameters on a normalized basis has been indicated with change in collector voltage. This set of curves was normalized at the same operating point of the transistor discussed in Fig. 5.14 so that a comparison between the two sets of curves can be made. Note that h_{ie} and h_{fe} are relatively steady in magnitude, while h_{oe} and h_{re} are much larger to the left and right of the chosen operating point. In other words, h_{oe} and h_{re} are much more sensitive to changes in collector voltage than are h_{ie} and h_{fe}.

In Fig. 5.16 the variation in h-parameters has been plotted for changes in junction temperature. The normalization value is taken to be room temperature: $T = 25°C$. The horizontal scale is a linear scale rather than a logarithmic scale as was employed for Figs. 5.14 and 5.15. In general, all the parameters increase in

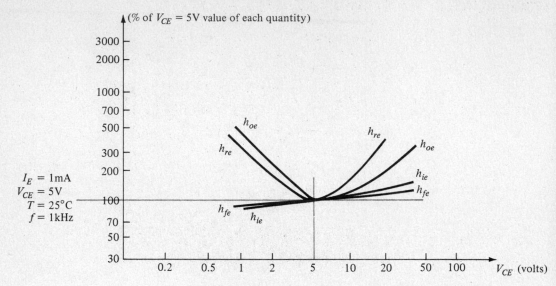

Figure 5.15. Hybrid parameter variations with collector-emitter potential.

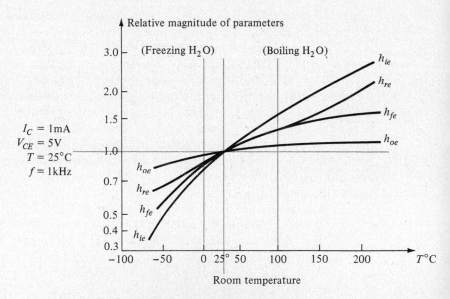

Figure 5.16. Hybrid parameter variations with temperature.

magnitude with temperature. The parameter least affected, however, is h_{oe}, while the input impedance h_{ie} changes at the greatest rate. The fact that h_{fe} will change from 50% of its normalized value at $-50°C$ to 150% of its normalized value at $+150°C$ indicates clearly that the operating temperature must be carefully considered in the design of transistor circuits.

5.5 SMALL-SIGNAL ANALYSIS OF THE BASIC TRANSISTOR AMPLIFIER USING THE HYBRID EQUIVALENT CIRCUIT

In this section the basic transistor amplifier will be examined in detail using the hybrid equivalent circuit. It will not be specified whether the transistor is in the common-emitter, base, or collector configuration. The results, therefore, are applicable to each configuration, requiring only that the proper h-parameters be used for the configuration of interest. All amplifiers are basically two-port devices as indicated in Fig. 5.17; that is, there are a pair of input terminals and a pair of output terminals. For an amplifier there are six quantities of general interest: current gain, voltage gain, input impedance, output impedance, power gain, and phase relationships, each of which will be discussed in detail in this section. The load impedance Z_L can be any combination of resistive and reactive elements. In Fig. 5.17 all voltages and currents refer to the effective value of the sinusoidally varying quantities. The resistor R_s represents, in total, the internal resistance of the source and any resistance in series with the source V_s. Before continuing, keep in mind that the analysis to follow is for small-signal inputs. There is no discussion of dc levels and biasing arrangement. For the resulting equations to be useful, however, the quiescent point must be established and the resulting h-parameters must be known.

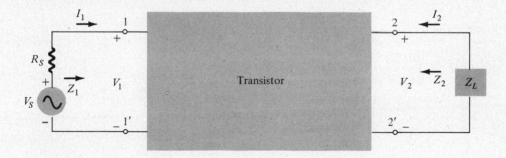

Figure 5.17. Basic transistor amplifier configuration.

The Current Gain $A_i = (I_2/I_1)$

Substituting the hybrid equivalent circuit for the transistor of Fig. 5.17 will result in the configuration of Fig. 5.18.

Applying Kirchhoff's current law to the output circuit, we get

$$I_2 = h_f I_1 + I = h_f I_1 + h_o V_2$$

Substituting $V_2 = -I_2 Z_L$, we get

$$I_2 = h_f I_1 - h_o Z_L I_2$$

The minus sign arises because the direction of I_2 as shown in Fig. 5.18 would result in a polarity across the load Z_L opposite to that shown in Fig. 5.18.

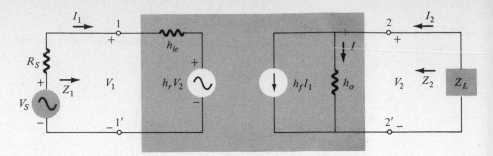

Figure 5.18. Transistor hybrid equivalent circuit substituted for the transistor of Fig. 5.17.

Rewriting the above equation, we get

$$I_2 + h_o Z_L I_2 = h_f I_1$$

and

$$I_2(1 + h_o Z_L) = h_f I_1$$

so that

$$A_i = \frac{I_2}{I_1} = \frac{h_f}{1 + h_o Z_L} \tag{5.10}$$

The Voltage Gain $A_v = (V_2/V_1)$

Applying Kirchhoff's voltage law to the input circuit results in

$$V_1 = I_1 h_i + h_r V_2$$

Substituting $I_1 = [(1 + h_o Z_L)I_2/h_f]$ from Eq. (5.10) and $I_2 = -(V_2/Z_L)$ from above results in

$$V_1 = \frac{-(1 + h_o Z_L)h_i}{h_f Z_L} V_2 + h_r V_2$$

Solving for the ratio V_2/V_1, we get

$$A_v = \frac{V_2}{V_1} = \frac{-h_f Z_L}{h_i + (h_i h_o - h_f h_r)Z_L} \tag{5.11}$$

The Input Impedance $Z_1 = (V_1/I_1)$

For the input circuit

$$V_1 = h_i I_1 + h_r V_2$$

Substituting

$$V_2 = -I_2 Z_L$$

we have

$$V_1 = h_i I_1 - h_r Z_L I_2$$

Since

$$A_i = \frac{I_2}{I_1}$$

$$I_2 = A_i I_1$$

so that the above equation becomes

$$V_1 = h_i I_1 - h_r Z_L A_i I_1$$

Solving for the ratio V_1/I_1, we get

$$Z_1 = \frac{V_1}{I_1} = h_i - h_r Z_L A_i$$

and substituting

$$A_i = \frac{h_f}{1 + h_o Z_L}$$

yields

$$\boxed{Z_1 = \frac{V_1}{I_1} = h_i - \frac{h_f h_r Z_L}{1 + h_o Z_L}}$$

(5.12)

The Output Impedance $Z_2 = (V_2/I_2)$

The output impedance of an amplifier is defined to be the ratio of the output voltage to the output current with the signal (V_s) set at zero.

For the input circuit with $V_s = 0$

$$I_1 = \frac{-h_r V_2}{R_s + h_i}$$

Substituting this relationship into the following equation obtained from the output circuit yields

$$I_2 = h_f I_1 + h_o V_2$$

$$I_2 = \frac{-h_f h_r V_2}{R_s + h_i} + h_o V_2$$

and the ratio

$$\boxed{Z_2 = \frac{V_2}{I_2}\bigg|_{V_s=0} = \frac{1}{h_o - \left(\dfrac{h_f h_r}{h_i + R_s}\right)}}$$

(5.13)

so that the output admittance

$$\boxed{Y_2 = \frac{I_2}{V_2}\bigg|_{V_s=0} = h_o - \frac{h_f h_r}{h_i + R_s}}$$

(5.14)

The Power Gain $A_p = (P_L/P_i)$

The power to the load is $V_L I_L \cos\theta$, which, for the situation being discussed, is $-V_2 I_2 \cos\theta$. The minus sign arises again for the same reason discussed in the derivation of the equation for A_i. It indicates that the load is absorbing power and not supplying it to the circuit. If we limit our discussion to *purely resistive loads*, then $\cos\theta = 1$ and $P_L = P_2 = -V_2 I_2$. The input power is $V_1 I_1$ so that

$$\boxed{A_p = \frac{P_L}{P_i} = \frac{-V_2 I_2}{V_1 I_1}}$$

(5.15)

but $$A_v = \frac{V_2}{V_1} \quad \text{and} \quad A_i = \frac{I_2}{I_1}$$

so that $$A_p = \left[-\frac{V_2}{V_1} \right]\left[\frac{I_2}{I_1} \right]$$

is also $$\boxed{A_p = -A_v A_i} \tag{5.16}$$

In terms of the h-parameters,

$$\boxed{A_p = \frac{h_f^2 Z_L}{(1 + h_o Z_L)[h_i + (h_i h_o - h_f h_r)Z_L]}} \tag{5.17}$$

If we consider that $$V_2 = -I_2 Z_L$$

and substitute $$I_2 = A_i I_1$$

then $$V_2 = -A_i I_1 Z_L$$

and $$A_v = \frac{V_2}{V_1} = \frac{-A_i I_1 Z_L}{V_1} = \frac{-A_i Z_L}{V_1/I_1} = \frac{-A_i Z_L}{Z_1}$$

so that $$A_p = -A_v A_i = -\left(-\frac{A_i Z_L}{Z_1} \right) A_i$$

and $$\boxed{A_p = \frac{A_i^2 Z_L}{Z_1}} \qquad (Z_L, Z_1 \text{—resistive}) \tag{5.18}$$

Phase Relationship

The phase relationship between the output current or voltage and the input current or voltage can be found by simply examining Eqs. (5.10) and (5.11), repeated below for convenience.

$$A_i = \frac{h_f}{1 + h_o Z_L}$$

$$A_v = \frac{-h_f Z_L}{h_i + (h_i h_o - h_f h_r)Z_L}$$

Recall from Section 5.3 that all the h-parameters are positive except h_f for the common-base and common-collector configurations. For the common-emitter configuration, therefore, it should be obvious from the above equations that the output current is in phase with the input current, while the output voltage, due to the negative sign, is 180° out of phase with the input voltage. For the common-collector and common-base configurations the opposite is true due to the difference in sign of h_f. Keep in mind, however, that the above discussion applies only to the current direction and voltage polarities as defined by Fig. 5.17.

EXAMPLE 5.1 Find the following for the fixed-bias transistor of Fig. 5.19:
  (a) Current gain $A_i = I_o/I_i$.
  (b) Voltage gain $A_v = V_o/V_i$.
  (c) Input impedance Z_i.
  (d) Output impedance Z_o.
  (e) Power gain A_p.

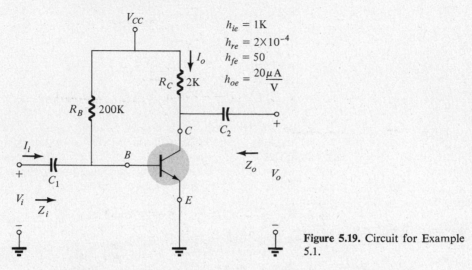

$$h_{ie} = 1K$$
$$h_{re} = 2\times10^{-4}$$
$$h_{fe} = 50$$
$$h_{oe} = \frac{20\mu A}{V}$$

Figure 5.19. Circuit for Example 5.1.

Solution: Replacing the dc supplies and capacitors by short circuits and substituting the transistor hybrid equivalent circuit will result in the configuration of Fig. 5.20.

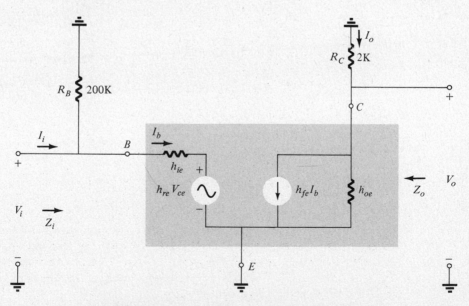

Figure 5.20. Circuit of Fig. 5.19 following the substitution of the small-signal hybrid equivalent circuit for transistor.

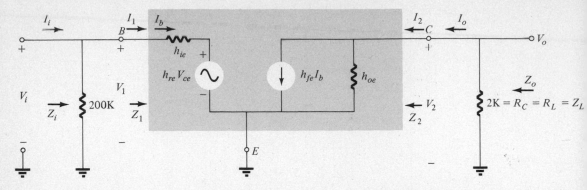

Figure 5.21. Redrawn circuit of Fig. 5.20.

Redrawing the circuit we obtain Fig. 5.21.

The basic hybrid equations will now be applied to the circuit of Fig. 5.21 to obtain the desired results. Note the similarities in the appearance of Fig. 5.21 as compared to the fundamental configuration gf Fig. 5.18. This is a basic requirement if the equations just derived from Fig. 5.18 are to be employed.

(a) To determine A_i, Z_1 must first be found

$$Z_1 = h_{ie} - \frac{h_{fe}h_{re}Z_L}{1 + h_{oe}Z_L} = 1 \times 10^3 - \frac{(50)(2 \times 10^{-4})(2 \times 10^3)}{1 + (20 \times 10^{-6})(2 \times 10^3)}$$

$$= 1 \times 10^3 - \frac{20}{1.04} \cong 981 \ \Omega$$

Since 200 K $\gg$ 0.963 K, $I_i \cong I_1$ and

$$A_i = \frac{I_o}{I_i} = \frac{I_o}{I_1} = \frac{I_2}{I_1} = \frac{h_{fe}}{1 + h_{oe}Z_L} = \frac{50}{1 + (20 \times 10^{-6})(2 \times 10^3)}$$

$$= \frac{50}{1.04} \cong \textbf{48.1}$$

(b) $\quad A_v = \dfrac{V_o}{V_i} = \dfrac{V_2}{V_1} = \dfrac{-h_{fe}Z_L}{h_{ie} + (h_{ie}h_{oe} - h_{fe}h_{re})Z_L}$

$$= \frac{-50(2 \times 10^3)}{1 \times 10^3 + [(1 \times 10^3)(20 \times 10^{-6}) - (50)(2 \times 10^{-4})]2 \times 10^3}$$

$$= \frac{-100 \times 10^3}{1 \times 10^3 + 20} \cong \textbf{-98}$$

(c) $\qquad\qquad\qquad Z_i = 200 \ \text{K} \, || \, Z_1 \cong Z_1 = \textbf{0.981 K}$

(d) $\qquad\qquad\qquad Z_o = 2 \ \text{K} \, || \, Z_2$

where $\quad Z_2 = \dfrac{1}{h_{oe} - \dfrac{h_{fe}h_{re}}{h_{ie} + R_s}} = \dfrac{1}{20 \times 10^{-6} - \dfrac{(50)(2 \times 10^{-4})}{1 \times 10^3 + 0}}$

$$= \frac{1}{(20 \times 10^{-6}) - (10 \times 10^{-6})} = \frac{1}{10 \times 10^{-6}} = 100 \ \text{K}$$

and $\qquad\qquad\qquad Z_o = 2 \ \text{K} \, || \, 100 \ \text{K} = \textbf{1.96 K}$

(e) $\qquad\qquad A_p = -A_v A_i = -(-98)(48.1) = \textbf{4713.8}$

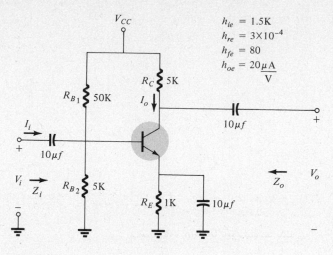

$$h_{ie} = 1.5K$$
$$h_{re} = 3 \times 10^{-4}$$
$$h_{fe} = 80$$
$$h_{oe} = 20\mu A \over V$$

Figure 5.22. Circuit for Example 5.2.

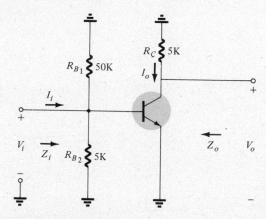

Figure 5.23. Circuit of Fig. 5.22 redrawn for small-signal ac analysis.

EXAMPLE 5.2 Find the following for the network of Fig. 5.22:

(a) $A_v = V_o/V_i$.

(b) $A_i = I_o/I_i$.

(c) Z_i.

(d) Z_o.

Solution: Replacing the dc supplies and capacitors by short circuits will result in the circuit of Fig. 5.23.

Substituting the hybrid equivalent circuit and redrawing the circuit will result in the following configuration (Fig. 5.24):

(a)

$$A_v = \frac{V_o}{V_i} = \frac{V_2}{V_1} = \frac{-h_{fe}Z_L}{h_{ie} + (h_{ie}h_{oe} - h_{fe}h_{re})Z_L}$$

$$= \frac{-80(5 \times 10^3)}{1.5 \times 10^3 + [(1.5 \times 10^3)(20 \times 10^{-6}) - (80)(3 \times 10^{-4})](5 \times 10^3)}$$

$$= \frac{-400 \times 10^3}{1.5 \times 10^3 + (6 \times 10^{-3})(5 \times 10^3)} = \frac{-400 \times 10^3}{1.5 \times 10^3 + 30}$$

$$= \frac{-400 \times 10^3}{1530} = \mathbf{-261.44}$$

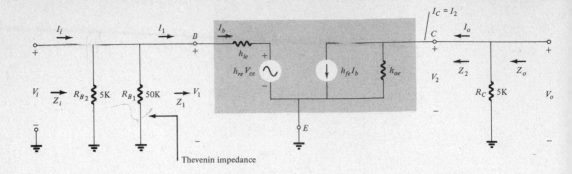

Thevenin impedance

Figure 5.24. Circuit of Fig. 5.23 following the substitution of the small-signal hybrid equivalent circuit for the transistor.

(b) Z_1 is required to determine the relationship between I_i and I_1

$$Z_1 = h_{ie} - \frac{h_{fe}h_{re}Z_L}{1 + h_{oe}Z_L} = 1.5 \times 10^3 - \frac{(80)(3 \times 10^{-4})(5 \times 10^3)}{1 + (20 \times 10^{-6})(5 \times 10^3)}$$

$$= 1.5 \times 10^3 - \frac{120}{1 + 0.1} = 1.5 \times 10^3 - \frac{120}{1.1}$$

$$= 1.5 \times 10^3 - 109.09$$

$$Z_1 \cong 1391 \ \Omega$$

The current divider rule can then be applied to the equivalent circuit of Fig. 5.25.

$$I_1 = \frac{4.55 \ \text{K} \ I_i}{4.55 \ \text{K} + 1.391 \ \text{K}} \cong 0.766 \ I_i$$

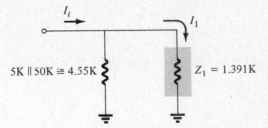

Figure 5.25. Determining the relationship between I_1 and I_i for the circuit of Fig. 5.24.

The current gain

$$A_i = \frac{I_o}{I_i} = \frac{I_2}{I_i} = \left[\frac{I_1}{I_i}\right]\left[\frac{I_2}{I_1}\right] = [0.766]\left[\frac{h_{fe}}{1 + h_{oe}R_L}\right]$$

$$= [0.766]\left[\frac{80}{1 + (20 \times 10^{-6})(5 \times 10^3)}\right] = [0.766]\left[\frac{80}{1 + 0.1}\right]$$

$$= \frac{61.28}{1.1} \cong \mathbf{55.71}$$

(c) $\qquad Z_i = 5 \ \text{K} \| 50 \ \text{K} \| Z_1 = 4.55 \ \text{K} \| 1.391 \ \text{K} \cong 1.065$

(d) R_s, as defined by Fig. 5.18, is required to determine Z_2. As in the previous example, it is simply the Thévenin impedance indicated in Fig. 5.24 and redrawn in Fig. 5.26.

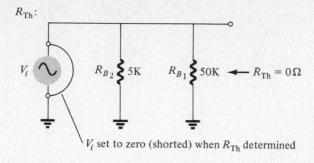

R_{Th}:

V_i set to zero (shorted) when R_{Th} determined

Figure 5.26. Determining R_{Th} for the portion of the circuit indicated in Fig. 5.24.

$$R_s = R_{Th} = 0$$

so that
$$Z_2 = \frac{1}{h_{oe} - \dfrac{h_{fe}h_{re}}{h_{ie} + R_s}} = \frac{1}{20 \times 10^{-6} - \dfrac{80(3 \times 10^{-4})}{1.5 \times 10^3}}$$

$$= \frac{1}{20 \times 10^{-6} - 16 \times 10^{-6}} = \frac{1}{4 \times 10^{-6}}$$

$$Z_2 = 250 \text{ K}$$

with
$$Z_o = Z_2 \| 5 \text{ K} = 250 \text{ K} \| 5 \text{ K} \cong 4.9 \text{ K}$$

EXAMPLE 5.3 Find the following quantities for the common-base configuration of Fig. 5.27:

(a) $A_i = I_o/I_i$.
(b) $A_{v1} = V_o/V_i$.
(c) Z_i.
(d) Z_o.

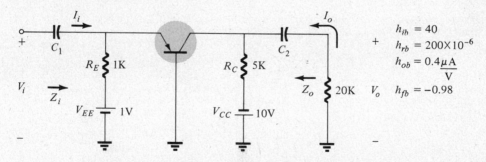

$h_{ib} = 40$
$h_{rb} = 200 \times 10^{-6}$
$h_{ob} = 0.4 \dfrac{\mu A}{V}$
$h_{fb} = -0.98$

Figure 5.27. Network for Example 5.3.

Solution: Replacing the dc supplies and capacitors by short circuits will result in the following configuration (Fig. 5.28).

(a) The equivalent load $Z_L = R_L = 5 \text{ K} \| 20 \text{ K} = 4 \text{ K}$

and
$$Z_1 = h_{ib} - \frac{h_{fb}h_{rb}Z_L}{1 + h_{ob}Z_L} = 40 - \frac{(-0.98)(200 \times 10^{-6})(4 \times 10^3)}{1 + (0.4 \times 10^{-6})(4 \times 10^3)}$$

$$= 40 - \frac{(-785 \times 10^{-3})}{1 + 1.6 \times 10^{-3}} = 40 + \frac{785 \times 10^{-3}}{1.0016}$$

$$Z_1 \cong 40.8 \ \Omega$$

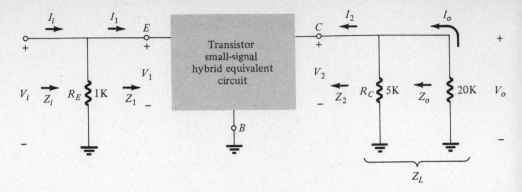

Figure 5.28. Circuit of Fig. 5.27 redrawn for small-signal ac analysis.

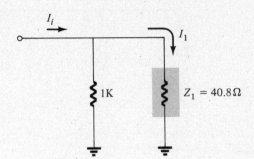

$Z_1 = 40.8\,\Omega$

Figure 5.29. Determining the relationship between I_1 and I_i for the circuit of Fig. 5.28.

The relationship between I_1 and I_i can then be determined by applying the current divider to the equivalent circuit of Fig. 5.29

$$I_1 = \frac{1\text{ K }(I_i)}{1\text{ K } + 40.8\ \Omega} = \frac{1\text{ K }(I_i)}{1040.8} \cong 0.96\,I_i$$

and the relationship between I_o and I_2 (Fig. 5.28).

Current divider rule:

$$I_o = \frac{5\text{ K }(I_2)}{5\text{ K } + 20\text{ K}} = 0.2\,I_2$$

so that

$$A_i = \frac{I_o}{I_i} = \left[\frac{I_o}{I_2}\right]\left[\frac{I_2}{I_i}\right] = \left[\frac{I_o}{I_2}\right]\left[\frac{I_1}{I_i}\right]\left[\frac{I_2}{I_1}\right]$$

$$= [0.2]\,[0.96]\left[\frac{h_{fb}}{1 + h_{ob}Z_L}\right] = [0.192]\left[\frac{-0.98}{1 + (0.4 \times 10^{-6})(4 \times 10^3)}\right]$$

$$= \frac{-0.188}{1 + 1.6 \times 10^{-3}} = \frac{-0.188}{1.0016} \cong \mathbf{-0.188}$$

(b)

$$A_{v1} = \frac{V_o}{V_i} = \frac{V_2}{V_1} = \frac{-h_{fb}Z_L}{h_{ib} + (h_{ib}h_{ob} - h_{fb}h_{rb})Z_L}$$

$$= \frac{-(-0.98)(4 \times 10^3)}{40 + [(40)(0.4 \times 10^{-6}) - (-0.98)(200 \times 10^{-6})]4 \times 10^3}$$

$$= \frac{3.92 \times 10^3}{40 + (16 \times 10^{-6} + 198 \times 10^{-6})(4 \times 10^3)}$$

$$= \frac{3.92 \times 10^3}{40 + (214 \times 10^{-6})(4 \times 10^3)} = \frac{3.92 \times 10^3}{40 + 0.856} = \mathbf{96}$$

(c) $$Z_i = 1\,K\,||\,Z_1 = 1\,K\,||\,40.8\,\Omega \cong 39.3\,\Omega$$

(d)

$$Z_2 = \cfrac{1}{h_{ob} - \cfrac{h_{fb}h_{rb}}{h_{ib} + R_s}} = \cfrac{1}{0.4 \times 10^{-6} - \cfrac{(-0.98)(200 \times 10^{-6})}{40 + 0}}$$

$$= \cfrac{1}{0.4 \times 10^{-6} - \cfrac{-196 \times 10^{-6}}{40}} = \cfrac{1}{0.4 \times 10^{-6} + 4.9 \times 10^{-6}}$$

$$= \frac{1}{5.3 \times 10^{-6}}$$

$$Z_2 \cong 189\,K$$

and $$Z_o = 5\,K\,||\,Z_2 = 5\,K\,||\,189\,K \cong 4.87\,K$$

5.6 APPROXIMATIONS FREQUENTLY APPLIED WHEN USING THE HYBRID EQUIVALENT CIRCUIT AND ITS RELATED EQUATIONS

On many occasions an exact theoretical solution requiring the use of the complete hybrid equivalent circuit is not required. Rather, an approximate solution, requiring considerably less time, is often desirable. In fact, considering that the h-parameters of a particular type of transistor will vary, even if only slightly, from transistor to transistor of the same manufacturing series with the operating point and with temperature, an approximate solution will often be as "accurate" as one involving all the parameters. To derive and demonstrate the validity of some of the approximation to be made, the transistor parameters indicated in Table 5.1 will be used. The common-emitter parameters are repeated here for convenience.

$$h_{fe} = 50, \qquad h_{ie} = 1000\,\Omega, \qquad h_{re} = 2.5 \times 10^{-4}, \qquad h_{oe} = 25\,\mu A/V$$

For the purposes of discussion we shall assume that the impedance of the source V_s is 1 K and that the load Z_L is 2 K (resistive).

The Current Gain A_i

$$A_i = \frac{h_{fe}}{(1 + h_{oe}Z_L)} \qquad \text{(Eq. 5.10)}$$

Substituting values:

$$(1 + h_{oe}Z_L) = [1 + (25 \times 10^{-6})(2 \times 10^3)] = (1 + 50 \times 10^{-3})$$
$$= (1 + 0.05) \cong 1$$

and $$A_i = \frac{h_{fe}}{(1 + h_{oe}Z_L)} \cong \frac{50}{1} \cong 50$$

resulting in

$$\boxed{A_i \cong h_{fe}} \qquad\qquad (5.19)$$

(Note that the term $h_{oe}Z_L$ in Eq. (5.10) is negligible because of the small value of h_{oe}.)

The Voltage Gain A_v

$$A_v = \frac{-h_{fe}Z_L}{h_{ie} + (h_{ie}h_{oe} - h_{fe}h_{re})Z_L} \qquad \text{(Eq. 5.11)}$$

Substituting values:

$$(h_{ie}h_{oe} - h_{fe}h_{re}) = [(1 \times 10^3)(25 \times 10^{-6}) - (50)(2.5 \times 10^{-4})]$$
$$= (25 \times 10^{-3} - 125 \times 10^{-4})$$
$$= 125 \times 10^{-4}$$

and $\quad h_{ie} + (h_{ie}h_{oe} - h_{fe}h_{re})Z_L = 1000 + (125 \times 10^{-4})(2 \times 10^3)$
$$= 1000 + 25 \cong 1000 = h_{ie}$$

and $\qquad A_v \cong \dfrac{-50(2 \times 10^3)}{1 \times 10^3} = -100$

so that

$$\boxed{A_v \cong \frac{-h_{fe}}{h_{ie}}Z_L} \qquad (5.20)$$

due to the fact that the term $Z_L(h_{ie}h_{oe} - h_{fe}h_{re})$ is negligible compared to h_{ie} in Eq. (5.11) because of the small values of h_{oe} and h_{re}.

The Input Impedance Z_1

$$Z_1 = h_{ie} - \frac{h_{fe}h_{re}Z_L}{1 + h_{oe}Z_L} \qquad \text{(Eq. 5.12)}$$

Substituting values in Eq. (5.12):

$$h_{fe}h_{re}Z_L = (50)(2.5 \times 10^{-4})(2 \times 10^3) = 25$$

and $\qquad \dfrac{h_{fe}h_{re}Z_L}{(1 + h_{oe}Z_L) \cong 1} \cong \dfrac{25}{1} = 25$

and $\qquad Z_1 = h_{ie} - \dfrac{h_{fe}h_{re}Z_L}{1 + h_{oe}Z_L} = 1000 - 25 \cong 1000 = h_{ie}$

so that

$$\boxed{Z_1 \cong h_{ie}} \qquad (5.21)$$

The Output Impedance Z_2

$$Z_2 = \frac{1}{h_{oe} - \dfrac{h_{fe}h_{re}}{h_{ie} + R_s}} \qquad \text{(Eq. 5.13)}$$

Equation (5.13) rewritten in a slightly different form:

$$Z_2 = \frac{h_{ie} + R_s}{h_{oe}(h_{ie} + R_s) - h_{fe}h_{re}}$$

Substituting values:

$$h_{oe}(h_{ie} + R_s) - h_{fe}h_{re} = 25 \times 10^{-6}(1000 + 1000) - 50(2.5 \times 10^{-4})$$
$$= 50 \times 10^{-3} - 12.5 \times 10^{-3}$$

For Z_2 the order of magnitude of $h_{oe}(h_{ie} + R_s)$ is too close to $h_{fe}h_{re}$ to warrant dropping the smaller quantity. Therefore, there is no clear-cut approximation that can be made for Z_2. Frequently, however, the relationship

$$\boxed{Z_2 > \frac{1}{h_{oe}}} \tag{5.22}$$

is used to obtain some idea of the order of magnitude for the output impedance.

Power Gain A_p

$$A_p = \frac{A_i^2 R_L}{R_i}$$

Through substitution in Eq. (5.18) of the complete expressions for A_i and $Z_i(R_i)$ and the typical values indicated for the parameters at the beginning of this section it can be shown that

$$\boxed{A_p \cong \frac{h_{fe}^2 R_L}{h_{ie}}} \tag{5.23}$$

In summary, the exact equation and the frequently used approximate form for each quantity discussed above are presented in Table 5.2.

TABLE 5.2 Exact vs. Approximate Equations for the
Quantities of Interest for the Basic Transistor
Amplifier

QUANTITY	EXACT	APPROXIMATE
A_i	$A_i = \dfrac{h_{fe}}{1 + h_{oe}Z_L}$	$A_i = h_{fe}$
A_v	$A_v = \dfrac{-h_{fe}Z_L}{h_{ie} + (h_{ie}h_{oe} - h_{fe}h_{re})Z_L}$	$A_v = \dfrac{-h_{fe}Z_L}{h_{ie}}$
Z_1	$Z_1 = h_{ie} - \dfrac{h_{fe}h_{re}Z_L}{1 + h_{oe}Z_L}$	$Z_1 = h_{ie}$
Z_2	$Z_2 = \dfrac{1}{h_{oe} - \dfrac{h_{fe}h_{re}}{h_{ie} + R_s}}$	$Z_2 > \dfrac{1}{h_{oe}}$
A_p	$A_p = \dfrac{A_i^2 R_L}{R_i}$	$A_p = \dfrac{h_{fe}^2 R_L}{h_{ie}}$

For comparison purposes the exact and approximate values for each quantity using the h-parameters employed in the above derivations have been listed in Table 5.3.

QUANTITY	EXACT	APPROXIMATE
A_i	47.62	50
A_v	−97.5	−100
Z_1	975 Ω	1000 Ω
Z_2	53.3 K	$Z_2 > 40$ K
A_p	4650	5000

Let us now turn our attention to the hybrid equivalent circuit and note the effect of these approximations on the circuit itself. In all cases, except for Z_2, the hybrid parameter h_{re} does not appear in the approximate equation (Table 5.2) for each quantity. On this basis, the voltage-controlled source $h_{re}V_2$ can be replaced by a short circuit whenever approximate values of A_i, A_v, and Z_i are desired. As far as Z_2 is concerned, we shall simply keep in mind that whenever $h_{re}V_2$ is eliminated, the resulting Z_2 will be somewhat smaller than the actual value. Removing $h_{re}V_2$ from the complete hybrid equivalent circuit will result in the approximate form of Fig. 5.30. For the circuit just discussed

$$\frac{1}{h_{oe}} = \frac{1}{25 \times 10^{-6}} = 40 \text{ K}$$

and $Z_L = 2$ K. The parallel combination of the two:

$$40 \text{ K} \| 2 \text{ K} = \frac{40 \text{ K} \times 2 \text{ K}}{40 \text{ K} + 2 \text{ K}} = \frac{80 \text{ K}}{42 \text{ K}} = 1.91 \text{ K} \cong 2 \text{ K} = Z_L$$

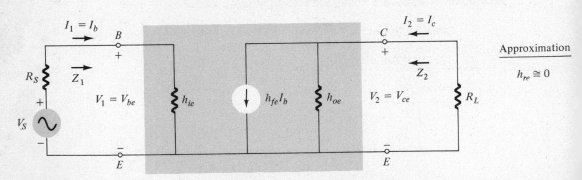

Figure 5.30. Circuit of Fig. 5.18 following the application of the approximation: $h_{re} \cong 0$.

For a number of applications the parallel combination of $1/h_{oe}$ and Z_L will result in an equivalent resistance close in magnitude to the load resistance. On this basis, the parameter h_{oe} can be eliminated as a second approximation from the circuit of Fig. 5.30, resulting in the reduced form of the hybrid equivalent circuit in Fig. 5.31.

The circuit of Fig. 5.31 would certainly require fewer calculations to obtain such

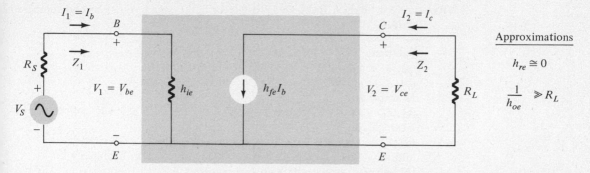

Figure 5.31. Circuit of Fig. 5.30 following the application of: $1/h_{oe} \gg R_L$.

quantities as the output voltage and current than the complete hybrid equivalent circuit. The conditions associated with the circuit of Fig. 5.31 must always be considered before it can be applied.

In the examples that follow we shall find that since h_{re} and h_{oe} have been removed from the equivalent circuit, $Z_2 = \infty$ Ω (open circuit). At first glance, this might appear to be a completely unacceptable approximation. We must keep in mind, however, that in this situation it is simply stating that compared to the other circuit parameters Z_2 is *sufficiently large* to be considered an open circuit.

There is one further approximation that can sometimes be made and that will certainly lead us to accept the fact that the transistor is a current-controlled device. In some cases, the resistance R_s, which was defined to be the "total" resistance of the source impedance and any resistance added in series with the source to limit the input current, is a bit larger in magnitude than h_{ie}. On this basis, if we drop h_{ie} from the picture, the result is Fig. 5.32.

It is obvious from Fig. 5.32 that the only link between the input and output circuits is a current source whose magnitude is controlled by the base current I_b. It would appear, however, that the number of conditions that must be satisfied before the circuit can be employed would substantially limit its use in transistor circuit analysis. It must be kept in mind, however, that in many manuals h_{fe} is the *only* parameter provided. This would require the use of the approximate equivalent circuit of Fig. 5.32 if some idea of the circuit's response is to be obtained.

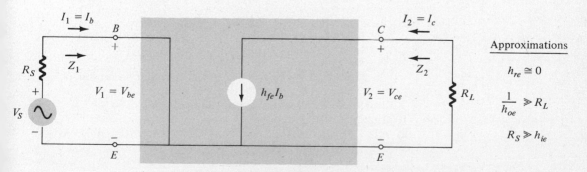

Figure 5.32. Circuit of Fig. 5.31 following the application of: $R_s \gg h_{ie}$.

SEC. 5.6 APPROXIMATIONS FREQUENTLY APPLIED

231

EXAMPLE 5.4 Repeat Example 5.1 using the approximate equations and compare results.

Solution: $(h_{fe} = 50, h_{re} = 2 \times 10^{-4}, h_{oe} = 20 \times 10^{-6}, h_{ie} = 1 \text{ K})$
(a) $A_i \cong h_{fe} = \mathbf{50}$ vs. 48.1 obtained in Example 5.1.
(b) $A_v \cong -h_{fe}Z_L/h_{ie} = -50(2 \times 10^3)/1 \text{ K} = \mathbf{-100}$ vs. -98 obtained in Example 5.1.
(c) $Z_i = 200 \text{ K} \| Z_1 \cong 200 \text{ K} \| h_{ie} \cong h_{ie} = \mathbf{1 \text{ K}}$ vs. 0.981 K obtained in Example 5.1.
(d) $Z_2 > 1/h_{oe} = 1/20 \times 10^{-6} = \mathbf{50 \text{ K}}$.
and $Z_o = 2 \text{ K} \| > 50 \text{ K} \cong \mathbf{2 \text{ K}}$ vs. 1.96 K obtained in Example 5.1.
(e) $A_p = A_v A_i \cong (-100)(50) = \mathbf{5 \times 10^3}$ vs. 4.7×10^3 obtained in Example 5.1.

In each case it is obvious that the approximate solution is in the "ball park" if a rough idea of a system's response or characteristics are required with a minimum of time and effort.

From Example 5.4 we can conclude for the fixed-bias circuit with typical values that

$$A \cong h_{fe}, \qquad A_v \cong \frac{-h_{fe}Z_L}{h_{ie}}, \qquad Z_i \cong h_{ie}, \qquad Z_2 > \frac{1}{h_{oe}}$$

and $Z_o \cong R_L$.

EXAMPLE 5.5 Repeat Example 5.2 using the approximate equivalent circuit and compare results.

Solution: Figure 5.22 will appear as shown in Fig. 5.33 following the substitution of the approximate equivalent circuit.

h_{oe} does not appear since

$$\frac{1}{h_{oe}} = \frac{1}{120 \ \mu\text{A/V}} = 50 \text{ K} \| 5 \text{ K} \cong 5 \text{ K} \ (10:1 \text{ ratio})$$

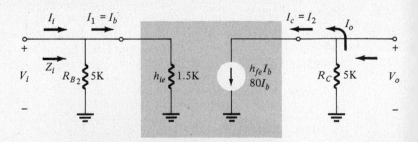

Figure 5.33. Approximate equivalent circuit for the network of Fig. 5.22.

In addition, R_{B_1}, does not appear since $R_{B_1} \| R_{B_2} = 50 \text{ K} \| 5 \text{ K} \cong 5 \text{ K}$ (10:1 ratio)
(a) A_v:

$$V_o = -I_o R_c = -h_{fe} I_b R_c$$

and

$$I_b = \frac{V_i}{h_{ie}}$$

so that

$$V_o = -h_{fe}\left(\frac{V_i}{h_{ie}}\right)R_c$$

and

$$A_v = \frac{V_o}{V_i} = \frac{-h_{fe}}{h_{ie}}R_c$$

as obtained for the fixed-bias network in Example 5.4.

Substituting values:

$$A_v = \frac{-(80)}{1.5\text{ K}}(5\text{ K}) \cong \mathbf{267}$$

as compared to 261.44 obtained earlier using the complete model.

(b) A_i:

$$I_o = h_{fe}I_b$$

Current divider rule:

$$I_b = \frac{5\text{ K}(I_i)}{5\text{ K} + 1.5\text{ K}} = 0.769I_i$$

and

$$I_o = h_{fe}(0.769I_i) = (80)(0.769)I_i = 61.52I_i$$

with

$$A_i = \frac{I_o}{I_i} \cong \mathbf{61.52}$$

as compared to 55.71 obtained earlier.

(c) Z_i:

$$Z_i \cong 5\text{ K}\,\|\,1.5\text{ K} = \mathbf{1.15\text{ K}}$$

as compared to 1.065 K obtained earlier.

(d) Z_o:

The output impedance is defined by the condition $V_i = 0$. Therefore,

$$I_b = \frac{V_i}{h_{ie}} = 0 \quad \text{and} \quad h_{fe}I_b = 0 \quad \text{(an open-circuit equivalent)}$$

and

$$Z_o = \mathbf{5\text{ K}}$$

as compared to 4.9 K obtained earlier. Take a moment to compare the degree of effort to obtain the results above as compared to using the complete model.

EXAMPLE 5.6 Determine $A_v, A_i, Z_i,$ and Z_o for the configuration appearing in

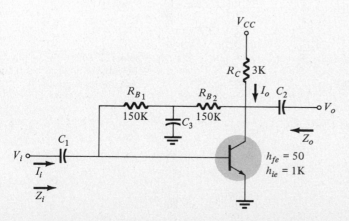

Figure 5.34. Network for Example 5.6.

Fig. 5.34 using the approximate equivalent circuit. Conditions are such that the effect of h_{re} and h_{oe} can be ignored.

Solution: Connecting the dc supplies to ground potential and replacing all capacitors by a short-circuit equivalent will result in the network of Fig. 5.35.

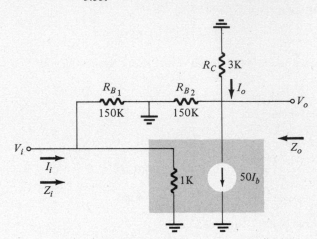

Figure 5.35. Network of Fig. 5.34 following the substitution of the approximate equivalent circuit and removing the dc levels.

Redrawing the circuit, we obtain Fig. 5.36 and, finally, Fig. 5.37.

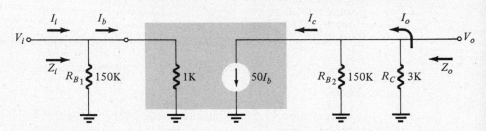

Figure 5.36. Reconstruction of Fig. 5.35.

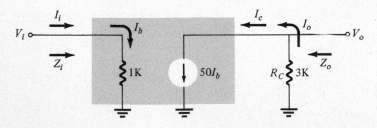

Figure 5.37. Network of Fig. 5.36 following the application of additional approximations.

From Fig. 5.37

$$V_o = -I_o R_L = -h_{fe} I_b R_L$$

but

$$I_b = \frac{V_i}{h_{ie}}$$

and

$$V_o = -h_{fe} \left(\frac{V_i}{h_{ie}} \right) R_L$$

with
$$A_v = \frac{V_o}{V_i} = \frac{-h_{fe}}{h_{ie}} R_L$$

as obtained for Examples 5.4 and 5.5.
Substituting numbers:

$$A_v = -\frac{50(3\ K)}{1\ K} = -150$$

A_i:

From Fig. 5.37
$$I_i = I_b$$

and
$$I_o = h_{fe} I_b$$

with result that

$$A_i = \frac{I_o}{I_i} = \frac{h_{fe} I_b}{I_b} = h_{fe} = 50$$

Z_i:

From Fig. 5.37

$$Z_i = h_{ie} = 1\ K$$

Z_o:

From Fig. 5.37

$$Z_o|_{V_i=0} = R_L = 3\ K$$

EXAMPLE 5.7 For the common-base circuit of Fig. 5.38 with a load applied, find the following:
(a) $A_i = I_o/I_i$.
(b) $A_v = V_o/V_i$.
(c) Z_i.
(d) Z_o.

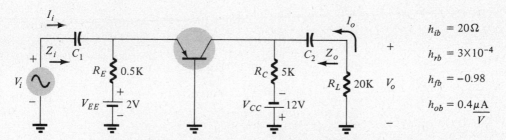

$h_{ib} = 20\Omega$

$h_{rb} = 3 \times 10^{-4}$

$h_{fb} = -0.98$

$h_{ob} = \dfrac{0.4\mu A}{V}$

Figure 5.38. Circuit for Example 5.7.

Solution: Replacing the dc supplies and coupling capacitors by short circuits will result in the configuration of Fig. 5.39.

Note, as mentioned in the above, the common-base equivalent circuit has exactly the same format as the common-emitter configuration but with the appropriate common-base parameters. Consider that now $h_{re}V_{ce}$ is $h_{rb}V_{cb}$ and $h_{fe}I_b$ is now $h_{fb}I_e$. Since the configuration is the same and $1/h_{ob}$ is greater than $1/h_{oe}$ with h_{re} usually close in magnitude to h_{rb}, the approximations ($1/h_{ob} \cong \infty\ \Omega$ and $h_{rb} \cong 0$) made for the common-emitter configuration are applicable here, also.

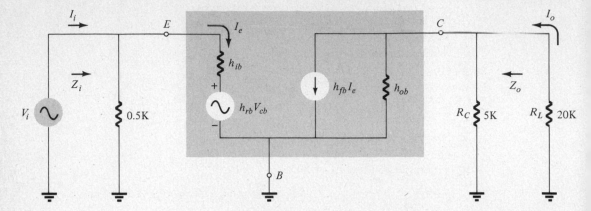

Figure 5.39. Circuit of Fig. 5.38 following the substitution of the hybrid equivalent circuit.

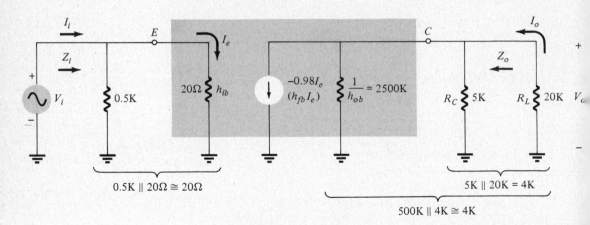

Figure 5.40. Redrawn circuit of Fig. 5.39 following the application of the approximation $h_{rb} \cong 0$.

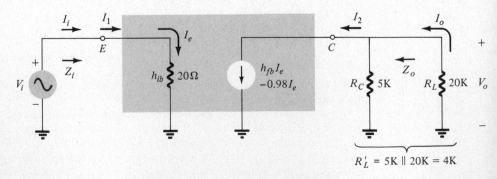

Figure 5.41. Circuit of Fig. 5.40 following the elimination (on an approximate basis) of certain parallel elements.

Applying $h_{rb} \cong 0$, we redraw the circuit (Fig. 5.40). Eliminating the 500-Ω and 500-K ($1/h_{ob}$) resistors due to the lower parallel resistor will result in the circuit of Fig. 5.41.

(a) $A_i = (I_o/I_i)$:

$$I_o = \frac{(5 \text{ K})(I_2)}{5 \text{ K} + 20 \text{ K}} = 0.2I_2 \qquad \text{(current divider rule)}$$

and $\quad I_2 = h_{fb}I_e = h_{fb}I_1 = h_{fb}I_i$ with A_i (transistor) $= \dfrac{I_2}{I_i} = h_{fb}$

Therefore, $\qquad\qquad\qquad I_o = 0.2I_2 = 0.2h_{fb}I_i$

and $\qquad\qquad A_i = \dfrac{I_o}{I_i} \cong 0.2(h_{fe}) = 0.2(-0.98) = -\mathbf{0.196}$

indicating as before that the current gain of the common-base configuration is always less than one.

(b) $A_v = \dfrac{V_o}{V_i}$:

Using $\qquad\qquad\qquad R'_L = 5 \text{ K} \,\|\, 20 \text{ K} = R_C \,\|\, R_L$

$$V_o = -I_2 R'_L = -h_{fb}I_e R'_L$$

and $\qquad\qquad\qquad\qquad I_e = I_i = \dfrac{V_i}{h_{ib}}$

so that $\qquad\qquad\qquad\quad V_o = -h_{fb}\left(\dfrac{V_i}{h_{ib}}\right)R'_L$

with $\qquad\qquad\qquad A_v = \dfrac{V_o}{V_i} = \dfrac{-h_{fb}}{h_{ib}}R'_L$

Note the similarities between this equation for voltage gain and that obtained for the common-emitter configurations. Consider also that the effect of the added load was only to change R_L to R'_L which is the parallel combination of R_C and the applied load R_L.

Substituting values, we get

$$A_v = -\frac{(-0.98)(4 \times 10^3)}{20} = \mathbf{196}$$

The voltage gain, therefore, can be significantly greater than one and results in an output that is *in phase* with the input (note the absence of the minus sign in the result).

(c) $Z_i \cong h_{ib} = \mathbf{20}\ \mathbf{\Omega}$.

(d) $Z_o|_{V_{i=0}} \cong R_C = \mathbf{5\ K}$.

EXAMPLE 5.8 Find the following for the circuit of Fig. 5.42:

(a) $A_i = I_o/I_i$.

(b) $A_v = V_o/V_i$.

(c) Z_i.

(d) Z_o.

Solution: Replacing the dc supplies and capacitors by short circuits and substituting the appropriate hybrid equivalent circuit will result in the configuration of Fig. 5.43. Note that the 2-K emitter resistor has been "shorted out" by the capacitor C_E.

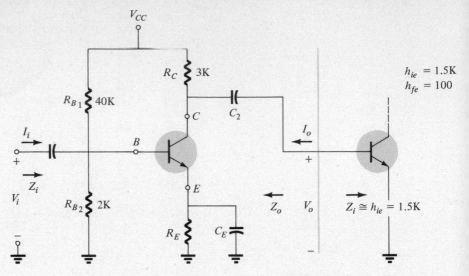

Figure 5.42. Circuit for Example 5.8.

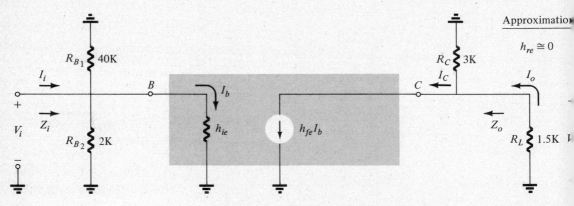

Figure 5.43. Circuit of Fig. 5.42 following the substitution of the approximate ($h_{re} \cong 0$) hybrid equivalent circuit.

Considering parallel elements will result in the configuration of Fig. 5.44.
(a) $A_i = (I_o/I_i)$:
From Fig. 5.44, $I_2 = 100I_b$. That is A_i (transistor) $= I_c/I_b = h_{fe}$. Applying the current divider rule to the input and output circuits, we get

$$I_b = \frac{2\,\text{K}\,I_i}{2\,\text{K} + 1.5\,\text{K}} = 0.571I_i$$

and

$$I_o = \frac{3\,\text{K}\,I_2}{3\,\text{K} + 1.5\,\text{K}} = 0.667I_2$$

Substituting, we get

$$A_i = \frac{I_o}{I_i} = \left[\frac{I_o}{I_2}\right]\left[\frac{I_2}{I_i}\right] = \left[\frac{I_o}{I_2}\right]\left[\frac{I_2}{I_b}\right]\left[\frac{I_2}{I_i}\right]$$

$$= [0.667][100][0.571]$$

$$= \mathbf{38.1}$$

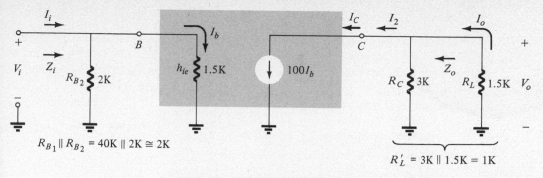

$R_{B_1} \| R_{B_2} = 40K \| 2K \cong 2K$

$R'_L = 3K \| 1.5K = 1K$

Figure 5.44. Circuit of Fig. 5.43 following the elimination (on an approximate basis) of certain parallel elements.

(b) $A_v = (V_o/V_i)$:

The configuration is similar to that obtained for both the CE and CB configurations and

$$A_v = \frac{V_o}{V_i} = \frac{-h_{fe}R'_L}{h_{ie}}$$

$$= -\frac{(100)(1 \times 10^3)}{1.5 \times 10^3}$$

$$= -66.7$$

(c) $Z_i \cong R_{B2} \| h_{ie} = 2\,\text{K} \| 1.5\,\text{K} = 0.86\,\text{K}$.

(d) $Z_o|_{V_{i=0}} \cong R_C = 3\,\text{K}$.

5.7 APPROXIMATE BASE, COLLECTOR, AND EMITTER EQUIVALENT CIRCUITS

In the following analysis it will prove very useful to know at a glance what the effect of loads and signal sources in another portion of the network will have on the base, collector, or emitter potential and current. In this section we shall find, on an approximate basis, the equivalent circuit "seen" looking into the base, collector, or emitter terminals of a transistor in the common-emitter configuration (Fig. 5.45).

The approximations $h_{re} \cong 0$ and $(1/h_{oe}) \cong \infty\,\Omega$ (open circuit), employed continually in Section 5.6, will be used throughout this section (they are valid only

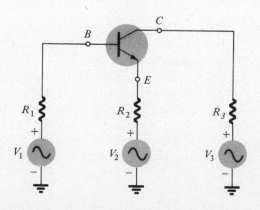

Figure 5.45. Representative circuit to be employed in the derivation of the base, collector, and emitter approximate equivalent circuits.

if $1/h_{oe} > R_3, R_2$). To include the effects of a supply and a load connected to each terminal the circuit of Fig. 5.45 will be employed. The full benefit of the circuits to be derived will be more apparent when the equivalent circuits have been obtained.

Substituting the approximate equivalent circuit for the transistor of Fig. 5.45 will result in the circuit of Fig. 5.46.

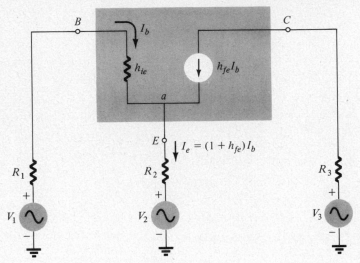

Figure 5.46. Circuit of Fig. 5.45 following the substitution of the approximate $(h_{re} \cong 0, \; h_{oe} \cong 0)$ hybrid equivalent circuit.

Base Circuit

The current through the resistor R_2 can be found by applying Kirchhoff's current law at node a:

$$I_{R_2} = I_b + h_{fe}I_b = (1 + h_{fe})I_b$$

Applying Kirchhoff's voltage law in the loop indicated in Fig. 5.46, we get

$$V_1 - I_bR_1 - h_{ie}I_b - (1 + h_{fe})I_bR_2 - V_2 = 0$$

Rewriting, we get

$$I_bR_1 + h_{ie}I_b + (1 + h_{fe})I_bR_2 = V_1 - V_2$$

and solving for I_b, we get

$$I_b = \frac{V_1 - V_2}{R_1 + h_{ie} + (1 + h_{fe})R_2} \tag{5.24}$$

The circuit that "fits" Eq. (5.24) appears in Fig. 5.47.

We must now take a moment to fully appreciate the benefits of having the base equivalent circuit of Fig. 5.47. It "tells" us that any signal (V_3) or load resistor from collector to ground (R_3) is not reflected to the equivalent "base" circuit of a transistor on an approximate basis. It also clearly indicates what effect the source V_2 and load resistor R_2 in the emitter leg will have on the base current I_b. Note that for the common-emitter circuit, when R_2 and $V_2 = 0$, if we substitute these conditions into the circuit of Fig. 5.47, the base circuit of Fig. 5.48 will result, which should by now be familiar.

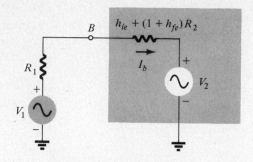

Figure 5.47. Approximate base equivalent circuit for the transistor.

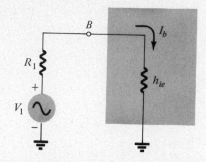

Figure 5.48. Circuit of Fig. 5.47 with the conditions $R_2 = 0$, $V_2 = 0$.

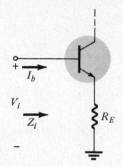

Figure 5.49. Effect of R_E on the input impedance.

A second glance at Fig. 5.47 will also reveal that any resistor in the emitter leg will appear as a much larger resistance in the base circuit due to the factor $(1 + h_{fe})$.

Typically $h_{fe} \gg 1$ and $(h_{fe} + 1) \cong h_{fe}$. In addition, it has been determined through practical experience that $(h_{fe} + 1)R_2 \cong h_{fe}R_2 \gg h_{ie}$, and h_{ie} can usually be dropped on an approximate basis. For the transistor configuration of Fig. 5.49, therefore, the input impedance Z_i is determined by

$$\boxed{Z_i \cong h_{fe}R_E} \tag{5.25}$$

with the result that

$$I_b \cong \frac{V_i}{Z_i} = \frac{V_i}{h_{fe}R_E}$$

Emitter Circuit

If Eq. (5.24) is multiplied by $(1 + h_{fe})$, the following equation will result:

$$(1 + h_{fe})I_b = (1 + h_{fe})\left[\frac{V_1 - V_2}{R_1 + h_{ie} + (1 + h_{fe})R_2}\right]$$

Rewritten:

$$(1 + h_{fe})I_b = I_e = \frac{V_1 - V_2}{\dfrac{R_1 + h_{ie}}{1 + h_{fe}} + R_2} \tag{5.26}$$

Constructing a circuit to "fit" Eq. (5.26), we obtain Fig. 5.50.

The "emitter circuit" also clearly indicates that a source (V_3) and load resistor (R_3) in the collector circuit are not reflected to the "emitter circuit." Note also that

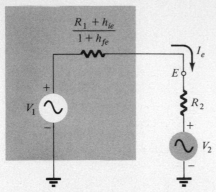

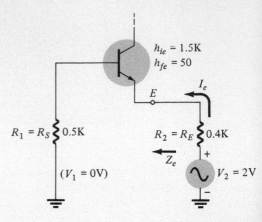

Figure 5.50. Approximate emitter equivalent circuit for the transistor.

Figure 5.51

V_1 will affect the current I_e but the reflected resistance is normally small in magnitude due to the division by the factor $(1 + h_{fe})$.

For the representative network of Fig. 5.51, if we use the approximation $(h_{fe} + 1) \cong h_{fe}$ and define $R'_s = R_S + h_{ie}$, then

$$Z_e = \frac{R_1 + h_{ie}}{1 + h_{fe}}$$

becomes

$$\boxed{Z_e \cong \frac{R'_s}{h_{fe}}}_{R'_s = R_S + h_{ie}}$$

(5.27)

and for the network of Fig. 5.51

$$I_e \cong \frac{V_2}{R_2 + Z_e} = \frac{V_2}{R_2 + \dfrac{R'_s}{h_{fe}}}$$

which for the values provided is

$$I_e = \frac{2}{400 + \dfrac{(0.5 + 1.5) \times 10^3}{50}} = \frac{2}{400 + 40} = \frac{2}{440}$$

$$= 4.55 \text{ mA}$$

Note the small magnitude of the transferred resistance Z_e. This is typical due to the division by the factor h_{fe}.

Collector Circuit

The current I_c is $h_{fe}I_b$ independent of the surrounding circuit. The collector voltage,

$$V_c = V_3 - I_cR_3 = V_3 - h_{fe}I_cR_3$$

is also independent of the external circuit. The resulting collector circuit is a

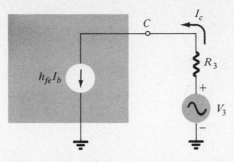

Figure 5.52. Approximate collector equivalent circuit for the transistor.

shown in Fig. 5.52. The collector current or potential, therefore, can be found directly after I_b has been determined.

The beneficial aspects of the equivalent circuits just derived will become obvious in the examples to follow and in later chapters. It would be wise to memorize these equivalent circuits for future use. They can be very powerful tools in analysis of transistor networks.

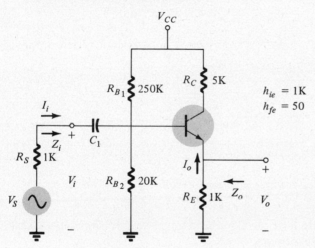

Figure 5.53. Circuit for Example 5.9.

EXAMPLE 5.9 The transistor configuration of Fig. 5.53, called the *emitter follower*, is frequently used for impedance matching purposes; that is, it presents a high impedance at the input terminals (Z_i) and a low output impedance (Z_o), rather than the reverse, which is typical of the basic transistor amplifier. The effect of the emitter follower circuit is much like that obtained using a transformer to match a load to the source impedance for maximum power transfer. The following analysis will reflect that the voltage gain of the emitter follower circuit is always less than one.

For the circuit of Fig. 5.53 calculate the following:

(a) Z_i.

(b) Z_o.

(c) $A_{v1} = V_o/V_s$ and $A_{v_2} = V_o/V_i$.

(d) $A_i = I_o/I_i$.

Solution: Eliminating the dc levels and replacing both capacitors by short circuits will result in the circuit of Fig. 5.54.

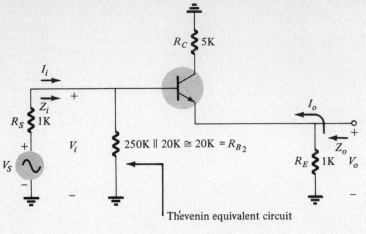

Figure 5.54. Circuit of Fig. 5.53 redrawn for small-signal ac analysis.

Thevenin equivalent circuit

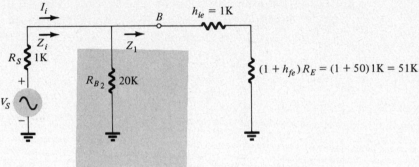

Figure 5.55. Substitution of the approximate base equivalent circuit into the circuit of Fig. 5.54.

(a) Using the base equivalent circuit (Fig. 5.55), we get

$$Z_1 = h_{ie} + (1 + h_{fe})R_2 = 52 \text{ K}$$

(certainly high compared to the typical input impedance $Z_i \cong h_{ie}$ for the basic transistor amplifier).

The resulting $Z_i = 20 \text{ K} \| 52 \text{ K} = \mathbf{14.5\ K}$

If we had used Eq. (5.25) and the conditions surrounding this approximation, the equivalent circuit would appear as shown in Fig. 5.56 and $Z_1 = h_{fe}R_E = 50 \text{ K}$ and $Z_i = 20 \text{ K} \| 50 \text{ K} \cong \mathbf{14.3\ K}$. This result will usually be so close to the result obtained using the complete approximate equivalent that we will use Eq. (5.25) throughout the following analysis unless otherwise noted.

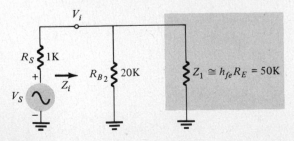

Figure 5.56. Approximation frequently used to determine the input impedance of a network with an unbypassed emitter resistor.

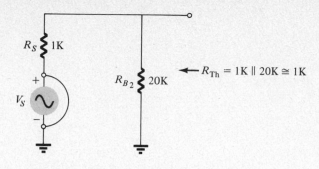

Figure 5.57. Determining R_{Th} for the portion of the circuit indicated in Fig. 5.54.

$$\leftarrow R_{\mathrm{Th}} = 1K \parallel 20K \cong 1K$$

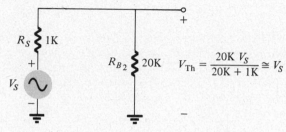

$$V_{\mathrm{Th}} = \frac{20K\ V_S}{20K + 1K} \cong V_S$$

Figure 5.58. Determining V_{Th} for the portion of the circuit indicated in Fig. 5.54.

(b) The Thévenin equivalent circuit of the portion of the network indicated in Fig. 5.54 will now be found so that the input circuit will have the basic configuration of Fig. 5.45. That is, R_1 and V_1 will be found to ensure that the proper values are substituted into the emitter equivalent circuit to be employed in the determination of Z_o.

R_{Th} (Fig. 5.57)
V_{Th} (Fig. 5.58)

Substituting the Thévenin equivalent circuit into the circuit of Fig. 5.54 will result in the configuration of Fig. 5.59.

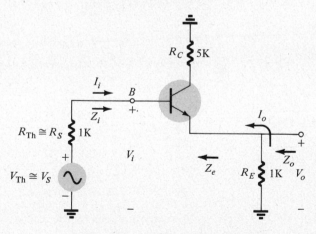

Figure 5.59. Circuit of Fig. 5.54 following the substitution of the Thévenin equivalent circuit.

Using the emitter equivalent circuit with $V_s = 0$ (as required by definition) to determine Z_o (Fig. 5.60), we get

$$Z_e = \frac{R'_S}{h_{fe}} = \frac{1\,K + 1\,K}{50} \cong \mathbf{40\ \Omega}$$

and

$$Z_o = Z_e \parallel R_L = 40 \parallel 1000 \cong \mathbf{40\ \Omega}$$

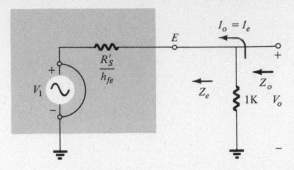

Figure 5.60. Substitution of the emitter equivalent circuit with V_1 set to zero.

(c) A careful examination of Fig. 5.53 will reveal that the output voltage V_o is separated from the input voltage V_i by only the voltage drop V_{be} across the base-to-emitter junction. For ac operations and an unbypassed (no capacitor present) emitter resistor the approximation indicated by Eq. (5.28) is frequently employed.

$$\boxed{V_{be} \cong 0V} {\scriptsize \begin{pmatrix} \text{ac operations and} \\ \text{unbypassed emitter resistor} \end{pmatrix}} \qquad (5.28)$$

Using this approximation, we get

$$V_o \cong V_i \quad \text{and} \quad A_{v2} = \frac{V_o}{V_i} \cong 1$$

For $A_{v1} = V_o/V_s$ we can turn to the network of Fig. 5.56 where

$$V_i = \frac{50\,\text{K}(V_s)}{50\,\text{K} + 1\,\text{K}} \cong 0.98 V_s$$

and

$$A_{o1} = \frac{V_o}{V_s} = \left[\frac{V_o}{V_i}\right]\left[\frac{V_i}{V_s}\right] = [\cong 1][0.98] = \mathbf{0.98}$$

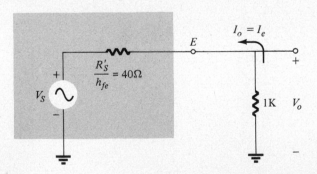

Figure 5.61. Circuit to be employed in determining A_{v_1}.

The validity of Eq. (5.28) can be demonstrated using the emitter equivalent circuit of Fig. 5.61 where

$$V_o = \frac{1\,\text{K}(V_s)}{1\,\text{K} + 40} = 0.96 V_s$$

and

$$A_{v2} = \frac{V_o}{V_s} = \mathbf{0.96} \cong 1$$

(d) From the emitter equivalent circuit:

$$I_o = I_e = -\frac{V_s}{1.040 \text{ K}}$$

and from Fig. 5.56

$$I_i = \frac{V_s}{R_s + Z_i} = \frac{V_s}{1 \text{ K} + 14.3 \text{ K}} = \frac{V_s}{15.3 \text{ K}}$$

or $V_s = I_i \, 15.3$ K

Substituting this result into the above equation, we get

$$I_o = -\frac{I_i 15.3 \text{ K}}{1.040 \text{ K}}$$

and

$$A_i = \frac{I_o}{I_i} = \frac{-15.3 \text{ K}}{1.040 \text{ K}} = -14.7$$

EXAMPLE 5.10 In Chapter 11 the *difference amplifier* will be discussed in detail. For the present, however, to demonstrate the usefulness of the base, collector, and emitter circuits, we shall consider a simple difference amplifier. In its basic form, a difference amplifier is simply a network that will produce a signal that is the difference of the two applied signals.

Figure 5.62 is such a circuit. Note that a signal has been applied to both the base and emitter leg of the transistor.

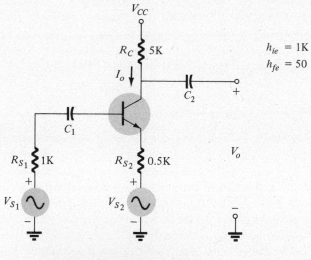

Figure 5.62. Circuit for Example 5.10—difference amplifier.

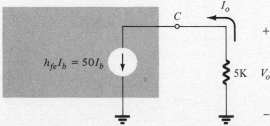

Figure 5.63. Application of the collector equivalent circuit to the circuit of Fig. 5.62.

Solution: Using the "collector equivalent circuit" after replacing the dc supply and capacitors by short circuits will result in the circuit of Fig. 5.63 where

$$V_o = -I_b 5 \text{ K} = -50 I_b 5 \text{ K}$$

Using the "base equivalent circuit" gives us Fig. 5.64 and

$$I_b \cong \frac{V_{s_1} - V_{s_2}}{R_{s_1} + h_{fe} R_{s_2}} = \frac{V_{s_1} - V_{s_2}}{1 \text{ K} + 25 \text{ K}} = \frac{V_{s_1} - V_{s_2}}{26 \text{ K}}$$

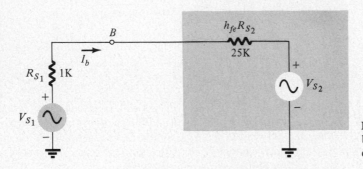

Figure 5.64. Application of the base equivalent circuit to the circuit of Fig. 5.62.

Substituting into the equation for V_o

$$V_o = -50 I_b 5 \text{ K} = -50 \frac{(V_{s_1} - V_{s_2})}{26 \text{ K}} 5 \text{ K} = \frac{-250(V_{s_1} - V_{s_2})}{26 \text{ K}}$$

and

$$V_o \cong 9.62(V_{s_1} - V_{s_2})$$

The collector potential V_o, therefore, is approximately 9.62 times the difference of the two applied signals. As mentioned earlier, difference amplifiers will be discussed in much greater detail in Chapter 12.

5.8 AN ALTERNATE APPROACH

In recent years there has been an increasing interest in an approximate equivalent circuit for the transistor in which one of the parameters is determined by the dc operating conditions. You will possibly recall from the transistor specification sheet provided in Chapter 3 that the hybrid parameter h_{ie} was specified at a particular operating point. Figure 5.14 revealed a significant variation in h_{ie} with $I_C (\cong I_E)$. The question then arises of what one would do with the provided value of h_{ie} if the conditions of operation (level of $I_C \cong I_E$) were different from those indicated on the specification sheet. The equivalent circuit derived below will permit the determination of an equivalent h_{ie} using the dc operating conditions of the network, thereby not limiting itself to the data on the device as provided by the manufacturer.

The derivation of the alternate equivalent circuit begins with a close examination of the input, and output characteristics, of the CB transistor configuration, as redrawn in Fig. 5.65, on an approximate basis. Note that straight line segments are used to represent the collector characteristics and a single diode characteristic for the emitter circuit (neglecting the variation in the input characteristics with the change in V_{CB}), resulting in an equivalent circuit such as shown in Fig. 5.66b. For ac conditions, therefore, the input impedance at the emitter of the CB transistor can be

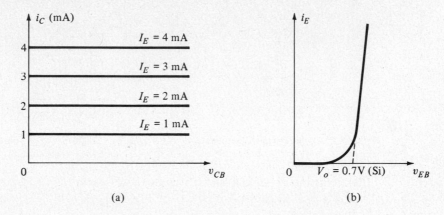

(a)

(b)

Figure 5.65. Approximate CB characteristics: (a) output; (b) input.

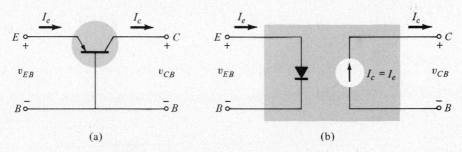

(a)

(b)

Figure 5.66. (a) CB configuration; (b) approximate CB equivalent circuit as defined by Fig. 5.65.

determined using Eq. (1.10) as introduced for the ac resistance of a diode. The factor r_B will be dropped to ensure that it does not affect the clarity of the introduction of this alternate technique. You will recall from Chapter 1 that in time it is conceivable that the factor r_B can be totally ignored with a negligible loss in accuracy if manufacturing techniques continue to improve. In time, when the alternate approach is developed, if you prefer to add a factor developed through experience for that transistor, then it can surely be introduced with little added confusion. For now we will define the input impedance for the CB configuration to be

$$r_e = \frac{26\ \text{mV}}{I_E\ (\text{mA})} \qquad \text{(ohms)} \qquad (5.29)$$

where I_E is the dc emitter current of the transistor. The emitter current is employed in Eq. (5.29) since it is the defined current of the diode in Fig. 5.65b. The result of the above discussion is the input equivalent circuit appearing in Fig. 5.67.

Figure 5.65a clearly indicates that the collector curves have been approximated to result in $I_C = I_E$ at any point on the characteristics. This would result in the output equivalent circuit appearing in Figs. 5.66 and 5.67. On an approximate

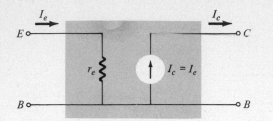

Figure 5.67. Approximate CB equivalent circuit.

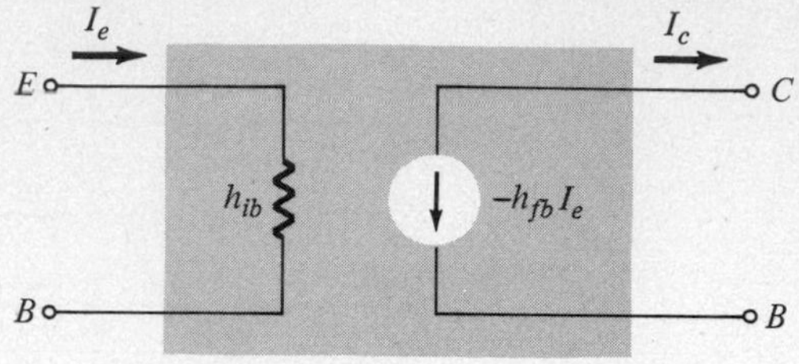

Figure 5.68. Approximate CB hybrid equivalent circuit.

basis, the alternate equivalent circuit is now defined. Note its similarities with the reduced hybrid equivalent circuit of Fig. 5.68. A comparison of the two clearly indicates that

$$\begin{aligned} h_{ib} &= r_e \\ h_{fb} &= 1 \end{aligned}$$

(5.30)

The following example will clarify the use of the alternate equivalent circuit.

EXAMPLE 5.11 For the network of Fig. 5.69, determine A_v, A_i, Z_i, and Z_o.

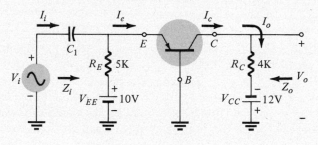

Figure 5.69. Network for Example 5.11.

Solution: dc conditions:

$$I_E = \frac{V_{EE} - V_{BE}}{R_E} = \frac{10 - 0.7}{5\,K} = \frac{9.3}{5\,K} = 1.86\,\text{mA}$$

and

$$r_e = \frac{26\,\text{mV}}{I_E} = \frac{26\,\text{mV}}{1.86\,\text{mA}} \cong 14\,\Omega$$

ac conditions:
The network redrawn (Fig. 5.70):

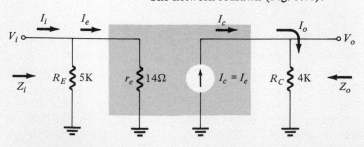

Figure 5.70. ac equivalent for the network of Fig. 5.69.

A_v:

$$V_o = I_c R_C = I_e R_C$$

and

$$I_e = \frac{V_i}{r_e}$$

so that

$$V_o = \left(\frac{V_i}{r_e}\right) R_C$$

and

$$\boxed{A_v = \frac{V_o}{V_i} = \frac{R_C}{r_e}}$$

Compare this result to that obtained in Example 5.7 where

$$|A_v| = \frac{h_{fb}}{h_{ib}} R_L'$$

but in this case $R_L' = R_C$, $h_{fb} = 1$, $h_{ib} = r_e$ which through substitution will result in Eq. 5.31. Substituting values, we get

$$A_v = \frac{R_C}{r_e} = \frac{4\,\text{K}}{14} = \mathbf{285.71}$$

A_i:

Since $R_E \| r_e \cong r_e$,

$$I_o = I_c = I_e = I_i$$

and

$$\boxed{A_i \cong 1 \cong h_{fb}}$$

as obtained in Example 5.7.

Z_i:

$$\boxed{Z_i \cong r_e = h_{ib}} = \mathbf{14\,\Omega}$$

Z_o:

$$\boxed{Z_o|_{V_i=0} = R_C} = \mathbf{4\,K}$$

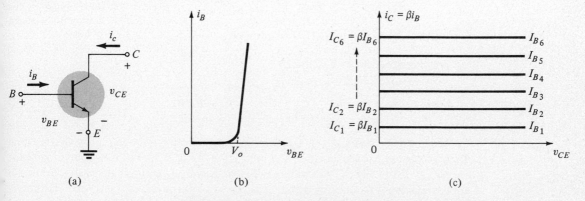

Figure 5.71. (a) CE configuration; (b) input characteristics; (c) output characteristics.

For the common-emitter configuration appearing in Fig. 5.71a the input and output characteristics have been approximated by the set appearing in Fig. 5.71b and 5.71c, respectively. The base characteristics are again approximated to be those of a diode (the effect of V_{CE} on the characteristics is ignored) and

$$r_{ac} = \frac{26 \text{ mV}}{I_B} \qquad (5.31a)$$

But $\qquad I_E \cong I_C = \beta I_B \quad$ and $\quad I_B \cong \dfrac{I_E}{\beta}$

so that $\qquad r_{ac} = \dfrac{26 \text{ mV}}{I_B} = \dfrac{26 \text{ mV}}{I_E/\beta} = \beta\left(\dfrac{26 \text{ mV}}{I_E}\right)$

or $\qquad \boxed{r_{ac} = \beta r_e} \qquad (5.31b)$

Equation (5.31) has the same format as Eq. (5.25) ($Z_i \cong h_{f_e}R_E$) used to reflect an emitter resistor to the base circuit. In this case Eq. (5.31) can be directly derived using the same equation [Eq. (5.25)] and Fig. 5.72a where r_e appears as a resistor in the emitter leg.

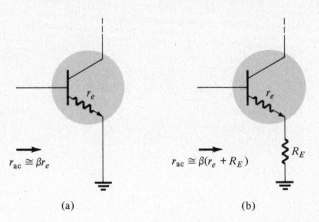

$$r_{ac} \cong \beta r_e \qquad\qquad r_{ac} \cong \beta(r_e + R_E)$$

(a) (b)

Figure 5.72. Determining r_{ac} for the CE configuration (a) bypassed R_E; (b) unbypassed R_E.

For the situation of Fig. 5.72b,

$$\boxed{r_{ac} = \beta(r_e + R_E) \cong \beta R_E} \qquad (5.32)$$

The input circuit for the CE configuration is approximated, for the reasons discussed above, by the diode circuit appearing in Fig. 5.73, but the input impedance appears as βr_e in Fig. 5.74a since r_e is determined by I_E and not I_B. In Fig. 5.71c the approximation was employed that β is the same throughout the device characteristics. This we know is absolutely untrue. However, its variation about the provided value for the typical application in the active region is assumed to be minimal and a

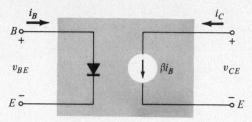

Figure 5.73. Approximate CE equivalent circuit.

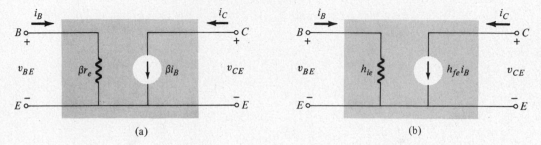

(a)

(b)

Figure 5.74. CE configuration (a) alternate equivalent circuit; (b) approximate hybrid equivalent circuit.

fixed value a valid first approximation. For our analysis, we will consider it to be a constant at the provided value resulting in the output equivalent circuits of Figs. 5.73 and 5.74a. From the equivalent circuits of Fig. 5.74 we can readily note that

$$\beta = h_{fe}$$
$$\beta r_e = h_{ie}$$

(5.33)

For the various configurations examined in detail earlier the resulting equations can be quickly converted to a set having only β and r_e in place of the hybrid parameters using the relationships of Eq. (5.33). It is also important to keep in mind that Eq. (5.33) permits a direct determination of h_{ie} at bias points different from the provided condition.

For the case indicated in Fig. 5.75,

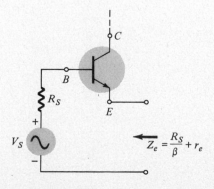

$$Z_e = \frac{R_S}{\beta} + r_e$$

Figure 5.75. Emitter impedance Z_e.

$$Z_e = \frac{R'_S}{h_{fe}} = \frac{R_S + h_{ie}}{h_{fe}} = \frac{R_S + \beta r_e}{\beta}$$

and
$$\boxed{Z_e = \frac{R_S}{\beta} + r_e} \qquad (5.34)$$

A few examples will clarify the use of the alternate CE equivalent circuit.

Example 5.12 Determine A_v, A_i, Z_i, and Z_o for the network of Fig. 5.76.

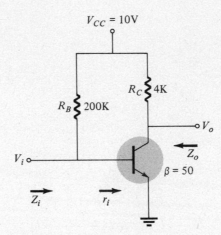

Figure 5.76. Network for Example 5.12.

Solution:

A_v:

For a network such as shown in Fig. 5.76 I am sure we have reached the point where it should be unnecessary to redraw the network for each calculation. Therefore, for dc conditions:

$$I_B = \frac{V_{CC} - V_{BE}}{R_B} = \frac{10 - 0.7}{200\ \text{K}} = \frac{9.3}{200\ \text{K}} \cong 46.5\ \mu\text{A}$$

and
$$I_E \cong I_C = \beta I_B = 50(46.5 \times 10^{-6}) = 2.325\ \text{mA}$$

so that
$$r_e = \frac{26\ \text{mV}}{I_E} = \frac{26}{2.325} \cong 11.2\ \Omega$$

ac conditions:
$$r_i = \beta r_e$$

and
$$I_b = \frac{V_i}{\beta r_e}$$

so that
$$V_o = -I_C R_C \cong \beta I_b R_C \cong \beta \left(\frac{V_i}{\beta r_e} \right) R_C = -\frac{R_C}{r_e} V_i$$

with
$$\boxed{A_v = \frac{V_o}{V_i} = -\frac{R_C}{r_e}} \qquad (5.35)$$

Substituting numbers, we get

$$A_v = -\frac{R_C}{r_e} = -\frac{4\,\text{K}}{11.2} \cong \mathbf{357.14}$$

A_i:
$$R_B \| r_i = R_B \| \beta r_e \cong \beta r_e$$

therefore,
$$I_b \cong I_i$$

and
$$I_o = h_{fe} I_b = h_{fe} I_i$$

with
$$\boxed{A_i = \frac{I_o}{I_i} = h_{fe}} \tag{5.36}$$

and
$$A_i = \mathbf{50}$$

Z_i:

$$\boxed{Z_i \cong \beta r_e} \tag{5.37}$$

$$= (50)(11.2) = \mathbf{560\,\Omega}$$

Z_o:

$$\boxed{Z_o \cong R_C} \tag{5.38}$$

$$= \mathbf{4\,K}$$

EXAMPLE 5.13 Determine A_v, A_i, Z_i, and Z_o for the network of Fig. 5.77.

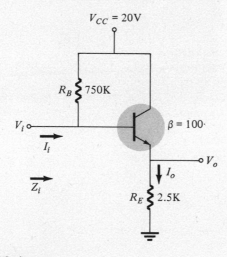

Figure 5.77. Network for Example 5.13.

Solution:
dc conditions:
From Chapter 4,

$$I_B = \frac{V_{CC} - V_{BE}}{R_B + \beta R_E} = \frac{20 - 0.7}{750\,\text{K} + (100)(2.5\,\text{K})} = \frac{19.3}{750\,\text{K} + 250\,\text{K}} = \frac{19.3}{1 \times 10^6}$$

$$= 19.3\ \mu\text{A}$$

$$I_E \cong I_C = \beta I_B = (100)(19.3 \times 10^{-6}) = 1.93 \text{ mA}$$

and
$$r_e = \frac{26 \text{ mV}}{I_E} = \frac{26}{1.93} = 13.47 \, \Omega$$

ac conditions:

A reduced ac equivalent circuit appears in Fig. 5.78.

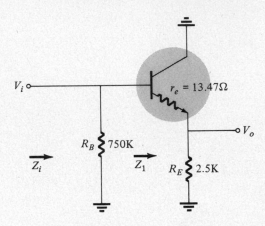

Figure 5.78. ac equivalent for the network of Fig. 5.77.

A_v:

In a previous section we introduced the approximation $V_{be} \cong 0$ V when a network had an unbypassed emitter resistor. With this in mind $V_o = V_i$

and
$$\boxed{A_v \cong 1}$$ (actually slightly less)   (5.39)

A_i:

The impedance Z_1:
$$Z_1 = \beta(r_e + R_E) = 100(13.47 + 2500)$$

Note here that r_e can realistically be ignored in comparison with R_E. Taking the approach, we get
$$Z_1 \cong \beta R_E = 100(2.5 \text{ K}) = 250 \text{ K}$$

and using Fig. 5.79, we get

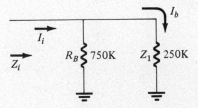

Figure 5.79. Determining the relationship between I_i and I_b.

$$I_b = \frac{750 \text{ K} I_i}{750 \text{ K} + 250 \text{ K}} = 0.75 I_i$$

and
$$A_i = \frac{I_o}{I_i} = \left[\frac{I_b}{I_i}\right]\left[\frac{I_o}{I_b}\right] = [0.75][\beta] = [0.75][100] = \mathbf{75}$$

Z_i:

From Fig. 5.79

$$\boxed{Z_i \cong R_B \| \beta R_E} \tag{5.40}$$

$$= 750 \text{ K } 250 \text{ K} = \mathbf{187.5 \text{ K}}$$

Z_o:

With V_i set to zero, R_B is effectively "shorted out" and in Eq. (5.34) $R_S = 0 \, \Omega$. Therefore,

$$Z_e = \frac{R_S}{\beta} + r_e = 0 + 13.47 = 13.47 \, \Omega$$

and

$$\boxed{Z_o = R_E \| r_e \cong r_e} = \mathbf{13.47 \, \Omega} \tag{5.41}$$

EXAMPLE 5.14   Determine A_v, A_i, Z_i, and Z_o for the network of Fig. 5.80.

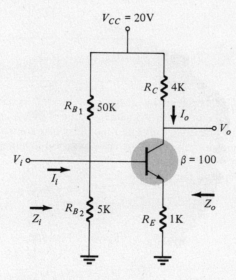

Figure 5.80. Network for Example 5.14.

Solution:

dc conditions:

From Chapter 4,

$$V_B \cong \frac{R_{B2}(V_{CC})}{R_{B2} + R_{B1}} = \frac{5 \text{ K}(20)}{5 \text{ K} + 50 \text{ K}} = \frac{5}{55}(20) = 18.18 \text{ V}$$

and

$$I_E = \frac{V_B - V_{BE}}{R_E} = \frac{18.18 - 0.7}{1 \text{ K}} = \frac{1.118}{1 \text{ K}} = 1.118 \text{ mA}$$

and

$$r_e = \frac{26 \text{ mV}}{I_E} = \frac{26}{1.118} \cong 23.26 \, \Omega$$

ac conditions:

The network is redrawn as shown in Fig. 5.81.

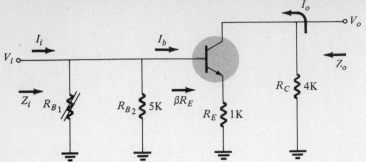

Figure 5.81. Network of Fig. 5.80 redrawn for ac conditions.

A_v:

$$V_o = -I_o R_C = -I_c R_C = -I_e R_C$$

but

$$I_e = \beta I_b$$

with

$$I_b \cong \frac{V_i}{\beta R_E}$$

so that

$$I_e = \beta\left(\frac{V_i}{\beta R_E}\right) = \frac{V_i}{R_E}$$

and

$$V_o = -\left(\frac{V_i}{R_E}\right) R_C$$

with

$$\boxed{A_v = \frac{V_o}{V_i} = -\frac{R_C}{R_E}} \tag{5.42}$$

The distinct advantage of this configuration is now obvious; it is β independent. It is not concerned with the value of $\beta(=h_{fe})$ which will vary depending on the operating point and the particular transistor of a series used. As a consequence, however, there is a significant loss in gain with an unbypassed emitter resistor.

Substituting numbers, we get

$$A_v \cong \frac{-R_C}{R_E} = -\frac{4\,\mathrm{K}}{1\,\mathrm{K}} = -4$$

A_i:

Since $R_{B1} \| R_{B2} \cong R_{B2}$ as shown in Fig. 5.81,

$$I_b \cong \frac{R_{B_2} I_i}{R_{B_2} + \beta R_E}$$

and

$$I_o = I_c = \beta I_b = \beta\left(\frac{R_{B_2} I_i}{R_{B_2} + \beta R_E}\right)$$

with

$$A_i = \frac{I_o}{I_i} = \frac{\beta R_{B_2}}{R_{B_2} + \beta R_E} = \frac{R_{B_2}}{\frac{R_{B_2}}{\beta} + R_E}$$

$$\boxed{A_i \cong \frac{R_{B_2}}{R_E}}\Bigg|_{R_{B1} \gg R_{B2}} \tag{5.43}$$

Substituting values, we get

$$A_i \cong \frac{5\,\mathrm{K}}{1\,\mathrm{K}} = 5$$

Z_i:

From Fig. 5.81

$$Z_i \cong R_{B_2} \| \beta R_E = 5 \text{ K} \| 100 \text{ K} \cong 5 \text{ K}$$

and

$$\boxed{Z_i \cong R_{B_2}} \tag{5.44}$$

Z_o:

From Fig. 5.81

$$Z_o = (Z_t + R_{\bar{E}}) \| R_C \cong (\infty \, \Omega + R_E) \| R_C \cong R_C$$

and

$$\boxed{Z_o \cong R_C} \tag{5.45}$$

For the following modifications (Fig. 5.82) of the network of Fig. 5.80 the results for $A_v, A_i, Z_i,$ and Z_o are provided. The derivations will appear as exercises at the end of the chapter.

$$\boxed{A_v \cong -\frac{R_C}{R_E}} \tag{5.46}$$

$$\boxed{A_i \cong \frac{\beta R_B}{R_B + \beta R_E}} \tag{5.47}$$

$$\boxed{Z_i \cong R_B \| \beta R_E} \tag{5.48}$$

$$\boxed{Z_o \cong R_C} \tag{5.49}$$

For the configuration of Fig. 5.83 the following results are obtained.

$$\boxed{A_v = -\frac{R_C}{R_{E_1}}} \tag{5.50}$$

$$\boxed{A_i \cong \frac{R_{B_2}}{R_E}}_{R_{B_1} \gg R_{B_2}} \tag{5.51}$$

$$\boxed{Z_i \cong R_{B_2} \| R_{E_1}} \tag{5.52}$$

$$\boxed{Z_o \cong R_C} \tag{5.53}$$

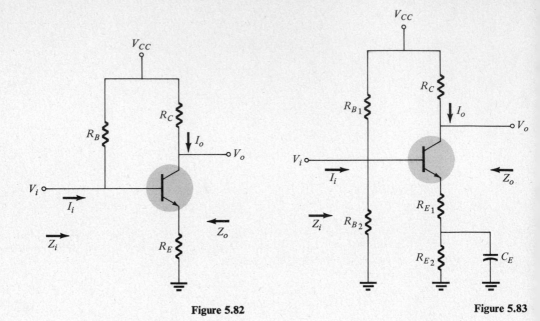

Figure 5.82

Figure 5.83

5.9 COLLECTOR FEEDBACK

The last transistor configuration to be analyzed in detail in this chapter appears in Fig. 5.84. The analysis will be in terms of the hybrid parameters but the results will appear in each form using the direct conversion equations of the previous section.

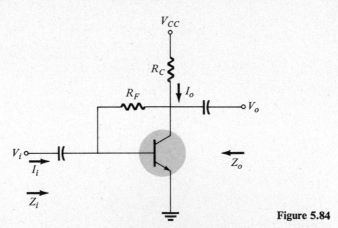

Figure 5.84

Note in this case that even though the collector is connected directly to the base through the feedback resistor R_F, it is not connected directly to V_{CC} so the output current I_o is not equal to the collector current of the transistor. This network provides a voltage feedback from the collector to the base to increase the stability of the system.

An approximate hybrid equivalent circuit has been substituted in Fig. 5.85.

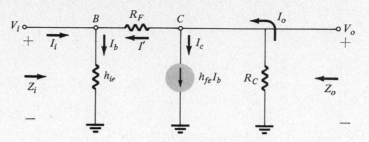

Figure 5.85. ac equivalent of Fig. 5.84.

A_v:

Applying Kirchhoff's current law at the collector terminal:

$$I_o = h_{fe}I_b + I'$$

and

$$V_o = -I_oR_C = -(h_{fe}I_b + I') = -(h_{fe}I_bR_C + I'R_C)$$

but

$$I_b = \frac{V_i}{h_{ie}} \quad \text{and} \quad I' = \frac{V_o - V_i}{R_F}$$

Substituting into the above equation for V_o results in

$$-V_o = h_{fe}\left(\frac{V_i}{h_{ie}}\right)R_C + \left(\frac{V_o - V_i}{R_F}\right)R_C$$

$$-V_o = \left(\frac{h_{fe}}{h_{ie}}R_C\right)V_i + \left(\frac{R_C}{R_F}\right)V_o - \left(\frac{R_C}{R_F}\right)V_i$$

or

$$-V_o\left(1 - \frac{R_C}{R_F}\right) = V_i\left(\frac{h_{fe}}{h_{ie}}R_C - \frac{R_C}{R_F}\right)$$

and

$$A_v = \frac{V_o}{V_i} = \frac{-\left(\frac{h_{fe}}{h_{ie}} - \frac{1}{R_F}\right)}{1 - \frac{R_L}{R_F}}R_C \cong \boxed{\frac{-h_{fe}}{h_{ie}}R_C} \tag{5.54}$$

For the conversion:

$$A_v = -\frac{\beta}{\beta r_e}R_C = \boxed{\frac{-R_C}{r_e}} \tag{5.55}$$

A_i:

At node B, $I_i + I' = I_b$

or

$$I' = I_b - I_i$$

At node C, $I_o = I' + I_c = I' + h_{fe}I_b$

or

$$I_o = (I_b - I_i) + h_{fe}I_b$$

$$I_o = (h_{fe} - 1)I_b - I_i$$

As an approximation: $(h_{fe} + 1) \cong h_{fe}$

and

$$I_b = \frac{I_o + I_i}{h_{fe}}$$

Applying Kirchhoff's voltage law around the outside network loop:

$$V_i + V_{RF} - V_o = 0$$

or
$$I_b h_{ie} + I'R_F + I_o R_C = 0$$

and
$$I_b h_{ie} + (I_b - I_i)R_F + I_o R_C = 0$$

gathering terms:

$$I_b(h_{ie} + R_F) - I_i R_F + I_o R_C = 0$$

Substituting $I_b = (I_o + I_i)/h_{fe}$ into the above and arranging terms, we get

$$I_o(h_{ie} + R_F + h_{fe}R_C) + I_i(h_{ie} + R_F - h_{fe}R_F) = 0$$

with
$$A_i = \frac{I_o}{I_i} = \frac{-(h_{ie} + R_F - h_{fe}R_E)}{(h_{ie} + R_F + h_{fe}R_C)}$$

$$= \frac{-\left(\dfrac{h_{ie}}{h_{fe}} + \dfrac{R_F}{h_{fe}} - R_F\right)}{\left(\dfrac{h_{ie}}{h_{fe}} + \dfrac{R_F}{h_{fe}} + R_C\right)} \cong \frac{R_F}{\dfrac{R_F}{h_{fe}} + R_C} = \boxed{\frac{h_{fe}R_F}{R_F + h_{fe}R_C}}$$

$$(5.56)$$

For $h_{fe}R_L \gg R_F$

$$A_i \cong \boxed{\frac{R_F}{R_C}} \qquad (5.57)$$

For the conversion

$$A_i = \frac{R_F}{R_C}\bigg|\beta R_C \gg R_F \qquad (5.58)$$

The above derivation could have been shortened significantly if we had assumed $I_o \cong h_{fe}I_b$. Then $I_b = (I_o/h_{fe})$ and not $(I_o + I_i)/h_{fe}$ and the mathematical manipulations would be substantially reduced.

Z_i:

$$V_i = I_b h_{ie}$$

and
$$I_b = I_i + I' = I_i + \frac{(V_o - V_i)}{R_F}$$

so that
$$V_i = \left(I_i + \frac{V_o - V_i}{R_F}\right)h_{ie}$$

with
$$V_i = I_i h_{ie} + \frac{h_{ie}}{R_F}V_o - \frac{h_{ie}}{R_F}V_i$$

but
$$V_o = A_v V_i$$

and
$$V_i = I_i h_{ie} + \frac{A_v h_{ie}}{R_F}V_i - \frac{h_{ie}}{R_F}V_i$$

with
$$V_i\left(1 - \frac{A_v h_{ie}}{R_F} + \frac{h_{ie}}{R_F}\right) = I_i h_{ie}$$

so that
$$Z_i = \frac{V_i}{I_i} = \frac{h_{ie}}{1 - \dfrac{A_v h_{ie}}{R_F}} = \frac{h_{ie}}{1 - \dfrac{h_{ie}}{\dfrac{R_F}{A_v}}}$$

Recall for parallel elements that

$$x \| y = \frac{x \cdot y}{x + y} = \frac{y}{1 + \dfrac{y}{x}}$$

In this case $y = h_{ie}$ and $x = (R_F / A_v)$

Therefore,
$$Z_i = \frac{V_i}{I_i} = \boxed{\frac{R_F}{A_v} \bigg\| h_{ie}} \tag{5.59}$$

or
$$Z_i = \frac{V_i}{I_i} = \boxed{\frac{R_F}{A_v} \bigg\| \beta r_e} \tag{5.60}$$

The resulting equations appear below for the configuration appearing in Fig. 5.86. The derivations will appear as exercises at the end of the chapter.

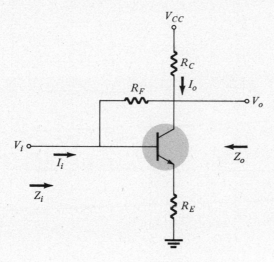

Figure 5.86

$$A_v \cong \boxed{-\frac{R_C}{R_E}} \tag{5.61}$$

$$A_i \cong \boxed{\frac{R_F}{R_E + R_C + \dfrac{R_F}{h_{fe}}}} \tag{5.62}$$

$$A_i \cong \boxed{\frac{-R_F}{R_E + R_C}}_{h_{fe}R_C \gg R_F} \tag{5.63}$$

$$Z_i \cong \boxed{\frac{R_F}{A_v} \bigg\| h_{fe} R_E} \tag{5.64}$$

(In each case $R_L' = R_L \parallel R_C$)	A_v	Z_i	Z_o
	$-\dfrac{h_{fe}}{h_{ie}} R_L'$ $-\dfrac{R_L'}{r_e}$	$R_B \parallel h_{ie}$ $R_B \parallel \beta r_e$	R_L'
	$-\dfrac{h_{fe} R_L'}{h_{ie}}$ $-\dfrac{R_L'}{r_e}$	$R_{B_1} \parallel R_{B_2} \parallel h_{ie}$ $R_{B_1} \parallel R_{B_2} \parallel \beta r_e$	R_L'
	$\cong 1$	$(R_E' = R_E \parallel R_L)$ $R_{B_1} \parallel R_{B_2} \parallel (h_{ie} + h_{fe} R_E')$ $R_{B_1} \parallel R_{B_2} \parallel \beta(r_e + R_E')$	$R_S' = R_S \parallel R_{B_1} \parallel R_{B_2}$ $R_E' \parallel \left(\dfrac{R_S' + h_{ie}}{h_{fe}}\right)$ $R_E' \parallel \left(\dfrac{R_S'}{\beta} + r_e\right)$
	$-\dfrac{h_{fb}}{h_{ib}} R_L'$ $\cong -\dfrac{R_L'}{r_e}$	$R_E \parallel h_{ib}$ $R_E \parallel r_e$	R_L'

CHAP. 5 SMALL-SIGNAL ANALYSIS

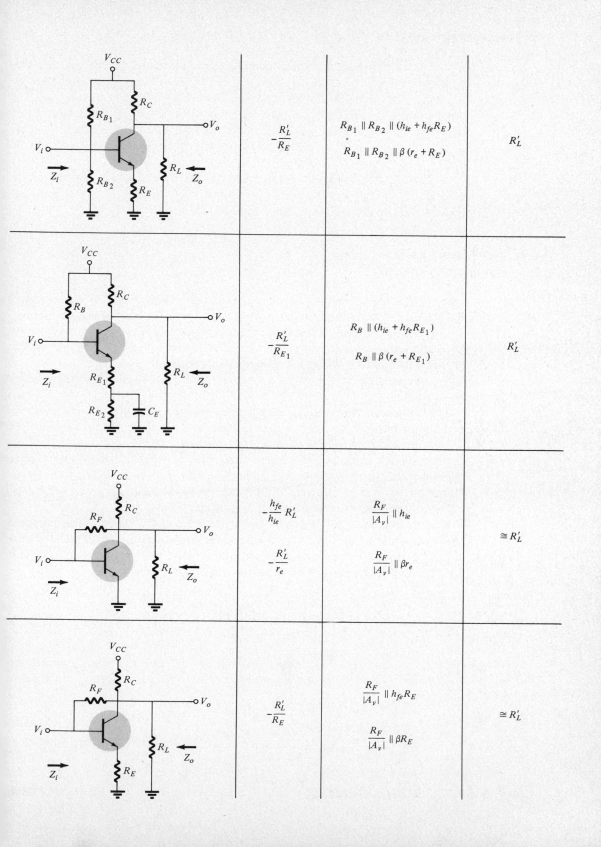

	A_v	Z_i	Z_o
	$-\dfrac{R'_L}{R_E}$	$R_{B_1} \parallel R_{B_2} \parallel (h_{ie} + h_{fe}R_E)$ $R_{B_1} \parallel R_{B_2} \parallel \beta(r_e + R_E)$	R'_L
	$-\dfrac{R'_L}{R_{E_1}}$	$R_B \parallel (h_{ie} + h_{fe}R_{E_1})$ $R_B \parallel \beta(r_e + R_{E_1})$	R'_L
	$-\dfrac{h_{fe}}{h_{ie}}R'_L$ $-\dfrac{R'_L}{r_e}$	$\dfrac{R_F}{|A_v|} \parallel h_{ie}$ $\dfrac{R_F}{|A_v|} \parallel \beta r_e$	$\cong R'_L$
	$-\dfrac{R'_L}{R_E}$	$\dfrac{R_F}{|A_v|} \parallel h_{fe}R_E$ $\dfrac{R_F}{|A_v|} \parallel \beta R_E$	$\cong R'_L$

5.10 SUMMARY TABLE

Table 5.4 summarizes the various configurations examined in this chapter. The approximation h_{re}, $h_{oe} \cong 0$ has been applied in each case. The latter case of $(1/h_{oe}) \cong \infty \; \Omega$ is one that should be checked for each application. If $(1/h_{oe})$ is close in magnitude to R_C, R_L, or their parallel combination $R_C \| R_L$, it cannot be dropped and its effects should be included with perhaps a hybrid equivalent circuit with just $h_{re} \cong 0$. If included, for the reasons above, it will cut both the voltage and current gain.

In addition, if a situation is encountered where R_E and r_e are comparable in value, the r_e cannot be dropped in equations such as $\beta(R_E + r_e)$. It must be included and the following equations modified.

5.11 THE TRIODE SMALL-SIGNAL EQUIVALENT CIRCUIT

Since the triode is still employed in a number of small-signal applications, we shall now derive, and demonstrate the use of, the triode small-signal equivalent circuit.

The resulting equivalent circuit for the triode has only two primary components like that of the reduced equivalent circuit of the transistor, making it appear somewhat similar to work with, but there are various facets of its use that must be firmly understood before it can be applied correctly. These will all be discussed in this section.

You will recall that in the *ideal* triode discussed in Chapter 3 there exists an open circuit between the grid and either the plate or the cathode. At high frequencies, however, capacitive effects are introduced that will couple the grid to both the plate and the cathode. Our present discussion, however, is for the open-circuit condition. This being the case, no matter what the grid-to-cathode potential, we will assume that the grid current will be zero, and a simple open circuit will suffice to represent the input, or grid circuit, of a triode.

If we assume that the curves of a triode are *linear* in the *region* of the quiescent point of operation, the small-signal equivalent circuit for the output circuit of a tube can be found by applying Thévenin's theorem. The Thévenin impedance can be shown to be the following using partial derivatives (calculus):

$$R_{Th} = r_p = \frac{\Delta v_{PK}}{\Delta i_p}\bigg|_{V_{GK}=\text{constant}} \tag{5.65}$$

where r_p is called the *plate resistance* of the triode; and $V_{Th} = \mu v_{GK}$, where the Greek letter μ (mu), called the *amplification factor*, has a magnitude defined by

$$|\mu| = \frac{\Delta v_{PK}}{\Delta v_{GK}}\bigg|_{I_P=\text{constant}} \tag{5.66}$$

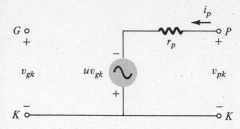

Figure 5.87. Triode piecewise linear equivalent circuit.

The resulting plate (output) circuit appears as shown in Fig. 5.87 with r_p and μ, defined by Eqs. (5.35) and (5.36). The method or procedure used in association with Eqs. (5.35) and (5.36) to determine r_p and μ is exactly the same as that employed for the transistor in the determination of the h-parameters.

A third quantity of interest for the triode is its *transconductance*, defined by Eq. (5.67).

$$g_m = \frac{\Delta_{ip}}{\Delta v_{GK}}\bigg|_{V_{PK}=\text{constant}} \tag{5.67}$$

The prefix *trans* from the word transfer is employed because it relates an output quantity (I_P) to an input quantity (V_{GK}).

By simply manipulating Eq. (5.66) in the following manner:

$$\mu = \frac{\Delta v_{PK}}{\Delta v_{GK}} = \frac{\Delta v_{PK}}{\Delta_{ip}} \times \frac{\Delta_{ip}}{\Delta v_{GK}}$$

and substituting the magnitude of the defined quantities for each ratio, we get

$$\mu = r_p g_m \tag{5.68}$$

so that if any two quantities are known, the third can be found using Eq. (5.68).

The magnitude of μ, r_p, and g_m will vary as indicated by Table 5.5.

TABLE 5.5 Typical Range of Values for μ, r_p, and g_m for the Triode

μ:	25–100
r_p:	0.5–100 K
g_m:	0.5–10 mA/V

The Norton small-signal equivalent circuit for the triode appears in Fig. 5.88.

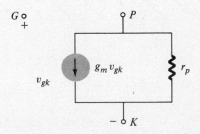

Figure 5.88. Norton form of the triode small-signal ac equivalent circuit.

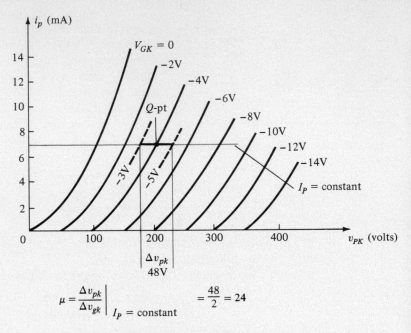

$$\mu = \left.\frac{\Delta v_{pk}}{\Delta v_{gk}}\right|_{I_P = \text{constant}} = \frac{48}{2} = 24$$

Figure 5.89. Determining μ from the plate characteristics.

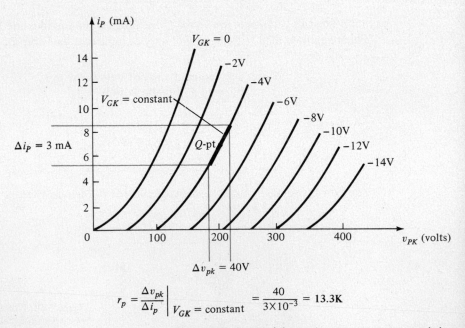

$$r_p = \left.\frac{\Delta v_{pk}}{\Delta i_p}\right|_{V_{GK} = \text{constant}} = \frac{40}{3 \times 10^{-3}} = 13.3\text{K}$$

Figure 5.90. Determining r_p from the plate characteristics.

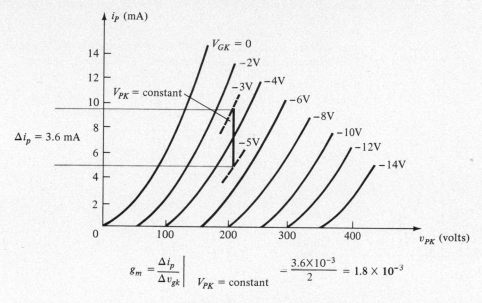

$$g_m = \frac{\Delta i_p}{\Delta v_{gk}}\bigg|_{V_{PK} = \text{constant}} = \frac{3.6 \times 10^{-3}}{2} = 1.8 \times 10^{-3}$$

Figure 5.91. Determining g_m from the plate characteristics.

EXAMPLE 5.15 (a) Find the magnitude of μ, r_p, and g_m at the indicated operating point on the characteristics of Fig. 5.89 and compare the value of g_m obtained graphically with that obtained using Eq. (5.68).

(b) Draw the Thévenin and Norton small-signal equivalent circuits for the triode using the data of part (a).

Solution: (a) See Figs. 5.89–5.91.

$$g_m = \frac{\mu}{r_p} = \frac{24}{13.3\ K} = 1.81 \times 10^{-3} \cong 1.8 \times 10^{-3} \qquad (5.69)$$

(b) Thévenin equivalent circuit: See Fig. 5.92.
  Norton equivalent circuit: See Fig. 5.93.

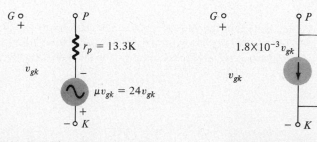

Figure 5.92. Resulting Thévenin equivalent circuit.

Figure 5.93. Resulting Norton equivalent circuit.

EXAMPLE 5.16 Find the voltage gain, $A_v = (V_o/V_i)$, for the basic triode amplifier of Fig. 5.94.

Solution: Eliminating dc levels and substituting the Thévenin equivalent circuit will result in the configuration of Fig. 5.95. In this case,

$$V_{gk} = V_i$$

and

$$V_o = \frac{-R_L(\mu V_{gk})}{R_L + r_p} = \frac{-R_L(\mu V_i)}{R_L + r_p}$$

so that

$$A_v = \frac{V_o}{V_i} = \frac{-\mu R_L}{R_L + r_p} \qquad (5.70)$$

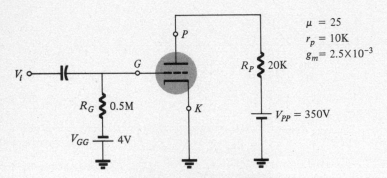

$\mu = 25$
$r_p = 10K$
$g_m = 2.5 \times 10^{-3}$

Figure 5.94. Circuit for Example 5.16.

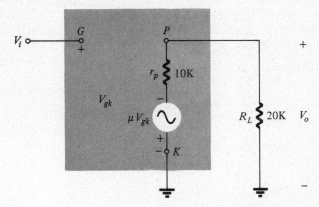

Figure 5.95. Circuit of Fig. 5.94 following the substitution of the small-signal ac equivalent circuit for the triode.

Note that the gain is equal to the amplification factor only when $r_p = 0$. For *maximum power transfer* to the load the load impedance R_L must equal the plate resistance (maximum power theorem); that is,

$$R_L = r_{p(\text{max. power})} \qquad (5.71)$$

for maximum power transfer to R_L.

The negative sign clearly indicates that the output (V_o) and input (V_i) voltages are 180° *out of phase*.

Substituting values for this example, we get

$$A_v = \frac{V_o}{V_i} = \frac{-25(20\text{ K})}{20\text{ K} + 10\text{ K}} = \frac{-500\text{ K}}{30\text{ K}} \cong -16.7$$

EXAMPLE 5.17 The circuit of Fig. 5.96, called the *cathode follower*, is the tube equivalent of the *emitter follower* circuit for the transistor. It is also used for impedance matching purposes since it has a high input impedance and low output impedance. The input and output voltages are also *in phase* as was true for the emitter follower.

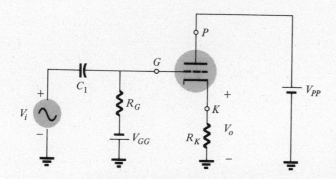

Figure 5.96. Cathode follower configuration to be explored in Example 5.17.

In this example the voltage gain $A_v = (V_o/V_i)$, and output impedance Z_o, will be found in terms of the Thévenin equivalent small-signal triode parameters.

Solution: Substituting the equivalent circuit and eliminating dc levels will result in the configuration of Fig. 5.97. Note in this case that V_{gk} is not the potential from grid to ground and therefore is not equal to the input signal V_i.

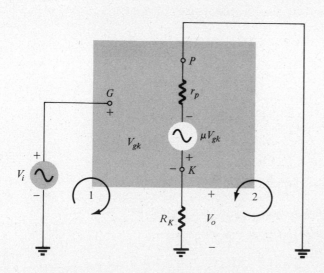

Figure 5.97. Circuit of Fig. 5.96 following the substitution of the small-signal ac equivalent circuit.

SEC. 5.11 THE TRIODE SMALL-SIGNAL EQUIVALENT CIRCUIT

271

Applying Kirchhoff's voltage law to the grid loop (1):

$$V_i - V_{gk} - I_p R_K = 0$$

and
$$V_{gk} = V_i - I_p R_K$$

Applying Kirchhoff's voltage law to the plate circuit:

$$\mu V_{gk} - V_o - I_p r_p = 0$$

or
$$V_o = \mu V_{gk} - I_p r_p$$

Substituting the above for V_{gk}, we get

$$V_o = \mu(V_i - I_p R_K) - I_p r_p$$

and
$$V_o = \mu V_i - I_p(\mu R_K + r_p)$$

but
$$I_p = \frac{V_o}{R_K}$$

and
$$V_o = \mu V_i - \frac{V_o}{R_K}(\mu R_K + r_p)$$

$$= \mu V_i - \mu V_o - \frac{r_p}{R_K} V_o$$

or
$$V_o + \mu V_o + \frac{r_p}{R_K} V_o = \mu V_i$$

and
$$\boxed{A_v = \frac{V_o}{V_i} = \frac{\mu}{1 + \mu + r_p/R_K} = \frac{\mu R_K}{(1 + \mu)R_K + r_p}}$$
(5.72)

For typical values:

$$\mu = 30, \qquad r_p = 10 \text{ K}, \qquad R_K = 20 \text{ K}$$

$$A_v = \frac{V_o}{V_i} = \frac{30(20 \text{ K})}{(31)(20 \text{ K}) + 10 \text{ K}} = \frac{600 \text{ K}}{620 \text{ K} + 10 \text{ K}} = \frac{600 \text{ K}}{630 \text{ K}} \cong 0.95$$

For the cathode follower, like the emitter follower, the gain is always less than one and as indicated above the ratio is always positive so that V_o and V_i are *in phase*.

Rewriting Eq. (5.72) in the following manner, we get

$$\frac{V_o}{V_i} = \frac{\dfrac{\mu R_K}{\mu + 1}}{R_K + \dfrac{r_p}{\mu + 1}}$$

or
$$V_o = \frac{R_K\left(\dfrac{\mu}{\mu + 1}\right)V_i}{R_K + \dfrac{r_p}{\mu + 1}}$$

and drawing the circuit to "fit" the above equation, we obtain Fig. 5.98.

We can now find the output impedance by setting $V_i = 0$ as defined by the conditions necessary to obtain the output impedance, and

$$\boxed{Z_o|_{V_i = 0} = \frac{r_p}{\mu + 1}}$$
(5.73)

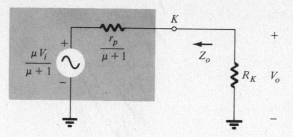

Figure 5.98. Circuit resulting from the rewriting of the fundamental gain equation (Eq. 5.71).

For the parameters just provided:

$$Z_o = \frac{10\text{ K}}{31} = 322\ \Omega$$

which is considerably less than the 10-K (r_p) output impedance that would result if the tube were used in the basic amplifier configuration of Example 5.16.

5.12 TRIODE PARAMETER VARIATION

The parameters of the small-signal equivalent circuit for the triode will vary with the operating conditions. Figure 5.99 illustrates the variations of μ, r_p, and g_m with plate current for various values of plate-to-cathode potential.

Note that μ remains fairly constant, while r_p and g_m vary considerably with change in plate current (I_p). Consider also that increasing values of plate-to-cathode potentials result in decreasing values of μ and increasing values of r_p.

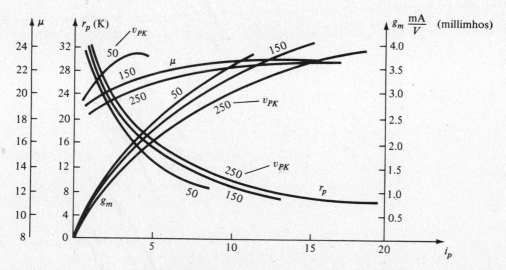

Figure 5.99. Variations in μ, r_p, and g_m with plate current for various values of plate to cathode potential.

5.13 THE PENTODE SMALL-SIGNAL EQUIVALENT CIRCUIT

The pentode small-signal equivalent circuit is the same in appearance as that discussed for the triode. The magnitude of the parameters, however, is very different from those of the triode. The pentode small-signal parameters can all be determined using the same equations and procedure described for the triode. The typical range of values for μ, r_p, and g_m are listed in Table 5.6.

TABLE 5.6

μ:	1000–5000
r_p:	0.5–2 M
g_m:	1000–9000 μA/V

Note that the magnitude of μ and r_p has increased considerably over those for the triode while the range of g_m has changed only slightly. For the pentode, the Norton equivalent circuit is the more frequently employed since the current through the load circuit is, for the majority of situations, unaffected by r_p and simply equal to the Norton current: $g_m V_{gk}$. It is a situation similar to that encountered with the transistor, where in general, $(1/h_{oe}) > R_L$. Example 5.18 will demonstrate the validity of the above statement.

EXAMPLE 5.18   Calculate the voltage gain, $A_v = (V_o/V_i)$ for the circuit of Fig. 5.100.

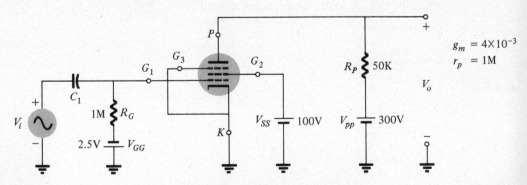

Figure 5.100. Basic pentode amplifier to be examined in Example 5.18.

Solution: Substituting the Norton small-signal equivalent circuit for the pentode and eliminating dc levels will result in the circuit of Fig. 5.101. Obviously, $V_{gk} = V_i$.

The parallel combination of $1 \text{ M} \| 50 \text{ K} \cong 50 \text{ K}$ (substantiating the discussion of the last few paragraphs) results in the plate circuit of Fig. 5.102 so that

$$V_o = (-4 \times 10^{-3} V_i)(50 \times 10^3)$$
$$= -200 V_i$$

and
$$A_v = \frac{V_o}{V_i} = -200$$

The minus sign again indicates a 180°-phase shift between V_o and V_i.

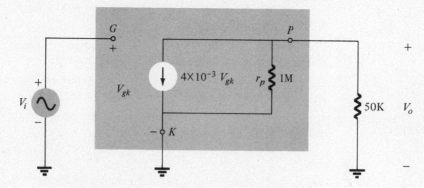

Figure 5.101. Circuit of Fig. 5.100 following the substitution of the small-signal ac equivalent circuit for the pentode.

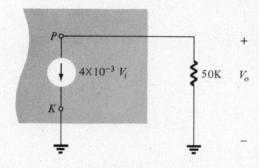

Figure 5.102. Plate circuit resulting from the removal of the 1M resistor (on an approximate basis).

PROBLEMS

§ 5.2

1. (a) Determine the hybrid parameters for the network of Fig. 5.103.
   (b) Sketch the hybrid equivalent circuit.

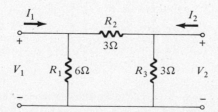

Figure 5.103

2. Sketch the complete hybrid equivalent circuit for the common-collector configuration and indicate the current directions as shown in Fig. 5.8a and b.

§ 5.3

3. Determine the hybrid parameters h_{fe} and h_{oe} from the collector characteristics of Fig. 5.9 at Q-pt of $V_{CE} = 5$ V and $I_C = 5$ mA and compare to those obtained in Section 5.3.

§ 5.4

4. For a change in I_C from 1 to 20 mA which hybrid parameter exhibits the greatest change in Fig. 5.14? Which exhibits the least change?

5. For the range of V_{CE} from 1 to 50 V which parameter in Fig. 5.15 exhibits the greatest change in value? Which exhibits the least change?

6. In Fig. 5.16 which parameter exhibits the least sensitivity to change in temperature? Which exhibits the most sensitivity?

§ 5.5

7. For the network of Fig. 5.104 determine the following:
   (a) Current gain $A_i = I_o/I_i$.
   (b) Voltage gain $A_v = V_o/V_i$.
   (c) Input impedance Z_i.
   (d) Output impedance Z_o.
   (e) Power gain A_p.

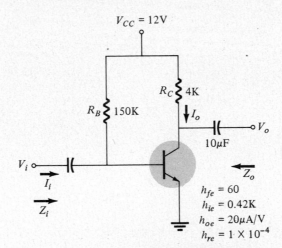

Figure 5.104

8. Repeat Problem 7 for the network of Fig. 5.105.

9. Repeat Problem 7 for the network of Fig. 5.106.

10. Repeat Problem 7 for the network of Fig. 5.107.

§ 5.6

11. Using an appropriate approximate hybrid equivalent circuit, determine A_i, A_v, Z_i, Z_o, and A_p for the network of Fig. 5.104 (compare with the earlier results of Problem 7 if available).

12. Repeat Problem 11 for the network of Fig. 5.105.

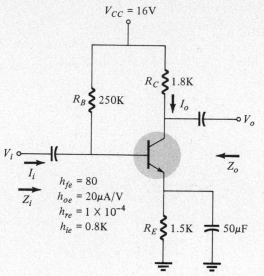

Figure 5.105

$V_{CC} = 16V$

R_B 250K

R_C 1.8K

I_o

V_o

V_i

I_i

Z_i

$h_{fe} = 80$
$h_{oe} = 20\mu A/V$
$h_{re} = 1 \times 10^{-4}$
$h_{ie} = 0.8K$

R_E 1.5K

$50\mu F$

Z_o

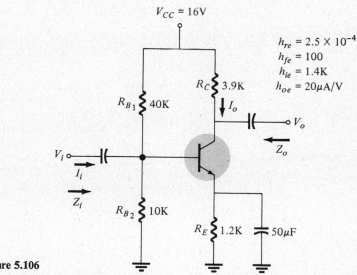

Figure 5.106

$V_{CC} = 16V$

$h_{re} = 2.5 \times 10^{-4}$
$h_{fe} = 100$
$h_{ie} = 1.4K$
$h_{oe} = 20\mu A/V$

R_{B1} 40K

R_C 3.9K

I_o

V_o

V_i

I_i

Z_i

R_{B2} 10K

R_E 1.2K

$50\mu F$

Z_o

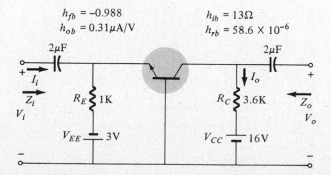

$h_{fb} = -0.988$
$h_{ob} = 0.31\mu A/V$

$h_{ib} = 13\Omega$
$h_{rb} = 58.6 \times 10^{-6}$

$2\mu F$

$2\mu F$

I_i

Z_i

V_i

R_E 1K

V_{EE} 3V

R_C 3.6K

V_{CC} 16V

I_o

Z_o

V_o

Figure 5.107

13. Repeat Problem 11 for the network of Fig. 5.106.

14. Repeat Problem 11 for the network of Fig. 5.107.

15. (a) Using an approximate equivalent circuit, determine the current gain $A_i = I_o/I_i$ and voltage gain $A_v = V_o/V_i$ for the network of Fig. 5.108.

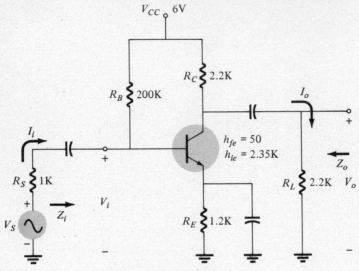

Figure 5.108

(b) Determine Z_i and Z_o.
(c) Determine $A_v = V_o/V_s$.

16. (a) Determine the new $A_v = V_o/V_i$ for Example 5.6 if a 5.6-K load is connected from collector to ground.
(b) Determine the new $A_i = I_o/I_i$ if I_o is now the current through the 5.6-K load.

§ 5.7

17. Determine the input impedance Z_i, the output impedance Z_o, and the voltage gain $A_v = V_o/V_i$ for the network of Fig. 5.109.

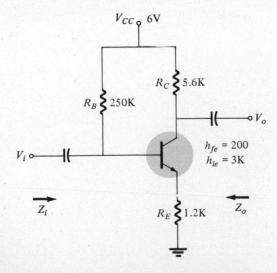

Figure 5.109

18. Determine Z_i, Z_o, A_v, A_t for the network of Fig. 5.110.

19. Determine Z_i, Z_o, A_v, A_t for the network of Fig. 5.111.

20. Determine V_o (in terms of V_1 and V_2) for the network of Fig. 5.112.

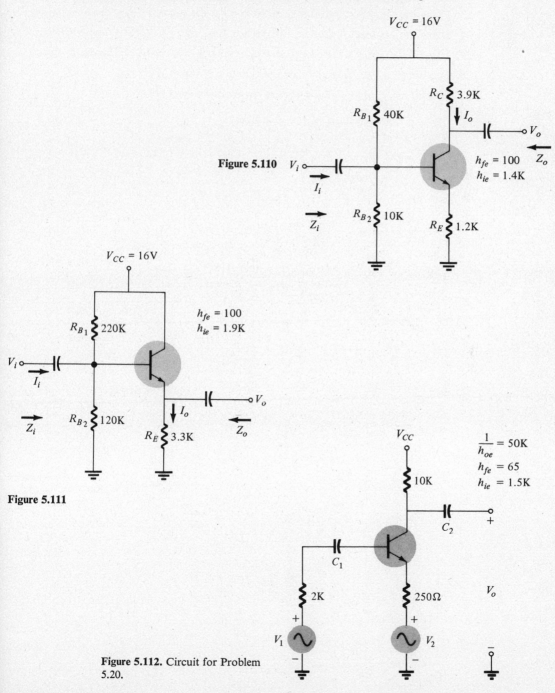

Figure 5.110

Figure 5.111

Figure 5.112. Circuit for Problem 5.20.

21. Using the approach introduced in Section 5.8 (use $r_B = 1.25\ \Omega$), repeat Problem 7.

22. Using the approach introduced in Section 5.8 (use $r_B = 1.7\ \Omega$), repeat Problem 10.

23. Using the approach introduced in Section 4.8 (use $r_B = 1.1\ \Omega$), repeat Problem 17.

24. Using the approach introduced in Section 5.8 (use $r_B = 0.5\ \Omega$), repeat Problem 18.

25. Using the approach introduced in Section 5.8 (use $r_B = 1.7\ \Omega$), repeat Problem 19.

26. Determine A_v, A_i, Z_i, and Z_o for the network of Fig. 5.113.

27. Determine A_v, A_i, Z_i, and Z_o for the network of Fig. 5.114.

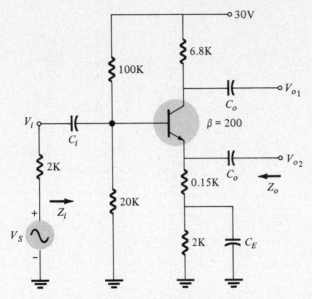

Figure 5.113

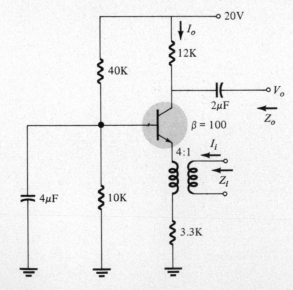

Figure 5.114

28. Derive the expressions for the network of Fig. 5.82.

29. Derive the expressions for the network of Fig. 5.83.

§ 5.9

30. Determine A_v, A_i, and Z_i for the network of Fig. 5.84 if $R_C = 5.6$ K, $R_F = 120$ K, $V_{CC} = 12$ V, and $\beta = 100$.

31. Determine A_v, A_i, and Z_i for the network of Fig. 5.86 if $R_C = 6.8$ K, $R_F = 180$ K, $R_E = 2.2$ K, $V_{CC} = 16$ V, and $\beta = 150$.

32. Determine the expressions for the network of Fig. 5.86.

§ 5.11

33. Determine the values of μ, r_p, and g_m at an operating point of 200 V and 5 mA on the triode plate characteristics of Fig. 5.115.

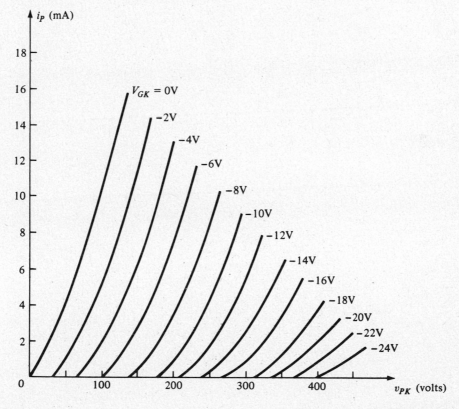

Figure 5.115

34. Calculate the voltage gain, $A_v = V_o/V_i$ for the triode amplifier circuit of Fig. 5.116.

35. Calculate A_v and Z_o for the cathode follower circuit of Fig. 5.117.

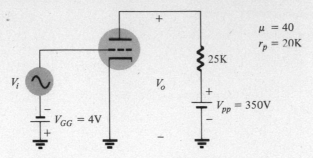

$\mu = 40$
$r_p = 20K$

V_i

V_o

25K

$V_{GG} = 4V$

$V_{pp} = 350V$

Figure 5.116. Circuit for Problem 5.34.

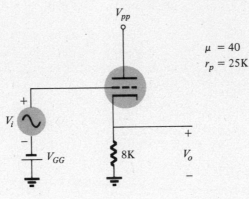

V_{pp}

$\mu = 40$
$r_p = 25K$

V_i

V_{GG}

8K

V_o

Figure 5.117. Circuit for Problem 5.35.

§ 5.12

36. Calculate the voltage gain, $A_v = V_o/V_i$, for the pentode amplifier of Fig. 5.118.

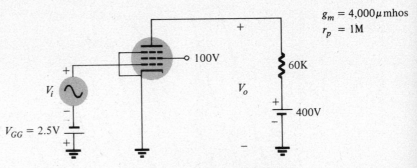

$g_m = 4,000\mu\,\text{mhos}$
$r_p = 1M$

100V

60K

V_i

V_o

$V_{GG} = 2.5V$

400V

Figure 5.118. Circuit for Problem 5.36.

field-effect transistors

6

6.1 GENERAL DESCRIPTION OF FET

A bipolar junction transistor (BJT) made as *npn* or as *pnp* is a current controlled device in which both electron current and hole current are involved. The field-effect transistor (FET) is unipolar. It operates as a voltage controlled device with either electron current in an *n*-channel FET or hole current in a *p*-channel FET. Either BJT or FET devices can be used to operate in an amplifier circuit (or other similar electronic circuits), with different bias considerations.

A few general comparisons between FET and BJT devices and resulting circuits can be made.

1. The FET has an extremely high input resistance with about 100 M typical.
2. The FET has no offset voltage when used as a switch (or chopper).
3. The FET is relatively immune to radiation, but the BJT is very sensitive (beta is particularly affected).
4. The FET is less "noisy" than a BJT and thus more suitable for input stages of low-level amplifiers (it is used extensively in hi-fi FM receivers).
5. The FET can be operated to provide greater thermal stability than a BJT.

Some disadvantages of the FET are the relatively small gain-bandwidth of the device compared to the BJT and the greater susceptibility to damage in handling the FET.

6.2 CONSTRUCTION AND CHARACTERISTIC OF JFET

The FET is a three-terminal device containing one basic *p-n* junction and can be built as either a Junction FET (JFET) or a Metal-Oxide-Semiconductor FET (MOSFET). Although the FET was one of the earliest solid-state devices proposed[1] for amplifier operation, until the mid-1960s the development of a commercially useful device lagged because of manufacturing construction limitations.

JFET Structure

The physical structure of a JFET is shown in Fig. 6.1. The *n*-channel JFET shown in Fig. 6.1a is constructed using a bar of *n*-type material into which a pair of *p*-type regions are diffused. A *p*-channel JFET is made using a bar of *p*-type material with *n*-type diffused regions as shown in Fig. 6.1b.

To examine how the device is operated, consider the *n*-channel JFET of Fig. 6.2, shown with applied bias voltage to operate the device. The supply voltage, V_{DD},

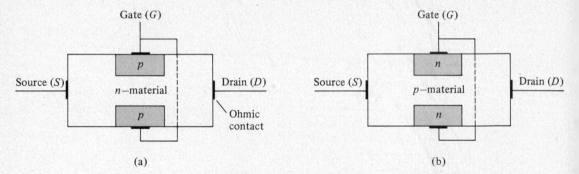

Figure 6.1. Physical structure of a JFET (a) *n*-channel; (b) *p*-channel.

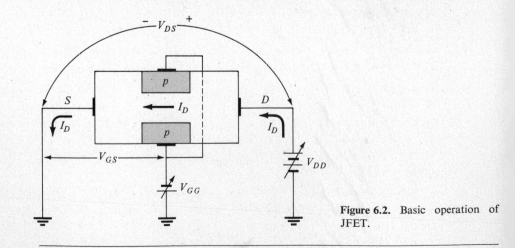

Figure 6.2. Basic operation of JFET.

[1]W. Shockley, *Electrons and Holes in Semiconductors* (New York: Van Nostrand, 1953).

CHAP. 6 FIELD-EFFECT TRANSISTORS

provides a voltage across drain-source, V_{DS}, which results in a current, I_D, from drain to source (electrons in an n-channel actually move from the source, hence name, to drain). This drain current passes through the *channel* formed by the p-type gate. A voltage between gate and source, V_{GS}, is shown here to be set by a voltage supply, V_{GG}. Since this gate-source voltage will reverse bias the gate-source junction, no gate current will result. The effect of the gate voltage will be to create a depletion region in the channel and thereby reduce the channel width to increase the drain-source resistance resulting in less drain current.

We shall first consider the device operation with $V_{GS} = 0$ V and then with the reverse-bias voltage, V_{GS}, increased (made more negative for an n-channel device). Figure 6.3a shows that drain current through the n-material of the drain-source

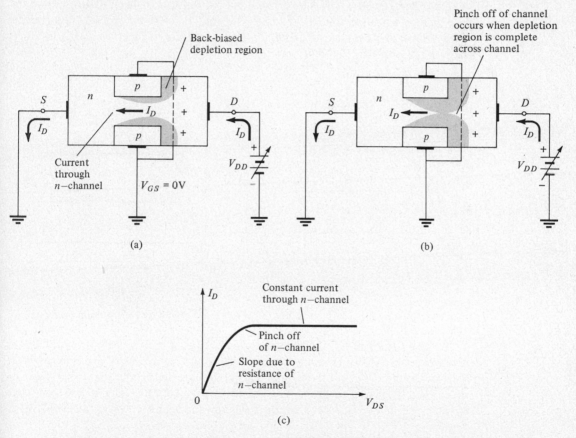

Figure 6.3. Pinch-off action due to channel current.

provides a voltage drop along the channel, which is more positive at the drain-gate junction than at the source-gate junction. This reverse-bias potential across the p-n junction causes a depletion region to form as shown in Fig. 6.3a. As the voltage, V_{DD}, is increased, the current, I_D, increases, resulting in a larger depletion region with increased channel resistance from drain to source. As the voltage V_{DD} is increased, the depletion region finally forms fully across the channel as shown in Fig. 6.3b. At this point any further increase in V_{DD} will result in further increasing the voltage across only the depletion region, leaving the voltage drop from depletion point d to ground constant, the current I_D then remaining constant. This operation is depicted in the $V_{GS} = 0$ characteristic curve of Fig. 6.3c. As V_{DS} increases, the current I_D increases until the depletion region is fully formed across the channel after which, as shown, the drain current saturates and remains a constant value for increased voltage, V_{DS}. This value of drain current, for $V_{GS} = 0$ V is an important parameter used to specify the operation of the JFET and is referred to as I_{DSS}, the drain to source current with gate-source shorted ($V_{GS} = 0$ V).

If V_{GS} is increased (more negative for an n-channel device), the channel will develop a depletion region so that the amount of current needed to close of the channel is less, as shown by the $V_{GS} = -1$ V curve of Fig. 6.4a. We then see that the gate voltage acts as a control, reducing the amount of drain current (at a specific voltage V_{DS}). For a p-channel JFET the drain current is reduced from I_{DSS} as V_{GS} is made more positive (Fig. 6.4b).

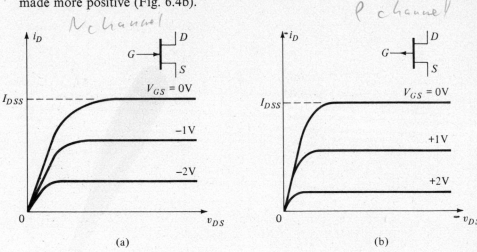

Figure 6.4. JFET drain characteristic.

As the value of V_{GS} is increased, with reduced drain current, a voltage is reached, after which no drain current will result, regardless of the voltage, V_{DS}. This gate-source pinch-off voltage, V_p, is also an important parameter used to specify the operation of the JFET. The drain characteristics of Fig. 6.4 show that for an n-channel FET, V_p is a negative voltage and that for a p-channel FET, V_P is positive.

Another form of device characteristic is the transfer characteristic which is a plot of drain current, i_D as a function of gate-source voltage, v_{GS}, for a constant value of drain-source voltage, v_{DS}. The transfer characteristic can be directly viewed on a curve tracer unit, obtained directly by measurement of device operation or drawn from the drain characteristic as shown in Fig. 6.5. Two important points of the transfer curve shown are the values I_{DSS} and V_p. When these points are fixed, the rest of the curve can be seen on the transfer characteristic or obtained

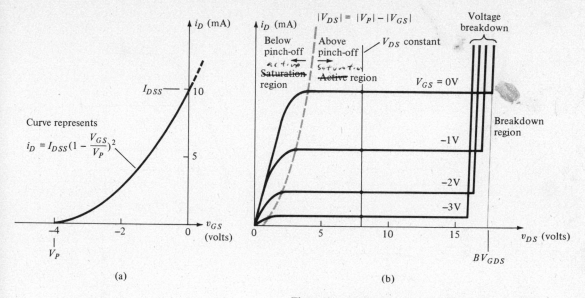

Figure 6.5. JFET drain and transfer characteristics.

from theoretical consideration of the physical process occurring in the JFET, leading to the relation[2]

$$i_D = I_{DSS}\left(1 - \frac{v_{GS}}{V_P}\right)^2 \qquad (6.1)$$

which represents the transfer characteristic curve of Fig. 6.5a. Notice that when $v_{GS} = 0$, $i_D = I_{DSS}$ and that when $i_D = 0$, $v_{GS} = V_P$ as seen on the transfer characteristic. The drain characteristic shows that the current saturates when the channel pinches off, the pinch off occurring at lower values of V_{DS} for more negative values of V_{GS} (as shown by the dashed line on the drain characteristic in Fig. 6.5b). The JFET device is normally biased to operate above pinch off in the region of current saturation. Within this region the device operation is more easily described by the transfer characteristic or by Shockley's equation.

6.3 dc BIAS OF JFET

An n-channel JFET amplifier circuit is shown in Fig. 6.6a using two supply voltages. We shall consider a number of methods to determine the dc bias voltages and current for the circuit. Since the voltage V_{GG} reverse biases the gate-source, there is no gate current and no voltage drop across R_G. The JFET drain characteristic is shown in Fig. 6.6b from which we see that $I_{DSS} = 2$ mA and $V_P = -4$ V. Since the gate-source voltage is set by the battery V_{GG}, we know that the curve $V_{GS} = -2$ V

---

[2]Shockley's equation applies above pinch-off region for JFET device.

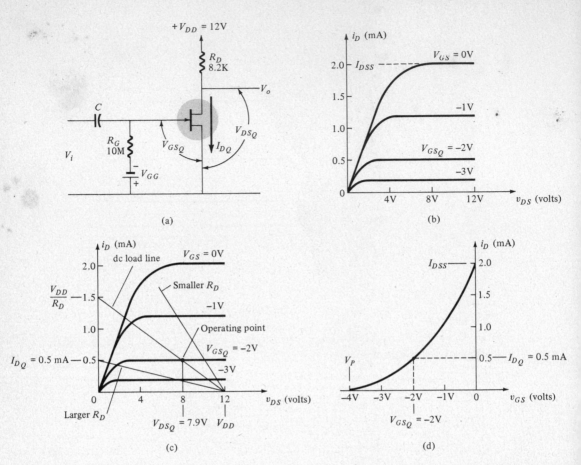

Figure 6.6. dc bias in JFET amplifier circuit.

represents one of the conditions of dc bias. To determine the exact bias point graphically we can solve by plotting a dc load line representing the Kirchhoff voltage equation around the drain-source loop

$$V_{DD} - I_{D_Q} \cdot R_D = V_{DS_Q}$$

This condition of circuit operation is shown graphically as a dc load line in Fig. 6.6c, drawn by interconnecting two points of the straight line:

1. At $I_D = 0$, $V_{DS} = V_{DD}$.

2. At $V_{DS} = 0$, $I_D = \dfrac{V_{DD}}{R_D}$.

As seen in Fig. 6.6c the intersection of the dc load line gate-source voltage curve occurs at approximately

$$V_{DS_Q} = 7.9 \text{ V}, \qquad I_{D_Q} = 0.5 \text{ mA}$$

and the operating point is established in the circuit of Fig. 6.6c. Decreasing V_{GS} will result in the bias point moving to lower values of V_{DS} and larger I_D along the dc load line. Using a smaller value of R_D will provide a steeper dc load line, as shown, resulting in a higher value of V_{DS_Q} at the same value of I_{D_Q} (for V_{GS_Q} remaining at -2 V).

The dc bias point can also be determined using the transfer characteristic as shown in Fig. 6.6d. At $V_{GS_Q} = -2$ V, the transfer curve shows $I_{D_Q} = 0.5$ mA. The value of V_{DS_Q} can then be calculated.

$$V_{DS_Q} = V_{DD} - I_{D_Q} \cdot R_D = 12 - (0.5 \text{ mA})(8.2 \text{ K}) = 7.9 \text{ V}$$

The transfer characteristic shows more clearly the result of varying V_{GS} on the bias current I_{D_Q}. As V_{GS} is increased toward $V_P = -4$ V, the device drains current is reduced, while reducing the magnitude of V_{GS} toward 0 V increases I_{D_Q} toward I_{DSS}.

The value of I_D at $V_{GS_Q} = -2$ V can also be calculated using Eq. (6.1).

$$I_{D_Q} = I_{DSS}\left(1 - \frac{V_{GS_Q}}{V_P}\right)^2 = 2 \text{ mA}\left(1 - \frac{-2}{-4}\right)^2 = 0.5 \text{ mA}$$

from which we calculate $V_{DS_Q} = 7.9$ V, as shown above.

EXAMPLE 6.1 Obtain the operating point for the fixed-bias circuit of Fig. 6.7a and n-channel JFET with drain characteristic shown in Fig. 6.7b.

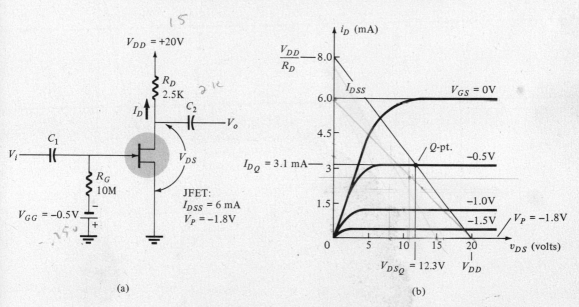

(a)   (b)

Figure 6.7. JFET circuit and drain characteristic for Example 6.1.

Solution: Drawing the dc load line between $V_{DD} = +20$ V and $V_{DD}/R_D = 20/2.5 \text{ K} = 8$ mA results in the Q-point (quiescent operating point) at

$$V_{DS_Q} = 12.3 \text{ V}, \qquad I_{D_Q} = 3.1 \text{ mA}$$

JFET Amplifier with Self-Bias

A more practical version of a JFET amplifier has only a single voltage supply using a self-bias resistor, R_S, to obtain the gate-source bias voltage, as shown in Fig. 6.8a. The presence of resistor R_S results in a positive voltage V_S due to the voltage drop $I_D R_S$. Since the gate voltage, V_G, is 0 V (no dc current flows through gate or resistor R_G), the net voltage measured from gate (0 V) to source ($+V_S$) is a negative voltage which is the gate-source bias voltage, V_{GS}. The gate circuit bias relation

$$V_{GS} = 0 - I_D R_S = -I_D R_S \qquad (6.2)$$

can be plotted on the i_d, v_{GS} characteristic as shown in Fig. 6.8b. To plot the straight line equation of Eq. (6.2) we can use $V_{GS} = 0$ at $I_D = 0$ and any other corresponding value. At $I_D = 3$ mA, for example,

$$V_{GS} = -I_D R_S = -(3 \text{ mA})(1.5 \text{ K}) = -4.5 \text{ V}$$

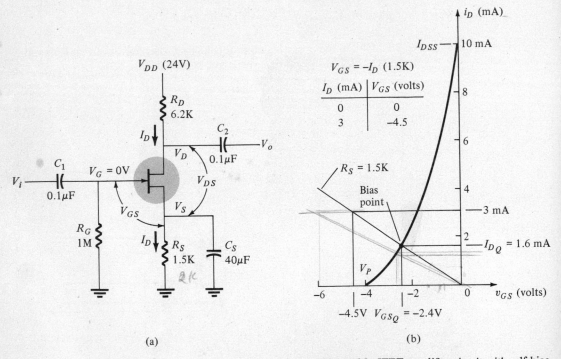

(a) (b)

Figure 6.8. JFET amplifier circuit with self-bias.

The self-bias line, as shown, intersects the device characteristic at $I_{D_Q} = 1.6$ mA, $V_{GS_Q} = -2.4$ V, the bias point set by R_S.

We should note that increasing R_S would lower the R_S line, resulting in a lower value of I_{D_Q} and corresponding larger value V_{GS_Q}. Reducing R_S would result in a steeper R_S line with higher I_{D_Q} and less voltage V_{GS_Q}.

EXAMPLE 6.2 Determine the bias point of the circuit shown in Fig. 6.8a for $R_S = 1$ K.

Solution: The self-bias line is drawn on the characteristic of Fig. 6.8b from the 0-axis ($V_{GS} = 0$, $I_D = 0$) passing through a point of selected value I_D, say $I_D = 4$ mA and corresponding voltage $V_{GS} = -I_D R_S = -(4\text{ mA})(1\text{ K}) = -4$ V. The intersection of this line with the device transfer characteristic is the bias point

$$I_{D_Q} = \textbf{2.2 mA}, \qquad V_{GS_Q} = \textbf{-2.2 V}$$

We can then calculate

$$V_{DS_Q} = V_{DD} - I_{DQ}R_D - I_{DQ}R_S = 24 - 2.2\text{ mA}(6.2\text{ K} + 1.5\text{ K})$$
$$= 24 - 16.9 = \textbf{7.1 V}.$$

EXAMPLE 6.3 Determine the operating point of the *n*-channel JFET amplifier shown in Fig. 6.9a.

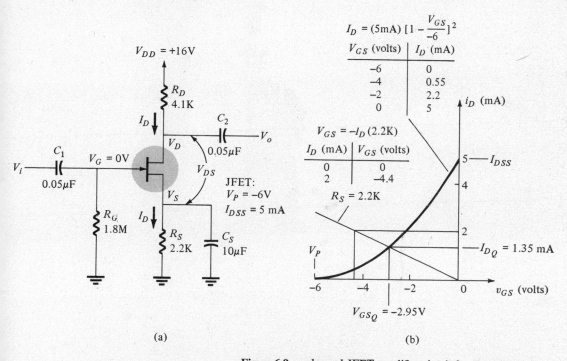

$$I_D = (5\text{mA}) \left[1 - \frac{V_{GS}}{-6}\right]^2$$

V_{GS} (volts)	I_D (mA)
-6	0
-4	0.55
-2	2.2
0	5

$$V_{GS} = -I_D (2.2\text{K})$$

I_D (mA)	V_{GS} (volts)
0	0
2	-4.4

$R_S = 2.2$K

(a)   (b)

Figure 6.9. *n*-channel JFET amplifier circuit for Example 6.3.

Solution: The transfer characteristic required to solve for the bias point need not be provided as long as V_P and I_{DSS} are specified. It is fairly easy to sketch the transfer characteristic using the values of V_P and I_{DSS} provided and the transfer curve formula [Eq. (6.1)].

$$I_D = I_{DSS}\left(1 - \frac{V_{GS}}{V_P}\right)^2$$

A sketch on a sheet of graph paper can be made for a few points of selected

V_{GS} and corresponding calculated values of I_D. For example, the present JFET would result in a transfer curve sketched from $V_P(= -6 \text{ V})$ along the x-axis through points

$$V_{GS} = -4 \text{ V}: \quad I_D = 5 \text{ mA}\left(1 - \frac{-4}{-6}\right)^2 = 0.55 \text{ mA}$$

$$V_{GS} = -2 \text{ V}: \quad I_D = 5 \text{ mA}\left(1 - \frac{-2}{-6}\right)^2 = 2.2 \text{ mA}$$

to the point $I_D = I_{DSS} = 5 \text{ mA}$ along the y-axis.

Using this transfer characteristic (Fig. 6.9b), we can then plot the self-bias line for $R_S = 2.2 \text{ K}$. Choosing a current, say $I_D = 2 \text{ mA}$, we obtain

$$V_{GS} = -I_D R_S = -(2 \text{ mA})(2.2 \text{ K}) = -4.4 \text{ V}$$

The self-bias line then intersects the transfer curve at about

$$V_{GS_Q} = -\textbf{2.95 V}, \qquad I_{D_Q} = \textbf{1.35 mA}$$

from which we calculate

$$V_{DS_Q} = V_{DD} - I_{D_Q}(R_S + R_D) = 16 - 1.35 \text{ mA}(2.2 \text{ K} + 4.1 \text{ K}) = \textbf{7.5 V}$$

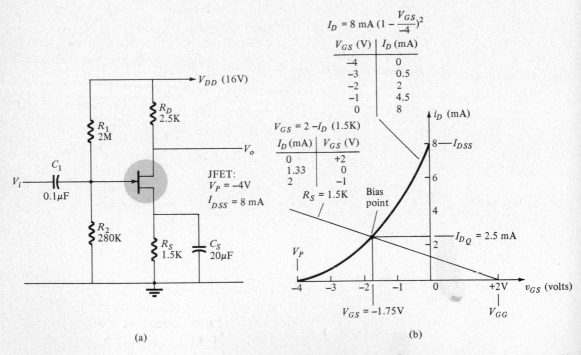

(a)

(b)

Figure 6.10. (a) JFET bias circuit with voltage divider gate bias; (b) transfer characteristic and self-bias line for Example 6.4.

Another form of dc bias circuit is that shown in Fig. 6.10a. Except for the gate voltage being set other than 0 V, the determination of bias voltage and current proceeds as discussed previously. The bias circuit of Fig. 6.10 provides a greater dc bias stability than the bias circuit of Fig. 6.8. The value of V_G obtained from the voltage divider network is

$$V_G = V_{GG} = \frac{R_2}{R_1 + R_2} \cdot V_{DD} \tag{6.3}$$

and the bias voltage V_{GS_Q} is then

$$V_{GS_Q} = V_{GG} - I_D R_S \tag{6.4}$$

EXAMPLE 6.4   Determine the dc bias of the JFET in the circuit of Fig. 6.10a.

Solution: The gate voltage is

$$V_G = V_{GG} = \frac{R_2}{R_1 + R_2} \cdot V_{DD} = \frac{280 \text{ K}}{2 \text{ M} + 280 \text{ K}} \cdot 16 \text{ V} = +2 \text{ V}$$

The result of a voltage drop $I_D R_S$ is a gate-source voltage

$$V_{GS} = V_G - V_S = +2 - I_D R_S$$

An R_S-bias line (see Fig. 6.10b) can be drawn corresponding to the above circuit equation. For the JFET with $V_P = -4$ V and $I_{DSS} = 8$ mA we can also plot the transfer characteristic as in Fig. 6.10b, the intersection of self-bias line and transfer characteristic providing a dc bias at

$$I_{D_Q} = 2.5 \text{ mA}, \qquad V_{GS_Q} = -1.75 \text{ V}$$

We can then calculate

$$V_{D_Q} = V_{DD} - I_{D_Q} R_s = 16 - 2.5 \text{ mA}(2.5 \text{ K}) = \textbf{9.75 V}$$

$$V_{S_Q} = I_{D_Q} R_D = 2.5 \text{ mA}(1.5 \text{ K}) = \textbf{3.75 V}$$

and $$V_{DS_Q} = V_{D_Q} - V_{S_Q} = 9.75 - 3.75 = \textbf{6 V}$$

(Note that $V_{GS_Q} = V_{G_Q} - V_{S_Q} = 2 - 3.75 = -1.75$ V, as expected.)

6.4 MOSFET CONSTRUCTION AND CHARACTERISTICS

A field-effect transistor can be constructed with the gate terminal insulated from the channel. The popular Metal-Oxide-Semiconductor FET (MOSFET)[3] is constructed as either a *depletion* MOSFET (Fig. 6.11a) or *enhancement* MOSFET (Fig. 6.11b). In the depletion mode construction a channel is physically constructed and current between drain and source will result from a voltage connected across the drain-source. The enhancement MOSFET structure has *no* channel formed when the device is constructed. Voltage must be applied at the gate to develop a channel of charge carriers so that a current results when a voltage is applied across the drain-source terminals.

---

[3]Also called Insulated Gate FET or IGFET.

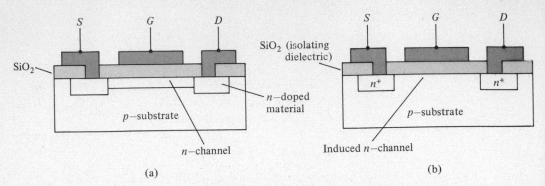

SiO₂

S G D

p–substrate

n–doped material

n–channel

(a)

SiO₂ (isolating dielectric)

S G D

n^+ n^+

p–substrate

Induced n–channel

(b)

Figure 6.11. MOSFET construction: (a) depletion; (b) enhancement.

Depletion MOSFET

The n-channel depletion MOSFET device of Fig. 6.11a is formed on a p-substrate (p-doped silicon material used as the starting material onto which the FET structure is formed. The source and drain are connected by metal (aluminum) to n-doped source and drain regions which are connected internally by an n-doped channel region. A metal layer is deposited above the n-channel on a layer of silicon dioxide (SiO_2) which is an insulating layer. This combination of a *metal* gate on an *oxide* layer over a *semi*conductor substrate forms the depletion MOSFET device. For the n-channel device of Fig. 6.11a negative gate-source voltages push electrons out of the channel region to deplete the channel and a large enough negative gate-source voltage will pinch off the channel. Positive gate-source

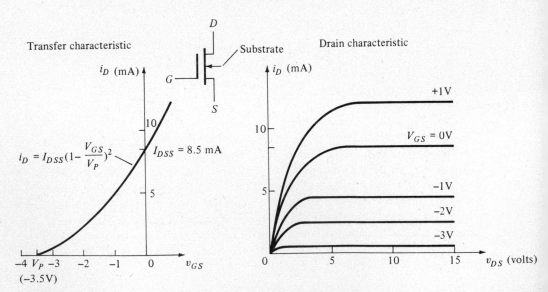

Transfer characteristic

i_D (mA)

$i_D = I_{DSS}(1 - \dfrac{V_{GS}}{V_P})^2$

I_{DSS} = 8.5 mA

10

5

−4 V_P −3 −2 −1 0 v_{GS}
(−3.5V)

D

G

Substrate

S

Drain characteristic

i_D (mA)

+1V

V_{GS} = 0V

−1V

−2V

−3V

10

5

0 5 10 15 v_{DS} (volts)

Figure 6.12. n-channel depletion MOSFET characteristic.

CHAP. 6 FIELD-EFFECT TRANSISTORS

294

voltage *on the other hand*, will result in an increase in the channel size (pushing away *p*-type carriers), allowing more charge carriers and therefore greater channel current to result.

An *n*-channel depletion MOSFET device characteristic is shown in Fig. 6.12. The device is shown to operate with either positive or negative gate-source voltage, negative values of V_{GS} reducing the drain current until the pinch-off voltage, after which no drain current occurs. The transfer characteristic is the same for negative gate-source voltages, but it continues for positive values of V_{GS}. Since the gate is isolated from the channel for both negative and positive values of V_{GS}, the device can be operated with either polarity of V_{GS}—no gate current resulting in either case. The device schematic symbol in Fig. 6.12 shows the addition of a substrate terminal (in addition to gate, source, and drain leads) on which the device type is indicated, the arrow here indicating a *p*-substrate and thus *n*-channel device. A *p*-channel depletion MOSFET characteristic is shown in Fig. 6.13.

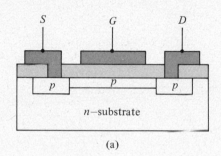

(a)

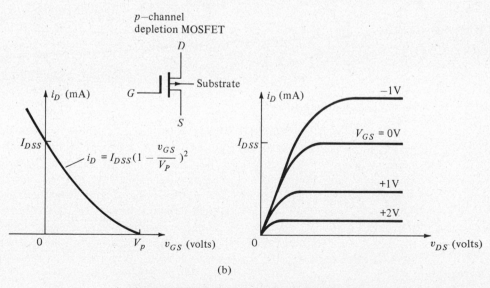

(b)

Figure 6.13. *p*-channel depletion MOSFET: (a) device structure; (b) device characteristic.

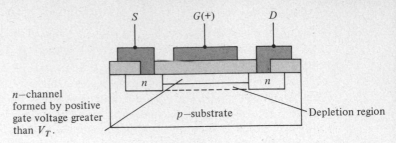

n—channel formed by positive gate voltage greater than V_T.

p—substrate

Depletion region

Figure 6.14. n-channel formed in enhancement MOSFET.

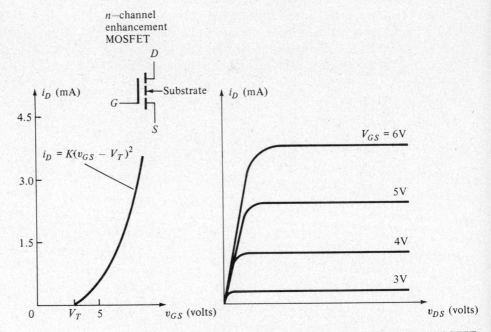

Figure 6.15. Device characteristic for n-channel enhancement MOSFET.

Enhancement MOSFET

The enhancement MOSFET of Fig. 6.14 has no channel between drain and source as part of the basic device construction. Application of a positive gate-source voltage will repel holes in the substrate region under the gate leaving a depletion region. When the gate voltage is sufficiently positive, electrons are attracted into this depletion region making it then act as an n-channel between drain and source. The resulting n-channel enhancement MOSFET characteristic is shown in Fig. 6.15. There is no drain current until the gate-source voltage exceeds the threshold value, V_T. Positive voltages above this threshold value result in increased drain current, the transfer characteristic being described by[4]

$$i_D = K(v_{GS} - V_T)^2 \qquad (6.5)$$

---

[4]Equation (6.5) is only valid for $|v_{GS}| > |V_T|$

where K, typically 0.3 mA/volt², is a property of the device construction. Note that no value I_{DSS} can be associated with an enhancement MOSFET because no drain current occurs with $V_{GS} = 0$ V. Although the enhancement MOSFET is more restricted in operating range than is the depletion device, the enhancement device is very useful in large-scale integrated circuits in which the simpler construction and smaller size make it a suitable device. The enhancement schematic symbol shows a broken line between drain and source indicating that there is no initial channel for the enhancement device. The substrate terminal arrow shows a p-substrate and an n-channel. P-channel enhancement MOSFETS can also be constructed, the device and characteristic being shown in Fig. 6.16.

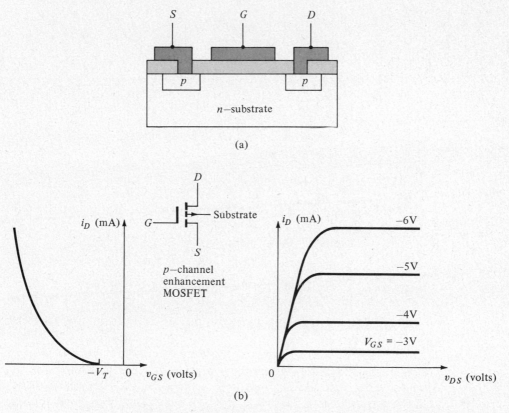

Figure 6.16. p-channel enhancement MOSFET: (a) device structure; (b) device characteristic.

6.5 MOSFET dc BIAS CIRCUITS

Depletion MOSFET Bias Circuit

The n-channel depletion MOSFET amplifier circuit shown in Fig. 6.17a is the same as that of a JFET device (except for the larger value of R_G possible with MOSFET devices). Since the gate-source voltage can go positive in this circuit, it is possible to bias the device at only slightly negative gate-source voltages. For the

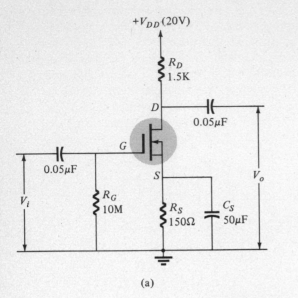

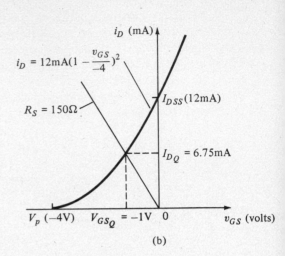

(a)

(b)

Figure 6.17. n-channel depletion MOSFET amplifier circuit and device characteristic.

circuit of Fig. 6.17 we obtain the dc bias values as follows:

Self-bias load line for $R_S = 150 \ \Omega$, drawn on Fig. 6.17, provides bias point at $V_{GS_Q} = -1$ V and $I_{D_Q} = 6.75$ mA.

The drain voltage is then

$$V_{D_Q} = V_{DD} - I_{D_Q}R_D = 20 - (6.75 \text{ mA})(1.5 \text{ K}) = 9.88 \text{ V}$$

and

$$V_{DS_Q} = V_{D_Q} - V_{S_Q} = 9.88 - 1 = 8.88 \text{ V}$$

Enhancement MOSFET bias Circuit

A popular dc bias arrangement for an enhancement device is shown in Fig. 6.18a. Using the drain-source voltage as gate-source bias voltage is achieved by connecting $R_G = 10$ M from drain to gate. Since there is no gate current, there is no voltage drop across R_G and $V_{DS} = V_{GS}$. The resulting bias point for $R_D = 2$ K (and $V_{DD} = 20$ V) can be obtained using the device transfer characteristic (see Fig. 6.18b). The device transfer characteristic can be drawn on graph paper using

$$I_D = K(V_{GS} - V_T)^2 = 0.3(V_{GS} - 3)^2$$

The circuit dc load line can also be drawn on the same graph, since

$$V_{GS} = V_{DS} = V_{DD} - I_D R_D = 20 - (2 \text{ K})I_D$$

The intersection of load line and device characteristic provides the operating point shown in Fig. 6.18b to be

$$V_{GS_Q} = V_{DS_Q} = 7.6 \text{ V}, \qquad I_{D_Q} = 6.2 \text{ mA}$$

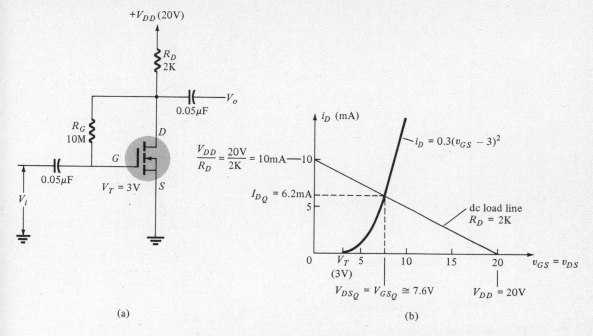

Figure 6.18. n-channel enhancement MOSFET dc bias circuit and characteristic.

The bias point can be adjusted by varying R_D. For example, R_D of 4 K would provide $V_{DS_Q} = V_{GS_Q} = 6.35$ V and $I_{D_Q} = 3.4$ mA (using a 4-K load line on Fig. 6.18b), while $R_D = 1$ K would result in $V_{DS_Q} = V_{GS_Q} = 9$ V and $I_{D_Q} = 11$ mA.

6.6 dc BIAS USING UNIVERSAL JFET BIAS CURVE

To reduce some of the effort in dc bias calculation with JFET (or depletion MOSFET) a normalized n-channel curve as shown in Fig. 6.19 may be used. The JFET transfer characteristic is plotted on normalized axes. To simplify the plotting of the R_S self-bias line an axis of normalized R_S values is plotted as the value m, where

$$m = \frac{|V_P|}{I_{DSS}R_S} \tag{6.6}$$

with V_P in volts, I_{DSS} in milliamperes and R_S in kilohms. For the voltage divider bias stabilized circuit the values of M and V_{GG} are used, as will be demonstrated shortly.

$$V_{GG} = \frac{R_2}{R_1 + R_2} \cdot V_{DD} \tag{6.7}$$

$$M = m \times \frac{V_{GG}}{|V_P|} \tag{6.8}$$

An example for each type of bias circuit will help show how this universal characteristic is used.

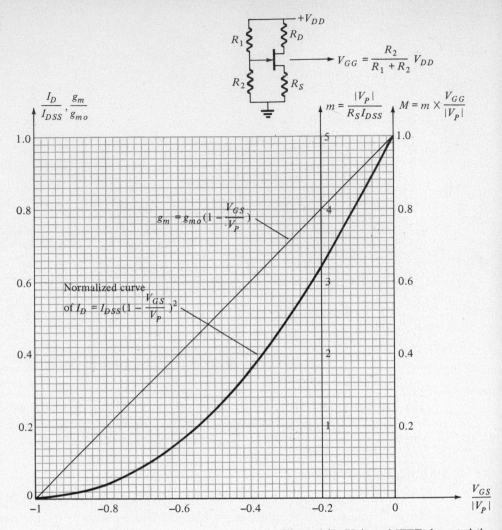

Figure 6.19. Universal JFET characteristic.

EXAMPLE 6.5 Determine the dc bias voltages and currents for the circuit of Fig. 6.20a.

Solution: Calculating the value of m.

$$m = \frac{|V_P|}{I_{DSS}R_S} = \frac{|-3|}{6(1.6)} = 0.31$$

we plot the R_S bias line from the point through the point $m = 0.31$ along the m-axis. The resulting bias point is seen in Fig. 6.20b to be

$$\frac{I_D}{I_{DSS}} = 0.18, \qquad \frac{V_{GS}}{|V_P|} = 0.575$$

from which we calculate

$$I_{D_Q} = 0.18(6 \text{ mA}) = \mathbf{1.08 \text{ mA}}$$
$$V_{GS_Q} = 0.575(-3) = \mathbf{-1.73 \text{ V}}$$

CHAP. 6 FIELD-EFFECT TRANSISTORS

300

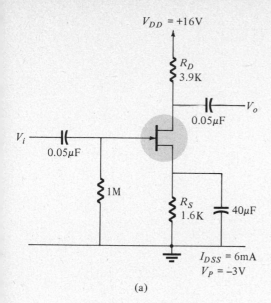

$V_{DD} = +16V$

R_D
3.9K

$\|\!($ — V_o
$0.05\mu F$

V_i — $\|\!($
$0.05\mu F$

1M

R_S
1.6K

$\|\!$ 40μF

$I_{DSS} = 6mA$
$V_P = -3V$

(a)

Figure 6.20

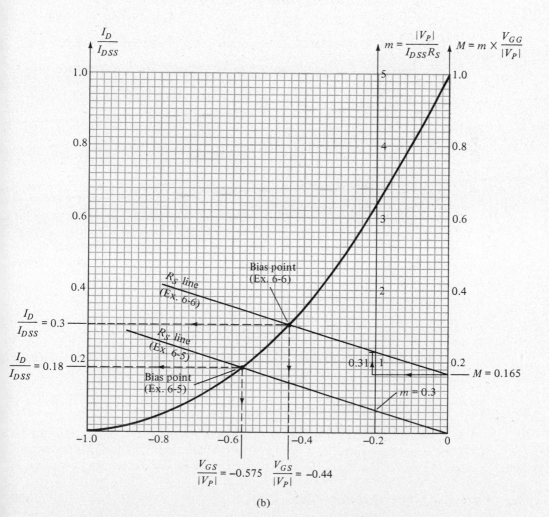

$\dfrac{I_D}{I_{DSS}}$

$m = \dfrac{|V_P|}{I_{DSS}R_S}$ $M = m \times \dfrac{V_{GG}}{|V_P|}$

1.0
0.8
0.6

Bias point
(Ex. 6-6)

0.4

R_S line
(Ex. 6-6)

$\dfrac{I_D}{I_{DSS}} = 0.3$

R_S line
(Ex. 6-5)

$\dfrac{I_D}{I_{DSS}} = 0.18$

0.2

0.31

Bias point
(Ex. 6-5)

$M = 0.165$

$m = 0.3$

-1.0 -0.8 -0.6 -0.4 -0.2 0

$\dfrac{V_{GS}}{|V_P|} = -0.575$ $\dfrac{V_{GS}}{|V_P|} = -0.44$

(b)

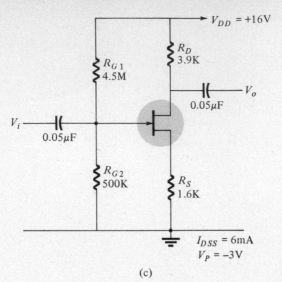

(c)   **Figure 6.20.** (continued)

Using the value of I_{D_Q} obtained, we then calculate

$$V_{DS_Q} = V_{DD} - I_{D_Q}(R_D + R_S) = 16 - 1.08(3.9 + 1.6) = \mathbf{10.06\ V}$$

EXAMPLE 6.6   Determine the dc bias condition for the circuit in Fig. 6.20.

Solution: For the voltage divider bias circuit of Fig. 6-20c we first calculate

$$m = \frac{|V_P|}{I_{DSS}R_S} = \frac{|-3|}{6(1.6)} = 0.31 \qquad \text{(same as in Fig. 6.20c)}$$

$$V_{GG} = \frac{R_{G_2}}{R_{G_1} + R_{G_2}} V_{DD} = \frac{500\ K}{4.5\ M + 500\ K} \cdot 16 = 1.6\ V$$

$$M = m \cdot \frac{V_{GG}}{|V_P|} = 0.31 \frac{1.6}{|-3|} = 0.165$$

To plot the bias line we must connect a line having the same slope (same R_S and JFET values) as in the previous example passing through the $M = 0.165$ point. This is directly accomplished by connecting a line between the point $M = 0.165$ along the M-axis and a point along the m-axis which is 0.31 higher. From Fig. 6.20b we see that this line gives a bias point

$$\frac{V_{GS}}{|V_P|} = 0.44 \qquad \frac{I_D}{I_{DSS}} = 0.3$$

from which we calculate

$$V_{GS_Q} = 0.44(-3\ V) = -\mathbf{1.32\ V}$$

$$I_{D_Q} = 0.3(6\ mA) = \mathbf{1.8\ mA}$$

We can then calculate

$$V_{DS_Q} = V_{DD} - I_{D_Q}(R_D + R_S) = 16 - 1.8\ mA(3.9\ K + 1.6\ K) = \mathbf{6.1\ V}$$

6.7 ac SMALL-SIGNAL AMPLIFIER OPERATION

Having biased the FET for operation in its linear region we can now investigate the voltage gain of the amplifier as well as the amplifier stage input and output impedances. Consider the basic JFET amplifier circuit of Fig. 6.21a. An ac voltage is

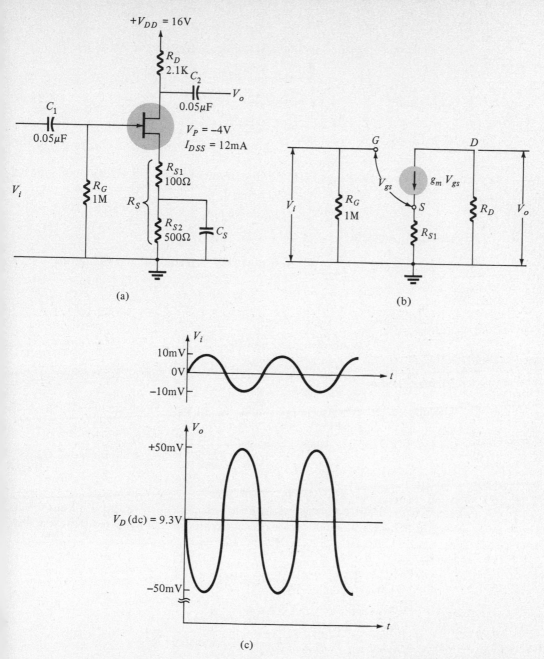

Figure 6.21. ac amplifier circuit.

applied through coupling capacitor C_1 which acts to block any signal dc level from affecting the stage bias. The ac signal V_i appears across resistor R_G. An amplified output voltage is developed across resistor R_D and is coupled through capacitor C_2 as the ac output voltage, V_o.

An ac equivalent circuit for the amplifier can be drawn as in Fig. 6.21b. The FET is represented by the current source $g_m V_{gs}$, where the value g_m is the FET transconductance measured in mhos. It can be shown[5] that at the dc bias point the value of g_m is

$$\boxed{g_m = g_{mo}\left(1 - \frac{V_{GS_Q}}{V_P}\right)} \quad (6.9)$$

where

$$g_{mo} = \frac{2I_{DSS}}{|V_P|} \quad (6.10)$$

is a constant for a particular FET. For the circuit of Fig. 6.21a we can obtain the dc bias, as previously covered, to be

$$V_{GS_Q} = -1.9\text{ V}, \qquad I_{D_Q} = 3.2\text{ mA}, \qquad V_{DS_Q} = 7.36\text{ V}$$

We can then calculate

$$g_{mo} = \frac{2I_{DSS}}{|V_P|} = \frac{2(12\text{ mA})}{|-4\text{ V}|} = 6 \times 10^{-3}\ \mho = 6\text{ mmhos} = 6000\ \mu\text{mhos}$$

and

$$g_m = g_{mo}\left(1 - \frac{V_{GS_Q}}{V_P}\right) = (6 \times 10^{-3})\left[1 - \frac{-1.9}{-4}\right] = 3.15 \times 10^{-3}$$

$$= 3.15\text{ mmhos} = 3150\ \mu\text{mhos}$$

Voltage Gain A_v

Calculation of the ratio V_o/V_i from the ac equivalent circuit of Fig. 6.21 results in the voltage gain[6]

$$\boxed{A_v = \frac{V_o}{V_i} = \frac{-g_m R_D}{1 + g_m R_{S1}}} \quad (6.11)$$

where R_D is the drain load resistor, g_m the transconductance of the JFET at the bias point, and R_{S1} is the value of source resistance which is *not* bypassed by a capacitor.

---

[5]Since

$$i_D = I_{DSS}\left(1 - \frac{v_{GS}}{V_P}\right)^2$$

and

$$g_m \triangleq \frac{i_D}{\partial v_{GS}}\Big|_{v_{DS}=\text{constant}}$$

we obtain

$$g_m = -\frac{2I_{DSS}}{V_P}\left(1 - \frac{v_{GS}}{V_P}\right) = g_{mo}\left(1 - \frac{v_{GS}}{V_P}\right)$$

where

$$g_{mo} \equiv \frac{2I_{DSS}}{|V_P|}$$

---

[6]

$$V_{gs} = V_i - (g_m V_{gs})R_{S1}$$

$$(1 + g_m R_{S1})V_{gs} = V_i$$

$$V_o = -g_m V_{gs}R_D$$

$$A_v = \frac{V_o}{V_i} = \frac{-g_m R_D}{1 + g_m R_{S1}} = \frac{-g_m R_D}{1 + g_m R_{S1}}$$

In the present example we obtain

$$A_v = \frac{-(3.15 \times 10^{-3})(2.1 \times 10^3)}{1 + (3.15 \times 10^{-3})(100)} = \frac{-6.615}{1 + 0.315} = -5.03$$

The value of A_v shows that the output ac signal will be 5.03 times as large as the input signal, the negative sign indicating that the output signal is inverted or 180° out of phase with the input signal, as shown in Fig. 6.21c.

A few examples will show how the voltage gain may be calculated for FET amplifier circuits.

EXAMPLE 6.7 Determine the voltage gain of the amplifier shown in Fig. 6.22, and the output voltage, V_o.

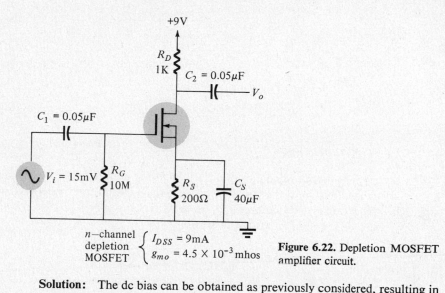

Figure 6.22. Depletion MOSFET amplifier circuit.

Solution: The dc bias can be obtained as previously considered, resulting in

$$V_{GS_Q} = -1 \text{ V}, \qquad I_{D_Q} = 5.1 \text{ mA}$$

Using Eqs. (6-9) and (6-10), we calculate g_m to be

$$g_m = 4.5 \times 10^{-3}\left(1 - \frac{-1}{-4}\right) = 3.38 \text{ mmhos}$$

The circuit voltage gain is then [using Eq. (6.11)]

$$A_v = \frac{-g_m R_D}{1 + g_m R_{S1}} = \frac{-(3.38 \times 10^{-3})(1 \times 10^3)}{1 + 0} = -\mathbf{3.38}$$

where $R_{S1} = 0$ (no unbypassed source resistance).
The output voltage is then

$$V_o = A_v V_i = -3.38(15 \text{ mV}) = -\mathbf{50.7 \text{ mV}}$$

EXAMPLE 6.8 Calculate the voltage gain of the enhancement MOSFET amplifier circuit of Fig. 6.23.

Solution: The dc bias point can be determined to be $V_{GS_Q} = V_{DS_Q} = 6.7$ V

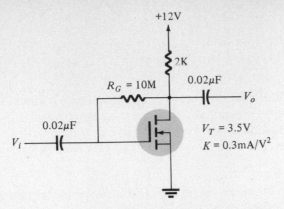

+12V

$R_G = 10M$

$0.02\mu F$

V_o

2K

$0.02\mu F$

V_i

$V_T = 3.5V$
$K = 0.3mA/V^2$

Figure 6.23. Enhancement MOSFET amplifier circuit.

at $I_{D_Q} = 3.1$ mA. The value of g_m at the bias point can be obtained from[7]

$$g_m = 2 K(V_{DS_Q} - V_T) \qquad (6.12)$$

$$g_m = 2(0.3 \times 10^{-3})(6.7 - 3.5) = 1.9 \text{ mmhos}$$

Using Eq. (6.11), we obtain

$$A_v = \frac{-g_m R_D}{1 + g_m(0)} = -g_m R_D = -1.9 \times 10^{-3}(2 \times 10^3) = \mathbf{-3.8}$$

Input Impedance, R_i

The input impedance of a FET amplifier at mid-frequency is simply the value of resistor R_G since the gate-source is an open circuit with the JFET input acting as a reverse biased diode junction, or MOSFET input being isolated by a dielectric. For the FET amplifier we can then use

$$R_i = R_G \qquad (6.13)$$

The JFET amplifier of Fig. 6.21 has an ac input impedance of $R_i = 1$ M; that of Fig. 6.22 has $R_i = 10$ M.

Output Impedance, R_o

The output impedance of a FET amplifier,[8] like that of a BJT amplifier, is the output resistance, R_D, for the FET circuit in parallel with the device output impedance. We can then use

$$R_o = R_D \| r_o \qquad (6.14a)$$

---

[7]Since $i_D = K(v_{GS} - V_T)^2$,

$$g_m = \frac{\partial i_D}{\partial v_{GS}}\bigg|_{V_{DS}=\text{constant}} = 2K(v_{GS_Q} - V_T)$$

---

[8]With input impedance assumed as short circuit.

where R_D is the drain resistance and r_o the FET output impedance given by the manufacturer. If we neglect the value of r_o ($r_o \gg R_D$), we then have

$$R_o = R_D \tag{6.14b}$$

Neglecting r_o for the JFET circuit of Fig. 6.21, we see that $R_o = R_D = 2.1$ K, for the depletion MOSFET circuit of Fig. 6.22, $R_o = R_D = 1$ K, and for the enhancement MOSFET circuit of Fig. 6.23 $R_o = R_D = 2$ K.

Source-Follower Amplifier (Output from Source Terminal)

If the output is taken from the source terminal (see Fig. 6.24), there is no phase inversion between output and input, and the voltage amplitude is reduced from the input value, as given by[9]

$$\boxed{A_v = \frac{V_o}{V_i} = \frac{+g_m R_s}{1 + g_m R_s}} \tag{6.15}$$

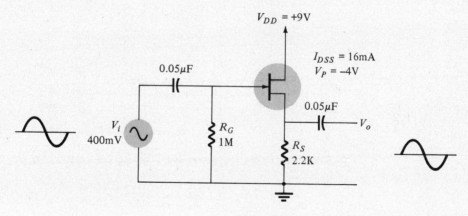

Figure 6.24. Source-follower amplifier circuit.

EXAMPLE 6.9 Calculate the voltage gain of the amplifier circuit of Fig. 6.24.

Solution: From dc bias calculations we obtain $V_{GS_Q} = -2.85$ V and $I_{D_Q} = 1.3$ mA, at which point we calculate

$$g_m = g_{mo}\left(1 - \frac{V_{GS_Q}}{V_P}\right) = \frac{2(16 \times 10^{-3})}{|-4|}\left(1 - \frac{-2.85}{-4}\right) = 2.3 \text{ mmhos}$$

---

[9]From ac equivalent circuit

$$V_o = g_m V_{gs} R_S, \qquad V_{gs} = V_i - V_o$$
$$V_{gs} = V_i - g_m V_{gs} R_S$$
$$(1 + g_m R_S)V_{gs} = V_i$$

so that
$$A_v = \frac{V_o}{V_i} = \frac{g_m R_S V_{gs}}{1 + g_m R_S V_{gs}} = \frac{g_m R_S}{1 + g_m R_S}$$

The voltage gain is then

$$A_v = \frac{V_o}{V_i} = \frac{g_m R_s}{1 + g_m R_s} = \frac{(2.3 \times 10^{-3})(2.2 \times 10^3)}{1 + (2.3 \times 10^{-3})(2.2 \times 10^3)} = 0.835$$

and
$$V_o = A_v V_i = 0.835(400 \text{ mV}) = 334 \text{ mV}$$

EXAMPLE 6.10 Determine V_{o1}, V_{o2}, R_i, and R_o for the amplifier circuit of Fig. 6.25.

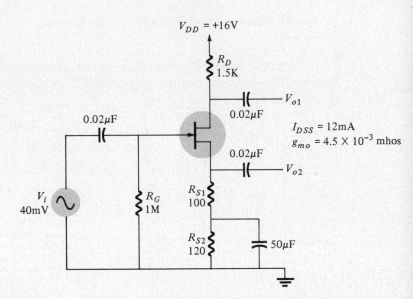

Figure 6.25. JFET amplifier circuit for Example 6.10.

Solution: Using the given JFET device parameters, we can determine V_P to be

$$V_P = -\frac{2I_{DSS}}{g_{mo}} = -\frac{2(12 \times 10^{-3})}{4.5 \times 10^{-3}} = -5.33 \text{ V}$$

Using a plot of the transfer characteristic or the universal JFET curve, we find the value of dc bias to be $V_{GS_Q} = -1.4$ V and $I_{D_Q} = 6.5$ mA, at which

$$g_m = g_{mo}\left(1 - \frac{V_{GS_Q}}{V_P}\right) = (4.5 \times 10^{-3})\left[1 - \frac{-1.4}{-5.3}\right] = 3.3 \text{ mmhos}$$

The voltage gains from input to each output are

$$A_{v1} = \frac{V_{o1}}{V_i} = \frac{-g_m R_D}{1 + g_m R_{s1}} = \frac{-(3.3 \times 10^{-3})(1.5 \times 10^3)}{1 + (3.3 \times 10^{-3})(100)} = -3.7$$

$$A_{v2} = \frac{V_{o2}}{V_i} = \frac{g_m R_{s1}}{1 + g_m R_{s1}} = \frac{(3.3 \times 10^{-3})(100)}{1 + (3.3 \times 10^{-3})(100)} = 0.25$$

so that
$$V_{o1} = A_{v1} V_i = -3.7(40 \text{ mV}) \cong \mathbf{150 \text{ mV}}$$
$$V_{o2} = A_{v2} V_i = +0.25(40 \text{ mV}) = \mathbf{10 \text{ mV}}$$

The input impedance of the amplifier stage is
$$R_i = R_G = \mathbf{1 \text{ M}}$$

and the output impedance at V_{o1} is

$$R_o = R_D = 1.5\,\text{K}$$

When a source having resistance drives an amplifier whose output is connected to a load, as in Fig. 6.26, the calculations can be made as shown in the following example.

EXAMPLE 6.11 (a) For the circuit of Fig. 6.26 calculate the output voltage, V_o, developed across a load of resistance $R_L = 10$ K. (b) Repeat if an impedance of 50 K is connected across the circuit output.

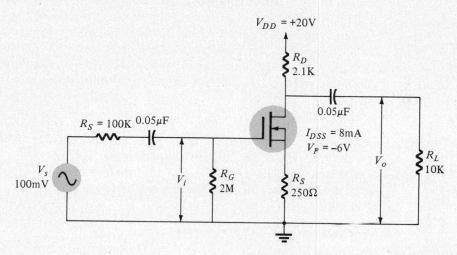

Figure 6.26. FET amplifier with loading.

Solution: The dc bias values can be found to be

$$V_{GS_Q} = -1.2\,\text{V}, \qquad I_{D_Q} = 5\,\text{mA}$$

at which $$g_m = \frac{2(8 \times 10^{-3})}{|-6|}\left(1 - \frac{-1.2}{-6}\right) = 2.13\,\text{mmhos}$$

The input voltage V_i is reduced from the source voltage ($V_S = 100$ mV) by the voltage divider of $R_S = 100$ K and $R_i = R_G = 2$ M

$$V_i = \frac{R_i}{R_i + R_S}V_S = \frac{2\,\text{M}}{2\,\text{M} + 100\,\text{K}}(100\,\text{mV}) = 95.24\,\text{mV}$$

(a) The output resistance $R_o = R_D = 2.1$ K is in parallel with the load resistance so that the output resistance used in gain calculation is

$$R_D \| R_L = 2.1\,\text{K} \| 10\,\text{K} = 1.74\,\text{K}$$

The ac voltage gain is then

$$A_o = \frac{V_o}{V_i} = \frac{-(2.13 \times 10^{-3})(1.74 \times 10^3)}{1 + (2.13 \times 10^{-3})(250)} = -2.42$$

and $$V_o = A_v V_i = -2.42(95.24\,\text{mV}) = -230.5\,\text{mV}$$

(b) If the output resistance, $r_o = 50$ K, of the FET is considered, then the output resistance used in the gain calculation is

$$R_D \| R_L \| r_o = 1.74\,\text{K} \| 50\,\text{K} = 1.68\,\text{K}$$

and
$$A_v = \frac{V_o}{V_i} = \frac{-(2.3 \times 10^{-3})(1.68 \times 10^3)}{1 + (2.13 \times 10^{-3})(250)} = -2.34$$

so that
$$V_o = A_v V_i = -2.34(95.24 \text{ mV}) = -222.86 \text{ mV}$$

6.8 HIGH- AND LOW-FREQUENCY EFFECTS IN FET

The ac analysis considered so far applies to mid-frequency operation. At high frequencies some device interterminal capacitances and circuit capacitances reduce the gain from the value at mid-frequency. For the amplifier of Fig. 6.27a a number of interterminal capacitances which, although not externally placed in the circuit, are present due to the physical construction of the FET device (or any other electronic device). Although we may not desire these capacitances, we cannot ignore their effect on the circuit operation at higher frequencies.

Miller Effect

The most important effect of these capacitances is due to the capacitance between input and output C_{gd} in the circuit of Fig. 6.27b.

The input capacitance for the FET can be obtained using the simpler representation in Fig. 6.27c. The input current is

$$I_i = V_i Y_i = I_1 + I_2 = V_i Y_{gs} + (V_i - A_v V_i) Y_{gd}$$

from which we get

$$Y_i = Y_{gs} + (1 - A_v) Y_{gd}$$

which can be expressed in terms of the capacitance values

$$\boxed{C_i = C_{gs} + \underbrace{(1 - A_v) C_{gd}}_{\text{Miller capacitance}}} \tag{6.16}$$

Notice that the capacitance from input to output C_{gd} is multiplied by the gain (actually, $1 - A_v$)—this being the Miller effect and $C_M = (1 - A_v) C_{gd}$ the Miller capacitance.

The total input capacitance across the load also may include the capacitance from drain to source C_{ds} and any stray wiring capacitance that exists, when FET amplifier stages are connected in cascade. The total high-frequency circuit effect is shown in Fig. 6.27d with

$$R_{\text{high}} = r_d \| R_D = R_o \tag{6.17}$$

$$C_{\text{high}} = C_{gs} + (1 - A_s) C_{gd} + C_{ds} + C_w \tag{6.18}$$

where C_W is the stray wiring capacitance and C_{ds}, C_{gd}, and C_{gs} are the interterminal capacitances of the FET.

EXAMPLE 6.12 Calculate the high-frequency capacitance for the FET amplifier of Fig. 6.27a.

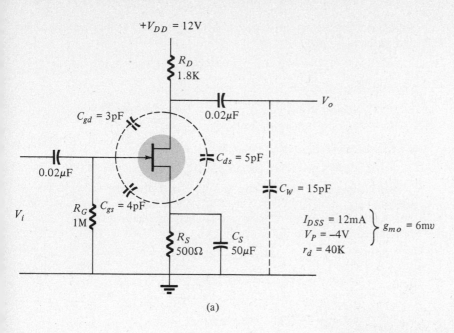

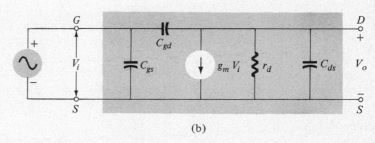

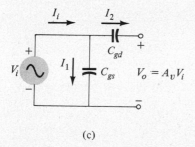

Figure 6.27. (a) FET amplifier showing high frequency capacitances; (b) FET high-frequency ac equivalent circuit; (c) simplified circuit for analysis; (d) simplified ac equivalent circuit.

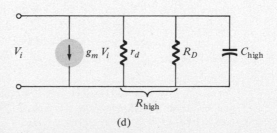

Solution: Calculating the dc bias to be $V_{GS_Q} = -1.8$ V, we then determine that at the bias point

$$g_m = (6 \times 10^{-3})\left(1 - \frac{-1.8}{-4}\right) = 3.3 \text{ mmhos}$$

The voltage gain is then

$$A_v = -g_m(R_D \| r_d) = -(3.3 \times 10^{-3})(1.8 \text{ K} \| 40 \text{ K})$$
$$= -(3.3 \times 10^{-3})(1.72 \times 10^3) \cong -5.7$$

We can calculate C_{high} using Eq. (6.18)

$$C_{\text{high}} = 4 \text{ pF } [1 - (-5.7)]3 \text{ pF} + 5 \text{ pF} + 15 \text{ pF} = 44.1 \text{ pF}$$

Notice that since the voltage gain A_v is negative, the value of $1 - A_v$ results in the sum $|A_v|$ *plus* 1. The effect of this high-frequency capacitance will be shown in Chapter 7 to reduce the amplifier gain at a high frequency determined when $R_o = X_{C_{\text{high}}}$, or

$$f_H = \frac{1}{2\pi R_o C_{\text{high}}} \tag{6.19}$$

Low Frequency

While the terminal capacitances of the FET have small effect at frequencies lower than mid-frequency, the coupling capacitor and input resistance determine the circuit gain at lower frequencies. When the magnitude of the coupling capacitor impedance equals the stage input resistance, the overall circuit gain is reduced from that at mid-frequency to a lower value. That is, when

$$X_{C_C} = R_i = R_G$$

$$\frac{1}{2\pi f_L C_C} = R_G$$

$$f_L = \frac{1}{2\pi R_i C_C} = \frac{1}{2\pi R_G C_C} \tag{6.20}$$

the gain of the amplifier stage drops to 0.707 of the mid-frequency gain.[10]

6.9 BOOTSTRAP SOURCE-FOLLOWER CIRCUIT

Although the FET amplifier has a high input impedance due to the large value of R_G, there are applications in which a much higher input impedance is desired. The bootstrap source-follower circuit of Fig. 6.28 provides this increased input impedance. The "boostrap" effect due to the 100 μF capacitor feeding a portion of the ac

---

[10]The voltage divider of R_G and X_{C_C}, when R_G and X_{C_C} are equal in value, results in $V_o = 0.707 \ V_i$ and the overall gain is reduced by this amount, which equals a drop of 3 dB. This material is covered more fully in Chapter 7.

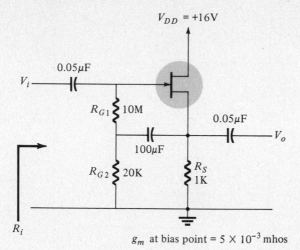

$V_{DD} = +16V$

$0.05\mu F$

V_i

R_{G1} 10M

$0.05\mu F$

V_o

$100\mu F$

R_{G2} 20K

R_S 1K

R_i

g_m at bias point $= 5 \times 10^{-3}$ mhos

Figure 6.28. Bootstrap source follower circuit.

output voltage back to R_G makes that resistor have an effective Miller impedance of

$$(R_{G1})_{\text{effective}} = R'_{G1} = \frac{R_{G1}}{1 - A_v} = R_i \qquad (6.21)$$

where

$$A_v = \frac{V_o}{V_i} \cong \frac{g_m(R_S \| R_{G2})}{1 + g_m(R_S \| R_{G2})}$$

As an example, the input impedance of the bootstrap circuit of Fig. 6.28 is

$$R_S \| R_{G2} = 1\,\text{K} \| 20\,\text{K} = 0.95\,\text{K}$$

$$A_v = \frac{(5 \times 10^{-3})(0.95\,\text{K})}{1 + (5 \times 10^{-3})(0.95\,\text{K})} = 0.83$$

so that

$$R_i = R'_{G1} = \frac{10\,\text{M}}{1 - 0.83} = 60\,\text{M}$$

An ideal buffer amplifier would have infinite input impedance, zero output impedance, and a voltage gain of unity. One practical form of such a circuit can be obtained using a bootstrap source-follower as shown in the circuit of Fig. 6.29. The following calculations show how the above circuit parameters can be calculated.

Since Q_2 has $V_{GS} = 0$ V, the drain current which is the current set for Q_1 is 2 mA. From the transfer characteristic of Q_1 it can be determined that at $I_{D_Q} = 2$ mA the value of V_{GS_Q} is $V_{GS_Q} = -2$ V. Since $V_Q = 0$ V, the value of V_S is

$$V_S = V_B = 2\,\text{V}$$

so that

$$V_E = 2 - 0.7 = 1.3\,\text{V}$$

and

$$I_E = \frac{V_E}{R_E} = \frac{1.3\,\text{V}}{1.2\,\text{K}} = 1.08\,\text{mA}$$

We can then calculate r_e

$$r_e = \frac{26}{1.08} + 2 \cong 26\,\Omega$$

The source impedance for Q_1 is then calculated from

$$R_S = 100\,\text{K} \| \beta(r_e + R_E) = 100\,\text{K} \| 200(1.2\,\text{K} + 26\,\Omega) \cong 71\,\text{K}$$

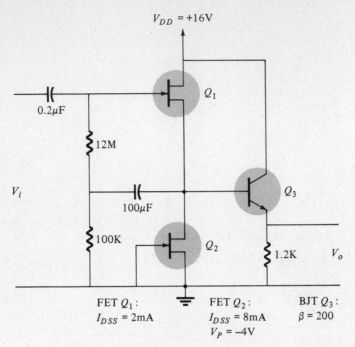

FET Q_1:
$I_{DSS} = 2mA$

FET Q_2:
$I_{DSS} = 8mA$
$V_P = -4V$

BJT Q_3:
$\beta = 200$

Figure 6.29. Buffer amplifier.

At bias point for Q_1 we calculate

$$g_m = g_{mo}\left(1 - \frac{V_{GS_Q}}{V_P}\right) = \frac{2(8 \times 10^{-3})}{1 - 41}\left(1 - \frac{-2}{-4}\right) = 2 \text{ mmhos}$$

and

$$A_v = \frac{g_m R_S}{1 + g_m R_S} = \frac{(2 \times 10^{-3})(71 \times 10^3)}{1 + (2 \times 10^{-3})(71 \times 10^3)} = 0.993$$

The input impedance is then determined to be

$$R_i = R'_{G1} = \frac{R_{G1}}{1 - A_v} = \frac{12 \text{ M}}{1 - 0.993} = 1.716 \times 10^9 = 1716 \text{ M}$$

an extremely large value, as desired.

The voltage gain of the BJT emitter follower is

$$A_{v2} = \frac{R_E}{R_E + r_e} = \frac{1.2 \text{ K}}{1.2 \text{ K} + 26 \, \Omega} = 0.9787$$

so that overall gain is

$$A_v = A_{v1}A_{v2} = (0.993)(0.9787) \cong 0.97$$

which is very close to the ideal value of 1.

Finally, the output resistance of the circuit taken from the emitter-follower output is

$$R_o \cong r_e = 26 \, \Omega$$

a low output resistance.

To complete our basic coverage of FET amplifier circuits we will consider the design of a few types of amplifiers. Starting with some description of FET and bias condition desired, the circuit resistor values are then determined.

EXAMPLE **6.13** Determine suitable resistor and capacitor values for the JFET amplifier of Fig. 6.30. Use a 2N4220 n-channel JFET having device parameters

$$BV_{GSS(\text{min})} = 30 \text{ V} \qquad\qquad g_{fs} = g_m \text{ is between 1000–4000 } \mu\text{mhos}$$

$$V_{GS(\text{off})} = -4 \text{ V} = V_p \qquad I_{DSS(\text{max})} = 3 \text{ mA}$$

to achieve a mid-frequency gain of at least 10.

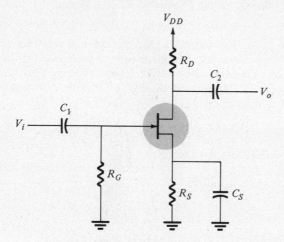

Figure **6.30.** JFET amplifier circuit for Example 6.13.

Solution: Our procedure will essentially work reverse from the analysis steps covered up to now.

For a voltage gain of 10 at the minimum g_m of 1000 μmhos we have

$$A_v = -g_m R_D$$

$$R_D = \frac{A_v}{-g_m} = \frac{-10}{-1000 \times 10^{-6}} = 10 \text{ K}$$

The value of $BV_{GSS(\text{min})}$ is the minimum voltage at which the drain-source voltage will cause breakdown to occur and should not be exceeded. Using a voltage supply of $V_{DD} = 25$ V in this design will then be acceptable. To achieve a bias voltage V_{D_Q} about midway in the range from 0 V to 25 V we select $V_{D_Q} = 10$ V as bias voltage, and with $R_D = 10$ K

$$V_{D_Q} = V_{DD} - I_{D_Q} R_D$$

$$I_{D_Q} = \frac{V_{DD} - V_{D_Q}}{R_D} = \frac{25 - 10 \text{ V}}{10 \text{ K}} = 1.5 \text{ mA}$$

To achieve the bias current of 1.5 mA we now determine the value of R_S using the device transfer characteristic or universal JFET characteristic.

From the universal characteristic the point

$$\frac{I_D}{I_{DSS}} = \frac{1.5 \text{ mA}}{3 \text{ mA}} = 0.5$$

corresponds to a point

$$\frac{V_{GS}}{|V_P|} = -0.3$$

so that

$$V_{GS_Q} = -0.3(4 \text{ V}) = -1.2 \text{ V}$$

We then calculate

$$V_{GS_Q} = -I_{D_Q} R_S$$

$$R_S = \frac{-V_{GS_Q}}{I_{D_Q}} = \frac{-(-1.2 \text{ V})}{1.5 \text{ mA}} = 800 \ \Omega$$

The value of R_G can be selected $R_G = 1$ M since

$$I_{GSS(\text{off})} R_G = (0.1 \text{ nA})(1 \times 10^6) = 10^{-4} \text{ V} = 0.1 \text{ mV}$$

where I_{GSS} the gate-source leakage current, will result in only a 0.1 mV dc voltage drop across a 1-M gate resistor, a negligible voltage in this circuit. Values of C_1 and C_2 are selected from frequency considerations. *At mid-frequency* ($f = 1000$ Hz) values of $C_1 = C_2 = 0.1 \ \mu\text{F}$ would provide an ac impedance magnitude of

$$X_C = \frac{1}{2\pi fC} = \frac{1}{2\pi(1000)(0.1 \times 10^{-6})} \cong 1.6 \text{ K}$$

which is small compared to $R_G = 1$ M

A value of C_S for which X_S is at least 10 times smaller than R_S to provide good ac bypass would be

$$X_C \leq \frac{1}{10} R_S = \frac{800}{10} = 80 \ \Omega$$

$$\frac{1}{2\pi fC_S} = 80 \ \Omega$$

$$C_S = \frac{1}{2\pi(1000)(80)} = 1.9 \times 10^{-6} \cong 2 \ \mu\text{F}$$

EXAMPLE 6.14 Design a FET amplifier as shown in Fig. 6.31 to operate from a 12-V supply with a gain of at least 5. Bias at $V_{D_Q} = 8$ V; $I_{D_Q} = 0.25$ mA.

Solution: To obtain the bias condition

$$V_{D_Q} = V_{DD} - I_{D_Q} R_D$$

$$R_D = \frac{V_{DD} - V_{D_Q}}{I_{D_Q}} = \frac{12 - 8 \text{ V}}{0.25 \text{ mA}} = 16 \text{ K}$$

Using universal characteristics, we can determine V_{GS} at bias using $I_D / I_{DSS} = 0.25 \text{ mA}/0.5 \text{ mA} = 0.5$ for which we get $(V_{GS_Q}/|V_P|) = -0.3$ so that

$$V_{GS_Q} = 0.3(-2) = -0.6 \text{ V}$$

The value of R_S needed is then

$$V_{GS_Q} = -I_{D_Q} R_S$$

$$R_S = -\frac{V_{GS_Q}}{I_{D_Q}} = \frac{-(-0.6 \text{ V})}{0.25 \text{ mA}} = 2.4 \text{ K}$$

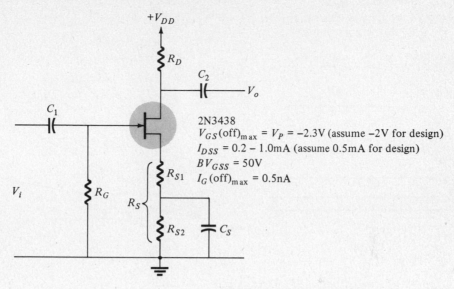

Figure 6.31. FET amplifier for Example 6.14.

At bias point we have

$$g_m = g_{mo}\left(1 - \frac{V_{GS_Q}}{V_P}\right) = \frac{2(0.5 \times 10^{-3})}{|-2|}\left(1 - \frac{-0.6}{-2}\right) = 0.35 \times 10^{-3}\,\text{V}$$

For the voltage gain we can write

$$A_v = \frac{-g_m R_D}{1 + g_m R_{S1}} = \frac{-0.35 \times 10^{-3}(16 \times 10^3)}{1 + 0.35 \times 10^{-3}(R_{S1})} \geq -5$$

which requires that $R_{S1} \leq 343\,\Omega$.

Choosing $R_{S1} = 270$ and $R_{S2} = 2.1\,\text{K}(R_{S1} + R_{S2} = 2.4\,\text{K})$ would provide the bias values specified with gain of

$$A_v = \frac{-(0.35 \times 10^{-3})(16 \times 10^3)}{1 + (0.35 \times 10^{-3})(270)} \cong -5.1$$

Choosing values of $R_G = 1\,\text{M}$ and $C_1 = C_2 = 0.1\,\mu\text{F}$ as in Example 6.13 would be satisfactory.

The value of C_S is chosen so that at $f = 1000\,\text{Hz}$

$$X_{C_s} \leq \frac{1}{10}R_{S2} = \frac{2.1\,\text{K}}{10} = 210\,\Omega$$

$$C_S = \frac{1}{2\pi(1000)(210)} = 0.76 \times 10^{-6} \cong 1\mu\,\text{F}$$

6.11 THE FET AS A VOLTAGE-VARIABLE RESISTOR (VVR)

The drain-source resistance of a FET can be varied as a function of applied gate-source voltage. This control is fairly linear and applies to the device operating region shown in Fig. 6.32a. Note that this is only a limited part of the FET

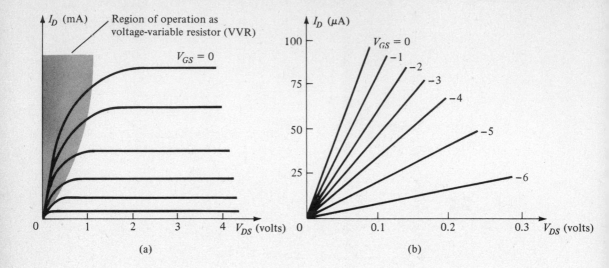

(a)

(b)

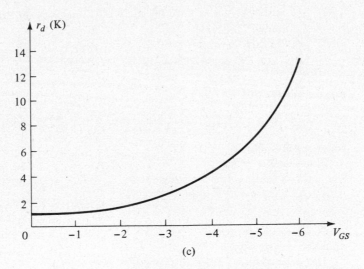

(c)

Figure 6.32. (a) and (b) operating action of FET as VVR; (c) resistance vs. control voltage.

operating region and is not the linear region of operation as an amplifier. The current range shown is limited to only about 100 μA and a corresponding voltage range of only a few hundred millivolts. Within this limited operating region the FET can be used as a voltage-variable resistor (VVR).

An enlarged view of the low-level region in which the FET can be used as a VVR is shown in Fig. 6.32b. The slope representing the device resistance is seen to vary as a function of gate-source control voltage. For example, we see that the slope is steepest, and therefore resistance least, for $V_{GS} = 0$ V whereas the slope is least, and resistance greatest, for $V_{GS} = -6$ V. A graph of device resistance versus control voltage obtained from Fig. 6.32b can be made as shown in Fig. 6.32c. Here we also see that the resistance increases with larger control voltage, although not in a linear manner. The change in device resistance is greatest at larger values of gate-source voltage.

Applications of VVR

One common application of a VVR is to vary the gain of an amplifier chain in order to achieve gain control. If this gain control results from a control voltage derived from the output voltage, then an automatic gain control (AGC) action is obtained. A simplified circuit diagram for such operation is shown in Fig. 6.33. The ac input signal is applied to an amplifier stage (Q_1) and then to succeeding amplifier stages to obtain an ac output signal. As a means of maintaining the output signal level constant the amplifier gain may be reduced as signal level increases. To achieve this gain control the output ac signal is rectified and filtered, thereby providing a dc voltage whose magnitude increases as the ac signal magnitude increases. This dc voltage is then applied to a FET used as a VVR. Notice that the FET is now used as a resistor, r_d, which effectively is in parallel (for ac operation) with resistor R_e. In this way the resistance of the FET from drain to source acts to vary the effective emitter degeneration resistance seen by amplifier Q_1. Capacitor C_2 serves only to prevent the operation of the VVR from affecting dc bias of stage Q_1.

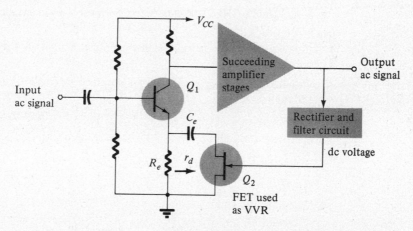

Figure 6.33. AGC amplifier using FET as VVR.

The gain of transistor Q_1 is then decreased as the output signal level increases. This occurs since the resulting increased dc control voltage to the FET causes its resistance as a VVR to increase, thereby allowing a larger effective emitter degeneration resistance for stage Q_1 and less voltage gain by that stage.

Other applications of the FET as a VVR might include voltage-controlled bandwidth in an LC circuit, electronically tuned RC filter, expander, and compressor circuits in hi-fi, and so on.

PROBLEMS

§ 6.3

1. Calculate the operating point of a FET circuit as in Fig. 6.7a for circuit values V_{DD} = 12 V, R_D = 2 K, V_{GG} = −1.5 V. Use the transistor characteristic of Fig. 6.7b.

2. Calculate the value of R_D needed to obtain bias values of V_{DS_Q} = 10 V, V_{GS_Q} = −1 V. Use the circuit and characteristic of Fig. 6.7.

3. Using V_{DD} = 16 V and R_D = 4 K, determine the value of V_{GG} = V_{GS} for operation at V_{DS_Q} = 8 V. Use circuit and characteristic of Fig. 6.7.

4. Determine the bias point of the circuit in Fig. 6.8 for R_S = 500 Ω.

5. Determine the operating point of the circuit in Fig. 6.9 for R_S = 1.2 K.

6. Determine the dc bias voltages at the JFET terminals for the circuit of Fig. 6.10. Use R_S = 1.2 K.

§ 6.6

7. Use the universal JFET curve to determine the bias point of the transistor in Fig. 6.20. Use R_S = 1.2 K.

§ 6.7

8. Calculate the ac voltage gain of a FET amplifier as in Fig. 6.22 for R_D = 2.7 K.

9. What value of transistor gain (g_m) is necessary for an amplifier as in Fig. 6.22 to have a gain of A_v = −4 if R_D = 2.7 K?

10. Calculate the voltage gain of the amplifier circuit in Fig. 6.22. Use R_S = 240 Ω and R_D = 1.5 K.

11. Calculate the voltage gain of the enhancement MOSFET amplifier circuit of Fig. 6.23 for R_D = 2.4 K.

12. Determine V_{o1}, V_{o2}, and R_i for the circuit of Fig. 6.25 for R_D = 2.1 K.

13. Calculate the output voltage, V_o, for the circuit of Fig. 6.26 with R_D = 2.4 K and R_L = 8.2 K.

§ 6.8

14. Calculate the input capacitance of a FET amplifier as in Fig. 6.27 having g_m = 1650 μmhos, C_{gs} = 1.5 pF, C_{gd} = 1 pF, C_{ds} = 5 pF, r_d = 40 K, R_D = 20 K, R_G = 1 M.

15. If the input capacitance is determined to be C_i = 44.5 pF calculate the input impedance of the FET amplifier of Problem 6.14 at frequencies (a) f = 20 kHz and (b) f = 1 MHz.

16. Calculate the output voltage of a FET amplifier having voltage gain of A_v = −80 at frequencies of (a) 20 kHz and (b) 1 MHz. Voltage is V_S = 2 mV from source impedance R_S = 10 K and input impedances are those of Problem 6.15.

17. Calculate the output impedance of the amplifier of Problem 6.14 (C_o = 20 pF) at frequencies (a) f = 20 kHz and (b) f = 1 MHz.

§ 6.9

18. Calculate the input and output impedances and voltage gain of the bootstrap FET source follower of Fig. 6.28 for circuit values $R_S = 750\ \Omega$ and $g_m = 4\ \text{m}\mho$ at bias point.

§ 6.10

19. Design an amplifier circuit as in Fig. 6.30 using a JFET and circuit values as follows, $V_{DD} = 20\ \text{V}$, $V_{GS\,\text{max}} = -20\ \text{V}$, $V_P = -6\ \text{V}$, $I_{DSS} = 12\ \text{mA}$, $I_{GSS} = 0.1\ \text{nA}$, $g_m = 2\ \text{mmhos}$, for a gain of 8.8.

20. Design an amplifier (as in Fig. 6.30) to have a voltage gain of 12 operating from a supply of 18 V.

21. Design a FET amplifier as shown in Fig. 6.31 for a gain of 10. (Design for $g_m = 0.4\ \text{m}\mho$). Use $V_{DD} = 22\ \text{V}$.

multistage systems, decibels (dB), and frequency considerations

7

7.1 INTRODUCTION

This chapter will include, under the heading of multistage systems, both the *cascaded* and *compound* configurations. The *cascaded* system, for the purposes of this text, is one in which each stage and the connections between each stage are very similar or identical. The *compound* system includes all other possible *multiple* active device configurations, each stage of which can be completely different in appearance with a variety of interconnections.

The first few sections of this chapter examine multistage systems employing the technique of analysis developed in earlier chapters. This is followed by a detailed discussion of decibels (dB) and the effect of frequency on the response of single and multistage systems.

7.2 GENERAL CASCADED SYSTEMS

A discussion of cascaded systems is best initiated by considering the block diagram representation of Fig. 7.1. The quantities of interest are indicated in the figure. The indicated A_v (voltage amplification) and A_i (current amplification) of each stage were determined with all stages connected as indicated in Fig. 7.1. In other words, A_v and A_i of each stage *do not* represent the gain of each stage on an independent basis. The loading effect of one stage on another was considered when these quantities were determined.

Rather than simply state the result for the overall gain of the system (voltage or current) a simple numerical example will clearly indicate the solution. If $A_{v_1} = -40$ and $A_{v_2} = -50$ with $V_{i_1} = 1 \text{ mV}$, then $V_{o_1} = A_{v_1} \times V_{i_1} = -40(1 \text{ mV}) = -40$

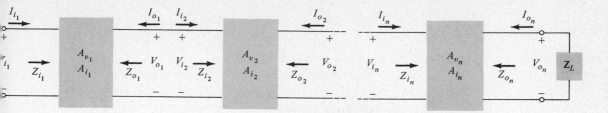

Figure 7.1. General cascaded system.

mV. Since $V_{o_1} = V_{i_2}$,

$$V_{o_2} = A_{v_2} V_{i_2} = -50(-40 \text{ mV}) = 2000 \text{ mV} = 2 \text{ V}$$

The overall gain is $A_{v_T} = 2000 \text{ mV}/1 \text{ mV} = 2000$.

Obviously, the total gain of the two stages is simply the product of the individual gains A_{v_1} and A_{v_2}. In general, for n stages,

$$A_{v_T} = A_{v_1} A_{v_2} A_{v_3} \cdots A_{v_n} \tag{7.1}$$

The same is true for the net current gain

$$A_{i_T} = A_{i_1} A_{i_2} A_{i_3} \cdots A_{i_n} \tag{7.2}$$

The input and output impedance of each stage as indicated in Fig. 7.1 are also those values obtained by considering the effects of each and every stage of the system. There is no generally employed equation, such as Eq. (7.2), for the input or output impedances of the system in terms of the individual values. However, in a number of situations (transistor, tube, or FET) the input (or output) impedance can normally be determined to an acceptable degree of accuracy by considering only one, or perhaps two, stages of the system.

The magnitude of the overall voltage gain of the representative system of Fig. 7.1 can be written as

$$|A_{v_T}| = \left|\frac{V_{o_n}}{V_{i_1}}\right| = \left|\frac{I_{o_n} Z_L}{I_{i_1} Z_{i_1}}\right|$$

so that

$$|A_{v_T}| = |A_{i_T}| \cdot \left|\frac{Z_L}{Z_{i_1}}\right| \tag{7.3}$$

Equation (7.3) will prove useful in the analysis to follow. To go a step further, if the product of the voltage and current gain is formed,

$$|A_{v_T} A_{i_T}| = \left|\frac{I_{o_n} Z_L}{I_{i_1} Z_{i_1}}\right| \cdot \left|\frac{I_{o_n}}{I_{i_1}}\right| = \left|\frac{I_{o_n}^2 Z_L}{I_{i_1}^2 Z_{i_1}}\right| = \frac{P_o}{P_i}$$

and
$$\boxed{|A_{p_T}| = |A_{v_T}| \cdot |A_{i_T}|} \qquad \text{(magnitude only)} \qquad (7.4)$$

this being the overall power gain of the system.

There are three types of coupling between stages of a system, such as in Fig. 7.1, that will be considered. The first to be described is the *RC-coupled* amplifier system, which is the most frequently applied of the three. This will be followed by the *transformer* and *direct-coupled* amplifier systems.

7.3 RC-COUPLED AMPLIFIERS

A cascaded RC-coupled transistor amplifier (two-stage) showing typical values and biasing techniques appears in Fig. 7.2. The terminology "RC-coupled" is derived from the biasing resistors and coupling capacitors employed between stages.

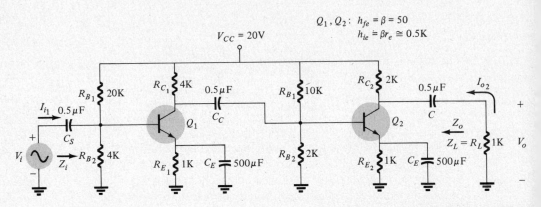

Figure 7.2. Two-stage RC-coupled amplifier.

The primary function of the approximate technique is to obtain a "ball-park" solution with a minimum of time and effort. A reduced time element obviously requires that the network be redrawn a minimum number of times. In fact, let us optimistically say that we can find the solutions for the network of Fig. 7.2 using only the original artwork.

Z_i

From past experience with single-stage amplifiers and the analysis just completed, it should be clear that for the ac response, both the 4-K and 20-K resistors will appear in parallel if the network is redrawn. They are also in parallel with the input impedance of Q_1, which is approximately $h_{ie} = \beta r_e = 0.5$ K since the emitter resistor is bypassed by C_E.

The parallel combination:

$$Z_{i_1} = 20\,\text{K} \,||\, 4\,\text{K} \,||\, 0.5\,\text{K} \cong 4\,\text{K} \,||\, 0.5\,\text{K} = \mathbf{0.444\,K} = R_{B_2} || h_{ie}$$

Z_o

Recall that the approximate collector-to-emitter equivalent circuit of a transistor is simply a current source $h_{fe}I_b$. This being the case, when $V_i = 0$, $I_{b_1} = 0$, $I_{b_2} = 0$ and $h_{fe}I_{b_2} = 0$, so that Z_o is simply R_{C_2} in parallel with the open-circuit representation of the controlled current source. That is,

$$Z_o|_{V_i=0} = R_{C_2} = 2\ \mathbf{K}$$

A_i

Applying the current divider rule: (Fig. 7.3)

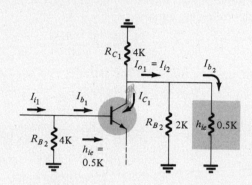

$$I_{b_1} = \frac{R_{B_2}I_{i_1}}{R_{B_2} + h_{ie}}$$

$$= \frac{4\,\mathrm{K}\,I_i}{4\,\mathrm{K} + 0.5\,\mathrm{K}}$$

and

$$I_{b_1} \cong 0.889I_{i_1}$$

Figure 7.3. Determining the relationship between I_{i_1} and I_{b_2}.

The collector current of the first stage $I_{c_1} \cong h_{fe}I_{b_1}$. However, I_{c_1} will divide between the 4-K resistor and the *loading* of the second stage (Fig. 7.3). A moment of reflection should indicate that, compared to the other parallel elements, the 10-K resistor can be dropped, in an approximate basis, from further consideration. The parallel combination of the 2-K and 0.5-K resistors as the loading of the next stage will result in an impedance of $= 0.4$ K.

Applying the current divider rule:

$$I_{o_1} = \frac{-R_{C_1}(I_{C_1})}{R_{C_1} + (R_{B_2}\|h_{ie})} = -\frac{4\,\mathrm{K}\,(I_{C_1})}{4\,\mathrm{K} + 0.4\,\mathrm{K}} = -\frac{4\,\mathrm{K}\,(h_{fe}I_{b_1})}{4.4\,\mathrm{K}} = -\frac{4\,\mathrm{K}\,(50)(0.889I_{i_1})}{4.4\,\mathrm{K}}$$

and

$$A_{i_1} = \frac{I_{o_1}}{I_{i_1}} \cong -\mathbf{40.409}$$

For the second stage:

$$I_{b_2} = \frac{R_{B_2}I_{i_2}}{R_{B_2} + h_{ie}} = \frac{2\,\mathrm{K}\,I_{i_2}}{2\,\mathrm{K} + 0.5} = 0.8I_{i_2}$$

and

$$I_{C_2} = h_{fe}I_{b_2} = 50(0.8I_{i_2}) = 40I_{i_2}$$

Applying the current divider rule to the output circuit:

$$I_{o_2} = \frac{R_{C_2} I_{C_2}}{R_{C_2} + R_L} = \frac{2\,\text{K}\,(h_{fe} I_{b_2})}{2\,\text{K} + 1\,\text{K}} = \frac{2\,\text{K}\,(40 I_{i_2})}{3\,\text{K}} = 26.667 I_{i_2}$$

and

$$A_{i_2} = \frac{I_{o_2}}{I_{i_2}} = 26.667$$

with

$$A_{i_T} = A_{i_1} \cdot A_{i_2} = (-40.409)(26.667) = -\mathbf{1077.6}$$

A_v

The direct connection under ac conditions clearly indicates in Fig. 7.2 that V_i appears directly at the base of the transistor of the first stage. Since the transistor has a grounded emitter terminal, the ac voltage gain can be obtained (on an approximate basis) using the following equation:

$$A_v \cong \frac{-h_{fe} R_L}{h_{ie}} = \frac{-R_L}{r_e}$$

R_L, the loading on the first stage, is the parallel combination of R_{C_1}, R_{B_2}, and $h_{ie}(= \beta r_e)$ which is $\cong 0.3636$ K. A_{v_1}, therefore, $= [-(50)(0.3636\,\text{K})]/0.5\,\text{K} = -36.36$. For the second stage,

$$A_{v_2} = \frac{-(50)(R_{C_2} \| R_L)}{0.5\,\text{K}} = \frac{-(50)(2\,\text{K} \| 1\,\text{K})}{0.5\,\text{K}} = \frac{-(50)(0.667\,\text{K})}{0.5\,\text{K}} = -66.7$$

The net gain is, therefore,

$$A_{v_T} = A_{v_1} A_{v_2} = (-36.36)(-66.7)$$

$$A_{v_T} \cong \mathbf{2425.2}$$

Using Eq. (7.3)

$$|A_{v_T}| = |A_{i_T}| \left| \frac{Z_L}{Z_{i_1}} \right| = \frac{(1077.6)(1\,\text{K})}{0.444\,\text{K}} = \mathbf{2427.03}$$

The very slight difference between the obtained values of A_{v_T} is due only to the decimal carry over in the individual solutions for A_{i_T} and A_{v_T}. In this case, the determination of A_{v_T} was markedly easier than that for A_{i_T}. In the future, therefore, there may be a savings in time if A_{v_T} is first determined, and A_{i_T} calculated from Eq. (7.3):

$$|A_{v_T}| \cdot \left| \frac{Z_{i_1}}{Z_L} \right|$$

EXAMPLE 7.1 We shall calculate the input and output impedance, voltage gain, and current gain of the two-stage amplifier of Fig. 7.4. Note that the second stage is an emitter-follower configuration.

Solution:

Z_i:

For ac conditions, R_E is bypassed by C_E and the input impedance to Q_1 is $\cong \beta r_e = (100)(11.61) = 1.161$ K.

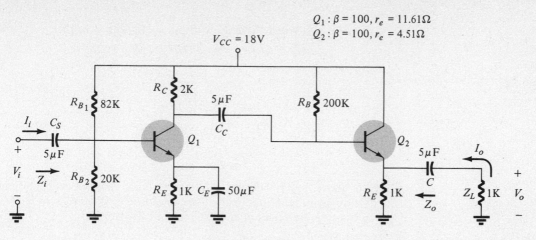

$$Q_1 : \beta = 100, r_e = 11.61\Omega$$
$$Q_2 : \beta = 100, r_e = 4.51\Omega$$

Figure 7.4. Two-stage transistor network to be examined in detail.

Then

$$Z_i = R_{B_1} \,\|\, R_{B_2} \,\|\, \beta r_e = 82\,\text{K} \,\|\, 20\,\text{K} \,\|\, 1.161\,\text{K} \cong \beta r_e = \mathbf{1.161\ K}$$

Z_o:

For ac conditions the network is redrawn as shown in Fig. 7.5. $Z_e = (R_s/\beta) + r_e$ where R_s is the source resistance connected to the base of the transistor. In this case $R_s = 2\,\text{K} \,\|\, 200\,\text{K} \cong 2\,\text{K}$ and

$$Z_e = \frac{2\,\text{K}}{100} + 4.51 = 20 + 4.51 - 24.51\ \Omega$$

$$Z_o = Z_e \,\|\, R_E = 24.51 \,\|\, 1\,\text{K} \cong \mathbf{24.51\ \Omega}$$

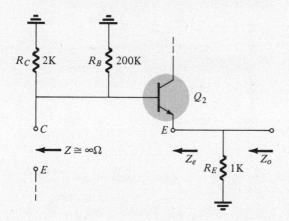

Figure 7.5. Determining Z_o for the network of Fig. 7.4.

$$V_i = V_{b_1}$$

and

$$A_{v_1} \cong \frac{-R_L}{r_e} = \frac{-(R_C \,\|\, R_B \,\|\, \beta(R_E \,\|\, Z_L))}{r_e} = \frac{-(2\,\text{K} \,\|\, 200\,\text{K} \,\|\, 100(1\,\text{K} \,\|\, 1\,\text{K}))}{r_e}$$

$$= -\frac{(2\,\text{K} \,\|\, 200\,\text{K} \,\|\, 50\,\text{K})}{11.61} \cong -\frac{2\,\text{K}}{11.61} = -172.27$$

$V_{be_2} \cong 0$ V and $V_{b_2} \cong V_o$ with $A_{v_2} = (V_{o_2}/V_{i_2}) = 1$. Then

$$A_{v_T} = A_{v_1} \cdot A_{o_2} = (-172.27)(1) = \mathbf{-172.27}$$

A_i:

$$|A_{i_T}| = |A_{v_T}| \left| \frac{Z_{i_1}}{Z_L} \right|$$

$$= \frac{(172.27)(1.161 \text{ K})}{1 \text{ K}}$$

$$\cong \mathbf{200}$$

Note above how rapidly the solutions to fairly complex configurations are developing using the approximations developed in Chapter 5. In the following example the values of r_e will have to be determined.

EXAMPLE 7.2 Determine Z_i, Z_o, A_v, A_i, and A_p for the network of Fig. 7.6.

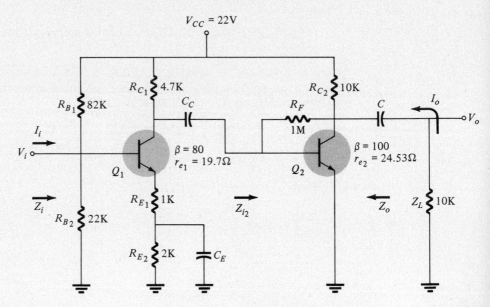

Figure 7.6

Solution: The values of r_e must be determined. For Q_1:

$$V_B = \frac{[R_{B_2} \| \beta(R_{E_1} + R_{E_2})]V_{CC}}{[R_{B_2} \| \beta(R_{E_1} + R_{E_2})] + R_{B_1}}$$

but

$$[R_{B_2} \| \beta(R_{E_1} + R_{E_2})] = 22 \text{ K} \| 80(3 \text{ K}) = 22 \text{ K} \| 240 \text{ K} \cong 22 \text{ K} = R_{B_2}$$

and

$$V_B \cong \frac{R_{B_2}V_{CC}}{R_{B_2} + R_{B_1}} = \frac{22 \text{ K} (22)}{22 \text{ K} + 82 \text{ K}} = \frac{484}{104} = 4.65 \text{ V}$$

and

$$V_E = V_B - V_{BE} = 4.65 - 0.7 = 3.95 \text{ V}$$

with
$$I_E = \frac{V_E}{R_{E_1} + R_{E_2}} = \frac{3.95}{3\ \text{K}} = 1.32\ \text{mA}$$

and
$$r_{e_1} = \frac{26\ \text{mV}}{I_E} = \frac{26}{1.32} = 19.70\ \Omega$$

For Q_2:
$$V_{CC} - (\beta + 1)I_B R'_C - R_F I_B - V_{BE} = 0$$
$$22 - (101)I_B\, 10\ \text{K} - 10^6 I_B - 0.7 = 0$$
$$21.3 = 2.01 \times 10^6 I_B$$

and
$$I_B = 10.6\ \mu\text{A}$$

with
$$I_E \cong I_C = \beta I_B = (100)(10.6\ \mu\text{A}) = 1.06\ \text{mA}$$

and
$$r_{e_2} = \frac{26\ \text{mV}}{I_E} = \frac{26}{1.06} = 24.53\ \Omega$$

Z_i:
$$Z_i = R_{B_1} \| R_{B_2} \| \beta R_{E_1} = 82\ \text{K} \| 22\ \text{K} \| 80\ \text{K} \cong \mathbf{14.26\ K}$$

Z_o:
$$Z|_{V_i=0} \cong R_C = \mathbf{10\ K}$$

A_v: $V_{b_1} = V_i$ and
$$A_{v_1} = \frac{-R_L}{(R_{E_1} + r_e)} = \frac{-(R_C \| Z_{i_2})}{R_{E_1} + r_e}$$

From Chapter 5,
$$Z_{i_2} = \frac{R_F}{A_v} \| \beta r_e$$

and
$$A_{v_2} = \frac{-R_L}{r_e} - \frac{-R_C \| Z_L}{r_e} = -\frac{5\ \text{K}}{24.53} = -203.83$$

$$Z_{i_2} = \frac{10^6}{203.83} \| 100(24.53) = 4.906\ \text{K} \| 2.453\ \text{K} = 1.6353\ \text{K}$$

so that
$$A_{v_1} = \frac{-(4.7\ \text{K} \| 1.6353\ \text{K})}{(1\ \text{K} + 0.0197\ \text{K})} = -\frac{1.213}{1.0197} \cong \mathbf{1.19}$$

and
$$A_{v_T} = A_{v_1} \cdot A_{v_2} = (-1.19)(-203.83) \cong \mathbf{242.56}$$

A_i:
$$|A_{i_T}| = |A_{v_T}| \left| \frac{Z_{i_1}}{Z_L} \right| = \frac{(242.56)(14.26\ \text{K})}{10\ \text{K}} \cong \mathbf{345.89}$$

A_p:
$$|A_p| = |A_{i_T}| \cdot |A_{v_T}| = (345.89)(242.56) \cong \mathbf{83.9 \times 10^3}$$

EXAMPLE 7.3   *FET RC-Coupled Amplifier.* RC coupling is not limited to transistor stages, as Fig. 7.7 indicates. Determine the total voltage gain.

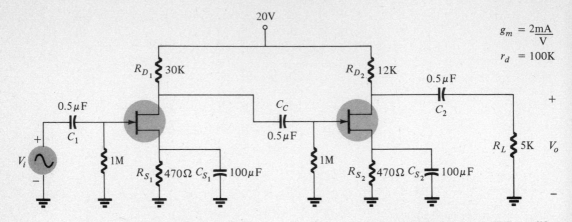

$g_m = \dfrac{2\text{mA}}{\text{V}}$

$r_d = 100\text{K}$

Figure 7.7. Two-stage FET amplifier.

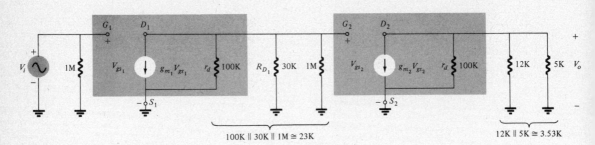

100K ∥ 30K ∥ 1M ≅ 23K   12K ∥ 5K ≅ 3.53K

Figure 7.8. Network of Fig. 7.7 following the substitution of the small-signal ac equivalent circuits.

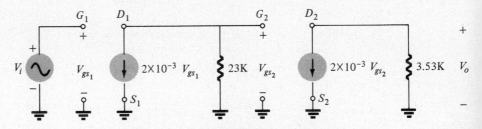

Figure 7.9. Network of Fig. 7.8 following the combination of parallel elements.

Solution: Substituting the small-signal equivalent circuit results in the configuration of Fig. 7.8. Combining parallel elements and eliminating those having no effect on the desired overall voltage gain will result in Fig. 7.9. Obviously,

$$V_{gs_1} = V_i$$

and

$$V_{gs_2} = -(2 \times 10^{-3} V_{gs_1})(23\text{ K})$$

so that

$$V_{gs_2} = -46 V_{gs_1}$$

The minus sign indicates that the polarity of the voltage across the 23-K resistor due to the current source is the reverse of the defined polarities for V_{gs_2}.

In conclusion:

$$V_o = -(2 \times 10^{-3} V_{gs_2})(3.53 \text{ K}) = -7.06 V_{gs_2}$$

so that

$$V_o = -7.06 V_{gs_2} = -7.06(-46 V_{gs_1}) = 324.8 V_{gs_1} = 324.8 V_i$$

and

$$A_v = \frac{V_o}{V_i} = \mathbf{324.8}$$

7.4 TRANSFORMER-COUPLED TRANSISTOR AMPLIFIERS

A two-stage transformer-coupled transistor amplifier is shown in Fig. 7.10. Note that step-down transformers are employed between stages while a step-up transformer is connected to the source V_i. The step-up transformer increases the signal level while the step-down transformer matches, as closely as possible, the loading of each stage to the output impedance of the preceding stage. This is done in an effort to be as close to maximum power transfer conditions as possible. The effect of this matching technique through the use of transformer coupling will be clearly demonstrated in the following analysis.

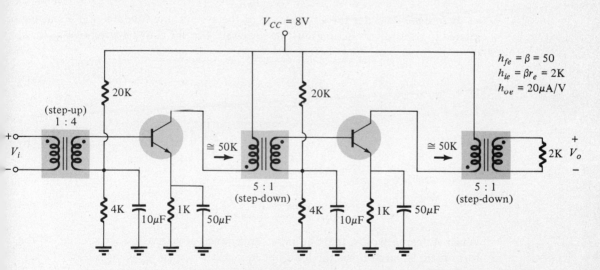

Figure 7.10. Two-stage transformer-coupled transistor amplifier.

Recall that a coupling capacitor was inserted to prevent any dc levels of one stage from affecting the bias conditions of another stage. The transformer provides this dc isolation very nicely.

The basic operation of this circuit is somewhat more efficient than the RC-coupled transistors due to the low dc resistance of the collector circuit of the transformer coupled system. The primary resistance of the transformer is seldom more than a few ohms as compared to the large collector resistance R_C of the RC-coupled system. This lower dc resistance results in a lower dc power loss under operating conditions. The efficiency, as determined by the ratio of the ac power out to the dc power in, is therefore somewhat improved.

There are some decided disadvantages, however, to the transformer-coupled system. The most obvious is the increased size of such a system (due to the transformers) compared to RC-coupled stages. The second is a poorer frequency response due to the newly introduced reactive elements (inductance of coils and capacitance between turns). A third consideration, frequently an important one, is the increased cost of the transformer-coupled (as compared to the RC-coupled) system.

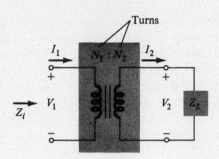

Figure 7.11. Basic transformer configuration.

Before we consider the ac response of the system, the fundamental equations related to transformer action must be reviewed. For the configuration of Fig. 7.11,

$$\frac{V_1}{V_2} = \frac{N_1}{N_2} = a \qquad \text{(transformation ratio)} \tag{7.5a}$$

$$\frac{I_1}{I_2} = \frac{N_2}{N_1} = \frac{1}{a} \tag{7.5b}$$

and
$$Z_i = a^2 Z_L \tag{7.5c}$$

which states, in words, that the input impedance of a transformer is equal to the turns ratio squared times the load impedance.

For the ac response, the circuit of Fig. 7.10 will appear as shown in Fig. 7.12. For maximum power transfer the impedances Z_2 and Z_4 should be equal to the output impedance of each transistor: $Z_o \cong 1/h_{oe} = 1/20 \ \mu\text{mhos} = 50 \ \text{K}$. This is one system where the effect of $1/h_{oe}$ must be considered. The hybrid parameters will therefore be employed in its solution. Applying $Z_i = a^2 Z_L$, $Z_4 = a^2 R_L = (5)^2 2 \ \text{K} = 50 \ \text{K}$. Z_2 is also 50 K since the input resistance to each stage (Z_1 and Z_3) is $\cong h_{ie} = 2 \ \text{K}$. Frequency considerations may not always permit Z_2 or Z_4 to be

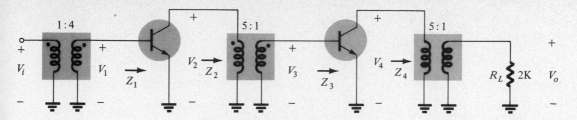

Figure 7.12. Cascaded transformer-coupled amplifiers of Fig. 7.10 redrawn to determine the small-signal ac response.

equal to $1/h_{oe}$. For situations of this type Z_2 and Z_4 are usually made as close as possible to $1/h_{oe}$ in magnitude.

Further analysis of the circuit of Fig. 7.12 results in

$$V_1 = \frac{N_2}{N_1}V_i = 4V_i$$

and

$$A_{v_1} = \frac{-h_{fe}Z_L}{h_{ie}} = \frac{-h_{fe}(\cong 1/h_{oe} \| Z_2)}{h_{ie}} = \frac{-50(50 \text{ K} \| 50 \text{ K})}{2 \text{ K}} = -625$$

so that

$$V_2 = -625V_1 = -625(4V_i) = -2500V_i$$

but

$$V_3 = \frac{N_2}{N_1}V_2 = \frac{1}{5}V_2 = \frac{1}{5}(-2500V_i) = -500V_i$$

and

$$A_{v_2} = \frac{-h_{fe}Z_L}{h_{ie}} = \frac{-(50)(25 \text{ K})}{2 \text{ K}} = -625 = \frac{V_4}{V_3}$$

so

$$V_4 = -625 \, V_3 = -625(-500 \, V_i)$$
$$= +312.5 \times 10^3 V_i$$

with

$$V_L = \frac{1}{5} V_4 = \frac{1}{5}.(312.5 \times 10^3 V_i)$$

and

$$A_{v_T} = \frac{V_L}{V_i} = \mathbf{62.5 \times 10^3}$$

7.5 DIRECT-COUPLED TRANSISTOR AMPLIFIERS

The third type of coupling between stages to be introduced in this chapter is *direct coupling*. The circuit of Fig. 7.13 is an example of a two-stage direct-coupled transistor system. Coupling of this type is necessary for very low-frequency applications. For a configuration of this type the dc levels of one stage are obviously related to the dc levels of the other stages of the system. For this reason the biasing arrangement must be designed for the entire network rather than for each stage independently. Although three separate 12-V supplies are indicated, only one is required if the three terminals of higher (positive) potential for each supply are paralleled.

One of the biggest problems associated with direct-coupled networks is stability. Any variation in dc level in one stage is transmitted on an amplified basis to the other stages. The addition of the emitter resistor aids as a stabilizing element in each stage.

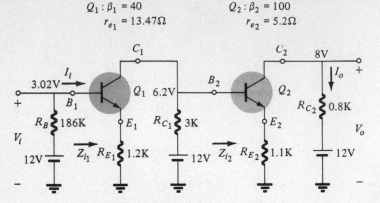

$$Q_1: \beta_1 = 40 \qquad\qquad Q_2: \beta_2 = 100$$
$$r_{e_1} = 13.47\Omega \qquad\qquad r_{e_2} = 5.2\Omega$$

Figure 7.13. Direct-coupled transistor stages.

(dc) Bias Conditions

For an output voltage $V_{C_2} = 8$ V as indicated in Fig. 7.13,

$$I_{0.8K} = \frac{12 - 8}{0.8 \text{ K}} = 5 \text{ mA}$$

therefore $\qquad\qquad I_{C_2} \cong I_{E_2} \cong 5 \text{ mA}$

and $\qquad\qquad V_{E_2} = (5 \text{ mA})(1.1 \text{ K}) = 5.5 \text{ V}$

For $\qquad\qquad V_{BE_2} = 0.7 \text{ V}$

$$V_{B_2} = V_{C_1} = 5.5 + 0.7 = 6.2 \text{ V}$$

as indicated. Applying

$$I_C \cong \beta_2 I_B$$

$$I_{B_2} \cong \frac{I_{C_2}}{\beta_2} = \frac{5 \text{ mA}}{100} = 50 \text{ } \mu A$$

and $\qquad\qquad I_{3K} = \dfrac{12 - 6.2}{3 \text{ K}} = \dfrac{5.8}{3 \text{ K}} = 1.93 \text{ mA}$

and since $\qquad\qquad I_{3K} \gg I_{B_2}$

assume $\qquad\qquad I_{C_1} \cong I_{3K} = 1.93 \text{ mA}$

and $\qquad\qquad I_{E_1} = 1.93 \text{ mA}$

so $\qquad\qquad V_{E_1} = (1.93 \text{ mA})(1.2 \text{ K}) = 2.32 \text{ V}$

and $\qquad\qquad V_{B_1} = V_{E_1} + V_{BE_1} = 2.32 + 0.7 = 3.02 \text{ V}$

as indicated.

The above verification of the potential levels appearing in Fig. 7.13 demonstrates clearly the close tie-in required between bias levels of a direct-coupled amplifier.

Now for the ac response. The approach uses the approximate technique introduced earlier in this chapter.

The input impedance to each emitter-follower configuration is $\cong \beta R_E$. Therefore,

$$Z_{i_1} \cong \beta_1 R_{E_1} = 40(1.2 \text{ K}) = 48 \text{ K}$$

and $\qquad\qquad Z_{i_2} \cong \beta_2 R_{E_2} = 100(1.1 \text{ K}) = 110 \text{ K}$

$$A_{v_1} = \frac{-R_{L_1}}{R_{E_1}} = \frac{-R_{C_1} \| \beta_2 R_{E_2}}{R_{E_1}} = -\frac{3\,\text{K} \| 110\,\text{K}}{1.2\,\text{K}} \cong -\frac{3\,\text{K}}{1.2\,\text{K}} = -2.5$$

$$A_{v_2} = \frac{-R_{L_2}}{R_{E_2}} = \frac{-R_{C_2}}{R_{E_2}} = -\frac{0.8\,\text{K}}{1.1\,\text{K}} = -0.7273$$

and

$$A_{v_T} = A_{v_1} A_{v_2} = (-2.5)(-0.7273) = \mathbf{1.818}$$

$$|A_i| = |A_v| \left| \frac{Z_{i_1}}{Z_L} \right| = \frac{(1.818)(48\,\text{K})}{0.8\,\text{K}} = \mathbf{109.08}$$

with

$$|A_{P_T}| = |A_v| \cdot |A_i| = (1.818)(109.08) = \mathbf{198.3}$$

7.6 CASCODE AMPLIFIER

For high-frequency applications the CB configuration has the most desirable characteristics of the three configurations. However, it suffers from a very low input impedance ($Z_i \cong h_{ib} = r_e$). The *cascode* configuration in Fig. 7.14 is designed to improve the input impedance level for the CB configuration through the use of a typical CE network. The gain of the CE configuration is low to ensure that the input Miller capacitance level is a minimum (recall the discussion of Miller capacitance in Chapter 6 for the FET) for high-frequency applications.

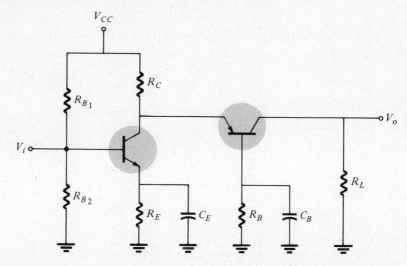

Figure 7.14. Cascode configuration.

A practical version of a cascode amplifier appears in Fig. 7.15. Note that the collector of the CE configuration is still tied directly to the emitter of the CB configuration.

For dc conditions:

$$I_{E_2} \cong I_{E_1} \quad \text{or} \quad I_{C_2} \cong I_{C_1}$$

or dividing each side by β since $\beta_1 = \beta_2 = \beta$

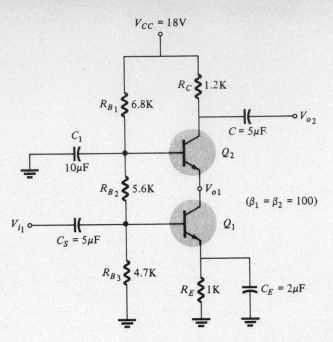

Figure 7.15. Practical cascode arrangement.

$$\frac{I_{C_2}}{\beta} \cong \frac{I_{C_1}}{\beta} \qquad \text{or} \qquad I_{B_2} \cong I_{B_1}$$

The current I_{B_1} will pass through βR_E of the parallel combination of R_{B_3} and βR_E. Since $\beta R_E = (100)(1\,\text{K}) = 100\,\text{K}$ and $R_{B_3} = 4.7\,\text{K}$, we shall assume that I_{B_1} is significantly smaller than $I_{4.7\text{K}}$ to permit ignoring its effect. Since this approximation is applied to I_{B_1}, it can also be applied to I_{B_2} (since $I_{B_2} \cong I_{B_1}$) and

$$V_{B_3} = \frac{R_{B_3}(V_{CC})}{R_{B_3} + R_{B_2} + R_{B_1}} = \frac{4.7\,\text{K}(18)}{4.7\,\text{K} + 5.6\,\text{K} + 6.8\,\text{K}} = \frac{84.6}{17.1}$$

$$= 4.95\,\text{V}$$

and

$$I_{E_1} = \frac{V_{E_1}}{R_E} = \frac{V_{B_3} - V_{BE}}{R_E} = \frac{4.95 - 0.7}{1\,\text{K}} = 4.25\,\text{mA}$$

with

$$r_{e_1} = \frac{26\,\text{mV}}{I_{E_1}} = \frac{26}{4.25} = 6.12\,\Omega$$

and since

$$I_{E_1} \cong I_{E_2},$$

$$r_{e_2} = 6.12\,\Omega$$

For ac conditions all capacitors assume their short-circuit equivalent and

$$A_{v_1} = \frac{V_{o_1}}{V_{i_1}} \cong \frac{-R_L}{r_{e_1}}$$

with $R_L = r_{e_2} = h_{ib_2}$ of Q_2, the input impedance of that connected stage and

$$A_{v_1} = \frac{-r_{e_2}}{r_{e_1}} \cong -1 \qquad \text{(low as desired because of the Miller effect)}$$

with

$$A_{v_2} = \frac{+R_L}{r_{e_2}} = \frac{+R_C}{r_{e_2}} = +\frac{1.2\,\text{K}}{6.12} \cong 196$$

and
$$A_{v_T} = \frac{V_{o_2}}{V_{i_1}} = A_{v_1} \cdot A_{v_2} = (-1)(+196) = -196$$

7.7 DARLINGTON COMPOUND CONFIGURATION

The Darlington circuit is a compound configuration that results in a set of improved amplifier characteristics. The configuration of Fig. 7.16 has a high input impedance with low output impedance and high current gain, all desirable characteristics for a current amplifier. We shall momentarily see, however, that the voltage gain will

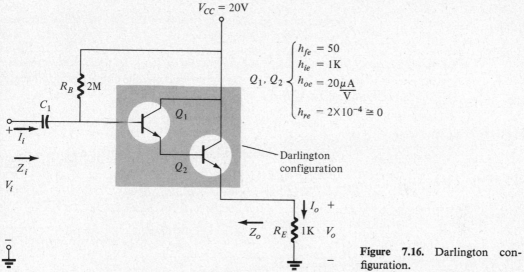

Figure 7.16. Darlington configuration.

be less than one if the output is taken from the emitter terminal. A variation in the configuration can result in a trade-off between the output impedance and voltage gain.

The description of the biasing arrangement is similar to that of a single-stage emitter-follower configuration with current feedback (Chapter 4). Note for the Darlington configuration that the emitter current of the first transistor is the base current for the second active device.

In its small-signal ac form the circuit will appear as shown in Fig. 7.17.

For the second stage:
$$Z_{i_2} \cong h_{fe_2} R_E$$
and
$$A_{i_2} = \frac{I_o}{I_2} = \frac{I_{e_2}}{I_{b_2}} \cong h_{fe_2}$$

On a *good* approximate basis, these equations cannot be applied to the first stage. The "fly in the ointment" is the closeness with which Z_{i_2} compares with $1/h_{oe_1}$. You will recall that $1/h_{oe_1}$ could be eliminated in the majority of situations because the load impedance $Z_L \ll 1/h_{oe_1}$. For the Darlington configuration the

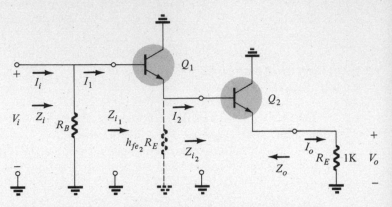

Figure 7.17. Darlington configuration of Fig. 7.16 redrawn to determine the small-signal ac response.

input impedance Z_{i_2} is close enough in magnitude to $1/h_{oe_1}$ to necessitate considering the effects of h_{oe_1}. In Chapter 5 it was found that for the single-stage grounded emitter transistor amplifier where $1/h_{oe}$ was considered,

$$A_i \cong \frac{h_{fe}}{1 + h_{oe}Z_L}$$

Applying the above equation to this situation, $Z_L = Z_{i_2} \cong h_{fe_2}R_E$ and

$$A_{i_1} = \frac{I_2}{I_1} = \frac{I_{c_1}}{I_{b_1}} \cong \frac{h_{fe_1}}{1 + h_{oe_1}(h_{fe_2}R_E)}$$

with

$$A_i = \frac{I_o}{I_1} = A_{i_1}A_{i_2} = \frac{h_{fe_1}h_{fe_2}}{1 + h_{oe_1}(h_{fe_2}R_E)} \qquad (7.6)$$

For $h_{fe_1} = h_{fe_2}$ and $h_{oe_1} = h_{oe_2} = h_{oe}$

$$A_i \cong \frac{h_{fe}^2}{1 + h_{oe}h_{fe}R_E} \qquad (7.7)$$

For $h_{oe}h_{fe}R_E \leq 0.1$ a fairly good approximation (within 10%) is

$$A_i \cong h_{fe}^2 = \beta^2 \qquad (7.8)$$

The current gain $A_i = I_o/I_i$, as defined by Fig. 7.17 can be determined through the use of the current divider rule

$$I_1 = \frac{R_B I_i}{R_B + Z_{i_1}}$$

Since $Z_{i_2} \cong h_{fe_2}R_E$ is the "emitter resistor" of the first stage (note Fig. 7.17), the input impedance to the first stages is $Z_{i_1} \cong h_{fe_1}(Z_{i_2} \| 1/h_{oe_1})$ since $Z_{i_2} = h_{fe_2}R_{E_1}$, and $1/h_{oe_1}$ will appear in parallel in the small-signal equivalent circuit. The result is

$$Z_{i_1} \cong h_{fe_1}\left(h_{fe_2}R_E \,\middle\|\, \frac{1}{h_{oe_1}}\right) = \frac{h_{fe_1}h_{fe_2}R_E(1/h_{oe_1})}{h_{fe_2}R_E + 1/h_{oe_1}}$$

and

$$Z_{i_1} = \frac{h_{fe_1}h_{fe_2}R_E}{h_{oe_1}h_{fe_2}R_E + 1} \tag{7.9}$$

which for $h_{fe_1} = h_{fe_2} = h_{fe}$ and $h_{oe_1} = h_{oe_2} = h_{oe}$

$$Z_{i_1} \cong \frac{h_{fe}^2 R_E}{1 + h_{oe}h_{fe}R_E} \tag{7.10}$$

For $h_{oe}h_{fe}R_E \leq 0.1$

$$Z_{i_1} \cong h_{fe}^2 R_E = \beta^2 R_E \tag{7.11}$$

Substituting the parameter values gives

$$h_{fe_1} = h_{fe_2} = h_{fe} = 50$$

and

$$h_{ie_1} = h_{ie_2} = h_{ie} = 1\,K$$

with

$$h_{oe_1} = h_{oe_2} - h_{oe} = 20 \ \mu A/V$$

$$A_i = \frac{I_o}{I_1} \cong \frac{(h_{fe})^2}{1 + h_{oe}h_{fe}R_E} = \frac{(50)^2}{1 + (20 \times 10^{-6})(50)(1\,K)}$$

$$= \frac{2500}{1 + 1} = \mathbf{1250}$$

and

$$Z_{i_1} \cong \frac{h_{fe}^2 R_E}{1 + h_{oe}h_{fe}R_E} = \frac{(50)^2 1\,K}{2} = 1250\,K = \mathbf{1.25\,M}$$

so that for $R_B = 2\,M$

$$\frac{I_1}{I_i} = \frac{R_B}{R_B + Z_{i_1}} = \frac{2\,M}{2\,M + 1.25\,M} = \frac{2}{3.25} = 0.615$$

and

$$A_{i_T} = \frac{I_o}{I_i} = \left[\frac{I_o}{I_1}\right]\left[\frac{I_1}{I_i}\right] = A_i \times \frac{I_1}{I_i}$$

$$= (1250)(0.615) = \mathbf{770}$$

with

$$Z_i = 2\,M \,\|\, Z_{i_1} = 2\,M \,\|\, 1.25\,M = \mathbf{770\,K}$$

Too frequently, the current gain of a Darlington circuit is assumed to be simply $A_i^2 \cong h_{fe}^2$ without any regard to the output impedance $1/h_{oe}$. In this case $A_i \cong (h_{fe})^2 = 2500$. Certainly, 2500 versus 1250 is *not* a good approximation. The effect of h_{oe_1} must therefore be considered when the current gain of the first stage is determined.

The output impedance Z_o can be determined directly from the emitter-equivalent circuits, as follows.

For the first stage

$$Z_{o_1} \cong \frac{R_{s_1} + h_{ie_1}}{h_{fe_1}} \tag{7.12}$$

$$= \frac{0 + 1\text{ K}}{50} \cong 20.0\ \Omega$$

and

$$Z_{o_2} \cong \frac{(Z_{o_1} \| 1/h_{oe_1}) + h_{ie_2}}{h_{fe_2}} \tag{7.13}$$

$$= \frac{(20.0 \| 50\text{ K}) + 1\text{ K}}{50} \cong \frac{20.0 + 1\text{ K}}{50} = \frac{1020}{50}$$

$$\cong 20.4\ \Omega$$

Note, as indicated in the introductory discussion, that the input impedance is high, output impedance low, and current gain high. We shall now examine the voltage gain of the system. Applying Kirchhoff's voltage law to the circuit of Fig. 7.16:

$$V_o = V_i - V_{be_1} - V_{be_2}$$

That the output potential is the input *less* the base-to-emitter potential of each transistor clearly indicates that $V_o < V_i$. It is closer in magnitude to one than to zero. On an approximate basis it is given by

$$A_v \cong \frac{1}{1 + \dfrac{h_{ie_2}}{h_{fe_2} R_E}} \tag{7.14}$$

as derived from the emitter equivalent circuit.

Substituting the numerical values of this general example, we get

$$A_v \cong \frac{1}{1 + \dfrac{1\text{ K}}{50\text{ K}}} = \frac{1}{1 + 0.2} = 0.98$$

7.8 CASCADED PENTODE AND TRIODE AMPLIFIERS

Many of the characteristics and general comments made about RC-coupled, transformer-coupled, and direct-coupled transistor amplifiers can also be applied to the pentode or tube amplifier employing the same coupling. Of the three, the RC-coupled pentode or triode amplifier is the most frequently applied, although the transformer-coupled generally results in the highest gain. This latter consideration, however, must be carefully weighed against the cost and desired frequency response.

The RC-coupled pentode amplifier of Fig. 7.18 will now be discussed. The

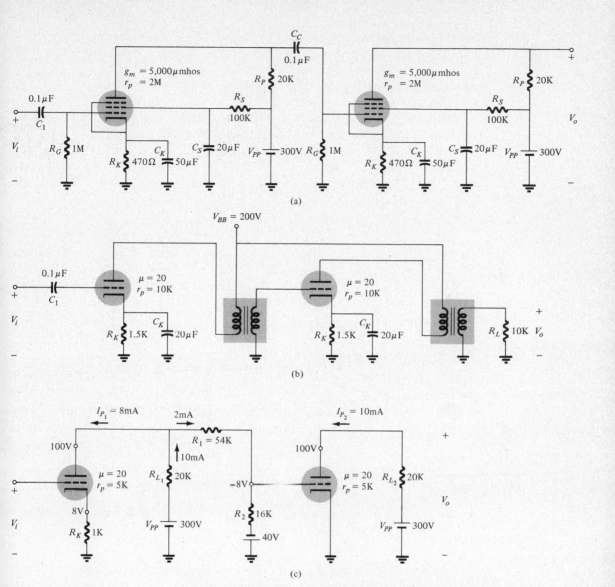

Figure 7.18. Vacuum-tube multistage systems: (a) RC-coupled; (b) transformer-coupled; (c) direct-coupled.

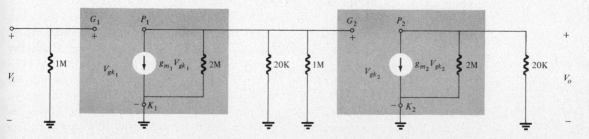

Figure 7.19. Small signal ac equivalent circuit for the cascaded system of Fig. 7.18.

analysis of the remaining transformer and direct-coupled pentode amplifier appearing in the same figure is similar to its transistor counterpart and therefore will be left as an exercise for the reader.

Substituting the small-signal ac equivalent circuit for the pentodes of Fig. 7.18a will result in the configuration of Fig. 7.19.

Obviously, $V_{gk_1} = V_i$, $2\,M \| 1\,M \| 20\,K \cong 20\,K$, and $2\,M \| 20\,K \cong 20\,K$, resulting in the reduced network of Fig. 7.20. From this circuit:

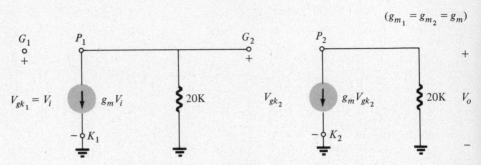

Figure 7.20. Network of Fig. 7.19 following the combination of parallel elements.

$$V_{gk_2} = V_{20K} = -(g_m V_i)(20\,K) = -(5000 \times 10^{-6})(20 \times 10^3 V_i)$$
$$= -100 V_i$$

and
$$g_m V_{gk_2} = -g_m(100 V_i) = -(5000 \times 10^{-6})(10^2 V_i)$$
$$= -0.5 V_i$$

so that
$$V_o = -(g_m V_{gk_2})(10\,K) = -(-0.5 V_i)(10\,K) = 5000 V_i$$

and
$$A_v = \frac{V_o}{V_i} = 5000$$

7.9 DECIBELS

The concept of the *decibel* (dB) and the associated calculations will become increasingly important in the remaining sections of this chapter. The background surrounding the term decibel has its origin in the old established fact that power and audio levels are related on a logarithmic basis. That is, an increase in power level, say 4 to 16 W, for discussion purposes, does not mean that the audio level will increase by a factor of $16/4 = 4$. It will increase by a factor of 2 as derived from the power of 4 in the following manner: $(4)^2 = 16$. For a change of 4 to 64 W the audio level will increase by a factor of 3 since $(4)^3 = 64$. In logarithmic form, the relationship can be written as

$$\log_4 64 = 3$$

In words, the equation states that the logarithm of 64 to the base 4 is 3. In general: $\log_b a = x$ relates the variables in the same manner as $b^x = a$.

Because of pressures for standardization, the *bel* (B) was defined by the following equation to relate power levels P_1 and P_2:

$$\text{bel} = \log_{10} \frac{P_2}{P_1} \tag{7.15}$$

Note that the common, or base 10, system was chosen to eliminate variability. Although the base is no longer the original power level, the equation will result in a basis for comparison of audio levels due to changes in power levels. The term bel was derived from the surname of Alexander Graham Bell.

It was found, however, that the bel was too large a unit of measurement for practical purposes, so the decibel (dB) was defined such that 10 decibels = 1 bel. Therefore,

$$\text{\# dB} = (10)(\text{\# bels}) = 10 \log_{10} \frac{P_2}{P_1}$$

and
$$\text{dB} = 10 \log_{10} \frac{P_2}{P_1} \tag{7.16}$$

The terminal rating of electronic communication equipment (amplifiers, microphones, etc.), is commonly rated in decibels. Equation (7.16) indicates clearly, however, that the decibel rating is a measure of the difference in magnitude between *two* power levels. For a specified terminal (output) power (P_2) there must be a reference power level (P_1). The reference level is generally accepted to be 1 mW although on occasion the 6 mW standard of earlier years is applied. The resistance to be associated with the 1-mW power level is 600 Ω, chosen because it is the characteristic impedance of audio transmission lines. When the 1-mW level is employed as the reference level, the decibel abbreviation frequently appears as dBm. In equation form

$$\text{dBm} = 10 \log_{10} \frac{P_2}{1 \text{ mW}}\bigg|_{600\Omega} \tag{7.17}$$

There exists a second equation for decibels that is applied frequently. It can be best described through the circuit of Fig. 7.21a. For V_i equal to some value V_1,

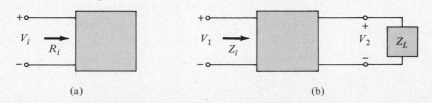

(a)   (b)

Figure 7.21. Configurations employed in the discussion of Eq. 7.16.

$P_1 = V_i^2/R_i$ where R_i is the input resistance of the system of Fig. 7.21a. If V_i should be increased (or decreased) to some other level, V_2, then $P_2 = V_2^2/R_i$. If we substitute into Eq. (7.16) to determine the resulting difference in decibels between the power levels,

$$dB = 10 \log_{10} \frac{P_2}{P_1} = 10 \log_{10} \frac{V_2^2/R_i}{V_1^2/R_i} = 10 \log_{10} \left(\frac{V_2}{V_1}\right)^2$$

and

$$dB = 20 \log_{10} \frac{V_2}{V_1} \tag{7.18}$$

Keep in mind, however, that this equation is only correct if the associated resistance for each applied voltage is the same. For the system of Fig. 7.21b, where output and input levels are being compared, $Z_i \neq Z_L$ and Eq. (7.18) will not result. Equation (7.16) should therefore be employed.

If $Z_i = Z_i \cos \theta_i$ and $Z_L = Z_L \cos \theta_L$ are substituted into Eq. (7.16) such that $P_L = \dfrac{V_L^2}{Z_L \cos \theta_L}$, etc., the following general equation will result

$$dB = 20 \log_{10} \frac{V_L}{V_i} + 10 \log_{10} \frac{Z_i}{Z_L} + 10 \log_{10} \frac{\cos \theta_i}{\cos \theta_L} \tag{7.19}$$

For resistive elements, which are most commonly encountered, $\cos \theta_i = \cos \theta_L$, and the last term, $\log_{10}(1) = 0$. In addition, if $Z_i = Z_L$, the second term will also drop out, resulting in Eq. (7.18).

Frequently the effect of different impedances ($Z_i \neq Z_L$) is ignored and Eq. (7.18) applied to simply establish a basis of comparison between levels—voltage or current. For situations of this type the decibel gain should more correctly be referred to as the *voltage or current gain in decibels* to differentiate it from the common usage of decibel as applied to power levels.

One of the advantages of the logarithmic relationship is the manner in which it can be applied to cascaded stages. For example, the overall voltage gain of a cascaded system is given by

$$A_{v_T} = A_{v_1} A_{v_2} A_{v_3} \cdots A_{v_n}$$

Applying the proper logarithmic relationship gives

$$20 \log_{10} A_{v_T} = 20 \log_{10} A_{v_1} + 20 \log_{10} A_{v_2}$$
$$+ 20 \log_{10} A_{v_2} + \cdots + 20 \log_{10} A_{v_n} \tag{7.20}$$

In words, the equation states that the decibel gain of a cascaded system is simply the sum of the decibel gains of each stage, that is,

$$dB_{A_{v_T}} = dB_{A_{v_1}} + dB_{A_{v_2}} + dB_{A_{v_3}} + \cdots + dB_{A_{v_n}} \tag{7.21}$$

The above equations can also be applied to current considerations. For $P_2 = I_2^2 R_o$ and $P_1 = I_1^2 R_o$,

$$dB = 20 \log_{10} \frac{I_2}{I_1} \qquad (7.22)$$

and
$$dB_{Ai_T} = dB_{Ai_1} + dB_{Ai_2} + dB_{Ai_3} + \cdots + dB_{Ai_n} \qquad (7.23)$$

Before considering a few examples, the fundamental operations associated with logarithmic functions will be considered. For many it will be simply a review. For some, an extended amount of time may be required to fully understand the material to follow.

Each equation introduced in this section employs the common or base 10 logarithmic system. As indicated in the introductory discussion, the logarithm of numbers that are powers of the chosen base are easily determined. For example,

$$\log_{10} \overset{a}{\underset{b}{10{,}000}} = x \Longrightarrow (10)^x = 10{,}000$$

and
$$x = 4$$

Similarly,

$$\log_{10} 1000 = \log_{10} (10)^3 = 3$$
$$\log_{10} 100 = \log_{10} (10)^2 = 2$$
$$\log_{10} 10 = \log_{10} (10)^1 = 1$$
$$\log_{10} 1 = \log_{10} (10)^0 = 0$$

For the logarithm of a number such as 24.8

$$\log_{10} 24.8 = x$$

or
$$10^x = 24.8$$

The unknown quantity x is obviously between 1 and 2, but a further determination would be purely a trial-and-error process if it were not for the logarithmic function. The procedure for determining the logarithm of a number requires that two components of the result be found separately. These two components are the *characteristic* and *mantissa*. The characteristic is simply the power of 10 associated with the number for which the logarithm is to be determined.

$$24.8 = 2.48 \times 10^1 \Longrightarrow 1 = \text{characteristic}$$
$$4860.0 = 4.860 \times 10^3 \Longrightarrow 3 = \text{characteristic}$$

The mantissa, or decimal portion of the logarithm, must be determined from a set of tables or the D and L scales of a slide rule. The result:

$$\log_{10} 24.8 = 1.3944$$
$$\log_{10} 4860.0 = 3.6870$$

Of course, calculators are available that can find the common logarithm directly.
There will be many occasions in which the antilogarithm of a number must be

determined; that is, for the example above, determine 24.8 and 4860.0 from the logarithm of these numbers. The process is simply the reverse of that applied to determine the logarithm. For example, find the antilogarithm of 2.140.

$$2.140 \begin{cases} & \text{characteristic} \Longrightarrow 10^2 \\ & \text{mantissa} \quad \Longrightarrow 138 \\ & \text{(L} \longrightarrow \text{D scale)} \end{cases} \quad 1.38 \times 10^2 = 138$$

For the calculator, keep in mind that

$$\log_{10} x = 2.140$$

is equivalent to

$$10^{2.140} = x$$

and 10^y is a common calculator function.

For ratios less than 1, the logarithm can be determined by simply inverting the ratio and introducing a negative sign.

$$\log_{10} \frac{1.6}{24} = -\log_{10} \frac{24}{1.6} = -\log_{10} 15 = -1.176$$

$$\log_{10} 0.788 = -\log_{10} \frac{1}{0.788} = -\log_{10} 1.27 = -0.4315$$

For power ratios, a negative decibel rating simply indicates a reduction in power level as compared to the initial or input power.

EXAMPLE 7.4 Find the magnitude gain corresponding to a decibel gain of 100.

Solution: By Eq. (7.16)

$$100 = 10 \log_{10} \frac{P_2}{P_1} \Longrightarrow \log_{10} \frac{P_2}{P_1} = 10$$

so that

$$\frac{P_2}{P_1} = 10^{10} = 10,000,000,000$$

This example clearly demonstrates the range of decibel values to be expected from practical devices. Certainly a future calculation giving a decibel result in the neighborhood of 100 should be questioned immediately. In fact, a decibel gain of 50 corresponds with a magnitude gain of 100,000, which is still very large.

EXAMPLE 7.5 The input power to a device is 10,000 W at a voltage of 1000 V. The output power is 500 W while the output impedance is 20 Ω.
(a) Find the power gain in decibels.
(b) Find the voltage gain in delibels.
(c) Explain why (a) and (b) agree or disagree.

Solution:

(a) $dB = 10 \log_{10} \frac{0.5 \times 10^3}{10 \times 10^3} = 10 \log_{10} \frac{1}{20} = -10 \log_{10} 20$

$= -10(1.301) = -13.01 \text{ dB}$

(b) $dB_v = 20 \log_{10} \dfrac{V_o}{V_i} = 20 \log_{10} \dfrac{\sqrt{PR}}{1000} = 20 \log_{10} \dfrac{\sqrt{500 \times 20}}{1000}$

$$= 20 \log_{10} \dfrac{100}{1000} = -20 \log_{10} \dfrac{1}{10} = -20 \log_{10} 10 = \mathbf{-20\ dB}$$

(c) $R_i = \dfrac{V^2}{P} = \dfrac{10^6}{10^4} = 10^2 \neq R_o = \mathbf{20\ \Omega}$

EXAMPLE 7.6   An amplifier rated at 40-W output is connected to a 10-Ω speaker. (a) Calculate the input power required for full power output if the power gain is 25 dB. (b) Calculate the input voltage for rated output if the amplifier voltage gain is 40 dB.

Solution:
(a) Eq. 7.16

$$25 = 10 \log_{10} \dfrac{40}{P_i} \Longrightarrow P_i = \dfrac{40}{\text{antilog } (2.5)} = \dfrac{40}{3.16 \times 10^2}$$

$$= \dfrac{40}{316} \cong \mathbf{126\ mW}$$

(b) $dB_v = 20 \log_{10} \dfrac{V_o}{V_i} \Longrightarrow 40 = 20 \log_{10} \dfrac{V_o}{V_i}$

$$\dfrac{V_o}{V_i} = \text{antilog } 2 = 100$$

$$V_o = \sqrt{PR} = \sqrt{40 \times 10} = 20\ V$$

$$V_i = \dfrac{V_o}{100} = \dfrac{20}{100} = \mathbf{200\ mV}$$

7.10 GENERAL FREQUENCY CONSIDERATIONS

The frequency of the applied signal can have a pronounced effect on the response of a single or multistage network. The analysis thus far has been for the mid-frequency spectrum. At low frequencies we shall find that the coupling and bypass capacitors can no longer be replaced by the short-circuit approximation because of the resulting change in reactance of these elements. The frequency-dependent parameters of the small-signal equivalent circuits and the stray capacitive elements associated with the active device and the network will limit the high-frequency response of the system. An increase in the number of stages of a cascaded system will limit both the high- and low-frequency response.

The magnitude gain of an RC-coupled, direct-coupled, and transformer-coupled amplifier system are provided in Fig. 7.22. Note that the horizontal scale is a logarithmic scale to permit a plot extending from the low- to the high-frequency regions. For each plot, a low-, high-, and mid-frequency region has been defined. In addition, the primary reasons for the drop in gain at low and high frequencies have also been indicated within the parentheses. For the RC-coupled amplifier the drop at low frequencies is due to the increasing reactance of C_C, C_S, or C_E, while its upper frequency limit is determined by either the parasitic capacitive elements of the

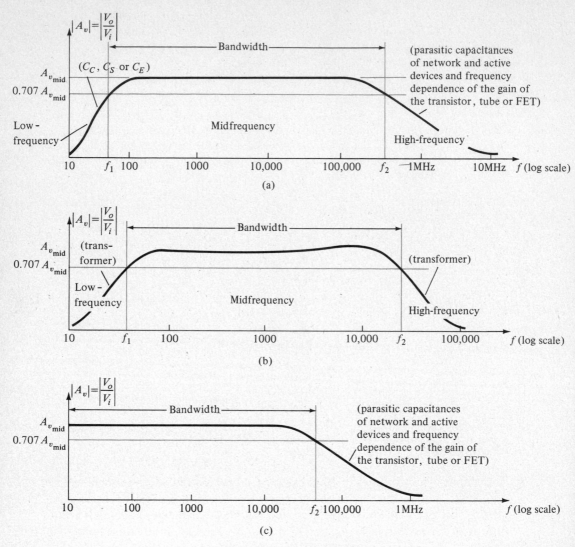

Figure 7.22. Gain vs. frequency for: (a) RC-coupled amplifiers; (b) transformer-coupled amplifiers; (c) direct-coupled amplifiers.

network and active device or the frequency dependence of the gain of the active device. An explanation of the drop in gain for the transformer-coupled system requires a basic understanding of "transformer action" and the transformer equivalent circuit. For the moment let us say that it is simply due to the shorting effect (across the input terminals of the transformer) of a magnetizing inductive reactance at low frequencies ($X_L = 2\pi f L$). The gain must obviously be zero at $f = 0$ since at this point there is no longer a changing flux established through the core to induce a secondary or output voltage. As indicated in Fig. 7.22, the high-frequency response is controlled primarily by the stray capacitance between the turns of the primary and secondary windings. For the direct-coupled amplifier, there are no coupling or by pass capacitors to cause a drop in gain at low frequencies. As the figure indicates, it is a flat response to the upper cutoff frequency which is determined by either the parasitic capacitances of the circuit and active device or the frequency dependence of the gain of the active device.

For each system of Fig. 7.22 there is a band of frequencies in which the gain is either equal or relatively close to the mid-band value. To fix the frequency boundaries of relatively high gain, $0.707A_{v_{mid}}$ was chosen to be the gain cutoff level. The corresponding frequencies f_1 and f_2 are generally called the corner, cutoff, band, break, or half-power frequencies. The multiplier 0.707 was chosen because of this level the output power is half the mid-band power output, that is, at mid-frequencies,

$$P_{o_{mid}} = \frac{|V_o^2|}{R_o} = \frac{|A_{v_{mid}}V_i|^2}{R_o}$$

and at the half-power frequencies,

$$P_{o_{HPF}} = \frac{|0.707\,A_{v_{mid}}V_i|^2}{R_o} = 0.5\frac{|A_{v_{mid}}V_i|^2}{R_o}$$

and

$$\boxed{P_{o_{HPF}} = 0.5\,P_{o_{mid}}} \tag{7.24}$$

The bandwidth (or pass band) of each system is determined by f_1 and f_2, that is,

$$\boxed{\text{bandwidth (BW)} = f_2 - f_1} \tag{7.25}$$

For applications of a communications nature (audio, video), a decibel plot of the voltage gain versus frequency is more useful than that appearing in Fig. 7.22. Before obtaining the logarithmic plot, however, the curve is generally normalized as shown in Fig. 7.23. In this figure the gain at each frequency is divided by the

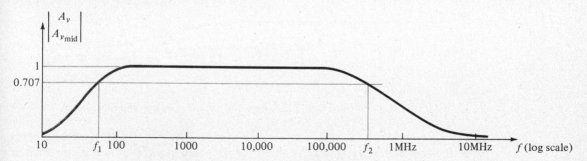

Figure 7.23. Normalized gain vs. frequency plot.

mid-band value. Obviously, the mid-band value is then 1 as indicated. At the half-power frequencies the resulting level is $0.707 = 1/\sqrt{2}$. A decibel plot can now be obtained by applying Eq. (7.18) in the following manner:

$$\left|\frac{A_v}{A_{v_{mid}}}\right|_{dD} = 20 \log_{10} \left|\frac{A_v}{A_{v_{mid}}}\right| \tag{7.26}$$

At mid-band frequencies, $20 \log_{10} 1 = 0$, and at the cutoff frequencies, $20 \log_{10} 1/\sqrt{2} = -3$ dB. Both values are clearly indicated in the resulting decibel plot of Fig. 7.24. The smaller the fractional ratio, the more negative the decibel level due to the inversion process discussed in Section 7.9.

For the greater part of the discussion to follow, a decibel plot will be made only for the low- or high-frequency regions. Keep Fig. 7.24 in mind, therefore, to permit a visualization of the broad system response.

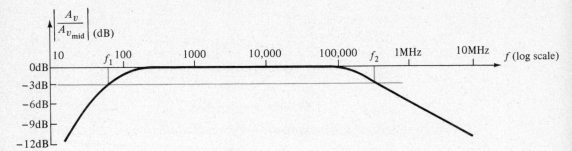

Figure 7.24. Decibel plot of the normalized gain vs. frequency plot of Fig. 7.23.

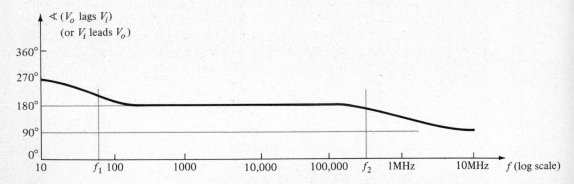

Figure 7.25. Phase plot for an RC-coupled amplifier system (for each stage).

It should be understood that an amplifier usually introduces a phase shift of 180° between input and output signals. This fact must now be expanded to indicate that this is only the case in the mid-band region. At low frequencies there is an additional phase shift such that V_o lags V_i by an increased angle. At high frequencies the phase shift will drop below 180°. Fig. 7.25 is a standard phase plot for an RC-coupled amplifier.

7.11 SINGLE-STAGE TRANSISTOR AMPLIFIER— LOW-FREQUENCY CONSIDERATIONS

Before considering the basic BJT amplifier let us first examine the series RC network appearing in Fig. 7.26 and the effect the applied frequency will have on the gain $|A_v| = |V_o/V_i|$. At mid-frequencies and high frequencies the capacitive reactance $X_C = (1/2\pi f C)$ is sufficiently small compared to R to permit a short-circuit approximation and $V_o = V_i$. However, at low frequencies the capacitive reactance can become sufficiently large to cause a significant voltage drop across C. The result is a drop in V_o and the gain $A_v = (V_o/V_i)$. When conditions are such that $|X_C| = R$, then

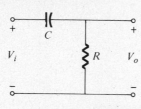

Figure 7.26

$$|V_o| = \frac{R}{\sqrt{R^2 + X_C^2}}|V_i| = \frac{R}{\sqrt{R^2 + R^2}}|V_i| = \frac{R}{\sqrt{2R^2}}|V_i| = \frac{R}{\sqrt{2}\,R}|V_i| = \frac{1}{\sqrt{2}}|V_i|$$

and

$$|A_v| = \left|\frac{V_o}{V_i}\right| = \frac{1}{\sqrt{2}} = 0.707$$

with

$$20\log_{10}\frac{1}{\sqrt{2}} = -3\,\text{dB}$$

We can conclude, therefore, that if an RC configuration is encountered, the condition $|X_C| = R$ will result in a 3-dB drop in voltage from the midfrequency value. A similar drop occurs for the current since at $X_C = R$,

$$|I| = \left|\frac{V_i}{Z}\right| = \frac{|V_i|}{\sqrt{R^2 + X_C^2}} = \frac{|V_i|}{\sqrt{2R^2}} = \frac{1}{\sqrt{2}}\frac{|V_i|}{R} = 0.707\frac{|V_i|}{R}$$

where $|V_i|/R$ is its mid-frequency value, if we replace X_C by its short-circuit equivalent.

Using the condition $X_C = R$, we can determine the frequency at which the 3-dB voltage drop will occur. That is,

$$|X_C| = R$$

$$\frac{1}{2\pi f C} = R$$

and

$$\boxed{f_1 = \frac{1}{2\pi R C}} \tag{7.27}$$

If the gain equation is written in the following manner

$$A_v = \frac{V_o}{V_i} = \frac{R}{R - jX_C} = \frac{1}{1 - j\frac{X_C}{R}} = \frac{1}{1 - j\frac{1}{\omega C R}} = \frac{1}{1 - j\frac{1}{2\pi f C R}}$$

and using the defined frequency above,

$$\boxed{A_v = \frac{1}{1 - j(f_1/f)}} \tag{7.28}$$

In the magnitude and phase form:

$$A_v = \frac{V_o}{V_i} = \underbrace{\frac{1}{\sqrt{1 + (f_1/f)^2}}}_{\text{magnitude of } A_v}\ \underbrace{\underline{/\tan^{-1}(f_1/f)}}_{\substack{\text{phase } \measuredangle \text{ by which} \\ V_o \text{ leads } V_i}}$$

For the magnitude, when $f_1 = f$,

$$|A_v| = \frac{1}{\sqrt{1 + (1)^2}} = \frac{1}{\sqrt{2}} = 0.707 \Longrightarrow -3\ \text{dB}$$

In the logarithimic form:

$$|A_v|_{\text{dB}} = 20 \log_{10} \frac{1}{\sqrt{1 + (f_1/f)^2}} = -20 \log_{10}(1 + (f_1/f)^2)^{1/2}$$

$$= -(\tfrac{1}{2})(20) \log_{10}(1 + (f_1/f)^2)$$

$$= -10 \log_{10}(1 + (f_1/f)^2)$$

For frequencies where $f_1 \gg f$ the above equation can be approximated by

$$= -10 \log_{10}(f_1/f)^2$$

and

$$\boxed{|A_v|_{\text{dB}} = -20 \log_{10} f_1/f}_{f_1 \gg f} \tag{7.29}$$

Ignoring the condition $f_1 \gg f$ for a moment, a plot of Eq. (7.29) on a frequency log scale will yield some results of a useful nature for future decibel plots.

At $f = f_1$, or $\frac{f_1}{f} = 1$, $\left(\frac{f}{f_1} = 1\right)$, $-20 \log_{10} 1 = 0\ \text{dB}$

At $f = 0.5 f_1$, or $\frac{f_1}{f} = 2$, $\left(\frac{f}{f_1} = 0.5\right)$, $-20 \log_{10} 2 = -6\ \text{dB}$

At $f = 0.25 f_1$, or $\frac{f_1}{f} = 4$, $\left(\frac{f}{f_1} = 0.25\right)$, $-20 \log_{10} 4 = -12\ \text{dB}$

At $f = 0.1 f_1$, or $\frac{f_1}{f} = 10$, $\left(\frac{f}{f_1} = 0.1\right)$, $-20 \log_{10} 10 = -20\ \text{dB}$

A plot of these points is indicated in Fig. 7.27 from $f/f_1 = 0.1$ to $f/f_1 = 1$. Note that this results in a straight line when plotted against a log scale. In the same figure a straight line is also drawn for the condition of 0 dB for $f \gg f_1$. As stated earlier, the straight-line segments (asymptotes) are only accurate for 0 db when $f \gg f_1$, and the sloped line when $f_1 \gg f$. We know, however, that when $f = f_1$, there is a 3-dB drop from the mid-band level. Employing this information in association with the straight-line segments permits a fairly accurate plot of the frequency response as indicated in the same figure. The piecewise linear plot of the asymtotes and associated breakpoints is called a *Bode plot*.

The above calculations and the curve itself demonstrate clearly that a change in frequency by a factor of 2 (equivalent to 1 octave) results in a 6-dB change in the ratio. For a 10:1 change in frequency (approximately equivalent to 1 decade) there is a 20-dB change in the ratio. In the future, therefore, a decibel plot can be easily

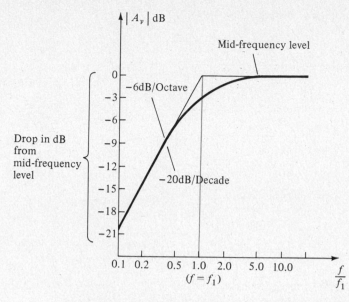

Figure 7.27. Bode plot for the low-frequency region.

obtained for a function having the format of Eq. (7.29). First simply find f_1 from the circuit parameters and then sketch two asymtotes—one along the 0 dB line and the other drawn through f_1 sloped at 6 dB/octave or 20 dB/decade. Then find the 3 dB point corresponding to f_1 and sketch the curve.

Let us now turn our attention to the basic BJT amplifier appearing in Fig. 7.28 with the capacitors that will affect the low-frequency response. We shall examine the effect of each independently.

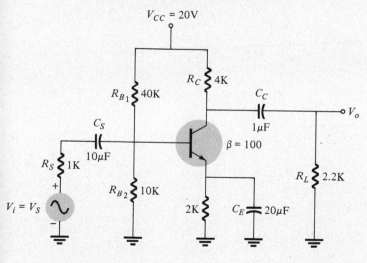

Figure 7.28

C_S

The reduced equivalent circuit surrounding C_S appears in Fig. 7.29. When conditions are such that $R_T = |X_{C_S}|$ or $R_S + R_i = |X_{C_S}|$, the input current I_i is reduced by -3 dB and a lower break frequency is defined by

$$R_S + R_i = \frac{1}{2\pi f C_S}$$

or

$$\boxed{f_{L_S} = \frac{1}{2\pi(R_S + R_i)C_S}}$$

(7.30)

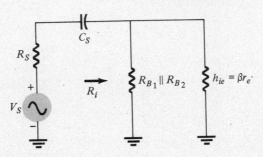

Figure 7.29. Localized ac equivalent for C_S.

Although not described in detail here, analysis will show that a 3-dB drop in input current will result in a similar drop in overall voltage gain.

C_C

The resulting equivalent circuit appearing in Fig. 7.30 defines a lower break frequency by

$$R_C + R_L = X_{C_C}$$

or

$$R_C + R_L = \frac{1}{2\pi f C_C}$$

and

$$\boxed{f_{L_C} = \frac{1}{2\pi(R_C + R_L)C_C}}$$

(7.31)

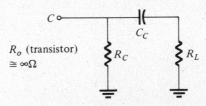

Figure 7.30. Localized ac equivalent for C_C.

C_E

The equivalent circuit external to the capacitor C_E appears in Fig. 7.31. The condition $R_e = R_E \,||\, ((R'_s/\beta) + r_e) = |X_{C_E}|$ defines the break frequency determined by the emitter bypass capacitor:

$$\boxed{f_{L_E} = \frac{1}{2\pi R_e C_E}}$$

(7.32)

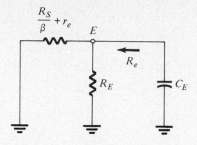

Figure 7.31. Localized ac equivalent for C_E.

At this frequency the emitter section of the network will have a sufficiently high impedance to reduce the emitter current and resulting voltage gain by 3 dB.

EXAMPLE 7.7 Determine the lower corner frequency for the network of Fig. 7.28 and sketch the Bode plot. Include an estimate of the actual gain versus frequency curve.

Solution: Determining r_e:

DC conditions:

$$\beta R_E = (100)(2\text{ K}) = 200\text{ K} \gg 10\text{ K}$$

and its effect ignored on an approximate basis and

$$V_B \cong \frac{R_{B_2}(V_{CC})}{R_{B_2} + R_{B_1}} = \frac{10\text{ K}(20)}{10\text{ K} + 40\text{ K}} = \frac{200}{50} = 4\text{ V}$$

with

$$I_E = \frac{V_E}{R_E} = \frac{4 - 0.7}{2\text{ K}} = \frac{3.3}{2\text{ K}} = 1.65\text{ mA}$$

so that

$$r_e = \frac{26\text{ mV}}{1.65\text{ mA}} \cong 15.76\ \Omega$$

C_S:

$$R_i = R_{B_1} \| R_{B_2} \| \beta r_e = 40\text{ K} \| 10\text{ K} \| 1.576\text{ K} \cong 1.32\text{ K}$$

$$f_{L_S} = \frac{1}{2\pi(R_S + R_i)C_S} = \frac{1}{(6.28)(1\text{ K} + 1.32\text{ K})(10 \times 10^{-6})}$$

$$= \frac{1}{(6.28)(2.32 \times 10^3)(10 \times 10^{-6})}$$

$$= \frac{1}{(6.28)(2.32)(10) \times 10^{-3}} = \frac{10^{+3}}{145.7}$$

$$f_{L_S} \cong 6.86\text{ Hz}$$

C_C:

$$f_{L_C} = \frac{1}{2\pi(R_C + R_L)C_C}$$

$$= \frac{1}{(6.28)(4\text{ K} + 2.2\text{ K})(1 \times 10^{-6})}$$

$$= \frac{1}{(6.28)(6.2 \times 10^3)(10^{-6})} = \frac{10^3}{(6.28)(6.2)} = \frac{10^3}{38.94}$$

$$\cong 25.68\text{ Hz}$$

C_E:

$$R_S' - R_S \| R_{B_1} \| R_{B_2} = 1\,\text{K} \| 40\,\text{K} \| 10\,\text{K} \cong 1\,\text{K}$$

$$R_e = R_E \left\| \left(\frac{R_S'}{\beta} + r_e \right) = 2\,\text{K} \left\| \left(\frac{1\,\text{K}}{100} + 15.76 \right) = 2\,\text{K} \| (10 + 15.76) \right.\right.$$

$$= 2\,\text{K} \| 25.76\ \Omega \cong 25.76\ \Omega$$

$$f_{L_E} = \frac{1}{2\pi R_e C_E} = \frac{1}{(6.28)(25.76)(20 \times 10^{-6})} = \frac{10^6}{3235.46} \cong 309.1\ \text{Hz}$$

As described earlier, each cutoff (corner, break) frequency will define a −3 dB point and an asymptote as appearing in Fig. 7.32. The highest frequency as determined by C_E (in this example) will determine the resulting lower corner frequency (−3 dB point) for the network. Note that the lower frequency determined by C_C increased the slope of the asymptote originating at 309.1 Hz by −6 dB/octave. A rough sketch of the lower frequency region appears on the same plot.

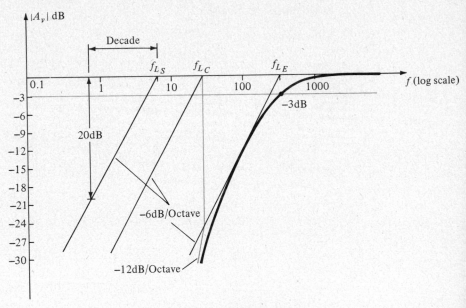

Figure 7.32. Low frequency plot for the network of Example 7.6.

The mid-band gain:

$$A_v = \frac{-R_L'}{r_e} = \frac{-R_C \| R_L}{r_e} = -\frac{(4\,\text{K}) \| (2.2\,\text{K})}{15.76} = -\frac{1419}{15.76}$$

$$\cong -90$$

At the −3 dB point the gain has dropped to 0.707 of its mid-band level and

$$|A_v|_{(-3\,\text{dB})} = (0.707)(90) = 63.63$$

as appearing in Fig. 7.32.

For an unbypassed emitter resistor there will obviously be only two cut off frequencies to be determined: that due to C_S and to C_C. The equation for f_{L_s} will be altered accordingly. There is an exercise at the end of the chapter on this very topic.

7.12 SINGLE-STAGE TRANSISTOR AMPLIFIER—HIGH-FREQUENCY CONSIDERATIONS

At the high frequency end there are two factors that will define the -3 dB point: the network capacitance (parasitic and introduced) and the frequency dependence of $h_{fe}(\beta)$.

In the high-frequency region the RC network of concern has the configuration appearing in Fig. 7.33. At increasing frequencies the reactance X_C will decrease in magnitude, resulting in a shorting effect across the output and a decrease in gain. The derivation leading to the corner frequency for this RC configuration follows along similar lines to that encountered for the low-frequency region. The most significant difference is in the general form of A_v appearing below:

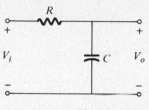

Figure 7.33

$$A_v = \frac{1}{1 + j(f/f_2)}$$ (7.33)

which results in an asymptotic plot such as shown in Fig. 7.34 that drops off at -6 dB/octave with increasing frequency.

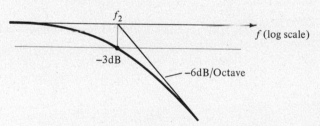

Figure 7.34. Asymptotic plot as defined by Eq. 7.33.

In Fig. 7.35 the various parasitic capacitances (C_{be}, C_{bc}, C_{ce}) of the transistor have been included with the wiring capacitance (C_{W1}, C_{W2}) introduced during construction. The high-frequency equivalent model for the network of Fig. 7.35 appears in Fig. 7.36. Note the absence of the capacitors C_S, C_C, and C_E, which are all assumed to be in the short-circuit state at these frequencies. Keep in mind that the capacitors that determined the low-frequency cutoff point appear between the input and output terminals of the RC network and not across the output as indicated in Fig. 7.33. The capacitance C_i includes the Miller capacitance C_M as introduced for the FET in Chapter 6, the capacitance C_{be}, and the input wiring capacitance C_{W1}. The capacitance C_o includes the output wiring capacitance and the collector capacitance.

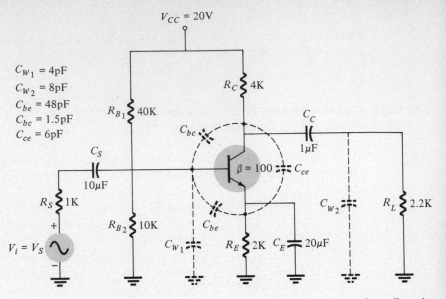

Figure 7.35. Network of Fig. 7.28 with the capacitors that affect the high-frequency response.

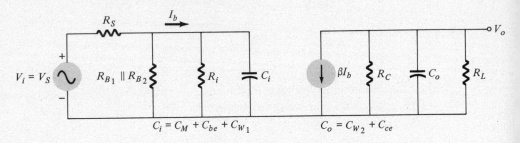

Figure 7.36. High-frequency model for the network of Fig. 7.35.

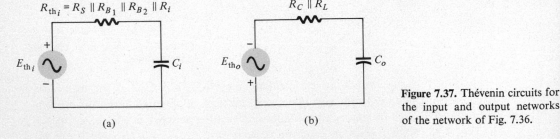

Figure 7.37. Thévenin circuits for the input and output networks of the network of Fig. 7.36.

Determining the Thévenin equivalent circuit of the input and output networks will result in the configurations of Fig. 7.37. For the input network the -3 dB frequency is defined by

$$f_{H_i} = \frac{1}{2\pi R_{Th_1} C_i} \qquad (7.34)$$

with
$$R_{Th_1} = R_S \| R_{B_1} \| R_{B_2} \| R_i$$
and
$$C_i = C_{W1} + C_{be} + C_M = C_{W1} + C_{be} + (1 + |A_v|)C_{bc}$$

For the condition $R_S = 0 \, \Omega$, $R_{Th_1} = 0 \, \Omega$ and f_{H_i} is infinite. In other words, with $R_S = 0 \, \Omega$, the applied voltage appears directly across C_i in the original network and an upper cutoff frequency for the input circuit is not defined.

Obviously, in the high-frequency range the voltage gain A_v will be a function of frequency due to the capacitive elements. It cannot be determined by a simple ratio of resistors as developed in Chapter 5. As a first approximation to the value of A_v, however, we shall use the mid-band value. As the frequency increases, the gain will obviously drop from the mid-band value due to the capacitive effects. If we therefore use the mid-band value, we have the maximum value of A_v and the maximum Miller capacitance. This will result in the maximum value for C_i and the minimum f_{H_i}. In essence, we have determined the lowest cutoff frequency due to C_i and defined a *worst-case* design situation. In other words, the actual cutoff due to C_i will always be higher than the level determined using the mid-band gain.

For the output network:

$$f_{H_o} = \frac{1}{2\pi R_{Th_2} C_o} \tag{7.35}$$

with
$$R_{Th_2} = R_C \| R_L$$
and
$$C_o = C_{W2} + C_{Ce}$$

For the network of Fig. 7.35:

$$R_{Th_1} = R_S \| R_{B_1} \| R_{B_2} \| R_i = 1 \, \text{K} \| 40 \, \text{K} \| 10 \, \text{K} \| 1.576 \, \text{K} \cong 0.575 \, \text{K}$$

$$C_i = C_{W1} + C_{be} + (1 + |A_v|)C_{bc}$$
$$= 4 \, \text{pF} + 48 \, \text{pF} + (1 + 90)(1.5 \, \text{pF})$$
$$= 188.5 \, \text{pF}$$

$$f_{H_i} = \frac{1}{2\pi R_{Th_1} C_i} = \frac{1}{(6.28)(0.575 \times 10^3)(188.5 \times 10^{-12})}$$
$$= \frac{1000 \times 10^6}{680.7} \cong \textbf{1.47 MHz}$$

$$R_{Th_2} = R_C \| R_L = 4 \, \text{K} \| 2.2 \, \text{K} = 1.419 \, \text{K}$$

$$C_o = C_{W2} + C_{ce} = 8 \, \text{pF} + 6 \, \text{pF} = 14 \, \text{pF}$$

$$f_{H_o} = \frac{1}{2\pi R_{Th_2} C_o} = \frac{1}{(6.28)(1.419 \times 10^3)(14 \times 10^{-12})}$$
$$= \frac{1000 \times 10^6}{124.76} \cong \textbf{8.02 MHz}$$

The h_{fe} or β variation with frequency must now be considered to ensure that it does not determine the high cutoff frequency of the amplifier. The variation of h_{fe} with frequency will approach, with some degree of accuracy, the following relationship:

$$h_{fe} = \frac{h_{fe_{mid}}}{1 + jf/f_\beta} \qquad (7.36)$$

The only undefined quantity, f_β, is determined by a set of parameters employed in the *hybrid π* or *Giacoletto* model frequently applied to best represent the transistor in the high-frequency region. It appears in Fig. 7.38. The various parameters warrant a moment of explanation. The resistance $r_{bb'}$ includes the base contact, base bulk, and base spreading resistance. The first is due to the actual connection

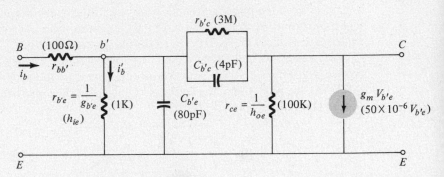

Figure 7.38. Giacoletto (or hybrid π) high-frequency transistor small-signal ac equivalent circuit.

to the base. The second includes the resistance from the external terminal to the active region of the transistors, while the last is the actual resistance within the active base region. The resistances $r_{b'e}, r_{ce},$ and $r_{b'c}$ are the resistances between the indicated terminals when the device is in the active region. The same is true for the capacitance $C_{b'c}$ and $C_{b'e}$, although the former is a transition capacitance while the latter is a diffusion capacitance. A more detailed explanation of the frequency dependence of each can be found in a number of readily available texts.

In terms of these parameters:

$$f_\beta(\text{sometimes appearing as } f_{hfe}) = \frac{g_{b'e}}{2\pi(C_{b'e} + C_{b'c})} \qquad (7.37)$$

or since the hybrid parameter h_{fe} is related to the hybrid parameter $g_{b'e}$ $g_m = h_{fe_{mid}} g_{b'e}$

$$f_\beta = \frac{1}{h_{fe_{mid}}}\left(\frac{g_m}{2\pi(C_{b'e} + C_{b'c})}\right) \qquad (7.38)$$

The basic format of Eq. (7.36) should suggest some similarities between it and the curves obtained for the low-frequency response. The most noticeable difference is the fact that f_β appears in the denominator while f_1 appears in the numerator of the frequency ratio. This particular difference will have the effect depicted in Fig.

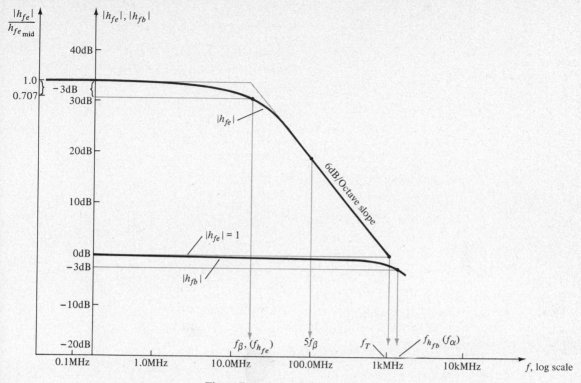

Figure 7.39. h_{fe} and h_{fb} vs. frequency in the high-frequency region.

7.39; the plot will drop off from the mid-band value rather than approach it with increase in frequency. The same figure has a plot of h_{fb} versus frequency. Note that it is almost constant for the frequency range. In general, the common-base configuration displays improved high-frequency characteristics over the common-emitter configuration. For this reason, common-base high-frequency parameters, rather than common-emitter parameters, are often specified for a transistor. The following equation permits a direct conversion for determining f_β if f_α and α are specified.

$$\boxed{f_\beta = f_\alpha(1 - \alpha)} \tag{7.39}$$

A quantity called the *gain-bandwidth product* is defined for the transistor by the condition

$$\left| \frac{h_{fe_{\text{mid}}}}{1 + jf/f_\beta} \right| = 1$$

so that $\qquad |h_{fe}|_{\text{dB}} = 20 \log_{10} \left| \dfrac{h_{fe_{\text{mid}}}}{1 + jf/f_\beta} \right| = 20 \log_{10} 1 = 0 \text{ dB}$

The frequency at which $|h_{fe}|_{\text{dB}} = 0$ dB is clearly indicated by f_T in Fig. 7.39. The magnitude of h_{fe} at the defined condition point ($f_T \gg f_\beta$) is given by

$$\frac{h_{fe_{\text{mid}}}}{\sqrt{1 + (f_T/f_\beta)^2}} \simeq \frac{h_{fe_{\text{mid}}}}{f_T/f_\beta} = 1$$

SEC. 7.12 HIGH-FREQUENCY CONSIDERATIONS

361

so that

$$\boxed{f_T \cong h_{fe_{\text{mid}}} \overbrace{f_\beta}^{(\cong \text{BW})} \text{ (Gain-bandwidth product)}} \qquad (7.40)$$

or

$$f_T \cong \beta f_\beta$$

with

$$\boxed{f_\beta = \frac{f_T}{\beta}} \qquad (7.41)$$

Substituting for f_β in Eq. (7.40) gives

$$f_T \cong h_{fe_{\text{mid}}} f_\beta = h_{fe}\left[\frac{1}{h_{fe}}\left(\frac{g_m}{2\pi(C_{b'e} + C_{b'c})}\right)\right]$$

$$\boxed{f_T \cong \frac{g_m}{2\pi(C_{b'e} + C_{b'c})}} \qquad (7.42)$$

Ignoring the effect of $r_{bb'}$ as a first approximation, $C_{b'e} = C_{be}$ and $C_{b'c} = C_{bc}$. For the network of Fig. 7.35.

$$f_\beta = \frac{g_{b'e}}{2\pi(C_{b'e} + C_{b'c})}$$

$$= \frac{1 \times 10^{-3}}{(6.28)(49.5 \times 10^{-12})} = \frac{1000 \times 10^6}{310.86}$$

$$f_\beta = \textbf{3.217 MHz}$$

and

$$f_T \cong h_{fe} f_\beta$$

$$= 100\,(3.217 \times 10^6)$$

$$= \textbf{321.7 M Hz}$$

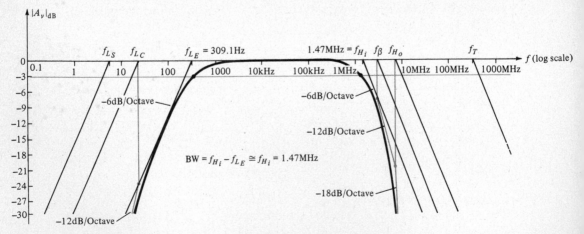

Figure 7.40. $|A_v|_{\text{dB}}$ versus frequency (log scale) for the network of Fig. 7.35.

For the low-, mid-, and high-frequency regions a Bode plot for the network of Fig. 7.35 appears in Fig. 7.40. Note in the high- and low-frequency regions that a cutoff frequency defines a —6 dB/octave asymptote and the slope increases by —6 dB/octave for the asymptote that will define the actual response each time it passes a cutoff frequency. A curve approximating the actual response also appears in the figure. Note the —3 dB drop at the highest low-frequency cutoff frequency and at the lowest high-frequency cutoff frequency.

7.13 MULTISTAGE FREQUENCY EFFECTS

For a second transistor stage connected directly to the output of a first stage there will be a significant change in the overall frequency response. In the high-frequency region the output capacitance C_o must now include the wiring capacitance (C_{w_1}), parasitic capacitance (C_{be}), and Miller capacitance (C_M) of the following stage. Further, there will be additional low-frequency cutoff levels due to the second stage that will further reduce the overall gain of the system in this region. For each additional stage the upper cutoff frequency will be determined primarily by that stage having the lowest cutoff frequency. The low-frequency cutoff is primarily determined by that stage having the highest cutoff frequency. Obviously, therefore, one poorly designed stage can offset an otherwise well-designed cascaded system.

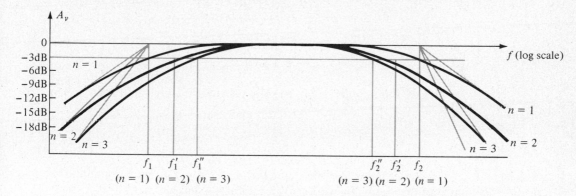

Figure 7.41. Effect of an increased number of stages on the cutoff frequencies and the bandwidth.

The effect of increasing the number of *identical* stages can be clearly demonstrated by considering the situations indicated in Fig. 7.41. In each case the upper and lower cutoff frequencies of each of the cascaded stages are identical. For a single stage the cutoff frequencies are f_1 and f_2 as indicated. For two identical stages in cascade the drop-off rate in the high- and low-frequency regions has increased to —12 dB/octave or —40 dB/decade. At f_1 and f_2, therefore, the decibel drop is now —6 dB rather than the defined band frequency gain level of —3 dB. The —3 dB point has shifted to f_1' and f_2' as indicated with a resulting drop in the bandwidth. A —18 dB/octave or —60 dB/decade slope will result for a three-stage system of identical stages with the indicated reduction in bandwidth (f_1'' and f_2'').

Assuming identical stages, an equation for each band frequency as a function of the number of stages (n) can be determined in the following manner:

For the low-frequency region,

$$A_{v\,\text{low, (overall)}} = A_{v_1\,\text{low}} A_{v_2\,\text{low}} A_{v_3\,\text{low}} \cdots A_{v_n\,\text{low}}$$

but since each stage is identical, $A_{v_1\,\text{low}} = A_{v_2\,\text{low}} = $ etc.

and

$$A_{v\,\text{low, (overall)}} = (A_{v_1\,\text{low}})^n$$

or

$$\frac{A_{v\,\text{low}}}{A_{v\,\text{mid}}}(\text{overall}) = \left(\frac{A_{v_1\,\text{low}}}{A_{v\,\text{mid}}}\right)^n = \frac{1}{(1 + jf_1/f)^n}$$

Setting the magnitude of this result equal to $1/\sqrt{2}$ (-3 db level) results in

$$\frac{1}{[\sqrt{1 + (f_1/f_1')^2}]^n} = \frac{1}{\sqrt{2}}$$

or

$$\left\{\left[1 + \left(\frac{f_1}{f_1'}\right)^2\right]^{1/2}\right\}^n = \left\{\left[1 + \left(\frac{f_1}{f_1'}\right)^2\right]^n\right\}^{1/2} = \{2\}^{1/2}$$

so that

$$\left[1 + \left(\frac{f_1}{f_1'}\right)^2\right]^n = 2$$

and

$$1 + \left(\frac{f_1}{f_1'}\right)^2 = 2^{1/n}$$

with the result

$$\boxed{f_1' = \frac{f_1}{\sqrt{2^{1/n} - 1}}} \tag{7.43}$$

In a similar manner, it can be shown that for the high-frequency region,

$$\boxed{f_2' = \sqrt{2^{1/n} - 1}\, f_2} \tag{7.44}$$

Note the presence of the same factor $\sqrt{2^{1/n} - 1}$ in each equation. The magnitude of this factor for various values of n is listed below.

n	$\sqrt{2^{1/n} - 1}$
1	1
2	0.64
3	0.51
4	0.44
5	0.39

For $n = 2$, consider that the upper cutoff frequency $f_2' = 0.64 f_2$ or 64% of the value obtained for a single stage, while $f_1' = (1/0.64)f_1 = 1.55 f_1$. For $n = 3$, $f_2' = 0.51 f_2$ or approximately $\frac{1}{2}$ the value of a single stage with $f_1' = (1/0.51)f_1 = 1.96 f_1$ or approximately *twice* the single-stage value.

Consider the example of the past few sections, where $f_2 = 1.47$ MHz and $f_1 = 309.1$ Hz. For $n = 2$,

$$f'_2 = 0.64f_2 = 0.64(1.47 \text{ MHz}) = 0.94 \text{ MHz}$$

and

$$f'_1 = 1.56f_1 = 1.56(309.1) \cong 482.2 \text{ Hz}$$

The bandwidth is now $0.94 \text{ MHz} - 482.2 \text{ Hz} \cong f'_2 \Rightarrow 64.4\%$ of its single-stage value, a drop of some significance.

For the RC-coupled transistor amplifier, if $f_2 = f_\beta$, or if they are close enough in magnitude for both to affect the upper 3-dB frequency, the number of stages must be increased by a factor of 2 when determining f'_2 due to the increased number of factors $1/(1 + jf/f_x)$.

A decrease in bandwidth is not always associated with an increase in the number of stages if the mid-band gain can remain fixed independent of the number of stages. For instance, if a single-stage amplifier produces a gain of 100 with a bandwidth of 10,000 Hz, the resulting gain-bandwidth product is $10^2 \times 10^4 = 10^6$. For a two-stage system the same gain can be obtained by having two stages with a gain of 10 since $(10 \times 10 = 100)$. The bandwidth of each stage would then increase by a factor of 10 to 100,000 due to the lower gain requirement and fixed gain-bandwidth product of 10^6. Of course, the design must be such as to permit the increased bandwidth and establish the lower gain level.

This discussion of the effects of an increased number of stages on the frequency response was included here, rather than at the conclusion of this chapter, to add a note of completion to the analysis of cascaded transistor amplifier systems. The results, however, can be applied directly to the discussion of FET and vacuum-tube cascaded systems to follow.

7.14 FREQUENCY RESPONSE OF CASCADED FET AMPLIFIERS

A representative cascaded system employing FET amplifiers appears in Fig. 7.42.

The equivalent circuit for the section a-a' indicated in Fig. 7.42 is presented in Fig. 7.43 for both the high- and low-frequency regions. It is assumed for the low-frequency response that the breakpoint frequency due to C_s is sufficiently less than that due to C_c, to result in C_c determining the lower-band frequency. For this reason C_s does not appear in the low-frequency model. For future reference, the breakpoint frequency determined by C_s is given by

$$f_S = \frac{1 + R_s(1 + g_m r_d)/(r_d + R_D)}{2\pi C_s R_s} \tag{7.45}$$

It was demonstrated in this chapter and in Chapter 6 that the mid-band voltage gain is given by

$$A_{v_{\text{mid}}} = \frac{V_o}{V_i} = -g_m R \qquad \text{where } R = r_{d_1} \| R_{D_1} \| R_{G_2}$$

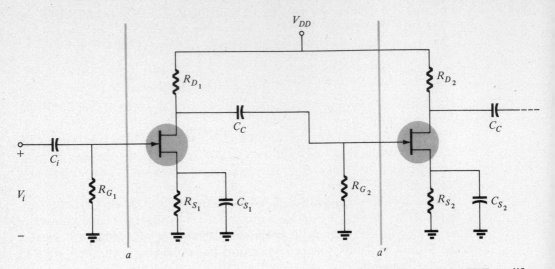

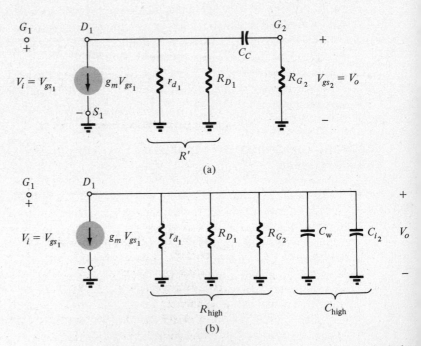

Figure 7.42. Cascaded FET amplifiers.

Figure 7.43. Small-signal ac equivalent circuit of section $a\text{-}a'$ of the network of Fig. 7.42: (a) low frequency; (b) high frequency.

The similarities between the circuits of Fig. 7.43 and those introduced in the last few sections on the transistor cascaded system are obvious. With this in mind, the steps leading to the following results should also be somewhat obvious.

$$\frac{A_{v_{\text{low}}}}{A_{v_{\text{mid}}}} = \frac{1}{1 - jf_1/f} \qquad (7.46)$$

where
$$f_1 = \frac{1}{2\pi C_C(R' + R_{G_2})}$$

and
$$R' = r_{d_1} \| R_{D_1}$$

In addition,
$$\frac{A_{v_{\text{high}}}}{A_{v_{\text{mid}}}} = \frac{1}{1 + jf/f_2} \qquad (7.47)$$

where
$$f_2 = \frac{1}{2\pi C_{\text{high}} R_{\text{high}}}$$

and
$$C_{\text{high}} = C_W + C_{i_2}$$

with
$$R_{\text{high}} = R = r_{d_1} \| R_{D_1} \| R_{G_2}$$

For FET amplifiers, the input capacitance of the succeeding stage, as derived in Chapter 6 is given by

$$C_i = C_{gs} + C_{gd}(1 + |A_v|) \qquad (7.48)$$

The gain-bandwidth product is

$$[\text{gain}][\text{BW}] = [A_{v_{\text{mid}}}][f_2 - f_1 \cong f_2] = [g_m R]\left[\frac{1}{2\pi R(C_W + C_i)}\right]$$

and
$$\text{GBW} = \frac{g_m}{2\pi(C_W + C_i)} = f_T = A_{v_{\text{mid}}} f_2 \qquad (7.49)$$

7.15 FREQUENCY RESPONSE OF CASCADED TRIODE AND PENTODE AMPLIFIERS

A cascaded vacuum triode amplifier system appears in Fig. 7.44. The high- and low-frequency equivalent circuits for the section a-a' of Fig. 7.44 appear in Fig. 7.45. The similarities between the circuits of Fig. 7.45 and those of Fig. 7.43 should be immediately obvious. The following discussion of cascaded vacuum-tube triode amplifiers will be a replica of that presented for the FET stages of the Section 7.14. The breakpoint due to C_K is again assumed to be sufficiently less than that due to C_C that it may be neglected in the low-frequency models. The breakpoint due to C_K is given by

$$f_K = \frac{1 + R_K(1 + \mu)/(r_p + R_P)}{2\pi C_K R_K} \qquad (7.50)$$

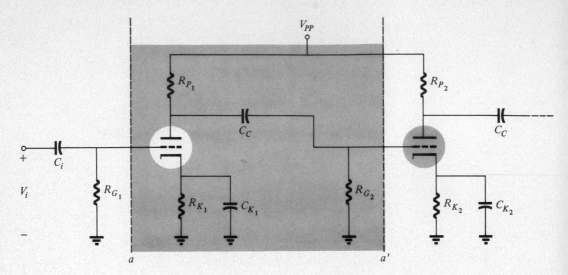

Figure 7.44. Cascaded vacuum tube triode stages.

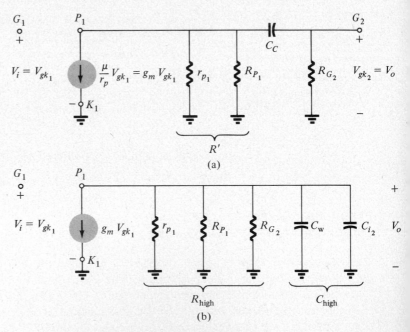

Figure 7.45. Small-signal ac equivalent circuit of section $a\text{-}a'$ of the network of Fig. 7.44: (a) low frequency; (b) high frequency.

The mid-band gain is

$$A_{v_{\text{mid}}} = \frac{V_o}{V_i} = -g_m R \qquad (7.51)$$

where
$$R = r_p \| R_P \| R_G$$

and
$$\boxed{\frac{A_{v_{\text{low}}}}{A_{v_{\text{mid}}}} = \frac{1}{1 - jf_1/f}} \tag{7.52}$$

where
$$f_1 = \frac{1}{2\pi C_C(R' + R_G)}$$

and
$$R' = r_p \| R_P$$

with
$$\boxed{\frac{A_{v_{\text{high}}}}{A_{v_{\text{mid}}}} = \frac{1}{1 + jf/f_2}} \tag{7.53}$$

where
$$f_2 = \frac{1}{2\pi C_{\text{high}} R_{\text{high}}}$$

$$C_{\text{high}} = C_W + C_{pk_1} + C_{i_2}$$

$$R_{\text{high}} = R = r_p \| R_P \| R_G$$

For triode amplifiers the input capacitance of the suceeding stage is given by

$$\boxed{C_i = C_{gk} + C_{gp}(1 + |A_v|)} \tag{7.54}$$

Through similarities with the FET circuit of Chapter 6. Eq. (7.54) can be derived directly through the use of the following triode equivalent circuit (Fig. 7.46). The gain-bandwidth product is

$$\boxed{\text{GBW} = \frac{g_m}{2\pi C_{\text{high}}} = f_T = |A_{v_{\text{mid}}}| f_2} \tag{7.55}$$

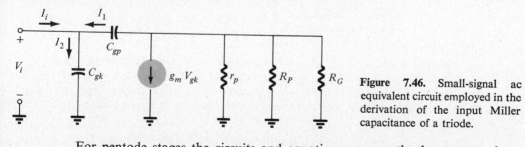

Figure 7.46. Small-signal ac equivalent circuit employed in the derivation of the input Miller capacitance of a triode.

For pentode stages the circuits and equations are exactly the same as those obtained for the triode system except for C_{high}. The action of the screen grid between plate and grid permits the approximation $C_{gp} \cong 0$, eliminating the high Miller feedback capacitance so that

$$\boxed{C_{\text{high}_{(\text{pentode})}} \cong C_W + C_{pk_1} + C_{gk_2}} \tag{7.56}$$

SEC. 7.15 CASCADED TRIODE AND PENTODE AMPLIFIERS

PROBLEMS

§ 7.2

1. (a) Determine A_{i_T} and A_{v_T} for a cascaded system if the power gain is 12.8×10^3 and $Z_L = 4\,\text{K}$ with the input impedance of the first stage $Z_{i_1} = 2\,\text{K}$.

   (b) If the system of part (a) is comprised of two identical stages, determine the voltage and current gain of each stage.

§ 7.3

2. A two-stage RC-coupled amplifier appears in Fig. 7.47.

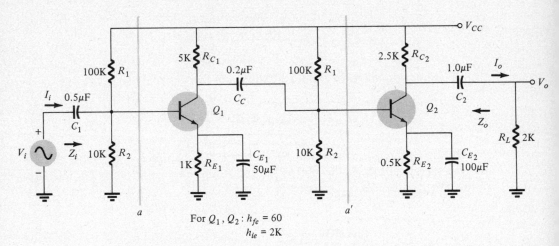

For Q_1, Q_2: $h_{fe} = 60$
$h_{ie} = 2\,\text{K}$

Figure 7.47. Two-stage RC-coupled amplifier.

   (a) Determine Z_i and Z_o.
   (b) Calculate the voltage gain $A_v = V_o/V_i$.
   (c) Determine the current gain $A_i = I_o/I_i$.

3. Determine Z_i, Z_o, A_{v_T}, A_{i_T}, and A_{p_T} for the RC-coupled amplifier of Fig. 7.2 if $R_{B_1} = 56\,\text{K}$ and $R_{B_2} = 5.6\,\text{K}$ for each transistor and $R_{C_1} = 6.8\,\text{K}$, $R_{C_2} = 3.3\,\text{K}$, $R_{E_1} = R_{E_2} = 0.5\,\text{K}$, $R_L = 2.2\,\text{K}$. All capacitor values are the same. For each transistor $\beta = 120$. Since h_{ie} is not provided, r_e will have to be determined for each transistor. Use all appropriate approximations.

4. Repeat Problem 3 if both emitter capacitors C_E are removed.

5. Repeat Example 7.1 if C_E is removed.

6. Repeat Example 7.1 if C_E is removed and the output load Z_L is connected through a capacitor to the collector of Q_2.

7. Repeat Example 7.2 if R_{E_1} is removed and Z_L is reduced to a significantly lower level of $0.5\,\text{K}$.

8. Design a two-stage RC-coupled amplifier to provide an overall gain of $\simeq 2000$. The circuit is to operate into a load of 10 K; the signal is supplied from a perfect voltage source $(R_S = 0\,\Omega)$. Show typical (commercially available) component values for each element and calculate the voltage gain of the resulting circuit as a check. The list of commercially available resistors can be found in any electronic products publication.

9. Repeat Example 7.3 if C_{S_1} and C_{S_2} are removed and $R_{S_1} = 2\,\text{K}$ with $R_{S_2} = 1\,\text{K}$.

§ 7.4

10. Calculate the impedance seen looking into the primary of a 5:1 step-down transformer connected to a load of 20 Ω.

11. Calculate the necessary transfomer turns ratio to match a 50-Ω load to a 20-K source impedance.

12. (a) Calculate the voltage gain (V_o/V_i) of the transformer-coupled amplifier of Fig. 7.48.

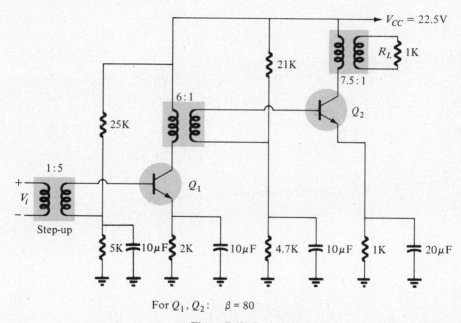

For Q_1, Q_2: $\beta = 80$

Figure 7.48. Two-stage transformer-coupled amplifier.

(b) What is the voltage gain of the circuit of Fig. 7.48 if the load is reduced to 0.5 K?

§ 7.5

13. Determine the new dc levels in Fig. 7.13 if the 12-V batteries are replaced by 16-V supplies. In addition, determine the new value of r_e for each transistor. How are the ac voltage and current gain affected? What are their new levels if affected by this change?

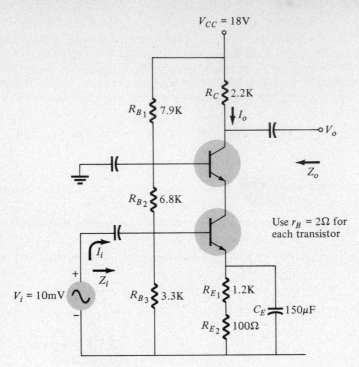

Figure 7.49

§ 7.6

14. Determine the following for the cascade amplifier of Fig. 7.49:
  (a) V_o.
  (b) Z_i, Z_o.
  (c) I_o, I_i, and A_i.
  (d) A_{PT}.

15. Determine the following for the cascade amplifier of Fig. 7.50:
  (a) r_{e_1} and r_{e_2} for $\beta = 50$.

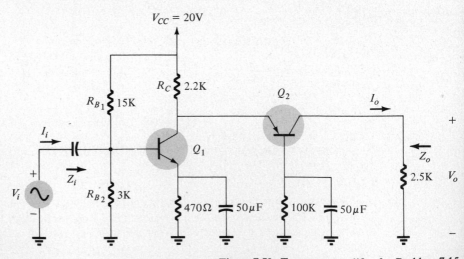

Figure 7.50. Two-stage amplifier for Problem 7.15.

(b) A_{V_T} and V_o if $V_i = 10$ mV.

(c) Z_i and Z_o.

§ 7.7

16. Determine A_i, Z_i, Z_o, and A_v for the Darlington configuration of Fig. 7.51.

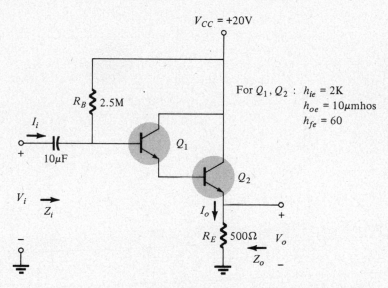

Figure 7.51. Amplifier circuit for Problem 7.16.

17. Repeat Problem 16 if a collector resistor of 2.2 K is added between the collector of Q_1 and V_{CC} and the output is taken off the collector of the Darlington configuration. I_o is the current through the added 2.2-K resistor.

18. Repeat Problem 16 if R_E is changed to 150 Ω.

19. Repeat Problem 17 if R_E is changed to 150 Ω.

§ 7.8

20. Calculate the voltage gain of a pentode amplifier of Fig. 7.18a for tube parameters $g_m = 8000$ μmhos, $r_p = 2$ M.

§ 7.9

21. Calculate the decibel power gain for the following:

(a) $P_o = 100$ W, $P_i = 5$ W.

(b) $P_o = 100$ mW, $P_i = 5$ mW.

(c) $P_o = 100$ μW, $P_i = 20$ μW.

22. Two voltage measurements made across the same resistance are $V_1 = 25$ V and $V_2 = 100$ V. Calculate the decibel power gain of the second reading over the first reading.

23. Input and output voltage measurements of $V_i = 10$ mV and $V_o = 25$ V are made. What is the voltage gain in decibels?

24. (a) The total decibel gain of a three-stage system is 120 dB. Determine the decibel gain of each stage if the second stage has twice the decibel gain of the first and the third has 2.7 times the decibel gain of the first.

(b) Determine the voltage gain of each stage.

§ 7.10–7.12

25. For the network of Fig. 7.52

(a) Determine the low cutoff frequencies f_{L_S}, f_{L_C}, and f_{L_E}.

(b) Calculate the mid-band voltage gain.

(c) Determine the high cutoff frequencies f_{H_i} and f_{H_o}.

(d) Sketch a rough plot of $A_v = V_o/V_i$ on a log plane.

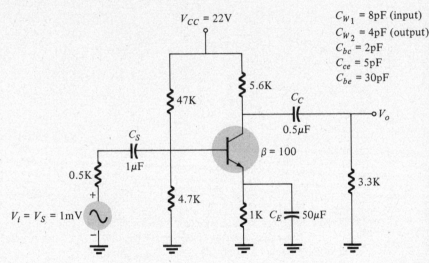

$C_{W_1} = 8\text{pF (input)}$
$C_{W_2} = 4\text{pF (output)}$
$C_{bc} = 2\text{pF}$
$C_{ce} = 5\text{pF}$
$C_{be} = 30\text{pF}$

Figure 7.52

26. Repeat Problem 25 if the capacitor C_E is removed.

27. Repeat Problem 25 if C_E is reduced to 1 μF.

§ 7.13

28. Calculate the overall voltage gain of four identical stages of an amplifier, each having a gain of 20.

29. Calculate the overall upper 3-dB frequency for a four-stage amplifier having an individual stage value of $f_2 = 2.5$ MHz.

30. A four-stage amplifier has a lower 3-dB frequency for an individual stage of $f_1 = 40$ Hz. What is the value of f_1 for ths full amplifier?

31. (a) Determine the low cutoff frequencies for the two stage amplifier of Fig. 7.53.

(b) Determine the high cutoff frequencies for the network of Fig. 5.73 ($f_\beta = 5$ MHz).

(c) Calculate the mid-band voltage gain and make a rough sketch of $A_v = V_o/V_i$ versus frequency (log plot).

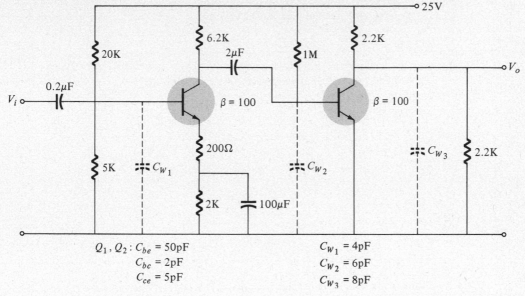

Q_1, Q_2: $C_{be} = 50\text{pF}$
$C_{bc} = 2\text{pF}$
$C_{ce} = 5\text{pF}$

$C_{W_1} = 4\text{pF}$
$C_{W_2} = 6\text{pF}$
$C_{W_3} = 8\text{pF}$

Figure 7.53

§ 7.14

32. Calculate the mid-frequency gain for a stage of a FET amplifier as in Fig. 7.42 for circuit values: $g_m = 6000\ \mu\text{mhos}$, $r_d = 50\ \text{K}$ (Q_1 and Q_2); $R_{D_1} = R_{D_2} = 10\ \text{K}$; $R_{G_1} = R_{G_2} = 1\ \text{M}$; $C_S = 10\ \mu\text{F}$, $R_S = 1\ \text{K}$, $C_C = 0.1\ \mu\text{F}$, $C_{gs} = C_{gd} = 4\ \text{pF}$, $C_W = 7\ \text{pF}$.

33. For the circuit and values of Problem 32 calculate f_1 and f_2.

34. Calculate the gain-bandwidth product for the circuit and values of Problem 32.

35. For the network of Fig. 7.54
(a) Calculate the low and high cutoff frequencies as determined by the provided levels of capacitance. Use $V_{CC} = 20\ \text{V}$, $r_B = 0\ \Omega$.

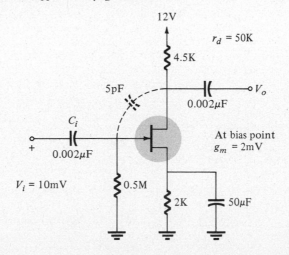

Figure 7.54

(b) What value of C_i will make the low cutoff frequency 10 Hz?

(c) Determine V_o at the high cutoff frequency.

§ 7.15

36. For the network of Fig. 7.44 and circuit values $r_p = 25$ K, $\mu = 50$, $R_p = 10$ K, $R_G = 0.5$ M, $R_K = 2$ K, $C_K = 1$ μF, $C_C = 0.5$ μF, $C_{gk} = C_{gp} = 5$ pF, calculate the following:

    (a) A_v (mid-band).

    (b) f_K.

    (c) f_T

large-signal amplifiers

8

8.1 GENERAL

An amplifier system consists of a signal pickup transducer, followed by a small-signal amplifier, a large-signal amplifier, and an output transducer device. The input transducer signal is generally small and must be amplified sufficiently to be used to operate some output device. The factors of prime interest in small-signal amplifier then are usually linearity and gain. Since the signal voltage and current from the input transducer is usually small, the amount of power handling capacity and power efficiency are of slight concern. Voltage amplifiers provide a large enough voltage signal to the large-signal amplifier stages to operate such output devices as speakers and motors. A large-signal amplifier must operate efficiently and be capable of handling large amounts of power—typically, a few watts to hundreds of watts. This chapter concentrates on the amplifier stage used to handle large signals, typically a few volts to tens of volts. The amplifier factors of greatest concern are the power efficiency of the circuit, the maximum amount of power that the circuit is capable of handling, and impedance matching to the output device.

A class-A series-fed amplifier stage is considered first to show some of the limitations in using such a simple circuit connection. The single-ended transformer-coupled stage is then discussed to show how impedance matching between driver stage and load (output transducer) is accomplished. The push-pull connection, a very popular connection for low distortion and efficient coupling of the signal to a speaker or motor device, is discussed next. Finally, complementary transistors for push-pull operation without a transformer are presented.

The simple fixed-bias circuit connection can be used as a large-signal class-A amplifier as shown in Fig. 8.1. The only difference between this circuit and the small-signal version considered previously is that the signals handled by the large-signal circuit are in the range of volts and the transistor used is a power transistor capable of operating in the range of a few watts. As will be shown, this circuit is not the best to use for a large-signal amplifier.

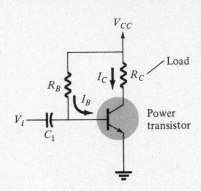

Figure 8.1. Series-fed class-A large signal amplifier.

Figure 8.2a shows a typical power transistor circuit with appropriate circuit values. The transistor-collector characteristic of Fig. 8.2b shows the load line and input and output signals. The input signal is an ac voltage of 0.5 V, peak amplitude. The circuit has an ac input resistance of 50 Ω. The output ac signal is to be developed across the 20-Ω load resistor. For the bias value of base current shown, the transistor is operated at a collector-emitter voltage of 10 V and a collector current of about 500 mA.

The input ac current variation is calculated to be

$$I_i = \frac{V_i}{r_i} = \frac{0.5\ \text{V}}{50\ \Omega} = 10\ \text{mA, peak}$$

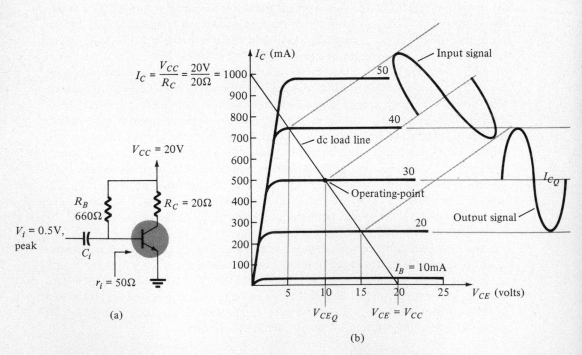

Figure 8.2. Operation of a series-fed circuit.

As shown on the transistor-collector characteristic (Fig. 8.2b) the corresponding current swing in the output (across the load resistor) is about 250 mA, peak. Corresponding to the indicated current swing is a voltage variation of 5 V, peak. From the present information we can calculate the current gain, voltage gain, and power gain of the circuit as follows:

$$A_v = \frac{V_o}{V_i} = \frac{5 \text{ V,peak}}{0.5 \text{ V, peak}} = 10$$

$$A_i = \frac{I_o}{I_1} = \frac{250 \text{ mA, peak}}{10 \text{ mA, peak}} = 25$$

$$A_p = \text{power gain} = A_v A_i = (10)(25) = 250$$

Up to now the power distribution in the amplifier has not been considered since the amount of power handled by small-signal circuits is small. In the large-signal circuit the amount of power in different parts of the amplifier is considerable and the efficiency of the amplifier circuit is of great interest. The dc battery is the source of power in a transistor amplifier. The dc power drawn from the battery is dissipated as heat lost in the load resistor and the transistor. The output power (ac signal developed across the load) is a part of the power taken from the dc battery. In fact, the operation of the circuit of Fig. 8.2a is to convert as much of the dc power drawn from the battery into ac power (output power) across the load (the 20-Ω resistor in this case). If the load were a speaker, the power delivered would result in audible sound.

In the present circuit we can calculate separately a number of different power terms as follows:

The average power taken from the dc battery is

$$\boxed{P_i(\text{dc}) = V_{CC} I_{C_Q}} \tag{8.1}$$

the product of the dc battery voltage times the *average* current drawn from the battery—the quiescent or average current for the nondistorted output signal, as shown in Fig. 8.2b.

The ac signal power developed across the load is

$$\boxed{P_o(\text{ac}) = I_C^2 (\text{rms}) R_C} \tag{8.2}$$

the square of the root-mean square value of the ac current *through* the load resistor times the value of collector resistance.

The dc power dissipated as heat by the collector resistor is equal to the square of the average or dc current through the collector resistor times the value of the resistance.

$$\boxed{P_{R_C}(\text{dc}) = I_{C_Q}^2 R_C} \tag{8.3}$$

The dc power dissipated by the transistor is

$$\boxed{P_t = P_i - P_o - P_{R_C}}$$ (8.4)

which is the input power minus that dissipated by the collector resistor (both ac and dc). This power is dissipated as heat with the transistor power rating indicating the maximum amount of power the particular device is capable of handling. (Transistor ratings and specifications are discussed in Section 8.7.)

The efficiency of the circuit in converting the dc power drawn from the battery into ac signal power across the load is calculated to be

$$\boxed{\% \text{ efficiency} = \eta = \frac{P_o\,(\text{ac})}{P_i\,(\text{dc})} \times 100}$$ (8.5)

The numerical calculations for the circuit of Fig. 8.2 are carried out in Example 8.1.

EXAMPLE 8.1 Calculate the various power terms for the circuit of Fig. 8.2a and transistor characteristic of Fig. 8.2b. Obtain the circuit power efficiency from these calculated values.

Solution:
(a) $P_i\,(\text{dc}) = V_{CC} I_{C_Q} = (20 \text{ V})(500 \text{ mA}) = \mathbf{10\ W}$

(b) $P_o\,(\text{ac}) = I_C^2\,(\text{rms})\,R_C = \left(\frac{0.25}{\sqrt{2}}\right)^2 20 = \mathbf{0.625\ W}$

(c) $P_{R_C}\,(\text{dc}) = I_{C_Q}^2 R_C = (0.5 \text{ A})^2(20\ \Omega) = \mathbf{5\ W}$

(d) $P_t\,(\text{dc}) = P_i - P_o - P_{R_C} = 10 - 0.625 - 5 = \mathbf{4.375\ W}$

(e) $\eta = \frac{P_o}{P_i} \times 100 = \frac{0.625}{10} \times 100 = \mathbf{6.25\%}$

The power efficiency of 6.25% is extremely poor and the circuit of Fig. 8.2 is therefore not a good amplifier for handling large amounts of power. Of the 10 W of power drawn from the battery, about half is dissipated in heat by the transistor and almost the same amount is wasted as heat in the load resistor, with only a very small amount of ac power being developed across the load. Although a 20-Ω speaker could be directly connected in the circuit as the load resistor, the operation of the circuit would be inefficient.

The maximum efficiency for the class-A series-fed amplifier can be seen to be 25% by considering the largest output power before distortion occurs is (see Fig. 8.2b)

$$V_o(p - p) = 20 \text{ V}, \qquad I_o(p - p) = 1000 \text{ mA} = 1 \text{ A}$$

which represents an output power of[1]

$$P_o = \frac{V_o(p - p) I_o(p - p)}{8} = \frac{(20 \text{ V})(1 \text{ A})}{8} = 2.5 \text{ W}$$

---

[1]$P_o\,(\text{ac}) = V_o\,(\text{rms})\,I_o\,(\text{rms}) = \frac{V_o(p)}{\sqrt{2}} \cdot \frac{I_o(p)}{\sqrt{2}} = \frac{V_o(p-p)/2}{\sqrt{2}} \cdot \frac{I_o(p-p)/2}{\sqrt{2}} = \frac{V_o(p-p)I_o(p-p)}{8}$

for an efficiency of

$$\eta = \frac{P_o\,(\text{ac})}{P_i\,(\text{dc})} \times 100 = \frac{2.5\,\text{W}}{10\,\text{W}} \times 100 = 25\%$$

For any input signal resulting in an output signal swing less than these maximum signal swings the efficiency is less than the maximum value of 25%. Note that the input power remains 10 W in this example, regardless of the amount of ac voltage swing, for the bias current value remains the same whether an input signal is applied or not.

8.3 TRANSFORMER-COUPLED AUDIO POWER AMPLIFIER

A more reasonable class-A amplifier connection uses a transformer to couple the load to the amplifier stage as shown in Fig. 8.3a. This is a simple version of the circuit for the presentation of a few basic concepts. More practical circuit versions will be covered shortly. Fig. 8.3b shows the output coupling transformer with voltage, current, and impedances indicated.

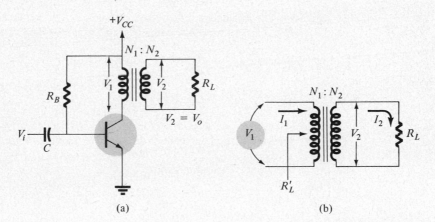

Figure 8.3. Transformer-coupled audio power amplifier.

Consider first the operation of the transformer as an impedance device. Typically, a low-impedance speaker is connected as the load. We have seen the result of connecting this speaker directly as the load in the circuit of Fig. 8.2. With the transformer used to connect the speaker load to the transistor-collector circuit, there results an impedance transformation.

Transformer Impedance Matching

The resistance seen looking into the primary of the transformer is related to the resistance connected across the secondary. The ratio of secondary resistance to primary resistance may be expressed as follows:

$$\frac{R'_L = V_1/I_1}{R_L = V_2/I_2} = \frac{R'_L}{R_L} = \frac{V_1}{I_1}\frac{I_2}{V_2} = \frac{V_1}{V_2}\frac{I_2}{I_1} = \frac{N_1}{N_2}\frac{N_1}{N_2} = \left(\frac{N_1}{N_2}\right)^2$$

where $V_1/V_2 = N_1/N_2$ and $I_2/I_1 = N_1/N_2$. Hence the ratio of the transformer input and output resistance varies directly as the *square* of the transformer turns ratio:

$$\boxed{\frac{R'_L}{R_L} = \left(\frac{N_1}{N_2}\right)^2 = a^2} \tag{8.6}$$

and

$$\boxed{R'_L = a^2 R_L = \left(\frac{N_1}{N_2}\right)^2 R_L} \tag{8.7}$$

where

R_L = resistance of load connected across the transformer secondary;

R'_L = effective resistance seen looking into primary of transformer;

$a = N_1/N_2$ is the step-down turns ratio needed to make the load resistance appear as a larger effective resistance seen from the transformer primary.

EXAMPLE 8.2 Calculate the effective resistance (R'_L) seen looking into the primary of a 15:1 transformer connected to an output load of 8 Ω.

Solution: Using Eq. (8.7), we get

$$R'_L = \left(\frac{N_1}{N_2}\right)^2 R_L = (15)^2 8 = 1800\ \Omega = \mathbf{1.8\ K}$$

EXAMPLE 8.3 What transformer turns ratio is required to match a 16-Ω speaker load to an amplifier so that the effective load resistance is 10 K?

Solution: Using Eq. (8.6), we get

$$\frac{R'_L}{R_L} = \left(\frac{N_1}{N_2}\right)^2 = \frac{10,000}{16} = 625$$

$$\left(\frac{N_1}{N_2}\right) = \sqrt{625} = \mathbf{25:1}$$

For the transformer-coupled amplifier of Fig. 8.3 the various power losses and output power developed across the load resistor will allow calculation of the circuit power efficiency. Figure 8.4 shows a transistor-collector characteristic, load lines, input, and output signals.

dc Load Line

The transformer dc (winding) resistance is used to obtain the dc load line for the circuit. Typically, this dc resistance is small and is shown in Fig. 8.4a to be 0 Ω, providing a straight (vertical) load line. This is the ideal load line for the trans-

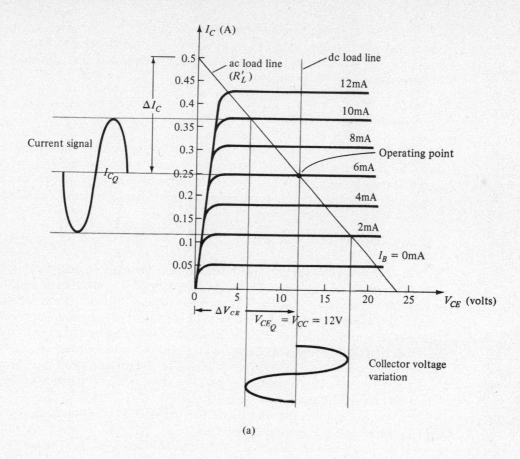

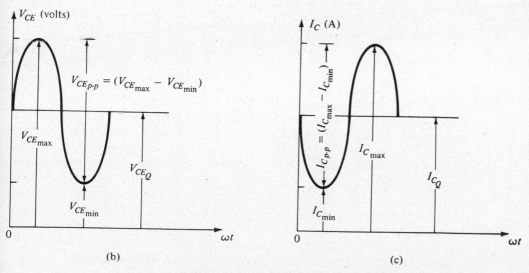

Figure 8.4. Graphical operation of transformer-coupled audio power amplifier.

former. Practical transformer windings would provide a slight slope for the load line, but only the ideal case will be considered in this discussion. There is no dc voltage drop across the dc load resistance in the ideal case and the load line is drawn straight vertically from the voltage point, $V_{CE_Q} = V_{cc}$.

Quiescent Operating Point

The operating point is obtained graphically as the point of intersection of the dc load line and the transistor base current curve. From the operating point the quiescent collector current I_{C_Q} is read. The value of the base current is calculated separately from the circuit as considered in the dc bias calculations of Chapter 4.

ac Load Line

In order to obtain the ac signal operation it is necessary to first calculate the ac load resistance seen looking into the primary side of the transformer and then to draw the ac load line on the transistor characteristic. The effective load resistance is calculated using Eq. (8.7) from the values of secondary load resistance and transformer turns ratio. Having obtained the value of R'_L the ac load line must be drawn so that it passes through the operating point and has a slope equal to $-1/R'_L$, the load line slope being the negative reciprocal of the ac load resistance. Since the collector signal passes through the operating point when no signal is applied, the load line must pass through the operating point.

In order to simplify drawing a load line of slope $-1/R'_L$ through the operating point the following technique may be used:

If the ac signal were to vary from the quiescent level to 0 V, it would also vary from the quiescent current level, I_{C_Q}, by an amount

$$\Delta I_C = \frac{\Delta V_{CE}}{R'_L} \tag{8.8}$$

Mark a point on the y-axis of the transistor characteristic ΔI_C units above the quiescent level and connect this point through the operating point to draw the ac load line desired (see Fig. 8.4a). Notice that the ac load line shows that the output signal swing can exceed the value of V_{CC}, the supply voltage. In fact, the voltage developed across the transformer primary can be large. One of the maximum operating values that will have to be checked carefully is that of $V_{CE_{max}}$, as specified by the transistor manufacturer, to see that the value of maximum voltage obtained after drawing the ac load line does not exceed the transistor rated maximum value. Assuming for the moment that maximum power and voltage ratings are not exceeded (the consideration of these maximum values is deferred until Section 8.7), ac signal swings of current and voltage are obtained as shown in Fig. 8.4a and redrawn in detail in Figs. 8.4b and 8.4c.

Signal Swing and Output ac Power

From the signal variations as shown in Figs. 8.4b and 8.4c the values of the peak-to-peak signal swings are obtained

$$V_{swing} = V_{CE}\,(\text{peak-to-peak}) = (V_{CE_{max}} - V_{CE_{min}}) \tag{8.9}$$

$$I_{swing} = I_C\,(\text{peak-to-peak}) = (I_{C_{max}} - I_{C_{min}}) \tag{8.10}$$

using the values as indicated (and defined) in Fig. 8.4. The ac power developed across the transformer primary can be calculated to be

$$P_o\,(\text{ac}) = V_{CE}\,(\text{rms})\,I_C\,(\text{rms})$$

$$= \frac{V_{CE}\,(\text{peak})}{\sqrt{2}}\,\frac{I_C\,(\text{peak})}{\sqrt{2}}$$

$$= \frac{V_{CE}\,(\text{peak-to-peak})/2}{\sqrt{2}} \times \frac{I_C\,(\text{peak-to-peak})/2}{\sqrt{2}}$$

$$\boxed{P_o\,(\text{ac}) = \frac{(V_{CE_{max}} - V_{CE_{min}})(I_{C_{max}} - I_{C_{min}})}{8}} \tag{8.11}$$

The ac power calculated is that developed across the primary of the transformer. Assuming a highly efficient transformer the power across the speaker is approximately equal to that calculated by Eq. (8.11). For our purposes an ideal transformer will be assumed so that the ac power calculated using Eq. (8.11) is also the ac power delivered to the load.

For the ideal transformer considered, the voltage across the secondary of the speaker can be calculated from

$$V_2 = V_S = V_L = \left(\frac{N_2}{N_1}\right) V_1 \tag{8.12}$$

where the secondary voltage (V_2) equals the speaker (V_S) or load voltage (V_L). The load voltage is related by the transformer turns ratio, N_2/N_1, to the voltage developed across the transformer primary (V_1). The voltage across the primary was previously labelled $V_{CE}\,(\text{rms})$, and for power considerations the rms values of voltage are usually used [unless otherwise stated, as in Eq. (8.11)]

From the calculated value of the secondary rms voltage the power across the load can be obtained

$$P_S = P_L = \frac{V_L^2}{R_L} \tag{8.13}$$

and equals the power calculated using Eq. (8.11). Thus the ac power can be calculated in a number of ways, including the following:

$$I_L\,(\text{rms}) = \frac{N_1}{N_2}\,I_C\,(\text{rms}) \tag{8.14}$$

where I_L is the rms value of the current through the load resistor (or speaker resistance) and the load current is related to the rms value of the ac component of collector current by the transformer turns ratio.

The ac power is then calculated from

$$P_S = P_L = I_L^2 R_L \qquad (8.15)$$

EXAMPLE 8.4 The circuit of Fig. 8.5a shows a transformer-coupled class-A audio power amplifier driving an 8-Ω speaker. The coupling transformer has a 3:1 step-down turns ratio. If the circuit component values result in a dc base current of 6 mA and the input signal (V_i) results in a peak base current swing of 4 mA, calculate the following circuit values using the transistor characteristic of Fig. 8.5b: $V_{CE_{max}}$, $V_{CE_{min}}$, $I_{C_{max}}$, $I_{C_{min}}$, the rms values of load current and voltage, and the ac power developed across the load. Calcu-

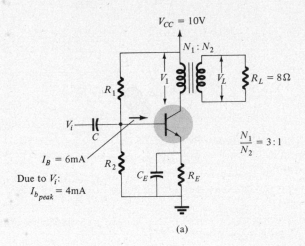

(a)

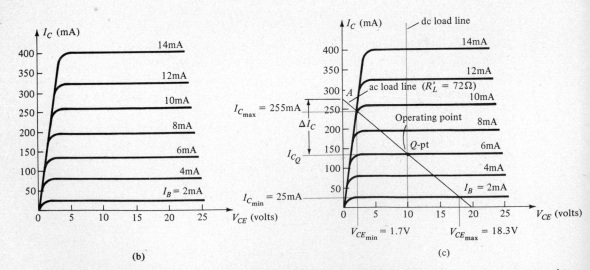

(b)   (c)

Figure 8.5. Transformer-coupled audio power amplifier and transistor characteristic for Example 8.4.

late the ac power using different equations as a check, that is, Eq. (8.11), Eq. (8.13), and Eq. (8.15). Draw the dc and ac load lines to obtain the voltages and currents in the collector side of the transformer.

Solution:

(a) The dc load line can be drawn vertically from the voltage point $V_{CE_Q} = V_{CC} = 10$ V (see Fig. 8.5c).

(b) For $I_B = 6$ mA the operating point on Fig. 8.5c is

$$V_{CE_Q} = 10 \text{ V} \qquad I_{C_Q} = 140 \text{ mA}$$

(c) The effective ac resistance R'_L is [use Eq. (8.7)]

$$R'_L = \left(\frac{N_1}{N_2}\right)^2 R_L = (3)^2 8 = 72 \text{ }\Omega.$$

(d) Draw the ac load line as follows: Use Eq. (8.8) to calculate the current swing above the operating current

$$\Delta I_C = \frac{\Delta V_{CE}}{R'_L} = \frac{10 \text{ V}}{72 \text{ }\Omega} = 139 \text{ mA}$$

Mark point A (Fig. 8.5c) $= I_{CE_Q} + \Delta I_C = 140 + 139 = 279$ mA along the y-axis. Connect point A to Q-point to draw ac load line.

(e) For the given peak base current swing of 4 mA, the maximum and minimum values of collector current and voltage obtained from Fig. 8.5c are

$$V_{CE_{min}} = 1.7 \text{ V} \qquad I_{C_{min}} = 25 \text{ mA}$$
$$V_{CE_{max}} = 18.3 \text{ V} \qquad I_{C_{max}} = 255 \text{ mA}$$

(f) Calculate the ac power across the transformer primary using Eq. (8.11)

$$P_o \text{ (ac)} = \frac{(V_{CE_{max}} - V_{CE_{min}})(I_{C_{max}} - I_{C_{min}})}{8}$$

$$= \frac{(18.3 - 1.7)(255 - 25) \times 10^{-3}}{8} = \mathbf{0.477 \text{ W}}$$

(g) Calculate the rms voltage across the primary

$$V_1 \text{ (rms)} = \frac{V_1 \text{ (peak-to-peak)}}{2\sqrt{2}} = \frac{V_{CE_{max}} - V_{CE_{min}}}{2\sqrt{2}}$$

$$= \frac{16.6}{2.828} = 5.87 \text{ V}$$

(h) The rms value of the load voltage is [using Eq. (8.12)]

$$V_L \text{ (rms)} = \left(\frac{N_2}{N_1}\right) V_1 \text{ (rms)} = \left(\frac{1}{3}\right)(5.87) = 1.96 \text{ V}$$

(i) Using Eq. (8.13) to calculate the ac power, we get

$$P_L \text{ (ac)} = \frac{V_L^2}{R_L} = \frac{(1.96)^2}{8} = \mathbf{0.480 \text{ W}}$$

(j) Using Eq. (8.14) to calculate the rms component of the load current, we get

$$I_L \text{ (rms)} = \left(\frac{N_1}{N_2}\right)\frac{(I_{C_{max}} - I_{C_{min}})}{2\sqrt{2}} = (3)\frac{(230 \times 10^{-3})}{2.828}$$

$$= 244 \text{ mA}$$

(k) Calculating the ac power using Eq. (8.15), we get

$$P_L(\text{ac}) = I_L^2 R_L = (244 \times 10^{-3})^2 8 = \textbf{0.434 W}$$

Power and Efficiency Calculations

So far we have considered calculating the ac power delivered to the load (the output ac power). We next consider the input power from the battery, power losses in the amplifier, and the overall power efficiency of the transformer-coupled class-A amplifier. The input dc power obtained from the battery is calculated from the values of dc battery voltage and average current drawn from the battery as in Eq. (8.1).

$$P_i(\text{dc}) = V_{CC} I_{C_Q}$$

For the transformer-coupled amplifier, as shown in Fig. 8.3, the power dissipated by the transformer is small and will be ignored in the present calculations. Thus for the transformer-coupled amplifier the only lost power is that dissipated by the power transistor as calculated by the following equation

$$\boxed{P_t = P_D = P_i - P_o} \tag{8.16}$$

when P_t is the power dissipated as heat by the active device (transistor in this case), also labelled P_D. The equation seems simple but is significant in operating a power amplifier. The amount of power dissipated by the transistor (which then sets the transistor power rating) is the difference between the average dc input power from the battery (which is a constant for a fixed battery and operating point) and the output ac power drawn by the load. If the output power is zero, then the transistor must handle the maximum amount, that set by the battery voltage and bias current. If the load does draw some of the power, then the transistor has to handle that much less (for the moment). In other words, the transistor has to work hardest (dissipate the most power) when the load is disconnected from the amplifier circuit and the transistor dissipates least power when the load is drawing maximum power from the circuit. Obviously, the safest rating of the transistor used is the maximum set when the load is disconnected. Since normal operation with the load connected requires the transistor to dissipate less power, it is always preferable to keep the load connected as long as the amplifier unit is turned on.

EXAMPLE 8.5 Calculate the efficiency of the amplifier circuit of Example 8.4. Also calculate the power dissipated by the transistor.

Solution: Using Eq. (8.1) to calculate the input power, we get

$$P_i = V_{CC} I_{C_Q} = (10)(140 \times 10^{-3}) = 1.4 \text{ W}$$

Using Eq. (8.16), we see that the power dissipated by the transistor is

$$P_D = P_i - P_o = 1.4 - 0.48 \cong \textbf{0.92 W}$$

$$\%\eta = \frac{P_o}{P_i} \times 100 = \frac{0.48}{1.4} \times 100 = \textbf{34.3\%}$$

Maximum Theoretical Efficiency

For a class-A amplifier the maximum theoretical efficiency for the series-fed circuit is 25% and for the transformer-coupled circuit it is 50%. From analysis of the operating range for a series-fed amplifier circuit the efficiency can be stated in the following form:

$$\% \, \eta = 25 \left[\frac{V_{CE_{\max}} - V_{CE_{\min}}}{V_{CC}} \right] \tag{8.17}$$

In practical operation the efficiency is less than 25%. In the circuit of Example 8.1 the efficiency was only 6.25%, indicating a poorly designed series-fed circuit.

The efficiency of a transformer-coupled class-A amplifier can be expressed by

$$\% \, \eta = 50 \left[\frac{(V_{CE_{\max}} - V_{CE_{\min}})}{2V_{CC}} \right]^2 = 50 \left[\frac{V_{CE_{\max}} - V_{CE_{\min}}}{V_{CE_{\max}} + V_{CE_{\min}}} \right] \tag{8.18}$$

The larger the value of $V_{CE_{\max}}$ and the smaller the value of $V_{CE_{\min}}$ the closer the efficiency approaches the theoretical limit of 50%. In the circuit of Example 8.4 the value obtained was 34.3%. Well-designed circuits can approach the limit of 50% closely so that the circuit of Fig. 8.5a would be considered average in operation. The larger the amount of power handled by the amplifier the more critical the efficiency becomes. For a few watts of power a simpler, cheaper circuit with less than maximum efficiency is acceptable (and sometimes desirable). For power levels in the tens to hundreds of watts, efficiency as close as possible to the theoretical maximum would be desired.

The value of 50% considered as the maximum for transformer-coupled amplifiers is only for class-A operation. There are additional ways of operating (biasing) the amplifier to obtain even higher efficiency, as will now be considered.

8.4 CLASSES OF AMPLIFIER OPERATION AND DISTORTION

Operating Classes

The only class of amplifier operation so far considered has been class A. By definition class-A operation provides collector (output) current flow during the complete cycle of the input signal (over a 360° interval). Figure 8.6a shows the output for class-A circuit operation. The bias level of current is I_{C_Q} and for the load line shown the output signal does not exceed values of $I_{C_{\max}}$ or $I_{C_{\min}}$, which would take the operation out of the linear region of device operation. Figure 8.6b shows *class-B* operation. The bias point is set at cutoff so that input signal variations driving the transistor into conduction will cause the output current to vary although input signal variations which tend to cut down the device conduction merely result

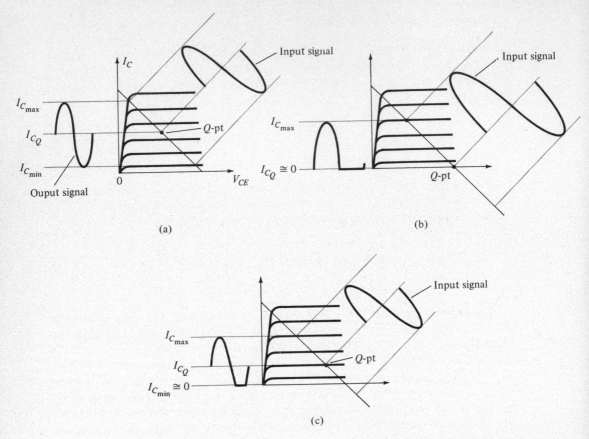

Figure 8.6. Various amplifier operating classes.

in the device remaining in cutoff. As shown in the figure, the output current flows for only about 180° of the cycle, this being the definition of class-B operation. Note that with no input signal the device is biased with no collector current flow and therefore no power dissipated by the transistor. Only when signal is applied does the transistor handle an average current which increases for larger input signals. Contrary to class-A operation, in which the worst condition occurs with no input signal and the least power is dissipated by the transistor for maximum input signal, the operation of a class-B circuit is to increase transistor dissipation for increased input signal. Since the average current in class-B operation is less than in class A, the amount of power dissipated by the transistor is less in class B.

A relation for the circuit efficiency of a class-B operating circuit is

$$\% \eta = 78.5\left(1 - \frac{V_{CE_{min}}}{V_{CC}}\right)$$

(8.19)

which shows that the efficiency approaches the theoretical maximum value of 78.5% as the value of $V_{CE_{min}}$ approaches 0 V and the supply voltage used is made larger.

In between class-A and class-B operation is *class-AB* operation, shown in Fig. 8.6c. The collector current flows for more than 180° of the signal cycle but less than 360°. The operating efficiency of class AB is between that of class A and class B, that is, between 50 and 78.5%, theoretic maximum.

Operation with the output conducting for less than 180° is called *class-C* operation and is found in resonant or tuned amplifier circuits, as, for example, in radio or television.

EXAMPLE 8.6 Determine the efficiency of the types of circuits indicated for the following operating conditions:
(a) Class-A operation with $V_{CE_{max}} = 30$ V and $V_{CE_{min}} = 3$ V.
(b) Class-B operation with $V_{CE_{min}} = 3$ V and $V_{CC} = 20$ V.

Solution: (a) Using Eq. (8.18), we get

$$\% \eta = 50 \left[\frac{V_{CE_{max}} - V_{CE_{min}}}{V_{CE_{max}} + V_{CE_{min}}} \right]^2 = 50 \left(\frac{30 - 3}{30 + 3} \right)^2 = \mathbf{33.5\%}$$

(b) Using Eq. (8.19), we get

$$\% \eta = 78.5 \left(1 - \frac{V_{CE_{min}}}{V_{CC}} \right) = 78.5 \left(1 - \frac{3}{20} \right) = \mathbf{66.7\%}$$

Distortion

Output signal variations of less than 360° of the signal cycle are considered to have *distortion*. This means that the output signal is no longer just an amplified version of the input signal but in some ways is distorted or changed from that of the input. The poor quality of music coming from a radio or hi-fi system with the music or voice no longer sounding like that which was originally recorded or transmitted is a result of distortion. Distortion can come from a number of different places in any audio system.

Distortion can occur because the device characteristic is not linear: *nonlinear or amplitude distortion*. This can occur with all classes of operation. In addition, the circuit elements and the amplifying device can respond to the signal differently at various frequency ranges of operation: *frequency distortion*.

When amplitude distortion occurs, the output signal no longer represents the input signal exactly. One technique of accounting for this change in the output signal is the method of *Fourier analysis*, which provides a means for describing a periodic signal in terms of its fundamental frequency component and frequency components at integer multiples—components called *harmonic components* or *harmonics*. For example, a signal which is originally 1000 Hz could result, after distortion, in a frequency component at 1000 Hz, and harmonic components at 2 kHz (2 × 1000 Hz), at 3 kHz (3 × 1000 Hz), 4 kHz (4 × 1000 Hz), and so on. The original frequency of 1000 Hz is called the fundamental frequency and those at integer multiples are the harmonics—that at 2 kHz is the second harmonic, the component at 3 kHz is the third harmonic, and so on. The fundamental signal is considered the first harmonic. (No harmonics at fractional amounts of the fundamental frequency exist using this technique.)

(a)

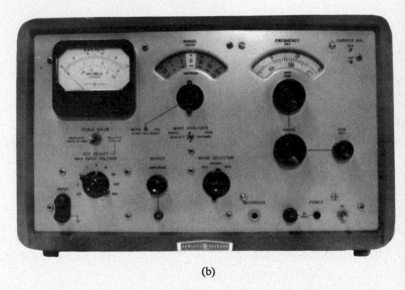

(b)

Figure 8.7. (a) Spectrum analyzer; (b) wave analyzer.

An instrument such as a spectrum analyzer would allow measurement of the harmonics present in the signal by providing display of the fundamental component of the signal and a number of its harmonics on a CRT screen as in Fig. 8.7a. Similarly, a wave analyzer instrument allows more precise measurement of the harmonic components of a distorted signal by filtering out each of these components and providing a calibrated dial reading of these components, one at a time (see Fig. 8.7b).

In any case the technique of considering any distorted signal as containing a fundamental component and harmonic components is practical and useful. For a signal occurring in class AB or class B the distortion may be mainly even harmonics, of which the second harmonic component is greatest. Thus, although the distorted signal contains all harmonic components from second harmonic on up, the most important in terms of the amount of distortion for the classes of operation we will consider is the second harmonic.

A current output waveform is shown in Fig. 8.8 with the quiescent, minimum, and maximum signal levels, and the times they occur, marked on the waveform. The signal shown indicates that some distortion is present. An equation which approximately describes the distorted signal waveform is

$$i_C \cong I_{C_Q} + I_O + I_1 \cos \omega t + I_2 \cos 2\omega t \qquad (8.20)$$

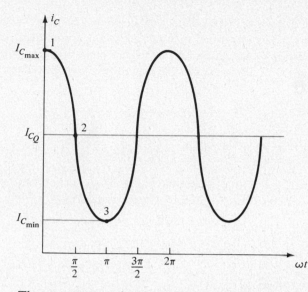

Figure 8.8. Waveform for obtaining second harmonic distortion.

The current waveform contains the original quiescent current I_{C_Q}, which occurs with zero input signal, an additional dc current I_O, due to the nonzero average of the distorted signal, the fundamental component of the distorted ac signal I_1, and a second harmonic component I_2, at twice the fundamental frequency. Although other harmonics are also present, only the second is considered here. Equating the resulting current from Eq. (8.20) at a few points in the cycle to that shown on the current waveform provides the following three relations:

At $\omega t = 0$

$$i_C = I_{C_{max}} = I_{C_Q} + I_O + I_1 \cos(0) + I_2 \cos(0)$$

$$I_{C_{max}} = I_{C_Q} + I_O + I_1 + I_2 \qquad (8.21)$$

At $\omega t = \pi/2$

$$i_C = I_{C_Q} = I_{C_Q} + I_O + I_1 \cos\left(\frac{\pi}{2}\right) + I_2 \cos\left(\frac{2\pi}{2}\right)$$

$$I_{C_Q} = I_{C_Q} + I_O - I_2 \qquad (8.22)$$

At $\omega t = \pi$

$$i_C = I_{C_{min}} = I_{C_Q} + I_O + I_1 \cos\left(\pi\right) + I_2 \cos\left(2\pi\right)$$

$$I_{C_{min}} = I_{C_Q} + I_O - I_1 + I_2 \qquad (8.23)$$

Solving Eq. (8.21), (8.22), and (8.23) simultaneously gives the following results

$$I_O = I_2 = \frac{I_{C_{max}} + I_{C_{min}} - 2I_{C_Q}}{4}$$

$$I_1 = \frac{I_{C_{max}} - I_{C_{min}}}{2} \qquad (8.24)$$

By definition the per cent of the second harmonic distortion is given by

$$\% D_2 \equiv \left| \frac{I_2}{I_1} \right| \times 100 \qquad (8.25)$$

The second harmonic distortion is the per cent of the second harmonic component present in the output current waveform with respect to the amount of the fundamental component. Obviously, 0% distortion is the ideal condition of no distortion.

Using the results in Eq. (8.24) to express the second harmonic distortion defined by Eq. (8.25) gives

$$\% D_2 = \left| \frac{\frac{1}{2}(I_{C_{\max}} + I_{C_{\min}}) - I_{C_Q}}{I_{C_{\max}} - I_{C_{\min}}} \right| \times 100 \qquad (8.26)$$

In a similar manner, the amount of second harmonic distortion can be related to the measured values of the distorted output voltage waveform

$$\% D_2 = \left| \frac{\frac{1}{2}(V_{CE_{\max}} + V_{CE_{\min}}) - V_{CE_Q}}{V_{CE_{\max}} - V_{CE_{\min}}} \right| \times 100 \qquad (8.27)$$

EXAMPLE 8.7 An output waveform displayed on a scope provides the following measured values:

(a) $V_{CE_{\min}} = 1\,V$, $\quad V_{CE_{\max}} = 22\,V$, $\quad V_{CE_Q} = 12\,V$
(b) $V_{CE_{\min}} = 4\,V$, $\quad V_{CE_{\max}} = 20\,V$, $\quad V_{CE_Q} = 12\,V$

For each set of values calculate the amount of the second harmonic distortion.

Solution: Using Eq. (8.27), we get

(a) $\%D_2 = \left| \dfrac{\frac{1}{2}(22 + 1) - 12}{22 - 1} \right| \times 100 = \mathbf{2.38\%}$

(b) $\%D_2 = \left| \dfrac{\frac{1}{2}(20 + 4) - 12}{20 + 4} \right| \times 100 = \mathbf{0\%}$ $\quad$ (no distortion)

The method used to obtain the amount of second harmonic distortion was called the three-point method since it involved equating the assumed form of the output voltage to the measured voltage at three points in the signal cycle. Using an assumed output signal equation containing more harmonic terms, along with choosing more points in the waveform, results in obtaining relations for the magnitude of the harmonic components at higher harmonic frequencies. Using a five-point method provides the dc component, first harmonic (fundamental), second harmonic, third harmonic, and fourth harmonic components. The harmonic distortion for each of these components is then defined as

$$D_2 = \left| \frac{I_2}{I_1} \right|, \qquad D_3 = \left| \frac{I_3}{I_1} \right|, \qquad D_4 = \left| \frac{I_4}{I_1} \right| \qquad (8.28)$$

The total distortion may be defined, in general, using the individual distortion components

$$D = \sqrt{D_2^2 + D_3^2 + D_4^2 + \cdots} \qquad (8.29)$$

When distortion does occur, the output power calculated by the undistorted case is no longer correct. Equation (8.11), for example, is true *only* for the nondistorted case. When distortion is present, the output power due to the fundamental component of the distorted signal is

$$P_1 = \frac{I_1^2 R_C}{2} \qquad (8.30)$$

The total output power due to all the harmonic components of the distorted signal is

$$P = (I_1^2 + I_2^2 + I_3^2 + \cdots)\frac{R_C}{2} \qquad (8.31)$$

The total power can also be expressed in terms of the total distortion

$$P = (1 + D_2^2 + D_3^2 + \cdots)I_1^2\frac{R_C}{2} = (1 + D^2)P_1 \qquad (8.32)$$

EXAMPLE 8.8 Using a five-point method to calculate harmonic components gives the following results: $D_2 = 0.1$, $D_3 = 0.02$, $D_4 = 0.01$, with $I_1 = 4$ A and $R_C = 8\ \Omega$. Calculate the total distortion, fundamental power component, and total power.

Solution: Total distortion, using Eq. (8.29), gives

$$D = \sqrt{D_2^2 + D_3^2 + D_4^2} = \sqrt{(0.1)^2 + (0.02)^2 + (0.01)^2} \cong 0.1$$

Fundamental power, using Eq. (8.30), gives

$$P_1 = \frac{I_1^2 R_C}{2} = \frac{(4)^2 8}{2} = 64 \text{ W}$$

Total power, using Eq. (8.32), gives

$$P = (1 + D^2)P_1 = [1 + (0.1)^2]64 = (1.01)64 = 64.64 \text{ W}$$

(Total power mainly due to fundamental component even with 10%, second harmonic distortion.)

Graphical Description of Harmonic Components of Distorted Signal

A demonstration of the use of harmonic components to represent a distorted signal is provided to help clarify the concept. As an example, a distorted waveform such as that resulting from class-B operation is shown in Fig. 8.9a. The signal is clipped on the negative half-cycle so that only the positive sinusoidal half-cycle provides an output signal.

Using Fourier analysis techniques, we can calculate a fundamental component of the distorted signal as shown in Fig. 8.9b. Figure 8.9b does not show the distorted waveform, only the fundamental component (which is a perfectly sinusoidal signal itself). Similarly, the second and third harmonic components can be obtained and are shown in Figs. 8.9c and 8.9d, respectively.

We now wish to check whether these components, each a purely undistorted

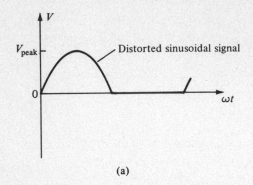

(a)

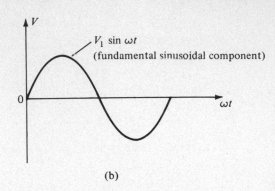

(b)

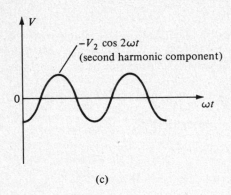

(c)

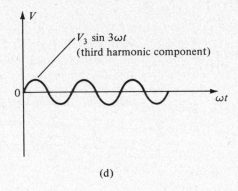

(d)

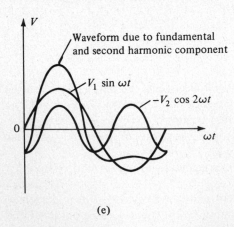

(e)

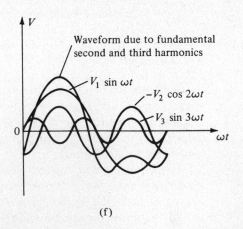

(f)

Figure 8.9. Graphical representation of a distorted signal through the use of harmonic components.

CHAP. 8 LARGE-SIGNAL AMPLIFIERS

sinusoidal signal, add up approximately to the original distorted signal. Figure 8.9e shows the resulting waveform when adding the fundamental and second harmonic components together. Note the flattening of the second half of the cycle. In Fig. 8.9f the third harmonic component is added to give a resulting waveform that comes fairly close to the original distorted signal. The addition of higher harmonic components of the right amplitude and correct phase will further alter the resulting waveform to approximate the original distorted signal. In a relatively simple manner we can observe that addition of a fundamental component and harmonic components can result in the original distorted waveform. In general, any periodic waveform can be represented by a fundamental component and harmonic components, each of varying amplitudes and at various phase angles.

The concept of harmonics is useful is both analyzing distorted (nonsinusoidal) waveforms and in providing a means of working with such signals. Since all the harmonic components are sinusoidal signals, we can separately consider the effect of each component on the circuit and obtain the total effect using superposition—adding together the voltages or currents being considered. Particular mention of harmonic content of a waveform will be made in the discussion of push-pull amplifiers, where it will be shown that the particular circuit connection eliminates the even harmonic components and leaves only the fundamental component and the odd harmonic components. Since the largest distortion components is the second harmonic, the elimination of that component will considerably reduce the total amount of distortion.

8.5 PUSH-PULL AMPLIFIER CIRCUIT

Previous discussion has shown that efficient operating conditions occur for class-AB or class-B operation. On the other hand, class-AB or class-B operation, as discussed so far, results in considerable distortion. To be forced to choose one or the other—more efficient operation or less distortion—is not the best of choices. Ideally, one wants to have the efficient operation of class-B but also the low distortion of class-A operation. This is not realistic, but a surprisingly low-distortion, high-efficiency operation can be obtained using the circuit connection known as *push-pull*. A typical transistor push-pull circuit connection is shown in Fig. 8.10. Similar circuits are also made using vacuum tubes or FETs.

The circuit of Fig. 8.10 requires an input transformer to produce opposite polarity signals to the two transistor inputs and an output transformer to drive the load in a push-pull mode of operation to be described. Figure 8.11a shows the input transformer with center tap at ground. As an example, the voltage across the secondary from the plus (+) to the minus (−) terminal is 100 V, peak. As shown in Fig. 8.11a, the voltages across each half of the transformer are 50 V, peak, adding up to the total of 100 V, peak across the transformer. With the center tap of the transformer connected to ground (0-V potential) the signal observed from the plus terminal to ground is, as shown in Fig. 8.11a, in phase with the voltage across the full transformer. However, since the voltage measured across the bottom half of the transformer is also in phase with the total signal when measured from center

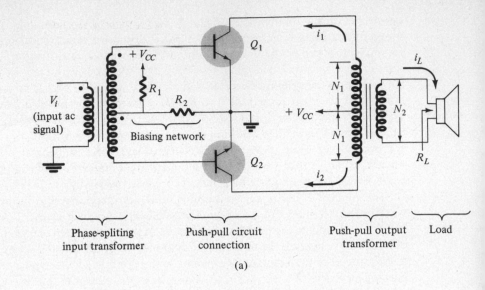

Phase-spliting
input transformer

Push-pull circuit
connection

Push-pull output
transformer

Load

(a)

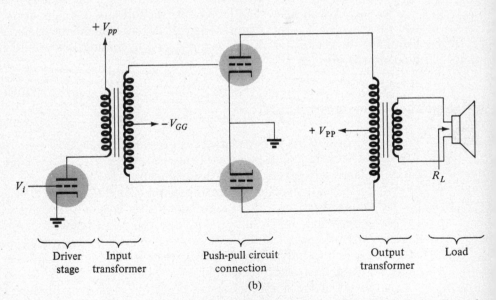

Driver
stage

Input
transformer

Push-pull circuit
connection

Output
transformer

Load

(b)

Figure 8.10. Push-pull circuit.

minus, it is opposite when measured from minus to center tap (or from minus to tap to ground). The signals at the plus and minus terminals measured with respect to ground are therefore opposite in phase as shown in Fig. 8.11a.

Having obtained opposite-phased input to the two transistor units the push-pull nature of the circuit operation can be considered as shown in the partial circuit diagram of Fig. 8.11b. Consider first the dc bias current (I_{C_Q}) for each transformer. Figure 8.11b shows that the bias currents for each transistor flow in opposite directions through the transformer winding. The magnetic flux set up by each of these currents results in opposite flux through the magnetic core so that the net flux, in the perfectly matched case, is zero. Thus the transformer need not handle a large flux due to the dc bias currents, resulting in a smaller size core, biased to operate near zero flux.

Considering only the first half-cycle of operation, transistor Q_1 is driven further into conduction whereas transistor Q_2 is driven less into conduction. The varying component of current for each transformer is marked I_1 in Fig. 8.11b. Since Q_1 is driven further into conduction, the varying component of current flows in the same direction as the dc bias current, resulting in a larger total current. The varying component of current flow in Q_2, however, flows in an opposite direction to the bias current for that transistor, resulting in a net decrease in current flow for transistor Q_2. Note from Fig. 8.11b that the overall operation results in a net current flow through the transformer. The input signals to each stage being of opposite phase results in a net output signal across the transformer. Had in-phase signals been

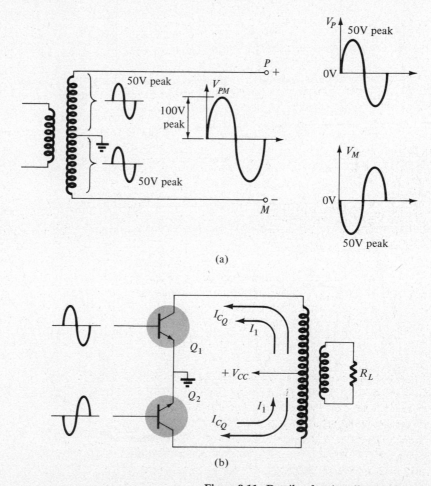

(a)

(b)

Figure 8.11. Details of push-pull operation.

applied to the transistor inputs, the net output signal for the varying components would have been zero. We now need to consider how the push-pull connection with class-AB or class-B operation of the transistors will provide low-distortion output.

Figure 8.12a shows the output signal for each section of the push-pull circuit when operated in class B. Note that since the two transistors are identically biased, they distort the output waveform on the negative half-cycle for each signal—the second half-cycle for stage 1 and the first half-cycle for stage 2, as shown, Intui-

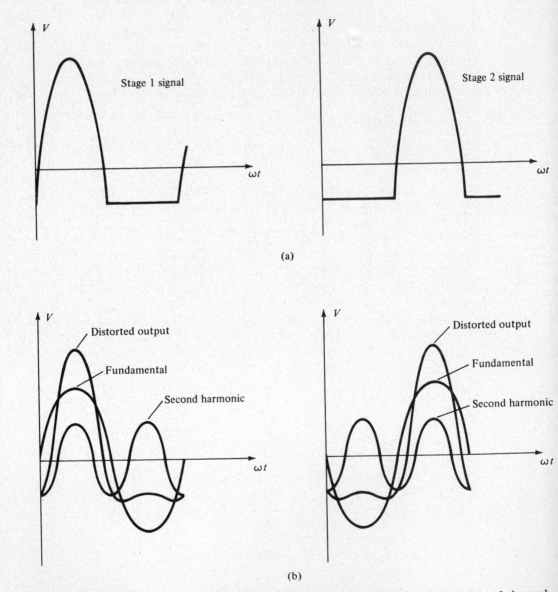

(a)

(b)

Figure 8.12. Distorted output waveforms for each section of the push-pull circuit operation.

tively, one might say that the positive half-cycle due to stage 1 provides the output signal during the first half-cycle and that the positive half-cycle signal operates stage 2 during the second half-cycle. Since the two stages would cause current flow in opposite directions, the two half-cycles, as indicated in Fig. 8.12a, would result in opposite half-cycles of signal in the transformer—thus providing a full or complete cycle of signal flow even with half-cycles of operation of each stage.

For a more detailed and meaningful discussion of the circuit operation the approximate distorted output waveforms in each stage of the circuit are shown in Fig. 8.12b as being made up of a fundamental component and a second harmonic component. Although other harmonic components would be necessary to describe more fully the actual distorted waveforms of Fig. 8.12a, consideration of only the fundamental and second harmonic components will provide the necessary information here.

Notice in Fig. 8.12b that the fundamental components of each stage are oppo-site in phase. For the push-pull circuit connection, opposite-phase output signals will result in a net voltage in the secondary which is the sum of the two component signals. The second harmonic components, however, are in phase and therefore cancel, providing a net voltage in the secondary of 0 V due to these signal compo-nents. In other words, the action of the push-pull connection would be to cancel out the second harmonic components of the signals applied to the primary of the out-put transformer, and the same cancellation can be shown to result for all other even harmonic components. From our previous consideration of harmonics we can now appreciate that a distorted signal applied to the push-pull transformer connec-tion, such as those shown in Fig. 8.12a, would result in cancellation of all even harmonic components so that the resulting output signal across the secondary can be considered to be made up of the fundamental component, third harmonic com-ponent, and all odd harmonic components of the distorted signal. Since the second harmonic is the largest component for the distorted signal of Fig. 8.12a, elimina-tion of that component will result in an output signal having considerably less distortion. In calculating the distortion of the output signal we need to consider the amount of the third, fifth, and so on, harmonic components, which we expect will not be too large, so that the total distortion as calculated by Eq. (8.30) will not be much more than the small amount of the third harmonic distortion.

Of course, the above conclusion of even harmonic cancellation is an ideal con-dition which will occur only if the circuit is perfectly balanced—transistors exactly matched, transformer center-tap connection perfect, and input signals exactly equal and opposite. This will never practically occur, but still the circuit can be operated class AB or class B for improved power efficiency with low distortion occurring (although more than for class-A operation).

To summarize, then, the advantages of the push-pull circuit connection are

1. The dc components of the collector currents oppose each other in the trans-former resulting in no net flux due to the bias current and, hence, smaller size cores can be used.

2. Class-AB or class-B operation is possible with small resulting distortion due to the cancellation of all the even harmonic components.

3. Ripple voltage in the voltage supply is cancelled out by the push-pull operation of the circuit.

4. High efficiency operation is possible by using class-AB or class-B biasing.

To complete the picture it must be pointed out that the power supply used in class-B operation must have good voltage regulation since the amount of current drawn is about zero for no signal operation and rises in value for larger amounts of signal level. Since the current drawn from the supply ranges considerably from not load to full load, the supply voltage regulation must be good. Also, hum voltages (60-Hz pickup), which are picked up on the input lines of the circuit, are *not* eliminated by the push-pull connection since these are brought along with the input signals to the driver transformer and connected out-of-phase as input to the push-pull circuit, thereby acting as proper input signals to drive the output device.

8.6 VARIOUS PUSH-PULL CIRCUITS INCLUDING TRANSFORMERLESS CIRCUITS

Although the circuit shown in Fig. 8.10 is the most common form of the push-pull circuit connection, a number of other circuit arrangements are possible We shall consider a few of these and their various advantages and disadvantages. It is important to keep in mind the overall operation of the circuit in order to appreciate the different methods of obtaining the advantages of push-pull operation. For the push-pull circuit it is necessary to develop the output voltage across the load in such a manner that two stages operating in class B will still provide a full cycle of signal by conducting on alternate half-cycles.

Starting with an input signal obtained from a driver amplifier stage, it is necessary to operate the two-stage push-pull circuit on alternate half-cycles for class-B operation. The opposite-phase input signals to the two stages of the push-pull circuit can be obtained in a number of ways. Figure 8.10 shows the use of an input transformer to provide the phase inversion between the two push-pull input signals. Another means of obtaining opposite-phase input signals is to use the paraphase circuit of Fig. 8.13. The input signal applied to the base appears at the collector, 180° out of phase. The output from the emitter is in phase with the input so that the two output signals are out of phase as shown in Fig. 8.13. Values of R_C, R_E, and h_{fe} can be chosen to make the voltage gain for the collector output signal to be 1. The gain for the signal taken from the emitter is 1 for the emitter-follower operation. Thus, the circuit provides no net gain but results in opposite-phase signals to drive the push-pull amplifier stage. The advantage of this driver connection is the savings on the use of a center-tapped transformer which is expensive and bulky and has a limited frequency operating range. A disadvantage is that the two signals do not come from similar impedance sources. The signal from the emitter provides a good driver connection since the source resistance viewed from the emitter is low. The collector circuit resistance, however, is relatively high and although the unloaded output signals are equal, they are different under load conditions. One possible improvement would be to add an additional emitter-follower stage to

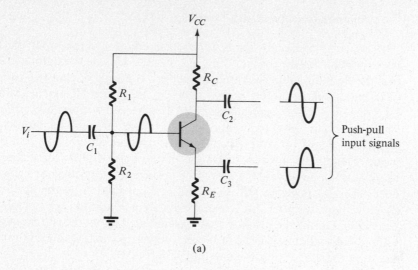

(a)

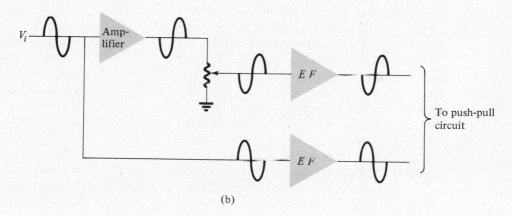

(b)

Figure 8.13. Phase-splitter circuits.

connect the output to the load since such stage would provide no additional voltage gain or phase inversion and would drive the push-pull stage from a low-resistance source.

Another means of obtaining opposite-phase signals to drive the push-pull stage is illustrated by the block diagram of Fig. 8.13b. The input signal is amplified and inverted by one amplifier stage and then attenuated for an overall gain of unity. The use of two emitter followers (possibly Darlington circuits) drives the push-pull stage from low-impedance sources.

One even more practical and popular circuit for obtaining the out-of-phase signals is using the difference amplifier discussed in Chapter 11. As will be shown in that chapter, the use of a difference amplifier requires only the output transformer to provide the complete push-pull circuit operation.

Complementary Symmetry Circuits

A number of circuits go beyond eliminating only the input phase inverting transformer from the circuit. These circuits also remove the output transformer so that the circuit is completely transformerless. A simple version of a transformerless push-pull amplifier circuit is shown in Fig. 8.14. Complementary transistors are used, that is, *npn* and *pnp* transistors are used instead of using two of the same type. The single input signal required is applied to both base inputs. However, since the transistors are of opposite type, they will conduct on opposite half-cycles of the input. During the positive half-cycle of the input signal, for example, the *pnp* transistor will be reverse biased by the positive half-cycle signal and will not conduct. The *npn* transistor will be biased into conduction by the positive half-cycle signal with a resulting half-cycle of output across the load resistor (R_L) as shown in Fig. 8.14b. During the negative half-cycle of input signal the *npn* transistor is biased off and the output half-cycle developed across the load is due to the operation of the *pnp* transistor at this time, as shown in Fig. 8.14c.

During a complete cycle of the input, a complete cycle of output signal is developed across the load. It should be obvious that one disadvantage of this circuit connection is the need for two supply voltages. Another, less obvious, but important, disadvantage with the complementary circuit as shown is the resulting *crossover* distortion in the output signal. Crossover distortion refers to the fact that during the signal crossover from positive to negative (or vice-versa) there is some nonlinearity in the output signal as indicated in Fig. 8.14d. This results from the fact that for the simple circuit shown in Fig. 8.14a the operation of the circuit does not provide exact switching of one transistor *off* and the other *on* at the zero voltage condition. Both may be off or partially conducting so that the output voltage is not exactly following the input and distortion occurs. This occurrence at the crossover point is of concern for the push-pull circuit of Fig. 8.10 as well, although not necessarily to the same degree. Bias of the transistors in class AB improves the operation by biasing the transistors so that each stays on for more than half of the cycle. For the circuit of Fig. 8.14a considerable effort must be made to reduce the crossover distortion. More practical circuit connections include additional biasing components in the base circuit to try to effect this improved operation.

Note that the load is driven as the output of an emitter-follower circuit so that the low resistance of the load is matched by low resistance from the driving source. Improved versions of the complementary circuit include the transistors, each connected in the Darlington arrangement, to provide even lower driver resistance than that with single transistors. The circuit of Fig. 8.14e shows a practical circuit connection using the Darlington connection of the transistors and additional emitter resistors for temperature bias stabilization.

Another transformerless push-pull circuit, shown in Fig. 8.15a, uses the same type of transistors rather than complementary transistors. However, the inputs applied to the circuit must be opposite in phase so that some phase-splitter circuit is required. Figures 8.15b and 8.15c show the operation of the transistors for alternate halves of the cycle and the resulting polarity of the signal across the load resistor.

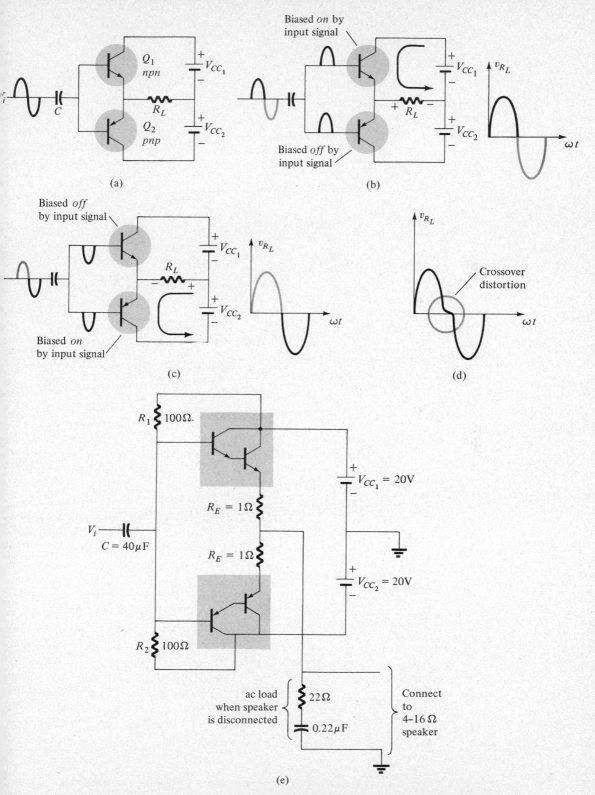

Figure 8.14. Complementary symmetry push-pull circuit.

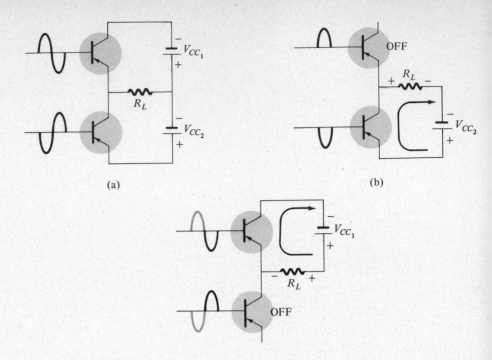

(a)

(b)

(c)

Figure 8.15. Transformerless push-pull circuit.

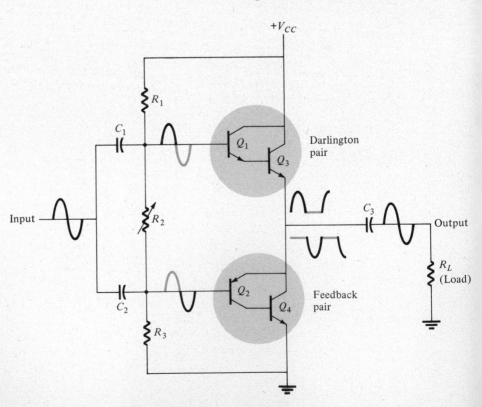

Figure 8.16. Quasi-complementary push-pull transformerless power amplifier.

Quasi-complementary Push-Pull Amplifier

The push-pull circuit form is achieved in the circuit of Fig. 8.16 by using complementary transistors (Q_1 and Q_2) before the power output, so that both power output transistors can be *npn* types. This is a practically preferable arrangement, for the *npn* power transistors are presently the best available. Notice that transistors Q_1 and Q_3 form a *Darlington connection* that provides output at a low-impedance level from the emitter. The connection of transistors Q_2 and Q_4 forms a *feedback pair*, which similarly provides a low-impedance drive to the load. Resistor R_2 can be adjusted to minimize crossover distortion. The single signal applied as input to the push-pull stage then results in a full cycle output to the load R_L, each half of the circuit operating class-B for efficient power operation.

Power and Efficiency Calculations in Class-B

The power and efficiency calculations of a variety of class-B power amplifiers should help in understanding how these circuits operate and provide some comparison between important circuit values.

Input dc Power

The power provided to the output speaker of a power amplifier circuit is drawn from the power supply (or power supplies) and is considered an input or dc power. The amount of this input power can be calculated from

$$P_i \,(\text{dc}) = V_{CC} I_{\text{dc}} \tag{8.33}$$

where I_{dc} is the average or dc current drawn from the power supply.

In class-B operation the current drawn from a single power supply is a full-wave rectified signal, while that drawn from a circuit having two power supplies is a half-wave rectified waveform from each supply. In either case the value of average power can be expressed as

$$I_{\text{dc}} = \frac{2}{\pi} I_{\text{peak}} \tag{8.34}$$

where I_{peak} is the peak value of the output current waveform.

Output ac Power

The power delivered to the load (usually referred to as a resistance, R_L) can be calculated from any one of a number of equal relations.

$$\boxed{P_o \,(\text{ac}) = \frac{V_L^2(p-p)}{8R_L} = \frac{V_L^2(p)}{2R_L} = \frac{V_L^2\,(\text{rms})}{R_L}} \tag{8.35}$$

Power Dissipated by Output Transistors

The power dissipated (as heat) by the output power transistors is the difference between the input dc power and the output ac power delivered to the load.

$$P_{2Q} = P_i \,(\text{dc}) - P_o(\text{ac}) \tag{8.36}$$

where P_{2Q} is the power dissipated by the *two* output power transistors.

The circuit's power efficiency is then calculated as

$$\% \, \eta = \frac{P_o \, (\text{ac})}{P_i \, (\text{dc})} \times 100 \%$$ (8.37)

Maximum Power Considerations

For class-B operation the maximum output power to the load is dependent on the value of the supply voltage.

$$\boxed{\text{maximum } P_o \, (\text{ac}) = \frac{V_L^2(p)}{2R_L} = \frac{V_{CC}^2}{2R_L}}$$ (8.38)

A corresponding ac current signal developed through the load varies to a peak value of

$$I_{\text{peak}} = \frac{V_{CC}}{R_L}$$

so that the average current from the power supply is

$$I_{\text{dc}} = \frac{2}{\pi} I_{\text{peak}} = \frac{2}{\pi} \cdot \frac{V_{CC}}{R_L}$$ (8.39)

The input power drawn by the circuit is then

$$\boxed{\text{maximum } P_i \, (\text{dc}) = V_{CC} I_{\text{dc}} = V_{CC}\left(\frac{2}{\pi} \cdot \frac{V_{CC}}{R_L}\right)}$$ (8.40)

The maximum circuit efficiency for class-B operation is then

$$\% \, \eta = \frac{P_o \, (\text{ac})}{P_i \, (\text{dc})} \, 100 = \frac{\dfrac{V_{CC}^2}{2R_L}}{V_{CC}\left(\dfrac{2}{\pi} \dfrac{V_{CC}}{R_L}\right)} \, 100 = \frac{\pi}{4} \cdot 100 = 78.54 \%$$ (8.41)

When the input signal results in less than the maximum output signal swing, the circuit efficiency is less than 78.5%. For class-B operation the maximum power dissipated by the output transistors *does not* occur at the maximum efficiency condition. The maximum power dissipated by the two output transistors occurs when the output voltage across the load is $0.636 V_{CC}$ $(= (2/\pi) V_{CC})$ and is

$$\boxed{\text{maximum } P_{2Q} = \frac{2}{\pi^2} \cdot \frac{V_{CC}^2}{R_L}}$$ (8.42)

EXAMPLE 8.9 For the circuit of Fig. 8.17
  (a) Calculate the input and output power handled by the circuit and the power dissipated by *each* output transistor for an input of 12 V, rms.
  (b) If the input signal is increased to provide the maximum undistorted output, calculate the values of maximum input and output power and the power dissipated by *each* output transistor for this condition.
  (c) Calculate the maximum power that *each* output transistor will have to handle.

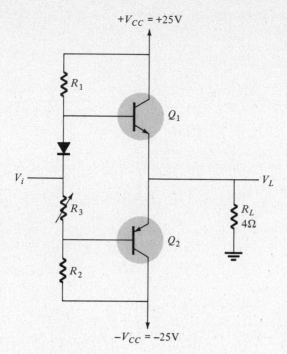

$+V_{CC} = +25\text{V}$

R_1

Q_1

V_i

R_3

R_L
4Ω

V_L

Q_2

R_2

$-V_{CC} = -25\text{V}$

Figure 8.17. Class-B power amplifier for Example 8.9.

Solution:

(a) The peak input voltage is

$$V_i(p) = \sqrt{2}\,V_i\,(\text{rms}) = \sqrt{2}\,(12) = 16.97 \cong 17\text{ V}$$

Since the resulting voltage across the load is ideally the same as the input signal (amplifier has, ideally, a voltage gain of unity),

$$V_L(p) = 17\text{ V}$$

$$P_o\,(\text{ac}) = \frac{V_L^2(p)}{2R_L} = \frac{(17)^2}{2(4)} = \textbf{36.125 W}$$

$$I_L(p) = \frac{V_L(p)}{R_L} = \frac{17\text{ V}}{4\ \Omega} = 4.25\text{ A}$$

The dc current drawn from the two power supplies is then

$$I_{dc} = \frac{2}{\pi}I_L(p) = \frac{2(4.25\text{ A})}{\pi} = 2.71\text{ A}$$

so that the power supplied to the circuit is

$$P_i\,(\text{dc}) = V_{CC}I_{dc} = (25\text{ V})(2.71\text{ A}) = \textbf{67.6 W}$$

The circuit efficiency (for an input of $V_i = 12$ V, rms) is

$$\%\,\eta = \frac{P_o\,(\text{ac})}{P_i\,(\text{dc})} \cdot 100 = \frac{36.125\text{ W}}{67.6\text{ W}} \cdot 100 = 53.4\%$$

and the power dissipated by each output transistor is

$$P_Q = \frac{P_{2Q}}{2} = \frac{P_i\,(\text{dc}) - P_o\,(\text{ac})}{2} = \frac{67.6 - 36.125\text{ W}}{2} = \textbf{14.76 W}$$

(b) If the input signal is increased to $V_i = 25$ V, *peak* ($V_i = 17.68$ V, rms) so that $V_L(p) = V_{CC} = 25$ V, the calculations are

$$\text{maximum } P_o \text{ (ac)} = \frac{V_{CC}^2}{2R_L} = \frac{(25)^2}{2(4)} = \textbf{78.125 W}$$

$$\text{maximum } P_i \text{ (dc)} = \frac{2}{\pi} \frac{V_{CC}^2}{R_L} = \frac{2}{\pi} \frac{(25)^2}{4} = \textbf{99.47 W}$$

and

$$\% \, \eta = \frac{P_o \text{ (ac)}}{P_i \text{ (dc)}} \cdot 100 = \frac{78.125}{99.74} \cdot 100 = 78.54\%$$

(the maximum circuit efficiency).

At this maximum signal condition the power dissipated by *each* output transistor is

$$P_Q = \frac{P_{2Q}}{2} = \frac{P_i \text{ (dc)} - P_o \text{ (ac)}}{2} = \frac{99.47 - 78.125}{2} = \textbf{10.67 W}$$

(Note that this is less than the power dissipated with a smaller input signal.)[2]

(c) The maximum power dissipation required of the output transistors is

$$\text{maximum } P_{2Q} = \frac{2}{\pi^2} \frac{V_{CC}^2}{R_L} = \frac{2}{\pi^2} \frac{(25)^2}{4} = 31.66 \text{ W}$$

so that

$$P_Q = \frac{P_{2Q}}{2} = \frac{31.66}{2} = \textbf{15.83 W}$$

8.7 POWER TRANSISTOR HEAT SINKING

Recent trends in electronics have been to replace individual transistors by complete integrated circuits for small-signal and low-power applications. Most high-power applications, however, still require individual power transistors. Improvements in production techniques have provided higher power ratings in smaller-sized packaging cases, have increased the maximum transistor breakdown voltage, and have provided faster switching power transistors.

Some discussion of power transistor rating occurred in Chapter 3. The maximum power handled by a particular device and the temperature of the transistor junctions are related since the power dissipated by the device causes an increase in temperature at the junctions of the device. Obviously, a 100-W transistor will provide more power capability than a 10-W transistor. On the other hand, proper heat sinking techniques will allow operation of a device closer to its maximum power rating.

We should note that of the two types of transistors—germanium and silicon—silicon transistors provide greater maximum temperature ratings. Typically, the maximum junction temperature of these types of power transistors is

germanium $100-110°C$

silicon $150-200°C$

---

[2]The power dissipation of the output transistors actually increases until the condition of $V_i = (2/\pi) V_{CC}$ after which the power dissipation decreases to a value of $(2/\pi - 1/2) \cdot V_{CC}/R_L$ at maximum circuit efficiency.

For many applications the average power dissipated may be approximated by

$$P_D = V_{CE}I_C$$

This power dissipation, however, is only allowed up to some maximum temperature. Above this temperature the device power dissipation capacity must be reduced (or *derated*) so that at higher case temperatures the power handling capacity is reduced—down to 0 W at the device maximum case temperature.

The greater the power handled by the transistor (dependent on power level set by the circuit) the higher the case temperature of the transistor. Actually, the limiting factor in power handled by a particular transistor is the temperature of the device collector junction. Power transistors are mounted in large metal cases to provide a large area from which the heat generated by the device may radiate. Even

so, operating a transistor directly into air (mounting it on a plastic board, for example) severely limits the device power rating. If, instead (as is usual practice), the device is mounted on some form of *heat sink*, its power handling capacity can approach the rated maximum value more closely. A few heat sinks are shown in Fig. 8.18. When the heat sink is used, the heat produced by the transistor dissipating power has a larger area from which to radiate the heat into the air, thereby holding the case temperature to a much lower value than would result without the heat sink. Even with an infinite heat sink (which, of course, is not available), for

Figure 8.18. Typical power heat sinks.

which the case temperature is held at the *ambient* (air) temperature, the junction will be heated above the case temperature and a maximum power rating must be considered.

Since even a good heat sink cannot hold the transistor case temperature at ambient (which, by the way could be more than 25°C if the transistor circuit is in a confined area where other devices are also radiating a good deal of heat), it is necessary to derate the amount of *maximum power* allowed for a particular transistor as a function of increased case temperature.

Figure 8.19 shows typical power derating curves for silicon transistors. The

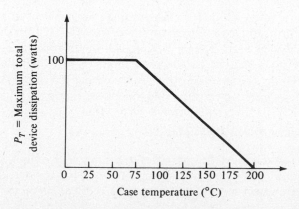

Figure 8.19. Typical power derating curve for silicon transistors.

curves show that the manufacturer will specify an upper temperature point (not necessarily 25°C) after which a linear derating takes place. For silicon the maximum power that should be handled by the device does not reduce to 0 W until a case temperature of 200°C (or 150° in some devices).

It is not necessary to provide a derating curve since the same information could be given simply as a listed derating factor on the device specification sheet. For example, a derating factor for a germanium transistor may be stated as follows:

Derate linearly to 100°C case temperature at the rate of 1 watt per degree above 50°C.

For a 50-W transistor (rated below 50°C) this means that at 80°C the maximum power rating of the device must be reduced by

$$(80°C - 50°C)\left(1\frac{W}{°C}\right) = 30 \text{ W}$$

so that the rated power dissipated should only be

$$50 \text{ W} - 30 \text{ W} = 20 \text{ W}$$

at 80° case temperature. Stated mathematically,

$$P_D(\text{temp}_1) = P_D(\text{temp}_0) - (\text{Temp}_1 - \text{Temp}_0)(\text{Derating factor})$$

where the value of Temp_0 is the temperature at which derating should begin, the value of Temp_1 is the particular temperature of interest (above the value Temp_0), $P_D(\text{temp}_0)$ and $P_D(\text{temp}_1)$ are the maximum power dissipations at the temperatures specified, and the derating factor is the value given by the manufacturer in units of watts (or milliwatts) per degree of temperature.

> EXAMPLE 8.10 Determine what maximum dissipation will be allowed for an 80-W silicon transistor (rated at 25°C) if derating is required above 25°C by a derating factor of 0.5 W/°C at a case temperature of 125°C.
>
> **Solution:**
> $$P_D(125°C) = P_D(25°C) - (125°C - 25°C)(0.5 \text{ W/°C})$$
> $$= 80 \text{ W} - 100(0.5) = 30 \text{ W}$$

It is interesting to note what power rating results using a power transistor without a heat sink. For example, a silicon transistor rated at 100 W at (or below) 100°C is rated at only 4 W at (or below) 25°C, free-air temperature. Thus, operated without a heat sink the device can handle a maximum of only 4 W at a room temperature of 25°C. Using a heat sink large enough to hold the case temperature to 100°C at 100 W allows operation at the maximum power rating.

Thermal Analogy of Power Transistor

Selection of a suitable heat sink requires a considerable amount of detailed determination which is not appropriate to our present basic considerations of the power transistor. However, more detail about the thermal characteristics of the transistor and its relation to the power dissipation of the transistor may help provide a clearer understanding of power as limited by temperature. The following discussion should provide some background information.

A picture of how the junction temperature (T_J), case temperature (T_C), and ambient (air) temperature (T_A) are related by the device heat handling capacity—a temperature coefficient usually called *thermal resistance*—is presented in the thermal-electrical analogy shown in Fig. 8.20.

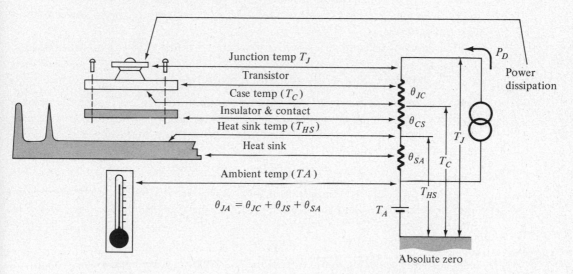

Figure 8.20. Thermal-to-electrical analogy.

In providing a thermal-electrical analogy the term thermal resistance is used to describe heat effects by an electrical term. The terms in Fig. 8.20 are defined as follows:

θ_{JA} = total thermal resistance (junction to ambient)
θ_{JC} = transistor thermal resistance (junction to case)
θ_{CS} = insulator thermal resistance (case to heat sink)
θ_{SA} = heat sink thermal resistance (heat sink to ambient)

Using the electrical analogy for thermal resistances, we can write

$$\theta_{JA} = \theta_{JC} + \theta_{CS} + \theta_{SA} \tag{8.43}$$

The analogy can also be used in applying Kirchhoff's law to obtain

$$T_J = P_D \theta_{JA} + T_A \tag{8.44}$$

The last relation shows that the junction temperature "floats" on the ambient temperature and that the higher the ambient temperature the lower the allowed value of device power dissipation.

The thermal factor θ provides information about how much temperature drop (or rise) results for a given amount of power dissipation. For example, the value of θ_{JC} is usually about 0.5°C/W. This means that for a power dissipation of 50 W the difference in temperature between case temperature (as measured by a thermocouple) and the inside junction temperature is only

$$T_J - T_C = \theta_{JC}P_D = (0.5°C/W)(50\ W) = 25°C$$

Thus, if the heat sink can hold the case at, say 50°C, the junction is then only at 75°C. This is a relatively small temperature difference, especially at lower power dissipation levels.

The value of thermal resistance from junction to free air (using no heat sink) is, typically,

$$\theta_{JA} = 40°C/W \qquad \text{(into free air)}$$

Note that in this case only 1 W of power dissipation results in a junction temperature 40°C greater than the ambient. A silicon transistor operating into an ambient temperature of 25°C could not dissipate 5 W without exceeding the junction temperature limit of 100°C.

$$T_J = T_A + \theta_{JA}P_D = 25°C + (40°C/W)(5\ W) = 225°C$$

A heat sink can now be seen to provide a low thermal resistance between case and air—much less than the 40°C/W value of the transistor case alone. Using a heat sink having

$$\theta_{SA} = 2°C/W$$

and, with an insulating thermal resistance (from case to heat sink) of

$$\theta_{CS} = 0.8°C/W$$

and, finally for the transistor,

$$\theta_{CJ} = 0.5°C/W$$

we can obtain

$$\theta_{JA} = \theta_{SA} + \theta_{CS} + \theta_{CJ}$$
$$= 2.0 + 0.8 + 0.5 = 3.3°C/W$$

So, with a heat sink, the thermal resistance between air and junction is only 3.3°C/W as compared to, say 40°C/W for the transistor operating directly into free air. Using the value of θ_{JA} above for a transistor operated at, say, 2 W we calculate

$$(T_J - T_A) = \theta_{JA}P_D = (3.3°C/W)(2\ W) = 6.6°C$$

In other words, the use of a heat sink in this example provided only a 6.6°C increase in junction temperature as compared to an 80°C rise without a heat sink.

EXAMPLE 8.11 A silicon power transistor is operated with a heat sink ($\theta_{SA} = 1.5°C/W$). The transistor, rated at 150 W (25°C), has $\theta_{JC} = 0.5\ °C/W$ and the mounting insulation has $\theta_{CS} = 0.6°C/W$. What maximum power can be dissipated if the ambient temperature is 40°C and $T_{J_{max}} = 200°C$.

Solution:

$$P_D = \frac{T_J - T_A}{\theta_{JC} + \theta_{CS} + \theta_{SA}} = \frac{200 - 40}{0.5 + 0.6 + 1.5} = \frac{160°C}{2.6°C/W} \cong \mathbf{61.5\ W}$$

PROBLEMS

§ 8.3

1. A class-A transformer-coupled amplifier uses a 25:1 transformer to drive a 4-Ω load. Calculate the effective ac load (seen by the transistor connected to the larger turns side of the transformer).

2. What turns ratio transformer is needed to couple to an 8-Ω load so that it appears as a 10-K effective load?

3. Calculate the transformer turns ratio required to connect four parallel 16-Ω speakers so that they appear as an 8-K effective load.

4. A transformer-coupled class-A amplifier drives a 16-Ω speaker through a $\sqrt{15}:1$ transformer. Using a power supply of 36 V (V_{CC}) the circuit delivers 2 W to the load. Calculate the following:
   (a) The ac power across the transformer primary.
   (b) The rms value of load voltage.
   (c) The rms value of primary voltage.
   (d) The rms values of load and primary current.

5. Calculate the efficiency of the circuit of Problem 4 if the bias current is $I_{C_Q} = 150$ mA.

6. Draw the circuit diagram of a class-B transformer-coupled amplifier using an *npn* transistor.

§ 8.4

7. Calculate the efficiency of the following amplifier classes and voltages:
   (a) Class-A operation with $V_{CE_{max}} = 24$, and $V_{CE_{min}} = 2$ V.
   (b) Class-B operation with $V_{CE_{min}} = 4$ V, and $V_{CC} = 22$ V.

8. For the following voltage values measured on a scope calculate the amount of the second harmonic distortion: $V_{CE_{max}} = 27$ V, $V_{CE_{min}} = 14$ V, $V_{CE_Q} = 20$ V.

§ 8.5

9. Draw the circuit diagram of a class-B *npn* push-pull power amplifier.

10. Sketch the waveforms of the ac signal in the circuit of Problem 9 at each point in the circuit for a sinusoidal input signal.

11. Draw the circuit diagram of an *npn* push-pull power amplifier operated class AB. Show the driver stage before the push-pull stage.

§ 8.6

12. Sketch the circuit diagram of a quasi-complementary amplifier, showing voltage waveforms in circuit.

13. List any advantages or disadvantages of a transformerless circuit over a transformer-coupled circuit.

14. For a class-B power amplifier as in Fig. 8.16 using a 30-V power supply and $R_L = 8\ \Omega$, calculate (a) maximum P_o (ac), (b) maximum P_i (dc), (c) maximum % η, and (d) maximum power dissipated by both output transistors. Use $R_L = 8\ \Omega$.

15. If the input voltage to the power amplifier in Fig. 8.16 is 15 V, rms using a 30-V power supply and $R_L = 8\,\Omega$, calculate (a) P_o (ac), (b) P_i (dc), (c) $\% \eta$, and (d) power dissipated by both output power transistors. Use $R_L = 8\,\Omega$.

16. For power amplifier as in Fig. 8.17 using a 40-V power supply and signal input of 18 V, rms calculate (a) P_o (ac), (b) P_i (dc), (c) $\% \eta$, and (d) power dissipated by both output power transistors.

§ 8.7

17. Determine the maximum dissipation allowed for a 100-W silicon transistor (rated at 25°C) for a derating factor of 0.6 W/°C, at a case temperature of 150°C.

18. A 160-W silicon power transistor operated with a heat sink ($\theta_{SA} = 1.5$°C/W) has $\theta_{JC} = 0.5$°C/W and mounting insulation of $\theta_{CS} = 0.8$°C/W. What maximum power can be handled by the transistor at an ambient temperature of 80°C? (Junction temperature should not exceed 200°C.)

19. What maximum power can a silicon transistor ($T_{J,\max} = 200$°C) dissipate at 80°C into free air at room temperature?

pnpn and other devices

9.1 INTRODUCTION

In this chapter we shall consider other important devices not discussed in detail in the previous chapters. Recall from Chapter 3 that the two-element vacuum tube led the way to the three-element triode, the four-element tetrode, and the five-element pentode. A similar natural sequence paved the way for a number of the three-layer and four-layer semiconductor devices to be introduced in this chapter. The family of four-layer *pnpn* devices will first be considered (SCR, SCS, GTO, LASCR, Shockley diode, DIAC, and TRIAC), followed by an increasingly important device—the UJT (unijunction transistor). The chapter will close with an introduction to the Power FET and a brief discussion of the phototransistor.

pnpn DEVICES

9.2 SILICON CONTROLLED RECTIFIER (SCR)

Within the family of *pnpn* devices the silicon controlled rectifier (SCR) is unquestionably of the greatest interest today. It was first introduced in 1956 by a group of Bell Telephone Laboratory engineers. A few of the more common areas of application for SCRs include relay controls, time delay circuits, regulated power suppliers, static switches, motor controls, choppers, inverters, cycloconverters, battery chargers, protective circuits, heater controls, and phase controls.

In recent years, SCRs have been designed to *control* powers as high as 10 MW with individual ratings as high as 2000 A at 1800 V. Its frequency range of application has also been extended to about 50 kHz, permitting some high-frequency applications such as induction heating and ultrasonic cleaning.

In some manuals and texts the SCR is called a *thyristor*, derived from the tube equivalent—the thyratron. The term thyristor is not limited only to the SCR but refers, in general, to all members of the *pnpn* family that have a control mechanism.

9.3 BASIC SILICON CONTROLLED RECTIFIER (SCR) OPERATION

As the terminology indicates the SCR is a rectifier constructed of silicon material which has a third terminal for control purposes. Silicon was chosen because of its high temperature and power capabilities. The basic operation of the SCR is different from the fundamental two-layer semiconductor diode in that a third terminal, called a *gate*, determines when the rectifier switches from the open-circuit to short-circuit state. It is not enough simply to forward bias the anode-to-cathode region of the device. In the conduction region the dynamic resistance of the SCR is typically 0.01–0.1 Ω. The reverse resistance is typically 100 K or more.

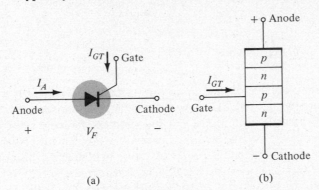

Figure 9.1. (a) SCR symbol and (b) basic construction.

The graphic symbol for the SCR is shown in Fig. 9.1 with the corresponding connections to the four-layer semiconductor structure. As indicated in Fig. 9.1a, if forward conduction is to be established, the anode must be positive with respect to the cathode. This is not, however, a sufficient criterion for turning the device on. A pulse of sufficient magnitude must also be applied to the gate to establish a turn-on gate current, represented symbolically by I_{GT}.

A more detailed examination of the basic operation of an SCR is best effected by splitting the four-layer *pnpn* structure of Fig. 9.1b into two three-layer transistor structures as shown in Fig. 9.2a and then considering the resultant circuit of Fig. 9.2b.

Note that one transistor for Fig. 9.2 is an *npn* device while the other is a *pnp* transistor. For discussion purposes, the signal shown in Fig. 9.3a will be applied to the gate of the circuit of Fig. 9.2b. During the interval $0 \rightarrow t_1$, $V_{gate} = 0$ V, the circuit of Fig. 9.2b will appear as shown in Fig. 9.3b ($V_{gate} = 0$ V is equivalent to the gate terminal being grounded as shown in the figure). For $V_{BE_2} = V_{gate} = 0$ V, the base current $I_{B_2} = 0$ and I_{C_2} will be approximately I_{CO}. The base current of

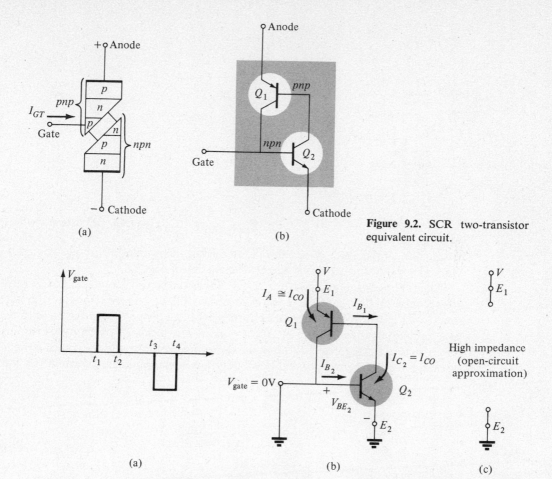

Figure 9.2. SCR two-transistor equivalent circuit.

Figure 9.3. OFF state of the SCR.

Q_1, $I_{B_1} = I_{C_2} = I_{CO}$ is too small to turn Q_1 on. Both transistors are therefore in the OFF state, resulting in a high impedance between the collector and emitter of each transistor and the open circuit representation for the controlled rectifier as shown in Fig. 9.3c.

At $t = t_1$ a pulse of V_G volts will appear at the SCR gate. The circuit conditions established with this input are shown in Fig. 9.4a. The potential V_G was chosen sufficiently large to turn Q_2 on ($V_{BE_2} = V_G$). The collector current of Q_2 will then rise to a value sufficiently large to turn Q_1 on ($I_{B_1} = I_{C_2}$). As Q_1 turns on, I_{C_1} will increase, resulting in a corresponding increase in I_{B_2}. The increase in base current for Q_2 will result in a further increase in I_{C_2}. The net result is a regenerative increase in the collector current of each transistor. The resulting anode to cathode resistance $[R_{SCR} = V/(I_A - \text{large})]$ is then very small, resulting in the short-circuit representation for the SCR as indicated in Fig. 9.4b. The regenerative action described above results in SCRs having typical turn-on-times of 0.1 to 1 μsec.

In addition to gate triggering, SCRs can also be turned on by significantly

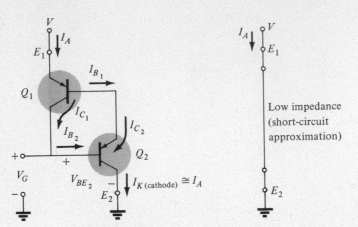

Figure 9.4. ON state of the SCR.

raising the temperature of the device or raising the anode-to-cathode voltage to the breakover value shown on the characteristics of Fig. 9.7.

The next question of concern is "How long is the turn-off time and how is turn-off accomplished?" An SCR *cannot* be turned off by simply removing the gate signal, and only a special few can be turned off by applying a negative pulse to the gate terminal as shown in Fig. 9.3a at $t = t_3$. The two general methods for turning off an SCR are categorized as the *anode current interruption* and the *forced commutation technique*. The two possibilities for current interruption are shown in Fig. 9.5.

Figure 9.5. Anode current interruption.

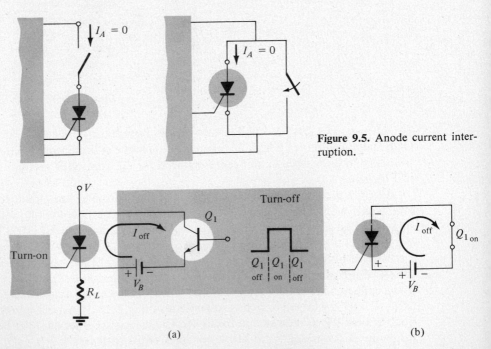

(a)

(b)

Figure 9.6. Forced commutation technique.

In Fig. 9.5a, I_A is zero when the switch is opened (series interruption) while in Fig. 9.5b the same condition is established when the switch is closed (shunt interruption). Forced commutation is the "forcing" of current through the SCR in the direction opposite to forward conduction. There are a wide variety of circuits for performing this function, a number of which can be found in the manuals of major manufacturers in this area. One of the more basic types is shown in Fig. 9.6. As indicated in the figure, the turn-off circuit consists of an *npn* transistor, a dc battery V_B, and a pulse generator. During SCR conduction the transistor is in the "off state," that is, $I_B = 0$ and the collector-to-emitter impedance is very high (for all practical purposes an open circuit). This high impedance will isolate the turn-off circuitry from affecting the operation of the SCR. For turn-off conditions, a positive pulse is applied to the base of the transistor, turning it heavily on, resulting in a very low impedance from collector to emitter (short-circuit representation). The battery potential will then appear directly across the SCR as shown in Fig. 9.6b, forcing current through it in the reverse direction for turn-off. Turn-off time of SCRs are typically 5–30 μsec.

9.4 SCR CHARACTERISTICS AND RATINGS

The characteristics of an SCR are provided in Fig. 9.7 for various values of gate current. The currents and voltages of usual interest are indicated on the characteristic. A brief description of each is listed below.

1. *Forward breakover voltage* $V_{(BR)F^*}$ is that voltage above which the SCR enters the conduction region. The asterisk (*) is a letter to be added that is dependent on the condition of the gate terminal as listed below

   O = open-circuit from G to K.
   S = short-circuit from G to K.
   R = resistor from G to K.
   V = fixed bias (voltage) from G to K.

2. *Holding current* (I_H) is that value of current below which the SCR switches from the conduction state to the forward blocking region under stated conditions.
3. *Forward and reverse blocking regions* are the regions corresponding to the open-circuit condition for the controlled rectifier which *block* the flow of charge (current) from anode to cathode.
4. *Reverse breakdown voltage* is equivalent to the Zener or avalanche region of the fundamental two-layer semiconductor diode.

It should be immediately obvious that the SCR characteristics of Fig. 9.7 are very similar to those of the basic two-layer semiconductor diode except for the horizontal offshoot before entering the conduction region. It is this horizontal jutting region that gives the gate control over the response of the SCR. For the characteristic having the solid line in Fig. 9.7 ($I_G = 0$) V_F must reach the largest required breakover voltage before the "collapsing" effect will result and the SCR

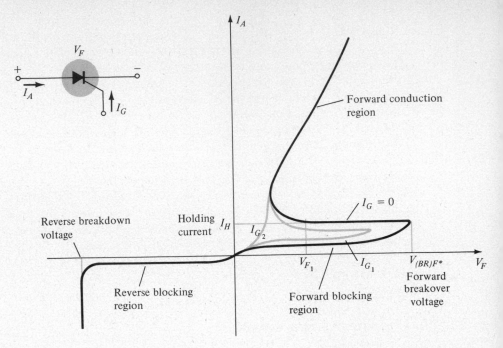

Figure 9.7. SCR characteristics.

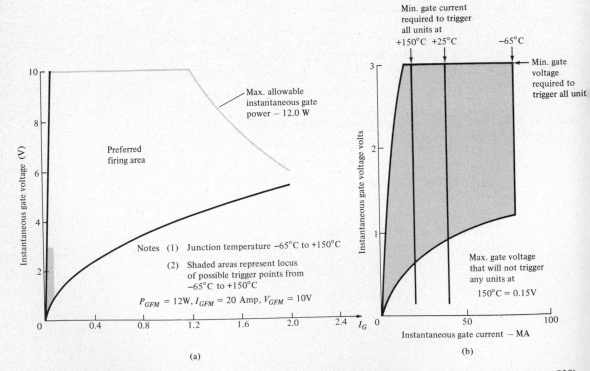

(a)

(b)

Figure 9.8. SCR gate characteristics (GE series—C38).

can enter the conduction region corresponding to the *on* state. If the gate current is increased to I_{G_1}, as shown in the same figure, by applying a bias voltage to the gate terminal the value of V_F required for the conduction is considerably less. Note also that I_H drops with increase in I_G. If increased to I_{G_2} the SCR will fire at very low values of voltage and the characteristics begin to approach those of the basic *p-n* junction diode. Looking at the characteristics in a completely different sense, for a particular V_F voltage, say V_{F_1} (Fig. 9.7), if the gate current is increased from $I_G = 0$ to $I_G = I_{G_1}$, the SCR will fire.

The gate characteristics are provided in Fig. 9.8. The characteristics of Fig. 9.8b are an expanded version of the shaded region of Fig. 9.8a. In Fig. 9.8a the three gate ratings of greatest interest, P_{GFM}, I_{GFM}, and V_{GFM} are indicated. Each is included on the characteristics in the same manner employed for the triode and transistor in Chapter 3. Except for portions of the shaded region any combination of gate current and voltage that falls within this region will fire any SCR in the series of components for which these characteristics are provided. Temperature will determine which sections of the shaded region must be avoided. At $-65°C$ the minimum current that will trigger the series of SCRs is 80 mA while at $+150°C$ only 20 mA are required. The effect of temperature on the minimum gate voltage is usually not indicated on curves of this type since gate potentials of 3 V or more are usually obtained easily. As indicated on Fig. 9.8b a minimum of 3 V is simply indicated for all units for the temperature range of interest. As an aid in the design of systems that will not trigger prematurely, an additional piece of information is usually provided as indicated in Fig. 9.8b. It states that at $150°C$, any gate voltages 0.15 V or less will not trigger the device into the ON state.

Other parameters usually included on the specification sheet of an SCR are the turn-on-time (t_{on}), turn-off time (t_{off}), junction temperature (T_J) and case temperature (T_C) all of which should by now be, to some extent, self-explanatory.

9.5 SCR CONSTRUCTION AND TERMINAL IDENTIFICATION

The basic construction of the four-layer pellet of an SCR is shown in Fig. 9.9a. The complete construction of a thermal-fatigue-free, high-current SCR is shown in Fig. 9.9b. Note the position of the gate, cathode, and anode terminals. The pedestal acts as a heat sink by transferring the heat developed to the chassis on which the SCR is mounted. The case construction and terminal identification of SCRs will vary with the application. Other case-construction techniques and the terminal identification of each are indicated in Fig. 9.10.

9.6 SCR APPLICATIONS

A few of the possible applications for the SCR are listed in the introduction to the SCR (Section 9.2). In this section we shall consider three: a static switch, a phase control system, and a battery charger.

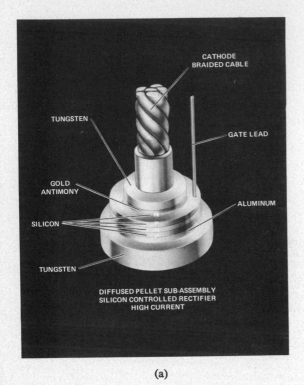

(a)

(b)

(Courtesy General Electric Company)

Figure 9.9. (a) Alloy-diffused SCR pellet; (b) thermal fatigue-free SCR construction.

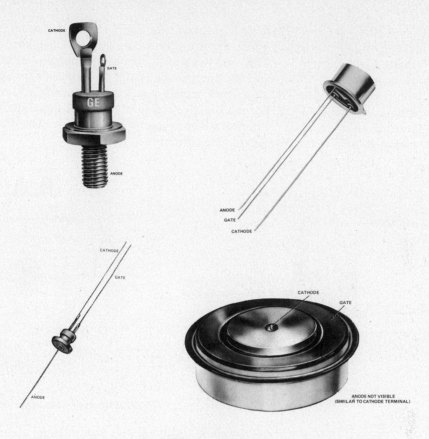

(Courtesy General Electric Company)

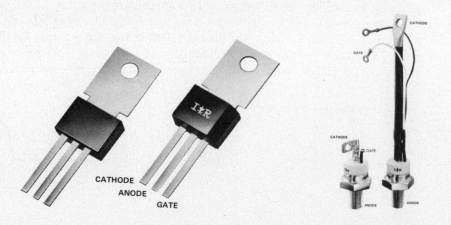

(Courtesy International Rectifier Corp. Inc.)

Figure 9.10. SCR case construction and terminal identification.

A half-wave *series static switch* is shown in Fig. 9.11a. If the switch is closed as shown in Fig. 9.11b, a gate current will flow during the positive portion of the input signal turning the SCR on. Resistor R_1 limits the magnitude of the gate current. When the SCR turns on, the anode-to-cathode voltage (V_F) will drop to the conduction value, resulting in a greatly reduced gate current and very little loss in the gate circuitry. For the negative region of the input signal the SCR will turn off since the anode is negative with respect to the cathode. The diode D_1 is included to prevent a reversal in gate current.

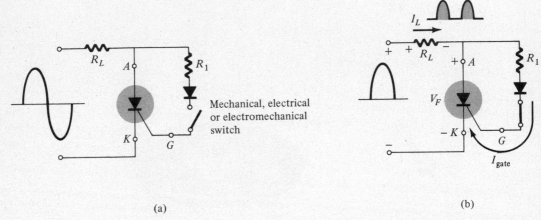

(a)

(b)

Figure 9.11. Half-wave series static switch.

The waveforms for the resulting load current and voltage are shown in Fig. 9.11b. The result is a half-wave rectified signal through the load. If less than 180° conduction is desired, the switch can be closed at any phase displacement during the positive portion of the input signal. The switch can be electronic, electromagnetic, or mechanical, depending on the application.

A circuit capable of establishing a conduction angle between 90° and 180° is shown in Fig. 9.12a. The circuit is similar to that of Fig. 9.11a except for the addition of a variable resistor and the elimination of the switch. The combination of the resistors R and R_1 will limit the gate current during the positive portion of the

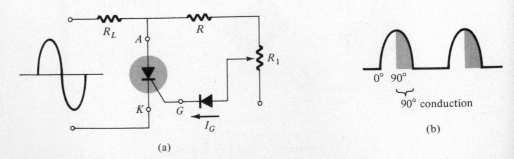

(a)

(b)

Figure 9.12. Half-wave variable-resistance phase control.

input signal. If R_1 is set to its maximum value, the gate current may never reach turn-on magnitude. As R_1 is decreased from the maximum the gate current will increase for the same input voltage. In this way, the required turn-on gate current can be established in any point between 0° and 90° as shown in Fig. 9.12b. If R_1 is low, the SCR will fire almost immediately, resulting in the same action as that obtained from the circuit of Fig. 9.11a (180° conduction). However, as indicated above, if R_1 is increased, a larger input voltage (positive) will be required to fire the SCR. The control as shown in Fig. 9.12b cannot be extended past a 90° phase displacement since the input is its maximum at this point. If it fails to fire at this and lesser values of input voltage on the positive slope of the input, the same response must be expected from the negatively sloped portion of the signal waveform. The operation here is normally referred to in technical terms as *half-wave variable resistance phase control*. It is an effective method of controlling the rms current and therefore power to the load.

A third popular application of the SCR is in a *battery charging regulator*. The fundamental components of the circuit are shown in Fig. 9.13. You will note that the control circuit has been blocked off for discussion purposes.

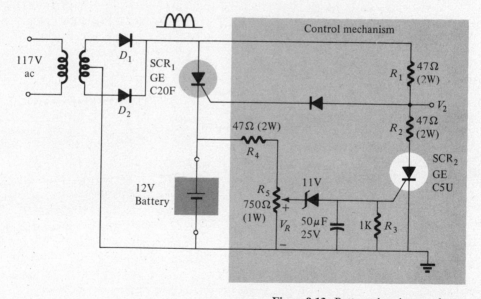

Figure 9.13. Battery charging regulator.

As indicated in the figure, D_1 and D_2 establish a full-wave rectified signal across SCR_1 and the 12-V battery to be charged. At low battery voltages SCR_2 is in the off state for reasons to be explained shortly. With SCR_2 open, the SCR_1 controlling circuit is exactly the same as the series static switch control discussed earlier in this section. When the full-wave rectified input is sufficiently large to produce the required turn-on gate current (controlled by R_1), SCR_1 will turn on and charging of the battery will commence. At the start of charging, the low battery voltage will result in a low voltage V_R as determined by the simple voltage divider circuit.

Voltage V_R is in turn too small to cause 11.0-V Zener conduction. In the off state, the Zener is effectively an open-circuit maintaining SCR_2 in the off state since the gate current is zero. The capacitor C_1 is included to prevent any voltage transients in the circuit from accidently turning on SCR_2. Recall from your fundamental study of circuit analysis that the voltage cannot instantaneously change across a capacitor. In this way C_1 prevents transient effects from affecting the SCR.

As charging continues, the battery voltage rises to a point where V_R is sufficiently high to both, turn on the 11.0-V Zener and fire SCR_2. Once SCR_2 has fired, the short-circuit representation for SCR_2 will result in a voltage divider circuit determined by R_1 and R_2 that will maintain V_2 at a level too small to turn SCR_1 on. When this occurs, the battery is fully charged and the open-circuit state of SCR_1 will cut off the charging current. Thus, the regulator recharges the battery whenever the voltage drops and prevents overcharging when fully charged.

9.7 SILICON CONTROLLED SWITCH (SCS)

The silicon controlled switch (SCS), like the silicon controlled rectifier is a four-layer *pnpn* device. All four semiconductor layers of the SCS are available due to the addition of an anode gate as shown in Fig. 9.14a. The graphic symbol and transistor equivalent circuit are shown in the same figure.

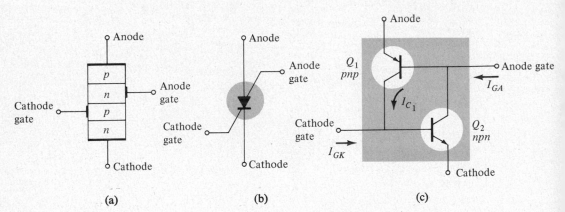

Figure 9.14. Silicon controlled switch (SCS): (a) basic construction; (b) graphic symbol; (c) equivalent transistor circuit.

The characteristics of the device are essentially the same as those for the SCR. The effect of an anode gate current is very similar to that demonstrated by the gate current in Fig. 9.7. The higher the anode gate current the lower the required anode-to-cathode voltage to turn the device on.

The anode gate connection can be used to either turn on or turn off the device. To turn on the device, a negative pulse must be applied to the anode gate terminal, while a positive pulse is required to turn off the device. The need for the type pulse indicated above can be demonstrated using the circuit of Fig. 9.14c. A negative pulse at the anode gate will forward bias the base-to-emitter junction of Q_1, turning it on. The resulting heavy collector current I_{C_1} will turn on Q_2, resulting in a regenerative action and the on state for the SCS device. A positive pulse at the anode gate will reverse bias the base-to-emitter junction of Q_1, turning it off, resulting in the open-circuit off state of the device. In general, the triggering (turn-on) anode gate current is larger in magnitude than the required cathode gate current. For one representative SCS device, the triggering anode gate current is 1.5 mA while the required cathode gate current is 1 μA. The required turn-on gate current at either terminal is affected by many factors. A few include the operating temperature, anode-to-cathode voltage, load placement, and type of cathode, gate-to-cathode or anode gate-to-anode connection (short-circuit, open-circuit, bias, load, etc.). Tables, graphs, and curves are normally available for each device to provide the type of information indicated above.

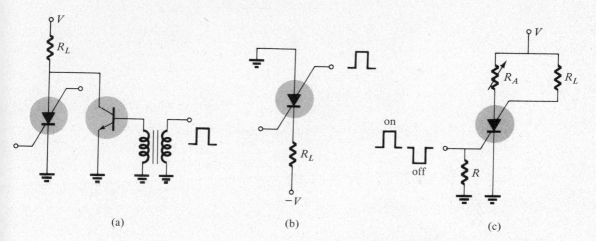

(a) (b) (c)

Figure 9.15. SCS turn-off techniques.

Three of the more fundamental types of turn-off circuits for the SCS are shown in Fig. 9.15. When a pulse is applied to the circuit of Fig. 9.15a, the transistor conducts heavily, resulting in a low impedance ($\cong$ short-circuit) characteristic between collector and emitter. This low-impedance branch diverts anode current away from the SCS, dropping it below the holding value and consequently turning it off. Similarly, the positive pulse at the anode gate of Fig. 9.15b will turn the SCS off by the mechanism described earlier in this section. The circuit of Fig. 9.15c can be turned either off *or* on by a pulse of the proper magnitude at the cathode gate. The turn-off characteristic is only possible if the correct value of R_A is employed. It will control the amount of regenerative feedback, the magnitude of which is critical for this type of operation. Note the variety of positions in which the load resistor R_L can be placed. There are a number of other possibilities that can be found in any comprehensive semiconductor handbook or manual.

An advantage of the SCS over a corresponding SCR is the reduced turn-off time, typically within the range 1–10 μsec for the SCS and 5–30 μsec for the SCR.

Some of the remaining advantages of the SCS over an SCR include increased control and triggering sensitivity and a more predictable firing situation. At present, however, the SCS is limited to low power, current, and voltage ratings. Typical maximum anode currents range from 100 to 300 mA with dissipation (power) ratings of 100 to 500 mW.

A few of the more common areas of application include a wide variety of computer circuits (counters, registers, and timing circuits) pulse generators, voltage sensors, and oscillators. One simple application for an SCS as a voltage-sensing device is shown in Fig. 9.16. It is an alarm system with *n* inputs from various stations. Any single input will turn that particular SCS on, resulting in an energized alarm relay and light in the anode gate circuit to indicate the location of the input (disturbance). The terminal identification of an SCS is shown in Fig. 9.17 with a packaged SCS.

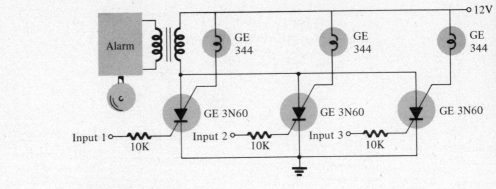

Figure 9.16. SCS alarm circuit.

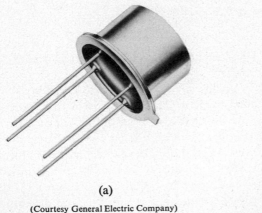

(a)

(Courtesy General Electric Company)

(b)

Figure 9.17. Silicon controlled switch (SCS): (a) device and (b) terminal identification.

9.8 GATE TURN-OFF SWITCH (GTO)

The gate turn-off switch is the third *pnpn* device to be introduced in this chapter. Like the SCR, however, it has only three external terminals as indicated in Fig. 9.18a. Its graphic symbol is also shown in the same figure (Fig. 9.18b). Although the graphic symbol is different from either the SCR or SCS, the transistor equivalent is exactly the same and the characteristics are similar.

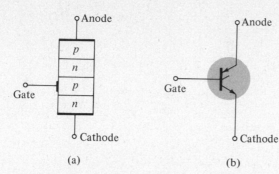

Figure 9.18. Gate turn-off switch (GTO): (a) basic construction; (b) symbol.

The most obvious advantage of the GTO over the SCR or SCS is the fact that it can be turned on *or* off by applying the proper pulse to the cathode gate (without the anode gate and associated circuitry required for the SCS). A consequence of this turn-off capability is an increase in the magnitude of the required gate current for triggering. For an SCR and GTO of similar maximum rms current ratings, the gate triggering current of a particular SCR is 30 μA while the triggering current of the GTO is 20 mA. The turn-off current of a GTO is slightly larger than the required triggering current. The maximum rms current and dissipation ratings of GTOs manufactured today is limited to about 3 A and 20 W, respectively.

A second very important characteristic of the GTO is improved switching characteristics. The turn-on time is similar to the SCR (typically 1 μsec), but the turn-off time of about the *same* duration (1 μsec) is much smaller than the typical turn-off time of an SCR (5–30 μsec). The fact that the turn-off time is similar to the turn-on time rather than considerably larger permits the use of this device in high-speed applications.

A typical GTO and its terminal identification are shown in Fig. 9.19. The GTO gate input characteristics and turn-off circuits can be found in a comprehensive manual or specification sheet. The majority of the SCR turn-off circuits can also be used for GTOs.

Some of the areas of application for the GTO include counters, pulse generators, multivibrators, and voltage regulators. Figure 9.20 is an illustration of a simple sawtooth generator employing a GTO and Zener diode.

When the supply is energized, the GTO will turn on, resulting in the short-circuit equivalent from anode to cathode. The capacitor C_1 will then begin to charge toward the supply voltage as shown in Fig. 9.20. As the voltage across the capacitor C_1 charges above the Zener potential a reversal in gate-to-cathode voltage will result, establishing a reversal in gate current. Eventually the negative gate current will be large enough to turn the GTO off. Once the GTO turns off, resulting in the open-circuit representation, the capacitor C_1 will dis-

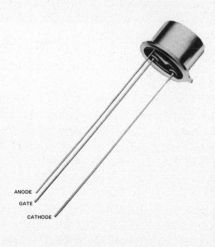

Figure 9.19. Typical GTO and its terminal identification.

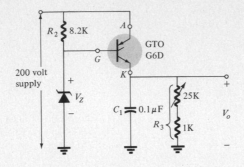
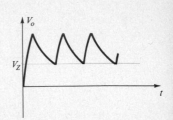

Figure 9.20. GTO sawtooth generator.

charge through the resistor R_3. The discharge time will be determined by the circuit time constant $\tau = R_3 C_1$. The proper choice of R_3 and C_1 will result in the sawtooth waveform of Fig. 9.20. Once the output potential V_o drops below V_Z, the GTO will turn on and the process will repeat.

9.9 LIGHT ACTIVATED SCR (LASCR)

The next in the series of *pnpn* devices is the light activated SCR (LASCR). As indicated by the terminology, it is an SCR whose state is controlled by the light falling upon a silicon semiconductor layer of the device. The basic construction of an LASCR is shown in Fig. 9.21a.

As indicated in Fig. 9.21a, a gate lead is also provided to permit triggering the device using typical SCR methods. Note also in the same figure that the mounting surface for the silicon pellet is the anode connection for the device.

The graphic symbols most commonly employed for the LASCR are provided in Fig. 9.21b. The terminal identification and typical LASCRs are shown in Fig. 9.22a.

Some of the areas of application for the LASCR include optical light controls, relays, phase control, motor control, and a variety of computer applications. The maximum current (rms) and power (gate) ratings for LASCRs commercially available today are about 3 A and 0.1 W. The characteristics (light-triggering) of a representative LASCR are provided in Fig. 9.22b. Note in this figure that an in-increase in junction temperature results in a reduction in light energy required to activate the device.

One interesting application of an LASCR is in the AND and OR circuits of Fig. 9.23. Only when light falls on LASCR$_1$ *and* LASCR$_2$ will the short-circuit representation for each be applicable and the supply voltage appear across the load. For the OR circuit, light energy applied to LASCR$_1$ *or* LASCR$_2$ will result in the supply voltage appearing across the load.

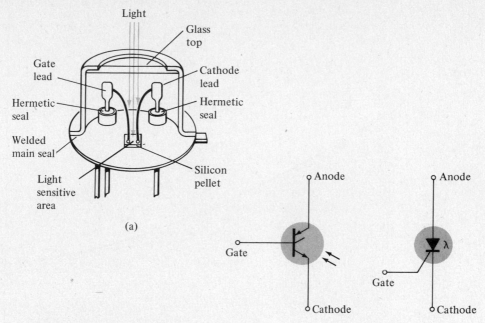

Figure 9.21. Light activated SCR (LASCR): (a) basic construction; (b) symbols.

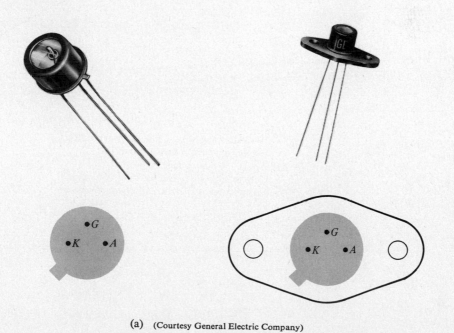

(a) (Courtesy General Electric Company)

Figure 9.22. LASCR: (a) appearance and terminal identification.

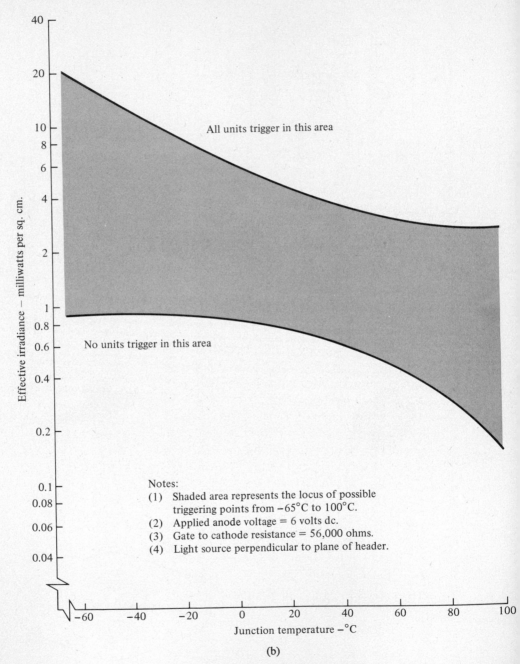

Figure 9.22. LASCR (continued): (b) light triggering characteristics.

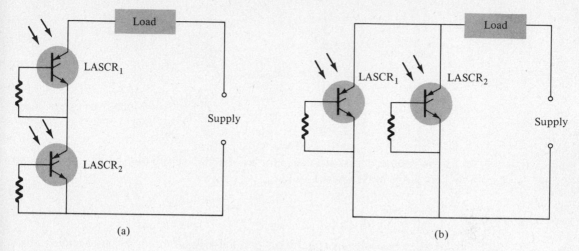

Figure 9.23. LASCR optoelectronic logic circuitry: (a) AND gate—input to LASCR$_1$ *and* LASCR$_2$ required for energization of the load; (b) OR gate—input to either LASCR$_1$ *or* LASCR$_2$ will energize the load.

9.10 SHOCKLEY DIODE

The Shockley diode is a four-layer *pnpn* diode with only two external terminals as shown in Fig. 9.24a, with its graphic symbol. The characteristics (Fig. 9.24b) of the device are exactly the same as those encountered for the SCR with $I_G = 0$. As indicated by the characteristics the device is in the off state (open-circuit representation) until the breakover voltage is reached, at which time avalanche conditions develop and the device turns on (short-circuit representation).

One common application of the Shockley diode is shown in Fig. 9.25 where it is employed as a trigger switch for an SCR. When the circuit is energized, the voltage across the capacitor will begin to change toward the supply voltage. Even-

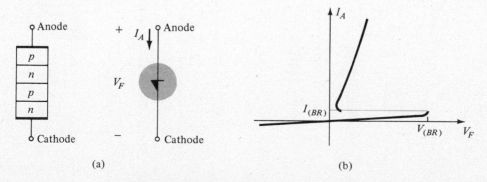

Figure 9.24. Shockley diode: (a) basic construction and symbol; (b) characteristics.

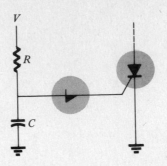

Figure 9.25. Shockley diode application—trigger switch for an SCR.

tually the voltage across the capacitor will be sufficiently high to first turn on the Shockley diode and then the SCR.

9.11 DIAC

The DIAC is basically a two-terminal parallel-inverse combination of semiconductor layers that permits triggering in either direction. The characteristics of the device, presented in Fig. 9.26a, clearly demonstrate that there is a breakover voltage

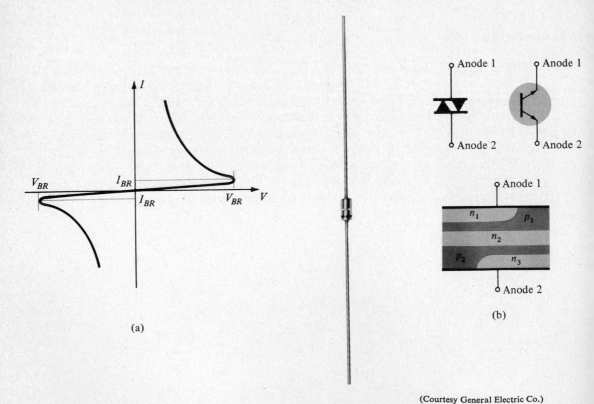

(Courtesy General Electric Co.)

Figure 9.26. DIAC: (a) characteristics; (b) symbols and basic construction.

in either direction. This possibility of an on condition in either direction can be used to its fullest advantage in ac applications.

The basic arrangement of the semiconductor layers of the DIAC is shown in Fig. 9.26b, along with its graphic symbol. Note that neither terminal is referred to as the cathode. Instead, there is an anode 1 (or electrode 1) and an anode 2 (or electrode 2). When anode 1 is positive with respect to anode 2, the semiconductor layers of particular interest are $p_1 n_2 p_3$ and n_3. For anode 2 positive with respect to anode 1 the applicable layers are $p_2 n_2 p_1$ and n_1.

9.12 TRIAC

The TRIAC is fundamentally a DIAC with a gate terminal for controlling the turn-on conditions of the bilateral device in either direction. In other words, for either direction the gate current can control the action of the device in a manner very similar to that demonstrated for an SCR. The characteristics, however, of the TRIAC in the first and third quadrants are somewhat different from those of the DIAC as shown in Fig. 9.27c. Note the holding current in each direction not present in the characteristics of the DIAC.

The graphic symbol for the device and the distribution of the semiconductor layers are provided in Fig. 9.27 with photographs of the device. For each possible

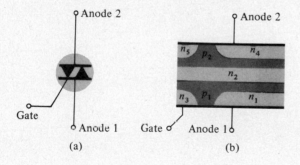

(a) (b)

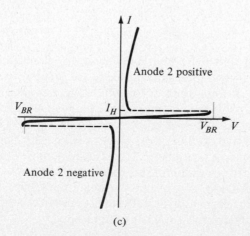

(c)

Figure 9.27. TRIAC: (a) symbol; (b) basic construction; (c) characteristics.

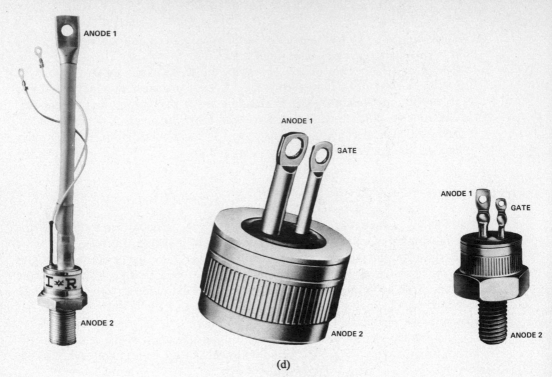

Figure 9.27. **Figure 9.27.** TRIAC (continued): (d) photographs.

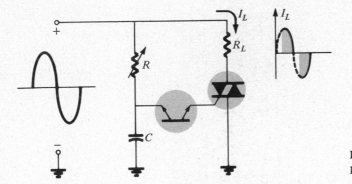

Figure 9.28. TRIAC application: phase (power) control.

direction of conduction there is a combination of semiconductor layers whose state will be controlled by the signal applied to the gate terminal.

One fundamental application of the TRIAC is presented in Fig. 9.28. In this capacity, it is controlling the ac power to the load by switching on and off during the positive and negative regions of input sinusoidal signal. The action of this circuit during the positive portion of the input signal is very similar to that encountered for the Shockley diode in Fig. 9.25. The advantage of this configuration is that during the negative portion of the input signal the same type of response will result since both the DIAC and TRIAC can fire in the reverse direction. The resulting waveform for the current through the load is provided in Fig. 9.28. By varying the resistor R the conduction angle can be controlled.

OTHER DEVICES

9.13 UNIJUNCTION TRANSISTOR

Recent interest in the unijunction transistor (UJT) has, like that for the SCR, been increasing at an exponential rate. Although first introduced in 1948, the device did not become commercially available until 1952. The low cost per unit combined with the excellent characteristics of the device have warranted its use in a wide variety of applications. A few include oscillators, trigger circuits, sawtooth generators, phase control, timing circuits, bistable networks, and voltage or current regulated supplies. The fact that this device is, in general, a low-power absorbing device under normal operating conditions is a tremendous aid in the continual effort to design relatively efficient systems.

The UJT is a three-terminal device having the basic construction of Fig. 9.29. A slab of lightly doped (increased resistance characteristic) n-type silicon material has two base contacts attached to both ends of one surface and an aluminum rod alloyed to the opposite surface. The p-n junction of the device is formed at the

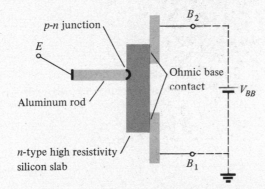

Figure 9.29. Unijunction transistor (UJT): basic construction.

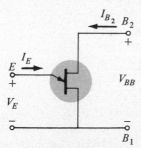

Figure 9.30. Symbol and basic biasing arrangement for the unijunction transistor.

boundary of the aluminum rod and the n-type silicon slab. The single p-n junction accounts for the terminology unijunction. It was originally called a duo (double) base diode due to the presence of two base contacts. Note in Fig. 9.29 that the aluminum rod is alloyed to the silicon slab at a point closer to the base 2 contact than the base 1 contact and that the base 1 terminal is made positive with respect to the base 2 terminal by V_{BB} volts. The effect of each will become evident in the paragraphs to follow.

The symbol for the unijunction transistor is provided in Fig. 9.30. Note that the emitter leg is drawn at an angle to the vertical line representing the slab of n-type material. The arrowhead is pointing in the direction of conventional current (hole) flow when the device is in the forward-biased, active, or conducting state.

The circuit equivalent of the UJT is shown in Fig. 9.31. Note the relative sim-

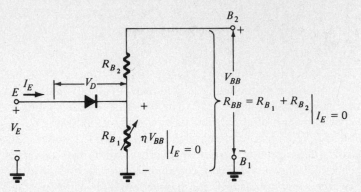

Figure 9.31. UJT equivalent circuit.

plicity of this equivalent circuit: two resistors (one fixed, one variable) and a single diode. The resistance R_{B_1} is shown as a variable resistor since its magnitude will vary with the current I_E. In fact, for a representative unijunction transistor, R_{B_1} may vary from 5 K down to 50 Ω for a corresponding change of I_E from 0 to 50 μA. The interbase resistance R_{BB} is the resistance of the device between terminals B_1 and B_2 when $I_E = 0$. In equation form

$$R_{BB} = (R_{B_1} + R_{B_2})|_{I_E=0} \tag{9.1}$$

(R_{BB} is typically within the range 4–10 K). The position of the aluminum rod of Fig. 9.29 will determine the relative values of R_{B_1} and R_{B_2} with $I_E = 0$. The magnitude of $V_{R_{B_1}}$ (with $I_E = 0$) is determined by the voltage divider rule in the following manner:

$$V_{R_{B1}} = \frac{R_{B_1} V_{BB}}{R_{B_1} + R_{B_2}} = \eta V_{BB}\bigg|_{I_E=0} \tag{9.2}$$

The Greek letter η (eta) is called the *intrinsic stand-off* ratio of the device and is given by

$$\eta = \frac{R_{B_1}}{R_{B_1} + R_{B_2}}\bigg|_{I_E=0} \tag{9.3}$$

For applied emitter potentials (V_E) greater than $V_{R_{B_1}} = \eta V_{BB}$ by the forward voltage drop of the diode, V_D (0.35 $\rightarrow$ 0.70 V) the diode will fire, assume the short-circuit representation (on an ideal basis), and I_E will begin to flow through R_{B_1}. In equation form the emitter firing potential is given by

$$V_P = \eta V_{BB} + V_D \tag{9.4}$$

The characteristics of a representative unijunction transistor are shown for $V_{BB} = 10$ V in Fig. 9.32. Note that for emitter potentials to the left of the peak

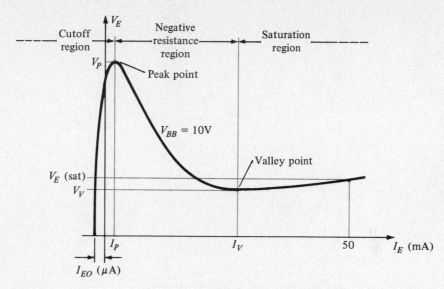

Figure 9.32. UJT static emitter characteristic curve.

point, the magnitude of I_E is never greater than I_{EO} (measured in microamperes). The current I_{EO} corresponds very closely with the reverse leakage current I_{CO} of the conventional bipolar transistor. This region, as indicated in the figure, is called the cutoff region. Once conduction is established at $V_E = V_P$, the emitter potential V_E will drop with increase in I_E. This corresponds exactly with the decreasing resistance R_{B_1} for increasing current I_E as discussed earlier. This device, therefore, has a *negative resistance* region which is stable enough to be used with a great deal of reliability in the areas of application listed earlier. Eventually, the valley point will be reached, and any further increase in I_E will place the device in the saturation region. In this region the characteristics approach that of the semiconductor diode in the equivalent circuit of Fig. 9.31.

The decrease in resistance in the active region is due to the holes injected into the *n*-type slab from the aluminum *p*-type rod when conduction is established. The increased hole content in the *n*-type material will result in an increase in the number of free electrons in the slab producing an increase in conductivity (G) and a corresponding drop in resistance ($R\downarrow = 1/G\uparrow$). Three other important parameters for the unijunction transistor are I_P, V_V, and I_V. Each is indicated on Fig. 9.32. They are all self-explanatory.

The emitter characteristics as they normally appear are provided in Fig. 9.33. Note that I_{EO} (μA) is not in evidence since the horizontal scale is in milliamperes. The intersection of each curve with the vertical axis is the corresponding value of V_P. For fixed values of η and V_D, the magnitude of V_P will vary as V_{BB}, that is,

$$V_P\uparrow = \underset{\text{fixed}}{\underline{\eta V_{BB}\uparrow}} + V_D\uparrow$$

A typical set of specifications for the UJT is provided in Fig. 9.34b. The discussion of the last few paragraphs should make each quantity readily recognizable.

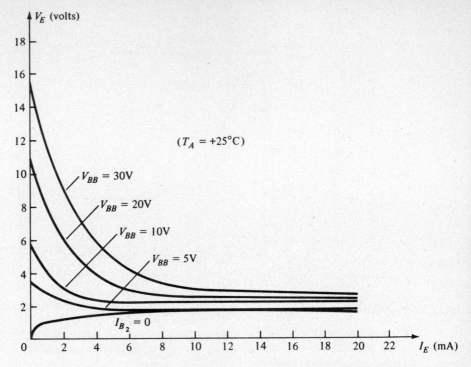

Figure 9.33. Typical static emitter characteristic curves for a UJT.

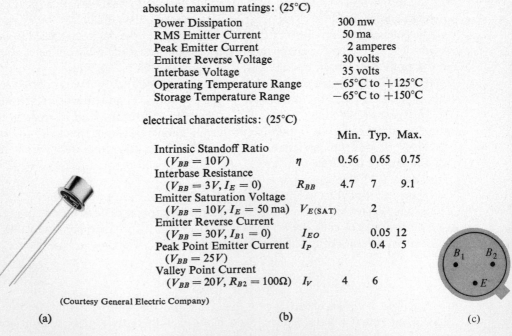

absolute maximum ratings: (25°C)

Power Dissipation	300 mw
RMS Emitter Current	50 ma
Peak Emitter Current	2 amperes
Emitter Reverse Voltage	30 volts
Interbase Voltage	35 volts
Operating Temperature Range	−65°C to +125°C
Storage Temperature Range	−65°C to +150°C

electrical characteristics: (25°C)

		Min.	Typ.	Max.
Intrinsic Standoff Ratio				
$(V_{BB} = 10V)$	η	0.56	0.65	0.75
Interbase Resistance				
$(V_{BB} = 3V, I_E = 0)$	R_{BB}	4.7	7	9.1
Emitter Saturation Voltage				
$(V_{BB} = 10V, I_E = 50 \text{ ma})$	$V_{E(SAT)}$		2	
Emitter Reverse Current				
$(V_{BB} = 30V, I_{B1} = 0)$	I_{EO}		0.05	12
Peak Point Emitter Current	I_P		0.4	5
$(V_{BB} = 25V)$				
Valley Point Current				
$(V_{BB} = 20V, R_{B2} = 100\Omega)$	I_V	4	6	

(Courtesy General Electric Company)

(a) (b) (c)

Figure 9.34. UJT: (a) appearance; (b) specification sheet; (c) terminal identification.

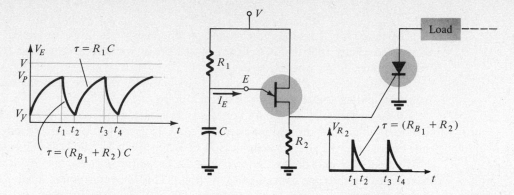

Figure 9.35. UJT triggering of an SCR.

The terminal identification is provided in the same figure with a photograph of a representative UJT. Note that the base terminals are opposite each other while the emitter terminal is between the two. In addition, the base terminal to be tied to the higher potential is closer to the extension on the lip of the casing.

One rather common application of the UJT is in the triggering of other devices such as the SCR. The basic elements of such a triggering circuit are shown in Fig. 9.35. The resistor R_1 must be chosen to ensure that the load line determined by R_1 passes through the device characteristics to the right of the peak point but to the left of the valley point. If the load line fails to pass to the right of the peak point, the device cannot turn on. An equation for R_1 that will ensure a turn-on condition can be established if we consider the peak point at which $I_P = I_{R_1}$ and $V_E = V_p$. (The equality $I_p = I_{R_1}$ is valid since the charging current of the capacitor, at this instant, is zero; that is, the capacitor is at this particular instant changing from a charging to a discharging state.) Then $V - I_P R_1 = V_P$ or $(V - V_P)/I_P = R_1$.

To ensure firing

$$\boxed{\frac{V - V_P}{I_P} > R_1}$$

(9.5)

At the valley point $I_E = I_V$ and $V_E = V_V$ so that to ensure turning off

$$\boxed{\frac{V - V_V}{I_V} < R_1}$$

(9.6)

For the typical values of $V = 30$ V, $\eta = 0.5$, $V_V = 1$ V, $I_V = 10$ mA, $I_P = 10$ μA, and $R_{BB} = 5$ K.

$$\frac{V - V_P}{I_P} = \frac{30 - [0.5(30) + 0.5]}{10 \times 10^{-6}} = \frac{14.5}{10 \times 10^{-6}} = 1.45 \text{ M} > R_1$$

and

$$\frac{V - V_V}{I_V} = \frac{30 - 1}{10 \times 10^{-3}} = 2.9 \text{ K} < R_1$$

Therefore 1.45 M $> R_1 >$ 2.9 K.

The range for R_1 is therefore extensive. The resistor R_2 must be chosen small enough to ensure that the SCR is not turned on by the interbase current I_{BB} that will flow through R_2 when $I_E = 0$.

The capacitor C will determine, as we shall see, the time interval between triggering pulses and the time span of each pulse.

At the instant the dc supply voltage V is applied, the voltage V_E will charge toward V volts since the emitter circuit of the UJT is in the open-circuit state. The time constant of the charging circuit is $R_1 C$. When $V_E = V_P$, the UJT will enter the conduction state and the capacitor C will discharge through R_{B_1} and R_2 at a rate determined by the time constant $(R_{B_1} + R_2)C$. This time constant is much smaller than the former, resulting in the patterns of Fig. 9.35. Once V_E decays to V_V, the UJT will turn off and the charging phase will repeat itself. Since I_{R_2} and V_{R_2} are related by Ohm's law (linear relationship), the waveform for I_{R_2} appears the same as for V_{R_2}. The positive pulse of V_{R_2} is designed to be sufficiently large to turn the SCR on. The operation of a UJT in an oscillator will be discussed further in Chapter 12.

9.14 V-FET

In recent years there has been increasing interest in raising the power limits of the FET. One technique that appears to be taking hold in the commercial market is the V-FET construction appearing in Fig. 9.36a. The basic construction of the typical MOSFET appears in Fig. 9.36b for comparison purposes. Most noticeable

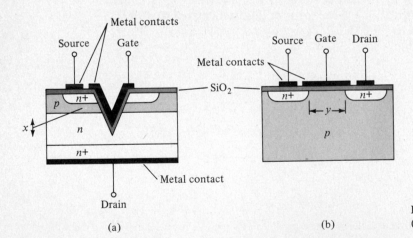

Figure 9.36. V-FET vs. MOSFET: (a) V-FET; (b) MOSFET.

are the four *diffused* layers in the V-FET as compared to the three regions of the MOSFET developed through *photolithographic* methods. The term V-FET is derived primarily from the fact that the drain-to-source current follows a "*vertical*" path rather than the horizontal path of the conventional MOSFET. Obviously, the V-type construction of the gate could also suggest this terminology. Increased currents are possible with the V-FET due to the significantly reduced channel length (1 : 3) (x versus y in Fig. 9.36), the availability of two current paths from the

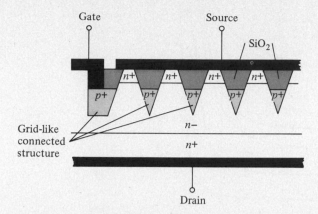

Figure 9.37. Saw-edge V-FET construction.

lower drain to the separated source, and the fact that many other Vs of construction can be introduced resulting in a saw-edge pattern in the top layer and a significant increase in the number of paths from drain to source as shown in Fig. 9.37. In the past FETs have been limited to the milliampere range with relatively low power ratings (milliwatts to few watts). Now the Siliconix Corporation of California has introduced to the commercial market an n-channel, enhancement mode, 60-W, 2-A V-FET.

In Fig. 9.36a the added n-type region will result in improved drain-to-source breakdown voltages and lower parasitic capacitance levels. Recall from Chapter 6 the nonlinear relationship between I_D and V_G. For the V-FET the short channel results in a straight line (linear) relationship.

Other important characteristics of the V-FET as compared to the bipolar transistor include a negative temperature coefficient to remove the concern about thermal runaway, a very low leakage current (a few nanoampere), and higher switching speeds (the VMP-1 produced by Siliconix can switch 2 A in 5 ns).

Further information on this commercial unit (VMP-1) include gate threshold voltages from 0.8 to 1.8 V, a maximum gate voltage of 10 V, and a drain-to-source breakdown voltage of 60 V. Its minimum g_m is 200 mmhos.

9.15 PHOTOTRANSISTORS

The fundamental behavior of photoelectric devices was introduced in Chapter 1 with the description of the photodiode. This discussion will now be extended to include the phototransistor, which has a photosensitive collector-base p-n junction. The current induced by photoelectric effects is the base current of the transistor. If we assign the notation I_λ for the photoinduced base current, the resulting collector current, on an approximate basis, is

$$I_C \cong h_{fe}I_\lambda \qquad (9.7)$$

A representative set of characteristics for a phototransistor is provided in Fig. 9.38 with the symbolic representation of the device. Note the similarities between

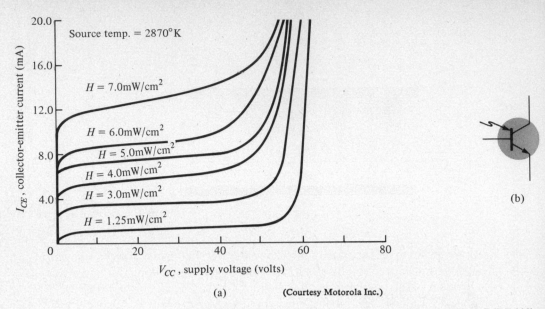

Figure 9.38. Phototransistor: (a) collector characteristics (MRD300); (b) symbol.

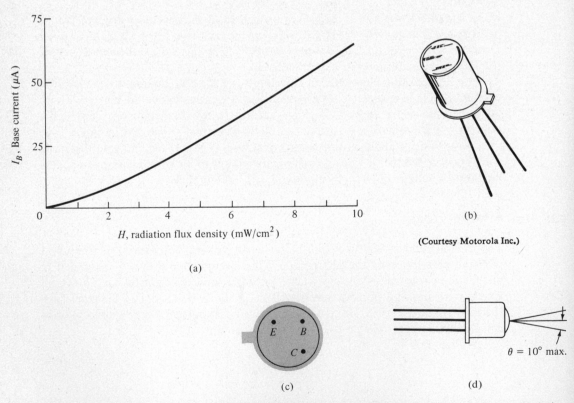

Figure 9.39. Phototransistor: (a) base current vs. flux density; (b) device; (c) terminal identification; (d) angular alignment.

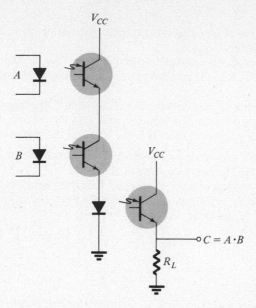

V_{CC}

A

B

V_{CC}

$C = A \cdot B$

R_L

Figure 9.40. High-isolation AND gate employing phototransistors and light emitting diodes (LEDs).

these curves and those of a typical bipolar transistor. As expected, an increase in light intensity corresponds with an increase in collector current. To develop a greater degree of familiarity with the light intensity unit of measurement, milliwatts per square centimeter, a curve of base current versus flux density appears in Fig. 9.39a. Note the exponential increase in base current with increasing flux density. In the same figure a sketch of the phototransistor is provided with the terminal identification and the angular alignment.

Some of the areas of application for the phototransistor include punch card readers, computer logic circuitry, lighting control (highways, etc.), level indication, relays, and counting systems.

A high-isolation AND gate is shown in Fig. 9.40 using three phototransistors and three LEDs (light emitting diodes). The LEDs are semiconductor devices that emit light at an intensity determined by the forward current through the device. With the aid of discussions in Chapter 1 the circuit behavior should be relatively easy to understand. The terminology "high-isolation" simply refers to the lack of an electrical connection between the input and output circuits.

PROBLEMS

§ 9.3

1. Describe in your own words the basic behavior of the SCR using the two-transistor equivalent circuit.

2. Describe two techniques for turning an SCR off.

3. Consult a manufacturer's manual or specification sheet and obtain a turn-off network. If possible, describe the turn-off action of the design.

§ 9.4

4. (a) At high levels of gate current the characteristics of an SCR approach those of what two-terminal device?
   (b) At a fixed anode-to-cathode voltage less than $V_{(BR)F*}$, what is the effect on the firing of the SCR as the gate current is reduced from its maximum value to the zero level.
   (c) At a fixed gate current greater than $I_{G=0}$, what is the effect on the firing of the SCR as the gate voltage is reduced from $V_{(BR)F*}$?
   (d) For increasing levels of I_G what is the effect on the holding current?

5. (a) In Fig. 9.8 will a gate current of 50 mA fire the device at room temperature (25°C)?
   (b) Repeat part (a) for a gate current of 10 mA.
   (c) Will a gate voltage of 2.6 V trigger the device at room temperature?
   (d) Is $V_G = 6$ V, $I_G = 800$ mA a good choice for firing conditions? Would $V_G = 4$ V, $I_G = 1.6$ A be preferred? Explain.

§ 9.6

6. In Fig. 9.11b why is there very little loss in potential across the SCR during conduction?

7. Fully explain why reduced values of R_1 in Fig. 9.12 will result in an increased angle of conduction.

8. Refer to the charging network of Fig. 9.13.
   (a) Determine the dc level of the full-wave rectified signal if a 1:1 transformer were employed.
   (b) If the battery in its uncharged state is sitting at 11 V, what is the anode-to-cathode voltage drop across SCR_1.
   (c) What is the maximum possible value of V_R?
   (d) At the maximum value of part (c) what is the gate potential of SCR_2?
   (d) Once SCR_2 has entered the short-circuit state, what is the level of V_2?

§ 9.7

9. Fully describe in your own words the behavior of the networks of Fig. 9.15.

§ 9.8

10. (a) In Fig. 9.20, if $V_Z = 50$ V, determine the maximum possible value the capacitor C_1 can charge to.
    (b) Determine the approximate discharge time (5τ) for $R_3 = 20$ K.
    (c) Determine the internal resistance of the GTO if the rise time is one-half the decay period determined in part (b).

§ 9.9

11. (a) Using Fig. 9.22, determine the minimum irradiance required to fire the device at room temperature (20°C).
    (b) What per cent reduction in irradiance is allowable if the junction temperature is increased from 0°C (32°F) to 100°C (212°F).

§ 9.10

12. For the network of Fig. 9.25, if $V_{(BR)} = 6$ V, $V = 40$ V, $R = 10$ K, $C = 0.2$ μF, and V_{GK} (firing potential) $= 3$ V, determine the time period between energizing the network and the turning on of the SCR.

§ 9.11

13. Using whatever reference you require, find an application of a DIAC and explain the network behavior.

§ 9.12

14. Repeat Problem 13 for the TRIAC.

§ 9.13

15. For the network of Fig. 9.35 in which $V = 40$ V, $\eta = 0.6$, $V_V = 1$ V, $I_V = 8$ mA, $I_P = 10$ μA, and $R_{BB} = 6$ K, determine the range of R_1 for the triggering network.

16. For a unijunction transistor with $V_{BB} = 20$ V, $\eta = 0.65$, $R_{B_1} = 2$ K ($I_E = 0$), $V_D = 0.7$ V, determine the following:
   (a) R_{B_2}.
   (b) R_{BB}.
   (c) $V_{R_{B_1}}$.
   (d) V_P.

§ 9.14

17. Describe the difference in construction between the standard FET and the V-FET.

18. Request information on a commercially available V-FET and discuss its characteristics.

§ 9.15

19. For a phototransistor having the characteristics of Fig. 9.39 determine the photo-induced base current for a radian flux density of 5 mW/cm². If $h_{fe} = 40$, find I_C.

20. Design a high isolation OR-Gate employing phototransistors and LEDs.

integrated circuits (ICs)

10

10.1 INTRODUCTION

During the last few years the excellent characteristics of (and advantages associated with) a relatively new type of electronic package have substantially altered the course of many areas of research and development. This product, called an *integrated circuit* (IC), has, through expanded usage and the various media of advertising, become a product whose basic function and purpose are now understood by the layman. The most noticeable characteristic of an IC is its size. It is typically hundreds and even thousands of times smaller than a semiconductor structure built in the usual manner with discrete components. In Fig. 10.1a all the circuit elements appearing in the circular pattern can be found in the IC appearing in the center of the figure. Figure 10.1b is an indication of the reduction in size for a 120-gate computer package.

Integrated circuits are seldom, if ever, repaired; that is, if a single component within an IC should fail, the entire structure (complete circuit) is replaced—a more economical approach. There are three types of ICs commercially available on a large scale today. They include the *monolithic*, *thin* (*or thick*) *film*, and *hybrid* integrated circuits.

10.2 MONOLITHIC INTEGRATED CIRCUIT

The term *monolithic* is derived from a combination of the Greek words *monos*, meaning single, and *lithos*, meaning stone, which in combination result in the literal translation, single-stone, or more appropriately, single-solid structure. As this descriptive term implies, the monolithic IC is constructed within a *single* wafer of

450

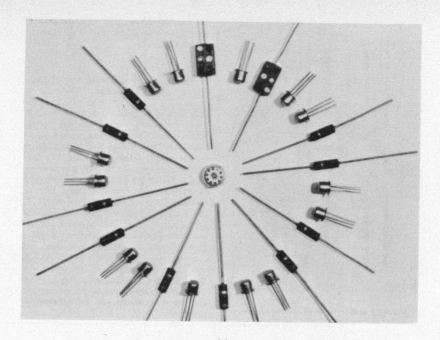

(a)

(Courtesy Motorola, Inc.)

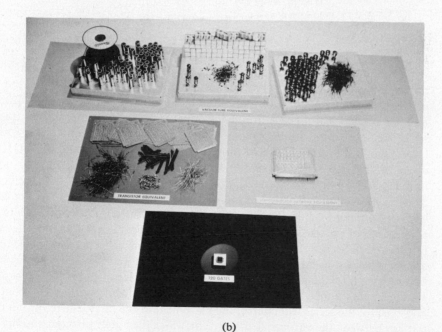

(b)

(Courtesy Texas Instruments, Inc.)

Figure 10.1. (a) An integrated circuit and the discrete elements required to build a circuit to perform the same function; (b) comparison of the resulting size of a 120 gate computer package manufactured by the indicated methods.

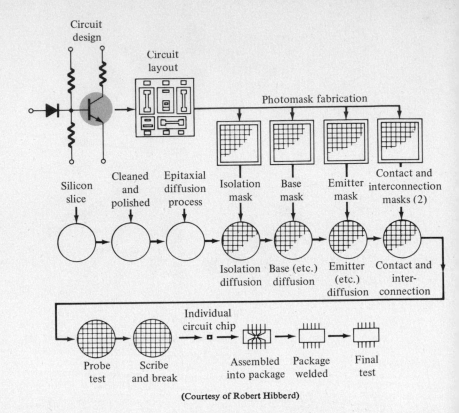

Circuit design

Circuit layout

Photomask fabrication

Silicon slice — Cleaned and polished — Epitaxial diffusion process — Isolation mask — Base mask — Emitter mask — Contact and interconnection masks (2)

Isolation diffusion — Base (etc.) diffusion — Emitter (etc.) diffusion — Contact and inter-connection

Probe test — Scribe and break — Individual circuit chip — Assembled into package — Package welded — Final test

(Courtesy of Robert Hibberd)

Figure 10.2. Monolithic integrated circuit fabrication.

semiconductor material. The greater portion of the wafer will simply act as a supporting structure for the very thin resulting IC. An overall view of the stages involved in the fabrication of monolithic ICs is provided in Fig. 10.2. The actual number of steps leading to a finished product is many times that appearing in Fig. 10.2. The figure does, however, point out the major production phases of forming a monolithic IC. The initial preparation of the semiconductor wafer of Fig. 10.2 was discussed in Chapter 3 in association with the fabrication of transistors. As indicated in the figure, it is first necessary to design a circuit that will meet the specifications. The circuit must then be laid out in order to ensure optimum use of available space and a minimum of difficulty in performing the diffusion processes to follow. The appearance of the mask and its function in the sequence of stages indicated will be introduced in Section 10.4. For the moment, let it suffice to say that a mask has the appearance of a negative through which impurities may be diffused (through the light areas) into the silicon slice. The actual diffusion process for each phase is similar to that applied in the fabrication of diffused transistors in Chapter 3. The last mask of the series will control the placement of the interconnecting conducting pattern between the various elements. The wafer then goes through various testing procedures, is scribed and broken into individual chips, packaged and assembled as indicated. A processed silicon wafer appears in Fig. 10.3. The original wafer can be anywhere from $\frac{1}{2}$ to 2 in. in diameter. The size of each chip will, of course, determine

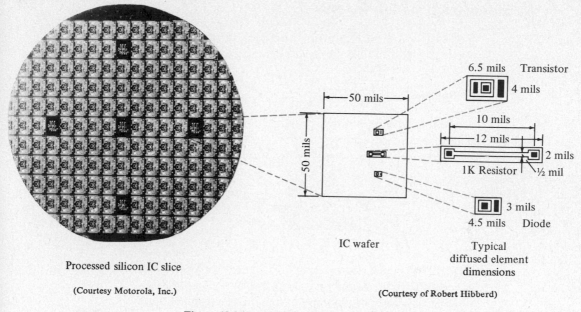

50 mils

50 mils

6.5 mils   Transistor

4 mils

10 mils

12 mils

2 mils

1K Resistor   ½ mil

3 mils

4.5 mils   Diode

IC wafer

Typical
diffused element
dimensions

Processed silicon IC slice

(Courtesy Motorola, Inc.)

(Courtesy of Robert Hibberd)

Figure 10.3. Processed monolithic IC wafer with the relative dimensions of the various elements.

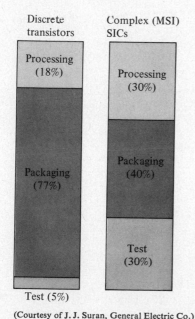

Discrete
transistors

Complex (MSI)
SICs

Processing
(18%)

Processing
(30%)

Packaging
(77%)

Packaging
(40%)

Test
(30%)

Test (5%)

(Courtesy of J. J. Suran, General Electric Co.)

Figure 10.4. Cost breakdown for the manufacturing of discrete transistors and complex silicon integrated circuits (SCIs).

the number of individual circuits resulting from a single wafer. The dimensions of each chip of the wafer in Fig. 10.3 are 50 × 50 mils. To point out the microminiature size of these chips, consider that 20 of them can be lined up along a 1-in. length. The average relative size of the elements of a monolithic IC appear in Fig. 10.3. Note the large area required for the 1-K resistor as compared to the other elements indicated. The next section will examine the basic construction of each of these elements.

A recent article indicated, by percentage, the relative costs of the various stages in the production of monolithic ICs as compared to discrete transistors. The resulting graphs appear in Fig. 10.4. The processing phase includes all stages leading up to the individual chips of Fig. 10.3. Note the high cost of packaging the integrated circuits and of testing the complex silicon integrated circuits (SIC). The cost of packaging has resulted in an increase (wherever feasible) in the number of IC chips within a single package. This multichip, hybrid type of integrated circuit will be considered in Section 10.7.

10.3 MONOLITHIC CIRCUIT ELEMENTS

The surface appearance of the transistor, diode, and resistor appear in Fig. 10.3. We shall now examine the basic construction of each in more detail.

Resistor

You will recall that the resistance of a material is determined by the resistivity, length, area, and temperature of the material. For the integrated circuit, each necessary element is present in the sheet of semiconductor material appearing in

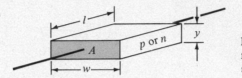

Figure 10.5. Parameters determining the resistance of a sheet of semiconductor material.

Fig. 10.5. As indicated in the figure, the semiconductor material can be either p- or n-type, although the p-type is most frequently employed.

The resistance of any bulk material is determined by

$$R = \rho \frac{l}{A}$$

For $l = w$, resulting in a square sheet,

$$R = \frac{\rho l}{yw} = \frac{\rho l}{yl}$$

and

$$\boxed{R_s = \frac{\rho}{y} \quad \text{(ohms)}} \qquad (10.1)$$

where ρ is in ohm-centimeters and y is in centimeters.

R_s is called the sheet resistance and has the units ohms per square. The equation clearly reveals that the sheet resistance is independent of the size of the square.

In general, where $l \neq w$,

$$\boxed{R = R_s \frac{l}{w} \quad \text{(ohms)}} \qquad (10.2)$$

For the resistor appearing in Fig. 10.3, $w = \frac{1}{2}$ mil, $l = 10$ mils, and $R_s = 100\ \Omega$/square:

$$R = R_s \frac{l}{w} = 100 \times \frac{10}{\frac{1}{2}} = 2\ \text{K}$$

A cross-sectional view of a monolithic resistor appears in Fig. 10.6 along with the surface appearance of two monolithic resistors. In Fig. 10.6a the sheet resistive material (p) is indicated with its aluminum terminal connections. The n-isolation

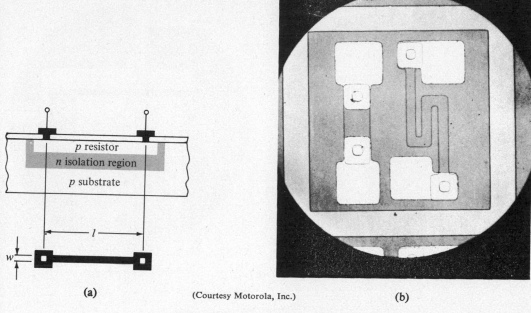

(a)

(Courtesy Motorola, Inc.)

(b)

Figure 10.6. Monolithic resistors: (a) cross section and determining dimensions; (b) surface view of two monolithic resistances in a single die.

region performs exactly that function indicated by its name; that is, it isolates the monolithic resistive elements from the other elements of the chip. Note in Fig. 10.6b the method employed to obtain a maximum l in a limited area. The resistors of Fig. 10.6 are called base-diffusion resistors since the p-material is diffused into the p-type substrate during the base-diffusion process indicated in Fig. 10.2.

Capacitor

Monolithic capacitive elements are formed by making use of the transition capacitance of a reverse-biased p-n junction. At increasing reverse-bias potentials, there is an increasing distance at the junction between the p- and n-type impurities. The region between these oppositely doped layers is called the depletion region (see Chapter 3) due to the absence of "free" carriers. The necessary elements of a capacitive element are therefore present—the depletion region has insulating characteristics that separate the two oppositely charged layers. The transition capacitance is related to the width (W) of the depletion region, the area (A) of the junction, and the permittivitiy (ϵ) of the material within the depletion region by

$$C_T = \frac{\epsilon A}{W}$$

(10.3)

The cross section and surface appearance of a monolithic capacitive element

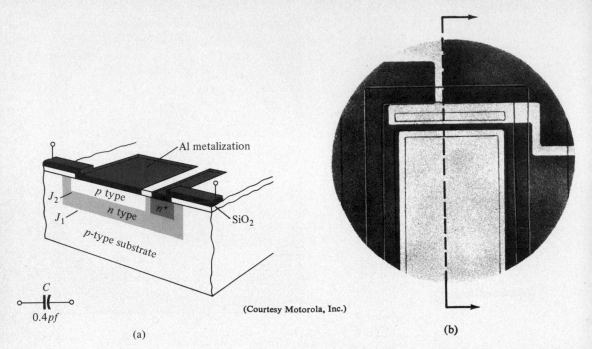

Figure 10.7. Monolithic capacitor: (a) cross section; (b) photograph.

appear in Fig. 10.7. The reverse-biased junction of interest is J_2. The undesirable parasitic capacitance at junction J_1 is minimized through careful design. Due to the fact that aluminum is a p-type impurity in silicon, a heavily doped n^+ region is diffused into the n-type region as shown to avoid the possibility of establishing an undesired p-n junction at the boundary between the aluminum contact and the n-type impurity region.

Inductor

Whenever possible, inductors are avoided in the design of integrated circuits. An effective technique for obtaining nominal values of inductances has so far not been devised for monolithic integrated circuits. In many instances, the need for inductive elements can be eliminated through the use of a technique known as RC synthesis. Thin (or thick) film or hybrid integrated circuits have an option open to them that cannot be employed in monolithic integrated circuits: the addition of discrete inductive elements to the surface of the structure. Even with this option, however, they are seldom employed due to their relatively bulky nature.

Transistors

The cross section of a monolithic transitor appears in Fig. 10.8a. Note again the presense of the n^+ region in the n-type epitaxial collector region. The vast majority of monolitic IC transistors are *npn* rather than *pnp* for reasons to be found in more advanced texts on the subject. Keep in mind when examining Fig. 10.8 that the *p*-substrate is only a supporting and isolating structure forming no part of the active device itself. The base, emitter, and collector regions are formed during the corresponding diffusion processes of Fig. 10.2.

The top view of a typical monolithic transistor appears in Fig. 10.8b. Note that two base terminals are provided while the collector has the outer rectangular aluminum contact surface.

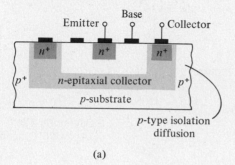

(a)

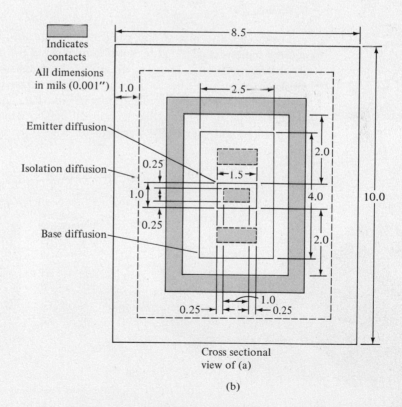

Cross sectional
view of (a)

(b)

(Courtesy Motorola Monitor)

Figure 10.8. Monolithic transistor: (a) cross section; (b) surface appearance and dimension for a typical monolithic transistor.

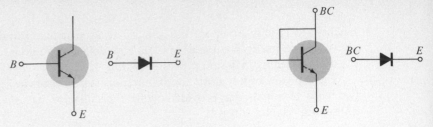

Figure 10.9. Transistor structure and connections employed in the formation of monolithic diodes.

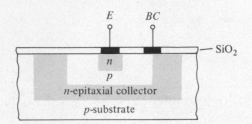

Figure 10.10. The cross-sectional view of a *BC-E* monolithic diode.

Diodes

The diodes of a monolithic integrated circuit are formed by first diffusing the required regions of a transistor and then masking the diode rather than transistor terminal connections. There is, however, more than one way of hooking up a transistor to perform a basic diode action. The two most common methods applied to monolithic integrated circuits appear in Fig. 10.9. The structure of a *BC-E* diode appears in Fig. 10.10 Note that the only difference between the cross-sectional view of Fig. 10.10 and that of the transistor of Fig. 10.8a is the position of the ohmic aluminum contacts.

(a)

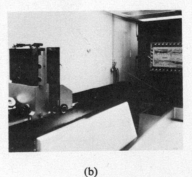

(b)

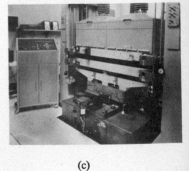

(c)

Figure 10.11. Mask preparation: (a) aristo handcutting of the mask pattern; (b) photo-reduction of the mask pattern; (c) step and repeat machine for the placement of a large number of the reduced mask pattern on a single production mask.

(d)

(Courtesy Motorola, Inc.)

Figure 10.11. Mask preparation (continued): (d) final mask.

10.4 MASKS

The selective diffusion required in the formation of the various active and passive elements of an integrated circuit is accomplished through the use of masks such as those appearing in Fig. 10.11. This figure depicts and describes the major steps involved in the production of these masks. We shall find in the next section that the light areas are the only areas through which donor and acceptor impurities can pass. The dark areas will block the diffusion of impurities somewhat because a shade will prevent sunlight from changing the pigment of the skin. The next section will demonstrate the use of these masks in the formation of a computer logic circuit.

10.5 MONOLITHIC INTEGRATED CIRCUIT—THE NAND GATE

This section is devoted to the sequence of production stages leading to a monolithic NAND-gate circuit (the operation of which is covered in Chapter 13). A detailed examination of each process would require many more pages than it is possible to include in this text. The description, however, should be sufficiently complete and informative to aid the reader in any future contact with this highly volatile area. The circuit to be prepared appears in Fig. 10.12a. The criteria of space allocation, placement of pin connection, and so on require that the elements be situated in the relative positions indicated in Fig. 10.12b. The regions to be isolated from one another appear within the solid heavy lines. A set of masks for the various diffusion processes must then be made up for the circuit as it appears in Fig. 10.12b.

We shall now slowly proceed through the first diffusion process to demonstrate the natural sequence of steps that must be followed through each diffusion process indicated in Fig. 10.2.

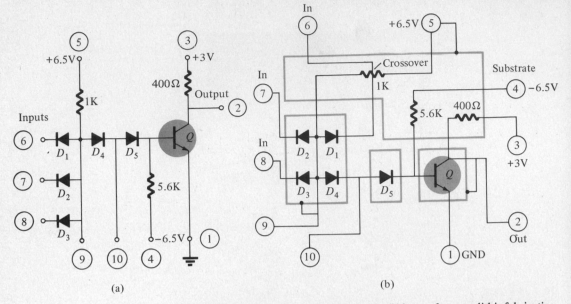

(a)

(b)

Figure 10.12. NAND gate: (a) circuit; (b) layout for monolithic fabrication.

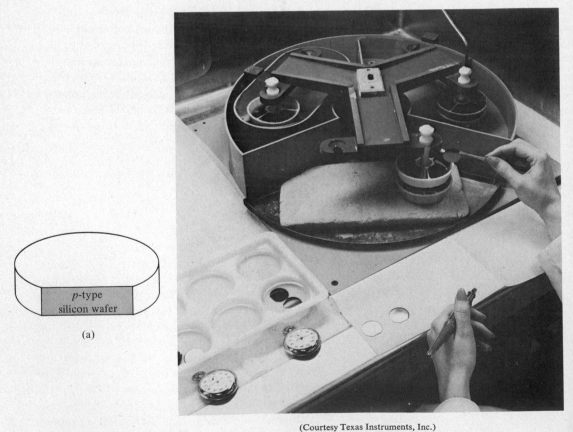

(a)

(Courtesy Texas Instruments, Inc.)

(b)

Figure 10.13. (a) p-Type silicon wafer; (b) polishing apparatus.

p-Type Silicon Wafer Preparation

After being sliced from the grown ingot, a *p*-type silicon wafer is polished and cleaned to produce the structure of Fig. 10.13.

n-Type Epitaxial Region

An *n*-type epitaxial region is then diffused into the *p*-type substrate as shown in Fig. 10.14. It is *in* this thin epitaxial layer that the active and passive elements will be diffused. The *p*-type area remaining is simply adding some thickness to the structure to give it increase strength and permit easier handling.

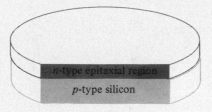

n-type epitaxial region
p-type silicon

Figure 10.14. *p*-Type silicon wafer after the *n*-type epitaxial diffusion process.

Silicon Oxidation (SIO$_2$)

The resulting wafer is then subjected to an oxidation process resulting in a surface layer of SiO$_2$ (silicon dioxide) as shown in Fig. 10.15. This surface layer will prevent any impurities from entering the *n*-type epitaxial layer. However, selective etching of this layer will permit the diffusion of the proper impurity into designated areas of the *n*-type epitaxial region of the silicon wafer.

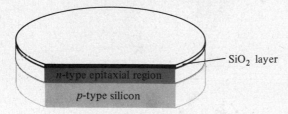

SiO$_2$ layer

n-type epitaxial region
p-type silicon

Figure 10.15. Wafer of Fig. 10.14 following the deposit of the SiO$_2$ layer.

Photolithographic Process

The selective etching of the SiO$_2$ layer is accomplished through the use of a photolithographic process. A mask is first prepared on a glass plate as explained in Section 10.4. The first mask will determine those areas of the SiO$_2$ layer to be removed in preparation for the isolation diffusion process. The wafer is first coated with a thin layer of photosensitive material, commonly called *photoresist*, as demonstrated in Fig. 10.16a. This new layer is then covered by the mask and an ultraviolet light is applied that will expose those regions of the photosensitive material not covered by the masking pattern (Fig. 10.16b). The resulting wafer is then subjected to a chemical solution that will remove the unexposed photosensitive material. A cross section of a chip (*S-S*) of Fig. 10.16 will then appear as indicated

(a)

(Courtesy Motorola, Inc.)

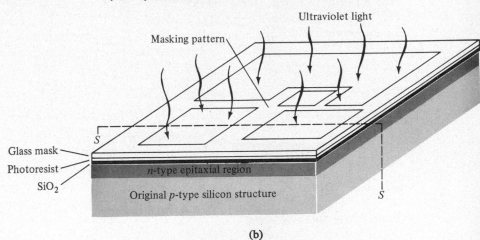

(b)

Figure 10.16. Photolithographic process: (a) applying the photoresist; the wafer is spun at a high speed to insure an even distribution of the photoresist; (b) the application of ultraviolet light after the mask is properly set; the structure is only one of the 200, 400, or even 500 individual NAND gate circuits being formed on the wafer of Figs. 10.13 through 10.15.

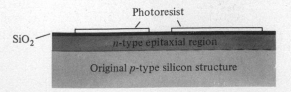

Figure 10.17. Cross section (*s-s*) of the chip of Fig. 10.16 following the removal of the unexposed photoresist.

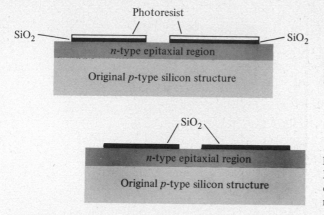

Figure 10.18. Cross section of Fig. 10.17 following the removal of the uncovered SiO₂ regions.

Figure 10.19. Cross section of Fig. 10.18 following the removal of the remaining photoresist material.

in Fig. 10.17. A second solution will then etch away the SiO₂ layer from any region not covered by the photoresist material (Fig. 10.18). The final step before the diffusion process is the removal, by solution, of the remaining photosensitive material. The structure will then appear as shown in Fig. 10.19.

Isolation Diffusion

The structure of Fig. 10.19 is then subjected to a p-type diffusion process resulting in the islands of n-type regions indicated in Fig. 10.20. The diffusion process ensures a heavily doped p-type region (indicated by p^+) between the n-type islands. The p^+ regions will result in improved *isolation* properties between the active and passive components to be formed in the n-type islands. In preparation for the next masking and diffusion process, the entire surface of the wafer is coated with a SiO₂ layer as indicated in Fig. 10.21.

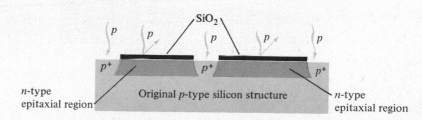

Figure 10.20. Cross section of Fig. 10.19 following the isolation diffusion process.

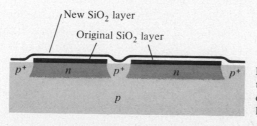

Figure 10.21. In preparation for the next diffusion process the entire wafer is coated with a SiO₂ layer.

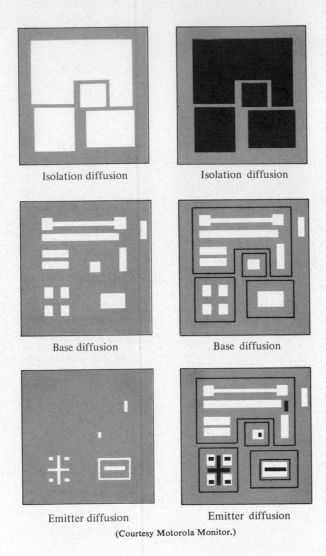

Isolation diffusion   Isolation diffusion

Base diffusion   Base diffusion

Emitter diffusion   Emitter diffusion

(Courtesy Motorola Monitor.)

Figure 10.22. The surface appearance of the monolithic NAND gate after the isolation, base, and emitter diffusion processes. The masks employed in each case are also included.

Base and Emitter Diffusion Processes

The isolation diffusion process is followed by the base and emitter diffusion cycles. The sequence of steps in either case is the same as that encountered in the description of the isolation diffusion process. Although "base" and "emitter" refer specifically to the transistor structure, necessary parts (layers) of each element (resistor, capacitor, and diodes) will be formed during each diffusion process. The surface appearance of the NAND gate after the isolation base and emitter diffusion processes appears in Fig. 10.22. The mask employed in each process is also provided next to each photograph.

The cross section of the transistor of Fig. 10.12 will appear as shown in Fig. 10.23 after the base and emitter diffusion cycles.

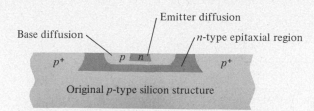

Base diffusion

Emitter diffusion

n-type epitaxial region

p^+ p n p^+

Original p-type silicon structure

Figure 10.23. Cross section of the transistor of Fig. 10.12 after the base and emitter diffusion cycles.

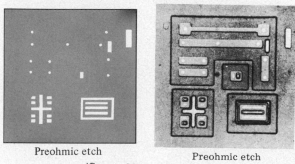

Preohmic etch

Preohmic etch

(Courtesy Motorola Monitor.)

Figure 10.24. Surface appearance of the chip of Fig. 10.22 after the preohmic etch cycle. The mask employed is also included.

Preohmic Etch

In preparation for a good ohmic contact, n^+ regions (see Section 10.3) are diffused into the structure as clearly indicated by the light areas of Fig. 10.24. Note the correspondence between the light areas and the mask pattern.

Metalization

A final masking pattern exposes those regions of each element to which a metallic contact must be made. The entire wafer is then coated with a thin layer of aluminum that after being properly etched will result in the desired interconnecting conduction pattern. A photograph of the completed metalization process appears in Fig. 10.25.

The complete structure with each element indicated appears in Fig. 10.26. Try to relate the interconnecting metallic pattern to the original circuit of Fig. 10.12a.

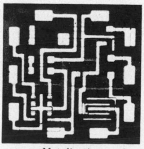

Metalization

(Courtesy Motorola Monitor.)

Figure 10.25. Completed metalization process.

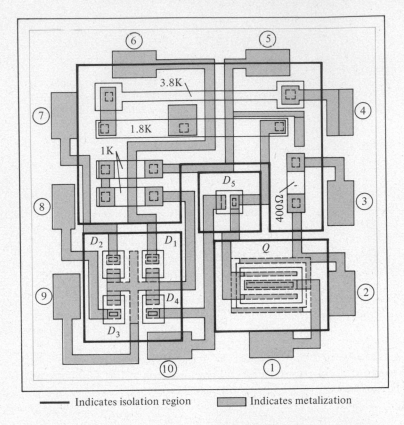

Figure 10.26. Monolithic structure for the NAND gate of Fig. 10.12.

—— Indicates isolation region   ▢ Indicates metalization

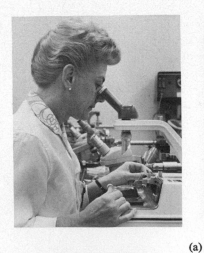

(Courtesy Autonetics, North American Rockwell Corp.)

(a)

(Courtesy Texas Instruments, Inc.)

(b)

(Courtesy Motorola, Inc.)

Figure 10.27. (a) Scribing and (b) breaking of the monolithic wafer into individual chips.

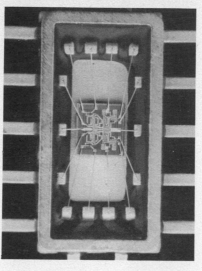

(a)

(b)

(c)

(Courtesy Texas Instruments, Inc.)

Figure 10.28. Monolithic packaging techniques: (a) flat package; (b) TO (top-hat)-type package; (c) dual in-line plastic package.

Packaging

Once the metalization process is complete, the wafer must be broken down into its individual chips. This is accomplished through the scribing and breaking processes depicted in Fig. 10.27. Each individual chip will then be packaged in one of the three forms indicated in Fig. 10.28. The name of each is provided in the figure.

SEC. 10.5 MONOLITHIC INTEGRATED CIRCUIT—THE NAND GATE

467

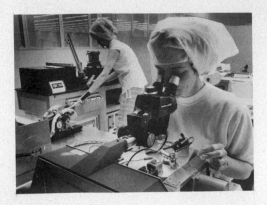

(a)

(Courtesy Autonetics, North American Rockwell Corp.)

(b)

(Courtesy Texas Instruments, Inc.)

(c)

(Courtesy Texas Instruments, Inc.)

Figure 10.29. Production testing.

Testing

The final production stage, as with every commercial electronic package, is the testing of the system. As indicated in Fig. 10.4 this can demand a good percentage of the manufacturing costs. Photographs of various testing procedures appear in Fig. 10.29.

10.6 THIN AND THICK FILM INTEGRATED CIRCUITS

The general characteristics, properties, and appearance of thin and thick film integrated circuits are similar although they both differ in many respects from the monolithic integrated circuit. They are not formed within a semiconductor wafer but *on* the surface of an insulating substrate such as glass or an appropriate ceramic material. In addition, *only* passive elements (resistors, capacitors) are formed through thin or thick film techniques on the insulating surface. The active elements (transistors, diodes) are added as *discrete* elements to the surface of the structure after the passive elements have been formed. The discrete active devices are frequently produced using the monolithic process.

Two thin film integrated circuits appear in Fig. 10.30. Note the active elements added on the surface between the proper aluminum contacts. The interconnecting conduction pattern and the passive elements are prepared through masking techniques.

The primary difference between the thin and thick film techniques is the process employed for forming the passive components and the metallic conduction pattern. The thin film circuit employs an evaporation or cathode-sputtering technique; the thick film employs silk-screen techniques. Priorities do not permit a detailed description of these processes here.

In general, the passive components of film circuits can be formed with a broader range of values and reduced tolerances as compared to the monolithic IC. The use of discrete elements also increases the flexibility of design of film circuits although, obviously, the resulting circuit will be that much larger. The cost of film circuits with a larger number of elements is also, in general, considerably higher than that of monolithic integrated circuits.

(Courtesy Autonetics, North American Rockwell Corp.)

Figure 10.30. Thin-film integrated circuits.

10.7 HYBRID INTEGRATED CIRCUITS

The term *hybrid integrated circuit* is applied to the wide variety of multichip integrated circuits and also those formed by a combination of the film and monolithic IC techniques. The multichip integrated circuit employs either the monolithic or film technique to form the various components, or set of individual circuits,

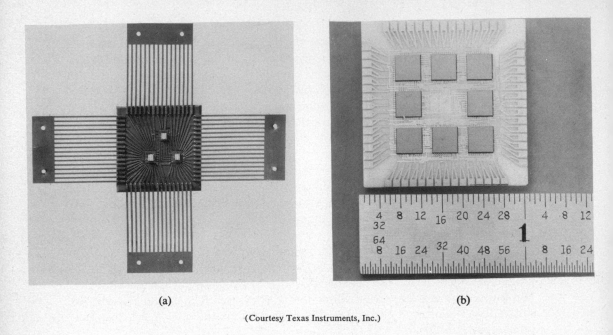

(a)   (b)

(Courtesy Texas Instruments, Inc.)

Figure 10.31. Hybrid integrated circuits.

which are then interconnected on an insulating substrate and packaged in the same container. Integrated circuits of this type appear in Fig. 10.31. In a more sophisticated type of hybrid integrated circuit the active devices are first formed within a semiconductor wafer which is subsequently covered with an insulating layer such as SiO_2. Film techniques are then employed to form the passive elements on the SiO_2 surface. Connections are made from the film to the monolithic structure through "windows" cut in the SiO_2 layer.

differential and operational amplifiers

11.1 BASIC DIFFERENTIAL AMPLIFIER

An amplifier is an electronic circuit containing transistors, FETs, or IC circuits that provides voltage gain. It may also provide current gain, or power gain, or allow impedance transformation. Since it is a basic part of practically every electronic application, the amplifier is an essential circuit. Amplifiers, as we have already discovered, may be classified in many ways. There are low-frequency amplifiers, audio amplifiers, ultrasonic amplifiers, radio-frequency (RF) amplifiers, wide-band amplifiers, video amplifiers, each type operating in a prescribed frequency range. We have considered small-signal and large-signal amplifiers and amplifiers that may be interconnected as either RC-coupled or transformer-coupled.

The *differential amplifier* is a special type of circuit that is used in a wide variety of applications. Let us consider a number of basic properties of differential amplifiers. Figure 11.1 shows a block symbol of a differential amplifier unit. As shown, there are two separate input (1 and 2) and two separate output (3 and 4) terminals. We must first consider the relation between these terminals to obtain an understanding of how the differential amplifier (D.A.) may be applied. Notice that in Fig. 11.1 a ground connection is shown separately to show that both input or output terminals may be different from ground. Voltages may be applied to either or both input terminals and output voltages may appear at both output terminals. However, there are some very specific phase relations between both input and both output terminals.

Figure 11.2 shows the block and circuit diagrams of a basic differential amplifier to be used in the following discussion. There are two inputs and outputs shown in the block diagram. Inputs are applied essentially to each base of the two separate

471

Figure 11.1. Block symbol of a differential amplifier.

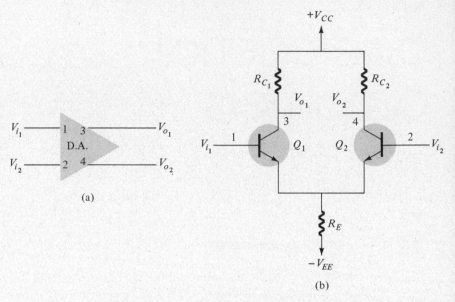

(a)

(b)

Figure 11.2. Basic differential amplifier: (a) block diagram; (b) circuit diagram.

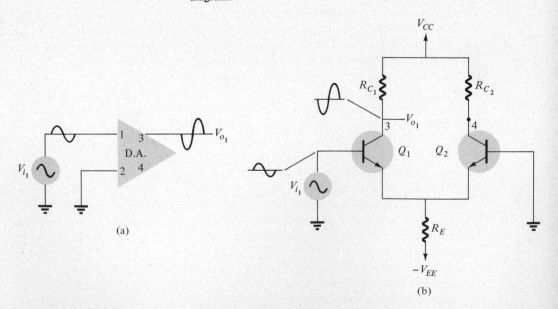

(a)

(b)

Figure 11.3. Single-ended operation of differential amplifier: (a) block diagram; (b) circuit diagram.

transistors. As shown, however, the transistor emitters are connected to a common-emitter resistor so that the two output terminals V_{o_1} and V_{o_2} are affected by either or both input signals. The outputs are taken from the collector terminals of each transistor. The input and output terminals are also numbered to facilitate reference. There are two supply voltages shown in the circuit diagram and it should be carefully noted that no ground terminal is indicated within the circuit although the opposite points of both positive- and negative-voltage supplies are understood to be connected to ground.

Single-Ended Input D.A.

Consider first the operation of the differential amplifier with a single input signal applied to terminal 1, with terminal 2 connected to ground (0 V). Only one output signal at either terminal 3 or terminal 4 is considered. Figure 11.3 shows the block and circuit diagrams for input signal V_{i_1} at terminal 1 and output V_{o_1} at terminal 3. The block diagram shows a sinusoidal input and an amplified, inverted output. The circuit diagram shows the sinusoidal input applied to the base of a transistor with the amplified output at the collector inverted, as we would expect from past knowledge of a single-stage transistor amplifier.

With input 2 grounded it might seem that there is no output at terminal 4, but this is incorrect. The block diagram of Fig. 11.4 shows the operation of the differential amplifier with the V_{o_2} output at terminal 4 resulting from an input V_{i_1} at terminal 1. The input at terminal 1 is shown as V_{i_1}, which is a small sinusoidal voltage measured with respect to ground. Since an emitter resistor is connected in common with both emitters, a voltage that is developed by V_{i_1} appears at the common-emitter point. This sinusoidal voltage, measured with respect to ground, is approximately one-half the magnitude and is in phase with V_{i_1} because it results from emitter-follower action of the circuit.

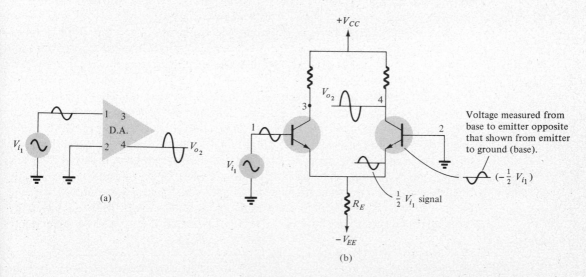

Figure 11.4. Single-ended operation of differential amplifier.

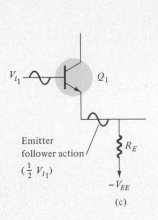

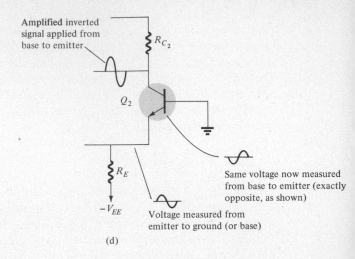

Figure 11.4. (continued)

To be sure that this is clearly understood, the part of the circuit, acting as an emitter follower is shown in Fig. 11.4c. An input applied to the base of Q_1 appears in phase and about one-half the magnitude at the emitter of Q_1 for the emitter-follower part of the circuit shown. Recall that for an emitter follower the gain is less than unity[1] (with no phase reversal). This emitter signal is measured with respect to ground. Figure 11.4d shows the part of the circuit with the emitter voltage affecting the operation of transistor Q_2. The voltage at the emitter of Q_2 is the same as that of Q_1 (since the emitters are connected together) and appears from the emitter of Q_2 to ground or to the base of Q_2 (since that is connected to ground). If the voltage measured from the emitter to base of Q_2 is in phase with input V_{i_1} as shown, the voltage measured from base to emitter of Q_2 is the same signal with opposite polarity. Thus, by measuring from base to emitter of Q_2 a voltage of about one-half the magnitude of V_{i_1} is obtained, but the signal is opposite in polarity to that of V_{i_1}. The amplifier action of transistor Q_2 and load resistor R_{C_2} provides an output at the collector of Q_2 that is amplified and inverted from the signal developed across base to emitter of Q_2.

In summary, an input V_{i_1} is applied to input 1 and an amplified, in-phase signal V_{o_2} results at output terminal 4. Because the input at terminal 2 is grounded this does not mean that no output occurs at terminal 4. It should be clear that internal connection (of common emitters) results in the input at terminal 1 causing an output at terminal 4. In fact, we can now see that the input at terminal 1 causes output signals at both terminals 3 and 4. In addition, these outputs are opposite in phase and of about the same magnitude. Finally, we should see (as in Fig. 11.5) that the output at terminal 4 is in phase with the input at terminal 1, while the output at terminal 3 is opposite in phase to the input at terminal 1. It should be understood

---

[1]Emitter-follower gain in this circuit connection is actually one-half since the load impedance equals the output impedance of the emitter follower.

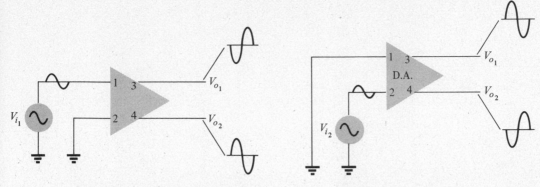

Figure 11.5. Single-input, opposite-phase outputs.

Figure 11.6. Single-ended input to terminal.

from the previous discussion that an input applied to terminal 2 with terminal 1 grounded will result in output voltages as shown in Fig. 11.6.

Differential Input (Double-Ended Input) Operation

In addition to using only one input to operate the differential amplifier circuitry, it is possible to apply signals to each input terminal, with opposite outputs appearing at the two output terminals. The usual use of the *double-ended* or *differential* mode of input is when the two input signals are themselves opposite in phase (180° out of phase), and about the same magnitude. Figure 11.7 shows such a situation.

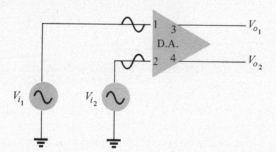

Figure 11.7. Operation with differential input signals.

We now must consider how each input affects the outputs and what the resulting output signal looks like. This can be done using the *superposition principle*, considering each input applied separately with the other at 0 V and summing the resulting output voltages at each terminal. Figures 11.8a and 11.8b show the result of each input acting alone and Fig. 11.8c shows the resulting overall operation. The input applied to terminal 1 results in an opposite-phase, amplified output at terminal 3 *and* an in-phase, amplified output at terminal 4. Assume that the inputs are about equal in magnitude and that the output magnitudes are about equal, of value V, for discussion purposes.

The input applied to terminal 2 results in an opposite-phase, amplified output at

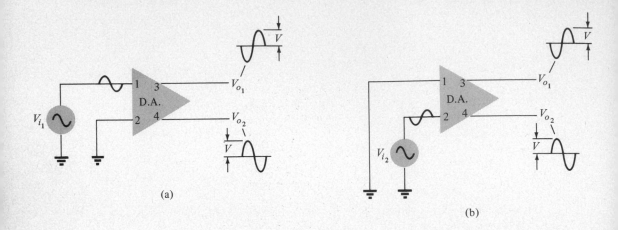

(a)

(b)

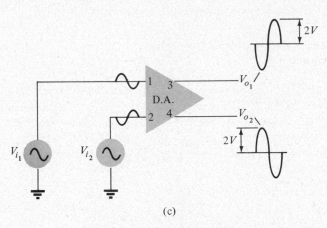

(c)

Figure 11.8. Differential operation of amplifier: (a) $V_{i_2} = 0$; (b) $V_{i_1} = 0$; (c) both inputs present.

terminal 4 *and* an in-phase, amplified output at terminal 3. The magnitudes of the outputs will both be V since the input magnitudes were assumed to be about the same. It is important to note that the outputs in each case are of the same phase at each output terminal. By superposition, the resulting signals at each output terminal are added, and we obtain the full operation of the circuit shown in Fig. 11.8c. The output at each terminal is twice that resulting from single-ended operation because the outputs due to each input are in phase. If the inputs applied were both in phase (or if the same input were applied to both input terminals), the resulting signals due to each input acting alone would be opposite in phase at each output and the resulting output would *ideally* be about 0 V, as shown in Fig. 11.9.

To bring the operation as single- and double-ended differential amplifier states into full perspective consider the connection of two differential amplifiers shown in Fig. 11.10. From the previous discussion, if the amplifiers had identical single-ended gains, then the outputs of stage 1 would be larger than the inputs by the

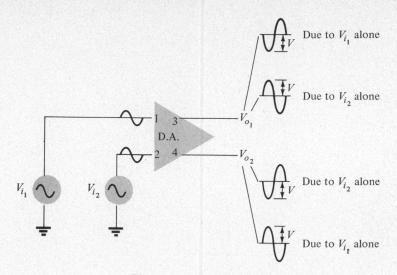

Figure 11.9. Operation with in-phase input signals.

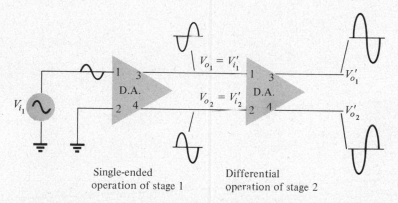

Single-ended
operation of stage 1

Differential
operation of stage 2

Figure 11.10. Single- and double-ended operation of differential amplifier stages.

amount of amplifier gain, while the outputs of stage 2 would be larger than the inputs to stage 2 by twice the amplifier gain. The initial signal, from an antenna of a radio, or a phonograph pickup cartridge, etc., is single-ended and is used as such. The second differential amplifier stage, however, could be operated double-ended to obtain twice the stage gain. Either output of stage 2 (or both) could then be used as amplified signals to the next section of the system. Although differential operation requires about equal and opposite phase signals, this is often available, especially after one single-ended stage of gain.

11.2 DIFFERENTIAL AMPLIFIER CIRCUITS

Having considered some features of a differential amplifier, we shall now look into some details of differential amplifier circuits. In particular, we shall consider the voltage gain of the stage and its input and output impedance. A basic circuit of a

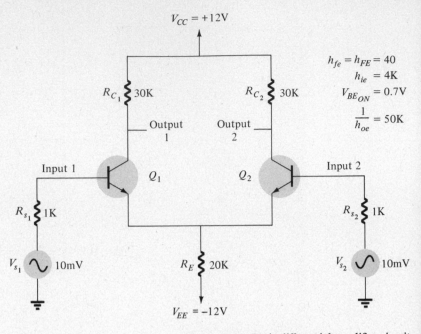

Figure 11.11. Basic differential amplifier circuit.

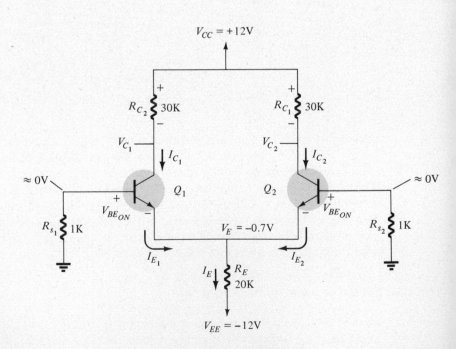

Figure 11.12. dc bias action of circuit.

differential amplifier is shown in Fig. 11.11. Input signals are shown as a voltage source with source resistance in the general case.

dc Bias Action of Circuit

Before considering the main action of the circuit as a voltage amplifier let us see how the circuit is biased to operate. A circuit diagram (Fig. 11.12) shows the main voltage and current features of the circuit for dc operation. No ac signal sources are present, the sources having been set to 0 V with only the source resistances present. The base emitter of Q_1 is forward biased by the $-V_{EE}$ battery from ground through resistor R_{S_1}, through the base emitter, through resistor R_E to $-V_{EE}$ (see Fig. 11.13a). The voltage drop across the forward-biased base emitter is about

$$V_{BE_{ON}} = 0.7\ \text{V}$$

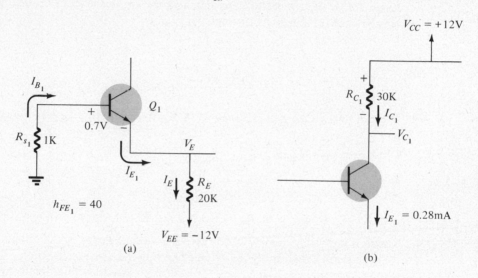

Figure 11.13. Partial circuits of differential amplifier: (a) input section; (b) output section.

We would have to write a number of equations to solve for the dc voltages and currents. However, it is possible to use good approximations to make the calculations more direct. For example, the dc voltage drop across source resistor R_{S_1} will be small as the following calculation indicates (assuming a typical base current in the order of microamperes)

$$I_{B_1} R_{S_1} = (100\ \mu\text{A})(1\ \text{K}) = 100\ \text{mV} = 0.1\ \text{V}$$

If the base current were only 10 μA, then the dc voltage drop across R_{S_1} would be 10 mV, which is negligible. On the other hand, a source resistance of 10 K with base current of 100 μA would result in a voltage drop of 1 V, which is not negligible. For our purposes we shall assume the voltage drop to be small (which is often correct) and check later in the calculations to be sure we were able to make such an assumption.

If we assume that[2]

$$V_{B_1} = 0 \text{ V}$$

then the emitter voltage is directly calculated to be

$$V_E = V_{E_1} = V_{B_1} - V_{BE_1} \tag{11.1}$$

$$= 0 - 0.7 \text{ V} = -0.7 \text{ V}$$

The current through resistor R_E is then calculated directly to be

$$I_E = \frac{V_E - V_{EE}}{R_E} \tag{11.2}$$

$$= \frac{-0.7 - (-12)}{20 \text{ K}} = \frac{11.3}{20} = 0.565 \text{ mA}$$

The current through resistor R_E is made up of the emitter currents from each transistor. If the transistors are matched, the emitter current of each transistor is one-half the total current through R_E.

$$I_{E_1} = I_{E_2} = \frac{I_E}{2} \tag{11.3}$$

$$= \frac{0.565 \text{ mA}}{2} = 0.2825 \text{ mA} \cong 0.28 \text{ mA}$$

We can now check our assumption of V_{B_1} by calculating I_{B_1} as follows:

$$I_{B_1} = \frac{I_{E_1}}{1 + h_{FE_1}} \tag{11.4}$$

$$= \frac{0.28 \text{ mA}}{1 + 40} = 6.8 \text{ } \mu\text{A}$$

$$V_{B_1} = I_{B_1} R_{S_1} = 6.8 \text{ } \mu\text{A} \times 1 \text{ K} = 6.8 \text{ mV}$$

which is negligible compared to the other voltage drops in the circuit.

The output section of the circuit can now be directly handled. Figure 11.13b shows a partial circuit diagram. The collector current is obtained from the calculation of emitter current.

$$I_{C_1} \cong I_{E_1} \tag{11.5}$$

$$= 0.28 \text{ mA}$$

and the collector voltage is

$$V_{C_1} = V_{CC} - I_{C_1} R_{C_1} \tag{11.6}$$

$$= 12 - (0.28 \text{ mA})(30 \text{ K})$$

$$= 12 - 8.4 = 3.6 \text{ V}$$

In summary, the base voltage being about 0 V, the emitter voltage is then fixed at about -0.7 V. The emitter currents of each transistor are then set by resistor R_E, this also setting the collector current, which is approximately equal to the emitter current of either transistor. For the collector current resulting the output dc

---

[2]This assumes that $(\beta + 1) 2R_2 \gg R_{S_1}$, a usually very good assumption.

voltage V_{C_1} is adjusted by choosing the value of R_{C_1}. These considerations hold true for the currents and voltages of transistor Q_2. The output voltage at V_{C_1} will probably be set about in the center of the potential voltage swing from 0 V to $+V_{CC}$. In the present circuit a lower value of R_{C_1} would result in a lower voltage drop with the value of V_{C_1} then higher, a value of about 6 V being desirable. It should be clearly noted that the value of R_{C_1} has almost no effect on the current I_{C_1}, which depends on the value of resistor R_E in setting I_E.

If an input ac signal causes the transistor to approach cutoff, the output voltage will approach V_{CC} in value. If, however, the input ac signal causes the transistor to turn on, the lowest the output voltage can go is about -0.7 V, the fixed dc voltage at the emitter. Thus, the largest possible voltage swing at V_{C_1} would be from near 0 to $+12$ V in the present circuit. Biasing the circuit in the center of this range will allow the largest voltage swing from dc bias before distortion (clipping) occurs.

ac Operation of Differential Amplifier Circuit

To consider the ac operation of the circuit all dc voltage supplies are set at zero and the transistors are replaced by small-signal ac equivalent circuits. Figure 11.14 shows the resulting ac equivalent circuit, with the transistors replaced by hybrid equivalent circuits. The circuit obviously appears complex, and analyzing the total

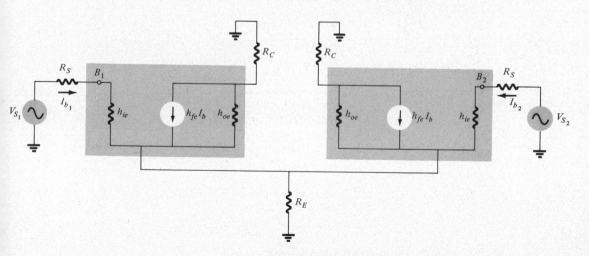

Figure 11.14. ac equivalent circuit of differential amplifier.

circuit would become involved. Again, we can break up the calculations by using some simplifying approximations so that smaller parts of the circuit can be analyzed separately. In addition,

$$h_{ie_1} = h_{ie_2} = h_{ie}, h_{fe_1} = h_{fe_2} = h_{fe}, h_{oe_1} = h_{oe_2} \cong 0$$

and
$$R_{C_1} = R_{C_2} = R_C, R_{s_1} = R_{s_2} = R_s$$

Figure 11.15a shows the partial ac equivalent circuit of the input for transistor Q_1. Looking into the emitter of transistor Q_2 a small ac equivalent resistance is present, equal in value to

$$R_{e_2} = \frac{R_s + h_{ie}}{1 + h_{fe}} \qquad (11.7)$$

As a general statement, the ac resistance seen looking into the emitter of a transistor circuit is approximately equal to the value of the transistor h_{ie} plus source resistances divided by the transistor h_{fe}. For the values of Fig. 11.11

$$R_{e_2} = \frac{1\,\text{K} + 4\,\text{K}}{1 + 40} = 122\,\Omega$$

The parallel combination of resistors R_E and R_{e_2} gives an equivalent ac resistance of

$$\frac{R_{e_2} R_E}{R_{e_2} + R_E} = \frac{122 \times 20{,}000}{122 + 20{,}000} \cong 122\,\Omega$$

Since the differential amplifier circuit generally has an R_E of large value, we can make the approximate statement that if

$$R_E \gg R_{e_2}$$

the parallel combination is approximately R_{e_2} in value, as shown in Fig. 11.15a. Using the resulting ac equivalent circuit, we see that the value of the ac base current is calculated to be

$$I_{b_1} = \frac{V_{s_1} - V_{s_2}}{R_s + h_{ie} + (1 + h_{fe})R_{e_2}} = \frac{V_{s_1} - V_{s_2}}{2(R_s + h_{ie})} \qquad (11.8a)$$

and defining $V_d \equiv V_{s_1} - V_{s_2}$ as the difference input voltage

$$I_{b_1} = \frac{V_d}{2(R_s + h_{ie})} \qquad (11.8b)$$

The output voltage can be written as

$$V_{o_1} = -I_{C_1} R_{C_1}$$

Using $I_{C_1} = h_{fe} I_{b_1}$ and I_{b_1} as expressed in Eq. (11.8b) results in

$$V_{o_1} = -\frac{h_{fe} R_{C_1}}{2(R_S + h_{ie})} V_d$$

The circuit ac difference gain is then

$$\boxed{A_{v_1} = \frac{V_{o_1}}{V_d} = -\frac{h_{fe} R_{C_1}}{2(R_S + h_{ie})}} \qquad (11.9)$$

Using the values of the circuit of Fig. 11.11 results in

$$A_{v_1} \cong -\frac{40 \times 30\,\text{K}}{2(1\,\text{K} + 4\,\text{K})} = -120$$

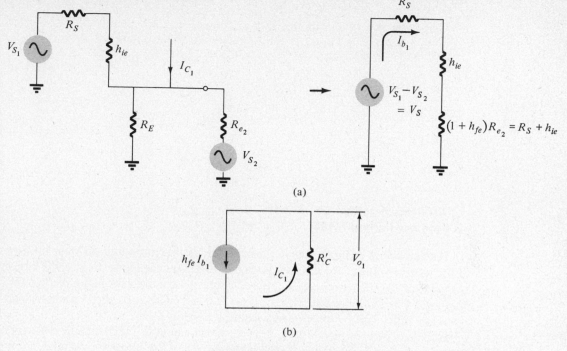

(a)

(b)

Figure 11.15. Partial ac equivalent circuit of difference amplifier.

Input Resistance

From the ac equivalent circuit of Fig. 11.15a the input resistance of the circuit seen from the source is

$$R_{i_1} = h_{ie} + (1 + h_{fe})R_{e_2} \tag{11.10a}$$

which can be expressed as

$$R_{i_1} = \frac{R_{S_2} + h_{ie}}{1 + h_{fe}} \tag{11.10b}$$

$$= h_{ie} + R_{S_2} + h_{ie} \tag{11.10c}$$

$$\boxed{R_{i_1} = 2h_{ie} + R_{S_2}} \tag{11.10d}$$

For the circuit of Fig. 11.11

$$R_{i_1} = 2(4\text{ K}) + 1\text{ K} = 9\text{ K}$$

Output Resistance

From the ac equivalent circuit of Fig. 11.15b the resulting approximate output resistance (assuming $h_{oe} \cong 0$) is

$$\boxed{R_{o_1} = R_C} \tag{11.11}$$

which has already been determined to be 30 K.

SEC. 11.2 DIFFERENTIAL AMPLIFIER CIRCUITS

EXAMPLE 11.1 Calculate the input and output resistance of a difference amplifier circuit as in Fig. 11.11 for the following circuit values: $R_{C_1} = R_{C_2} = 15$ K, $R_E = 10$ K, $h_{fe} = 60$, $h_{ie} = 2.5$ K, and $R_{s_1} = R_{s_2} = 600$ Ω.

Solution:

$$R_{i_1} = 2h_{ie} + R_{s_2}$$
$$= 2(2.5 \text{ K}) + 0.6 \text{ K}$$
$$= 5.6 \text{ K}$$
$$R_{o_1} = R_C = 15 \text{ K}$$

Difference Amplifier Circuit with Constant-Current Source

One important thing to note in the previous circuit considerations was that with $R_{e_2} \ll R_E$, the value of R_E was very large and therefore negligible. In fact, the larger the value of R_E, the better certain desirable aspects of a difference amplifier circuit. The main reason for R_E being very large is a circuit factor called *common-mode rejection*, which will be discussed in detail in Section 11.3.

However, dc bias calculations showed that the emitter (and thus the collector) current is determined partly by the value of R_E. For a fixed negative-voltage supply of, say, $V_{EE} = -20$ V a value of R_E of 10 K would limit the emitter resistor current to about

$$I_E \cong \frac{V_{EE}}{R_E} = \frac{20 \text{ V}}{10 \text{ K}} = 2 \text{ mA}$$

If a preferably larger value of $R_E = 100$ K were used, the value of dc emitter resistor current would then be

$$I_E \cong \frac{V_{EE}}{R_E} = \frac{20 \text{ V}}{100 \text{ K}} = 0.2 \text{ mA} = 200 \text{ } \mu\text{A}$$

and if a very large value of $R_E = 1$ M were used,

$$I_E = \frac{20 \text{ V}}{1 \text{ M}} = 20 \text{ } \mu\text{A}$$

We see that as larger values of R_E are used the resulting dc emitter current becomes much too small for proper operation of the transistors since the emitter and collector current of each transistor is one-half the already very small emitter resistor current.

One way to achieve high ac resistance while still allowing reasonable dc emitter currents is to use a constant-current source as shown in Fig. 11.16. The value of I_E could be set by the constant-current circuit to any desired value—1, 10, 20 mA, and so on. The ac resistance of a constant-current source is ideally infinite and practically from 100 K to about 1 M.

A practical difference amplifier circuit containing a constant-current source is shown in Fig. 11.17. To determine the dc currents and voltages let us first consider the details of the constant-current circuit.

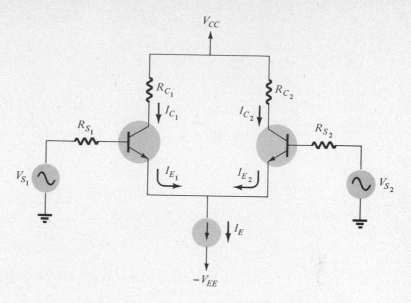

Figure 11.16. Difference amplifier with constant-current source.

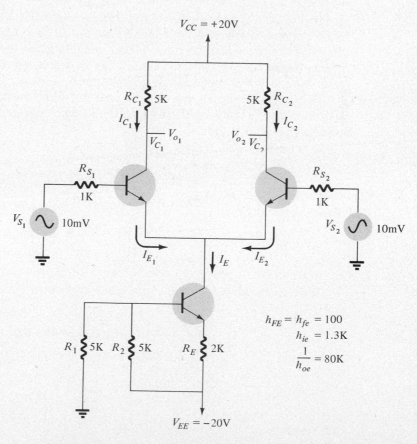

Figure 11.17. Practical difference amplifier circuit with constant-current source.

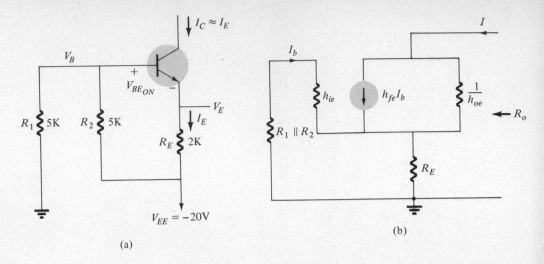

$$I_C \approx I_E$$

(a)

(b)

Figure 11.18. Constant-current circuit: (a) dc considerations; (b) ac considerations.

dc Operation

The constant-current section of the difference amplifier is shown in Fig. 11.18a. No connection to the collector is shown since the collector current is determined by the value of emitter current set by the base-emitter section of the circuit. To a great degree the amount of collector current can be set at any desired value without regard to the circuit connected to the collector.

To simplify the calculation of I_E we note that if the resistance looking into the transistor base is much larger than R_2, the base voltage can be calculated by the voltage divider of R_1 and R_2. That is, if

$$(1 + h_{FE})R_E \gg R_2$$

then

$$V_B = \frac{R_1}{R_1 + R_2}(-V_{EE}) \qquad (11.12)$$

For the present circuit

$$(1 + 100)(2\ \text{K}) = 202\ \text{K} \gg 5\ \text{K}$$

so that

$$V_B = \frac{5\ \text{K}}{5\ \text{K} + 5\ \text{K}}(-20) = -10\ \text{V}$$

The emitter voltage is less than the base voltage by the voltage drop $V_{BE_{ON}}$

$$V_E = V_B - V_{BE_{ON}} \qquad (11.13)$$

$$= -10 - (0.7) = -10.7\ \text{V}$$

The emitter current is then calculated to be

$$I_E = \frac{V_E - V_{EE}}{R_E} \qquad (11.14)$$

$$= \frac{-10.7 - (-20)\ \text{V}}{2\ \text{K}} = 4.65\ \text{mA}$$

The emitter current is held constant fairly well and variations in the difference amplifier section have almost no effect on the value of I_E. Once I_E is determined, the remaining dc bias calculations are the same as those previously considered.

The emitter current of each transistor is then

$$I_{E_1} = I_{E_2} = \frac{I_E}{2} \tag{11.15}$$

$$= \frac{4.65}{2} \text{ mA} \cong 2.3 \text{ mA}$$

and

$$I_{C_1} = I_{C_2} \cong I_{E_1} = I_{E_2} \tag{11.16}$$

$$= 2.3 \text{ mA}$$

The collector voltage is, as before,

$$V_C = V_{CC} - I_C R_C \tag{11.17}$$

$$= 20 - (2.3 \text{ mA})(5 \text{ K}) = 20 - 11.5 = 8.5 \text{ V}$$

ac Operation

The ac action of the constant-current source is that of a very high resistance—ideally infinite. An ac equivalent of the constant-current circuit is shown in Fig. 11.18b. From the equivalent circuit an expression for the circuit ac output impedance can be derived

$$R_o \cong \frac{1}{h_{oe}} \left[1 + \frac{h_{fe} R_E}{R_E + h_{ie} + R_1 \| R_2} \right] \tag{11.18}$$

For the circuit of Fig. 11.17 the calculation of R_o is

$$R_o = 80 \text{ K} \left[1 + \frac{100(2\text{K})}{2 \text{ K} + 1.3 \text{ K} + 2.5 \text{ K}} \right] = 2.84 \text{ M}$$

Thus, the circuit provides a high ac impedance of 2.84 M while providing a dc bias current around 5 mA.

Improved Constant-Current Circuit

An improved version of the constant-current circuit is shown in Fig. 11.19. A Zener diode is used to ensure that the current remains constant. The Zener diode conducts when the reverse-bias voltage exceeds the Zener breakdown voltage, V_Z. The Zener diode will then conduct keeping the voltage across the diode fixed at V_Z, for a wide range of current values. In the circuit, the emitter current is calculated from the voltage drops around the loop containing the Zener diode and emitter resistor

$$+V_Z - V_{BE_{ON}} - I_E R_E = 0$$

$$I_E = \frac{V_Z - V_{BE_{ON}}}{R_E} \tag{11.19}$$

For a Zener voltage of $V_Z = 10$ V, we calculate

$$I_E = \frac{V_Z - V_{BE}}{R_E} = \frac{10 - 0.7 \text{ V}}{2\text{K}} = 4.65 \text{ mA}$$

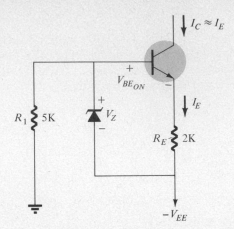

Figure 11.19. Constant-current circuit using Zener diode.

All other dc calculations then remain the same.

The ac output impedance for Fig. 11.19 can then be calculated as

$$R_o = \frac{1}{h_{oe}}(h_{fe} + 1)$$

which is

$$R_o = 80\text{ K }(100 + 1) = 8.08\text{ M}$$

We can thus bias the circuit with either bias resistors or using a Zener diode. The improvement with the Zener diode comes from the higher value of R_o. Another circuit form used to provide a constant current is shown in Fig. 11.20. Dc bias calculations to obtain the emitter current, I_E, result in

$$V_B = \frac{R_1}{R_1 + R_2}(-V_{EE} + V_{BE}) = \frac{3.3\text{ K}}{3.3\text{ K} + 3.3\text{ K}}(-15 + 0.7) = -7.5\text{ V}$$

$$I_E = \frac{V_E - V_{BE} - V_{EE}}{R_E} = \frac{-7.15 - 0.7 - (-15)\text{ V}}{1.8\text{ K}} = 4\text{ mA}$$

The dc collector currents are then

$$I_C = \frac{I_E}{2} = \frac{4\text{ mA}}{2} = 2\text{ mA}$$

so that

$$V_C = V_{CC} - I_C R_C = 15\text{ V} - 2\text{ mA }(3.9\text{ K}) = 7.2\text{ V}$$

The ac voltage gain is then (neglecting $1/h_{oe}$)

$$A_v = \frac{V_o}{V_i} = \frac{h_{fe}R_C}{2(h_{ie} + h_{fe}r_e)} = \frac{80(3.9\text{ K})}{2(1.2\text{ K} + 80(50))} = 30$$

where no phase inversion takes place between the output taken from Q_2 with input applied to Q_1. Although the gain is reduced by the inclusion of resistors r_E in the circuit, the gain is more stable. In addition, the input impedance is increased because of r_E.

$$R_i = 2(h_{ie} + h_{fe}r_E) = 2(1.2\text{ K} + 80(50)) = 10.4\text{ K}$$

The ac output impedance can be obtained from Eq. (11.18)

$$R_o = 120\text{ K}\left(1 + \frac{80(1.8\text{ K})}{1.8\text{ K} + 1.2\text{ K} + 1.65\text{ K}}\right) = 3.84\text{ M}$$

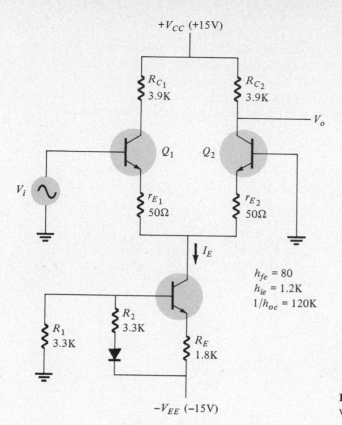

$+V_{CC}$ (+15V)

R_{C_1} 3.9K

R_{C_2} 3.9K

V_o

Q_1 Q_2

V_i

r_{E_1} 50Ω

r_{E_2} 50Ω

I_E

$h_{fe} = 80$
$h_{ie} = 1.2K$
$1/h_{oc} = 120K$

R_1 3.3K

R_2 3.3K

R_E 1.8K

$-V_{EE}$ (−15V)

Figure 11.20. Difference amplifier with emitter resistance.

This value can be further increased by removing resistor R_2 for which we obtain

$$R_o = 120\ K\left(1 + \frac{80(1.8\ K)}{1.8\ K + 1.2\ K}\right) = 5.88\ M$$

11.3 COMMON-MODE REJECTION

One of the more important features of a difference amplifier is its ability to cancel out or reject certain types of unwanted voltage signals. These unwanted signals are referred to as "noise" and can occur as voltages induced by stray magnetic fields in the ground or signal wires, as voltage variations in the voltage supply. What is important in this consideration is that these noise signals are not the signals that are desired to be amplified in the difference amplifier. Their distinguishing feature is that the noise signal appears equally at both inputs of the circuit.

We can say then that any unwanted (noise) signals that appear in phase, or common to both input terminals, will be greatly rejected (cancelled out) at the output of the difference amplifier. The signal that is to be amplified appears at only one input or opposite in phase at both inputs. What we wish to consider in this section is, if undesirable noise does occur, how much does the amplifier reject or cancel out this noise? A measure of this rejection of signals common to both inputs is called the amplifier's *common-mode rejection* and a numerical value is assigned, which is called the *common-mode rejection ratio* (CMRR).

Figure 11.21a shows an amplifier with two input signals. These signals can, in general, be considered to contain components that are exactly opposite in phase *and* components that are exactly in phase. For ideal operation we would want the difference amplifier to provide high gain for the out-of-phase components of the signals and zero gain for the in-phase components of the signals.

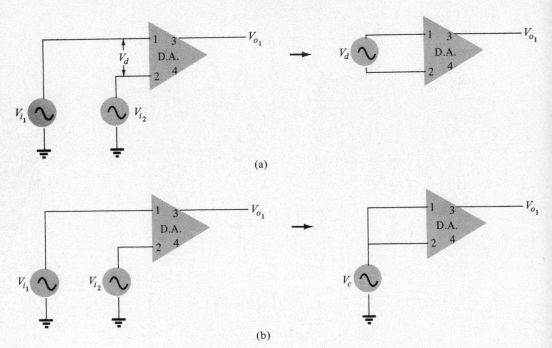

(a)

(b)

Figure 11.21. Difference and common operation: (a) ideal difference-mode operation; (b) ideal common-mode operation.

The voltage measured *from* terminal 1 *to* terminal 2 can be considered as a difference voltage

$$\boxed{V_d = V_{i_1} - V_{i_2}} \tag{11.20}$$

If, as in the ideal case, $V_{i_1} = -V_{i_2}$, we note that

$$V_d = V_{i_1} - (-V_{i_1}) = 2V_{i_1} = -2V_{i_2}$$

In general, there may also exist common components to the input signals. We can define a common input as

$$\boxed{V_c = \tfrac{1}{2}(V_{i_1} + V_{i_2})} \tag{11.21}$$

The ideal case (shown in Fig. 11.21b) is $V_{i_1} = V_{i_2}$, for which

$$V_c = \tfrac{1}{2}(V_{i_1} + V_{i_2}) = V_{i_1} = V_{i_2}$$

From Eqs. (11.20) and (11.21) we can obtain expressions for V_{i_1} and V_{i_2} based on V_c and V_d

$$V_{i_1} = V_c + \frac{V_d}{2} \tag{11.22a}$$

$$V_{i_2} = V_c - \frac{V_d}{2} \tag{11.22b}$$

The output voltages can then be expressed as

$$V_{o_1} = A_1 V_{i_1} + A_2 V_{i_2} \tag{11.23a}$$

$$V_{o_2} = A_2 V_{i_1} + A_1 V_{i_2} \tag{11.23b}$$

where $A_1 =$ negative voltage gain from input terminal 1 to output terminal 3 (with input terminal 2 grounded), and

$A_2 =$ positive voltage gain from input terminal 2 to output terminal 3 (with input terminal 1 grounded).

It is more important to consider the difference and common-mode operation of the amplifier since this allows determining the common-mode rejection of the circuit. This second way of considering the operation of the amplifier provides an output voltage as

$$V_{o_1} = A_d V_d + A_c V_c \tag{11.24a}$$

$$V_{o_2} = -A_d V_d + A_c V_c \tag{11.24b}$$

where $A_d =$ difference-mode gain of the amplifier,

$A_c =$ common-mode gain of the amplifier, and

V_d and V_c are defined in Eqs. (11.20) and (11.21), respectively.

Opposite-Phase Inputs: If the inputs are equal and opposite, $V_{i_1} = V_s$ and $V_{i_2} = -V_s$, then from Eq. (11.20)

$$V_d = V_{i_1} - V_{i_2} = V_s - (-V_s) = 2V_s$$

and from Eq. (11.21)

$$V_c = \tfrac{1}{2}(V_{i_1} + V_{i_2}) = \tfrac{1}{2}[V_s + (-V_s)] = 0$$

so that in Eq. (11.24)

$$V_{o_1} = A_d V_d + A_c V_c = A_d(2V_s) + A_c(0)$$
$$V_{o_1} = 2A_d V_s$$

which shows that only differential-mode operation occurs (and that the overall gain is twice the value of A_d).

In-Phase Inputs: If the inputs are equal and in phase, $V_{i_1} = V_s = V_{i_2}$, then from Eq. (11.20)

$$V_d = V_{i_1} - V_{i_2} = V_s - V_s = 0$$

and from Eq. (11.21)

$$V_d = \tfrac{1}{2}(V_{i_1} + V_{i_2}) = \tfrac{1}{2}(V_s + V_s) = V_s$$

so that in Eq. (11.24)

$$V_{o_1} = A_d V_d + A_c V_c = A_d(0) + A_c V_s$$
$$= A_c V_s$$

which shows that only common-mode operation occurs.

Common-Mode Rejection Ratio (CMRR)

The above calculations indicate how A_d and A_c can be measured in a difference amplifier circuit.

To measure A_d: Set $V_{i_1} = -V_{i_2} = 0.5$ V so that $V_d = 1$ V and $V_c = 0$ V. Under these conditions the output voltage is $A_d \times (1)$ so that the output voltage equals A_d.

To measure A_c: Set $V_{i_1} = V_{i_2} = 0.5$ V so that $V_d = 0$ V and $V_c = 1$ V. Then the output voltage measured equals A_c.

Having measured A_d and A_c for the amplifier we can now calculate a common-mode rejection ratio, which is defined as

$$\boxed{\text{CMRR} = \frac{A_d}{A_c}} \tag{11.25}$$

It should be clear that the desired operation will have A_d very large with A_c very small. That is, the signals appearing opposite in phase will appear greatly amplified at the output terminal, whereas the in-phase signals will mostly cancel out so that the common-mode gain A_c is very small. Ideally, A_d is very large and A_c is zero so that the value of CMRR is infinite. The larger the value of CMRR the better the common-mode rejection of the circuit.

It is possible to obtain an expression for the output voltage as follows:

$$\boxed{V_{o_1} = A_d V_d \left(1 + \frac{1}{\text{CMRR}} \frac{V_c}{V_d}\right)} \tag{11.26}$$

Even if both V_c and V_d components of voltage exist at the inputs, the value of $(1/\text{CMRR})(V_c/V_d)$ will be very small, for CMRR very large, and the output voltage will be approximately $A_d V_d$. In other words, the output will be almost completely due to the difference signal with the common-mode input signals rejected (or cancelled out). Some practical examples should help clarify these ideas.

EXAMPLE 11.2 Determine the output voltage of a difference amplifier for input voltages of $V_{i_1} = 150\ \mu$V and $V_{i_2} = 100\ \mu$V. The amplifier has a difference-mode gain of $A_d = 1000$ and the value of CMRR is
(a) 100
(b) 10^5

Solution:

$$V_d = V_{i_1} - V_{i_2} = 150 - 100 = 50 \ \mu V$$

$$V_c = \tfrac{1}{2}(V_{i_1} + V_{i_2}) = \frac{150 + 100}{2} = 125 \ \mu V$$

Note that the common signal is more than twice as large as the difference signal.

(a)
$$V_o = A_d V_d \left(1 + \frac{1}{\text{CMRR}} \frac{V_c}{V_d}\right)$$

$$= A_d V_d \left(1 + \frac{1}{100} \times \frac{125}{50}\right)$$

$$= A_d V_d (1.025)$$

$$= (1000)(50 \ \mu V)(1.025) = \mathbf{51.25 \ mV}$$

The output is only 0.025 or 2.5% more than the output, due only to a difference signal of 50 μV.

The common-mode signal, even larger than the difference component, has been rejected so that only 1.25 mV appear as output.

(b)
$$V_o = A_d V_d \left(1 + \frac{1}{10^5} \times \frac{125}{50}\right)$$

$$= A_d V_d (1.00025)$$

$$\cong 100 \times 50 \ \mu V = \mathbf{50 \ mV}$$

The output in this case is larger than that due to only the difference signal by 0.025%.

Example 11.2 shows that the larger the value of CMRR the better the circuit rejects common-input signals. Thus, one of the important difference amplifier factors to consider is the circuit's common-mode rejection ratio.

As summarized in Fig. 11.22 the difference gain between any input and output terminal is

$$A_d = -\frac{h_{fe} R_C}{2h_{ie}} \tag{11.27}$$

the phase relation between input and output depending on which set of terminals is used.

A common-mode gain can also be calculated as given by

$$A_C = -\frac{R_C}{R_E} \tag{11.28}$$

using a circuit with emitter resistor, or

$$A_C = -\frac{R_C}{R_o} \tag{11.29}$$

for a circuit with constant-current source having output resistance, R_o.

EXAMPLE 11.3 Calculate the difference and common-mode gains for the circuits of Fig. 11.22 and also the respective values of CMRR.

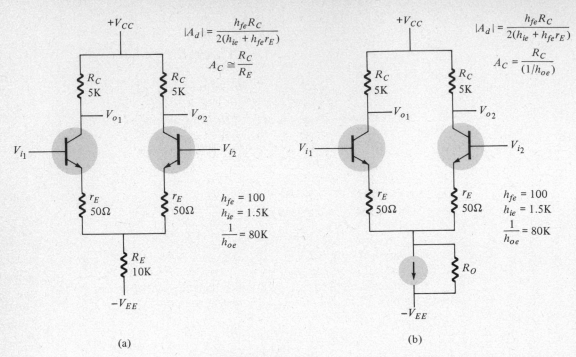

Figure 11.22. Difference amplifiers showing differential and common-mode gains.

Solution:
(a) For Fig. 11.22a

$$A_d = -\frac{h_{fe}R_C}{2h_{ie}} = \frac{-(100)(5\text{ K})}{2(1.5\text{ K})} = -166.7$$

$$A_c = -\frac{R_C}{2R_E} = \frac{-5\text{ K}}{2(10\text{ K})} = -0.25$$

$$\text{CMRR} = \frac{A_d}{A_c} = \frac{-166.7}{-0.25} = 666.8 \ (= \mathbf{56.5\ dB})$$

(b) For the circuit of Fig. 11.22b $A_d = -166.7$ (as in part (a)) using Eq. (11.20) to calculate R_o

$$R_o = 80\text{ K}(101) = 8.08\text{ M}$$

so that A_c is

$$A_c = -\frac{R_C}{R_o} = \frac{-5\text{ K}}{8.08\text{ M}} = -6.2 \times 10^{-4}$$

We then calculate

$$\text{CMRR} = \frac{A_d}{A_c} = \frac{-166.7}{-6.2 \times 10^{-4}} = 2.69 \times 10^5 \ (= \mathbf{108.6\ dB})$$

11.4 PRACTICAL DIFFERENTIAL AMPLIFIER UNITS—IC CIRCUITS

Differential amplifiers are versatile and useful in many areas of electronic operation. They are some of the most widely used linear IC devices. Since it is easier, cheaper, and thus more desirable to use an IC circuit than to build an equivalent circuit using discrete components, we shall consider some typical IC units in this section.

As an example, Fig. 11.23 shows the schematic and block diagrams of an RCA CA3000 IC differential amplifier. The manufacturer lists some of the possible uses of this unit as communications, telemetry, instrumentation, and data processing. Some of the specific applications include RC-coupled feedback amplifier, crystal oscillator, sense amplifier, comparator, and modulator. The manufacturer also lists a number of specifications for the unit, including

> Input impedance: 195 K, typical
> Voltage gain: 37 dB, typical
> CMRR: 98 dB, typical
> Frequency capability: dc to 30 MHz
> Differential input and push-pull output

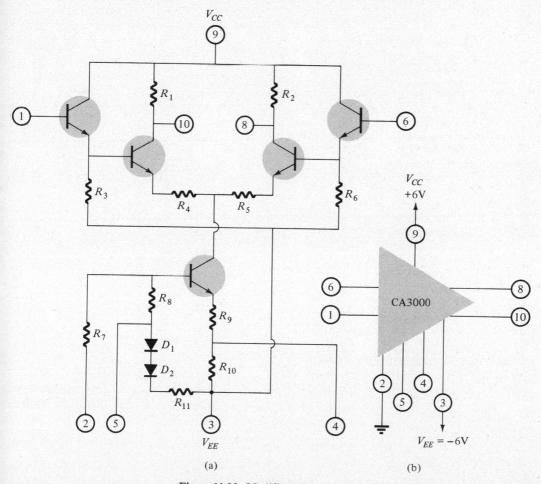

(a)   (b)

Figure 11.23. IC differential amplifier: (a) schematic diagram; (b) block diagram.

Notice how complex the circuit appears. However, we need not consider the details of the circuit at all in order to use it. The external features that will be considered next are sufficient to allow using the amplifier. It is helpful to consider the operation of a basic version of the circuit so that the external operation of the unit makes some sense. The additional complexity in most IC units improves certain of these circuit features but the basic operation remains the same.

Table 11.1 is part of the manufacturer's listing of electrical characteristics and definitions of terms. The following discussion will elaborate on a number of the more important characteristics and give some examples.

TABLE 11.1 Electorical Characteristics, at T = 25°C, V_{CC} = +6 V, V_{EE} = −6 V, f = 1 KHz

CHARACTERISTICS	SYMBOLS	Min.	Typ.	Max.	UNITS
Dynamic Characteristics					
Differential voltage gain, single-ended input	A_{diff}	28	32		dB
Single-ended input impedance	Z_i	70 K	195 K		Ω
Single-ended output impedance	Z_o	5.5 K	8 K	10.5 K	Ω
Common-mode rejection ratio	CMR	80	98		dB
Maximum output voltage swing	$V_o(p-p)$		6.4		$V(p-p)$
Bandwidth at −3-dB point	BW		650		kHz
Static Characteristics					
Input offset voltage	V_{IO}		1.4	8	mV
Input offset current	I_{IO}		1.2	10	μA
Quiescent operating voltage	V_8 or V_{10}		2.6		V
Device dissipation	P_T		30		mW

DEFINITIONS OF TERMS FOR CA3000

INPUT OFFSET VOLTAGE: The difference in the dc voltages that must be applied to the input terminals to obtain equal quiescent operating voltages (zero-output offset voltage) at the output terminals

INPUT OFFSET CURRENT: The difference in the currents at the two input terminals

QUIESCENT OPERATING VOLTAGE: The dc voltage at either output terminal, with respect to ground

DC DEVICE DISSIPATION: The total power drain of the device with no signal applied and no external load current

COMMON-MODE VOLTAGE GAIN: The ratio of the signal voltages developed at either of the two output terminals to the common signal voltage applied to the two input terminals connected in parallel for ac

DIFFERENTIAL VOLTAGE GAIN—SINGLE-ENDED INPUT-OUTPUT: The ratio of the change in output voltage at either output terminal with respect to ground, to difference in the input voltages

COMMON-MODE REJECTION RATIO: The ratio of the full differential voltage gain to the common-mode voltage gain

BANDWIDTH AT—3-dB POINT (BW): The frequency at which the voltage gain of the device is 3 dB below the voltage gain at a specified lower frequency

MAXIMUM OUTPUT VOLTAGE V_o(p-p): The maximum peak-to-peak output voltage swing, measured with respect to ground, that can be achieved without clipping of the signal waveform

SINGLE-ENDED INPUT IMPEDANCE (Z_i): The ratio of the change in input voltage to the change in input current measured at either input terminal with respect to ground

SINGLE-ENDED OUTPUT IMPEDANCE (Z_o): The ratio of the change in output voltage to the change in output current measured at either output terminal with respect to ground

Differential Voltage Gain—Single-Ended Input-Output

The typical value of 32 dB is the gain from one input terminal to either output terminal. This was considered the gain A_1 or A_2 in Sections 11.1–11.3. The manufacturer lists the gain in units of decibels (dB). The relation of decibels and the gain as numerical ratio of output voltage (V_o) to input voltage (V_i) is

$$A_{dB} = 20 \log |A_v| = 20 \log \left| \frac{V_o}{V_i} \right| \qquad (11.30)$$

As an example, a gain of $A_v = 100$ is the same as

$$A_{dB} = 20 \log 100 = 20(2) = 40 \text{ dB}$$

and a gain of $A_v = 10$ is the same as

$$A_{dB} = 20 \log 10 = 20(1) = 20 \text{ dB}$$

A gain of 32 dB is then the same as a gain between 10 and 100 and can be calculated exactly as

$$32 = 20 \log A_v$$
$$1.6 = \log A_v$$
$$A_v = \text{antilog } 1.6 \cong 4 \times 10 = 40$$

Single-Ended Input Impedance (Z_i)

The input impedance is measured at either input terminal. A listed value of 195 K indicates a relatively high value. It should be recalled from Chapter 7 how important input impedance values are when interconnecting amplifier stages or driving the amplifier with a practical voltage source. If the input impedance is not much larger than the source impedance, loading will cause the input voltage to be less than that of the unloaded source signal, resulting in less output voltage.

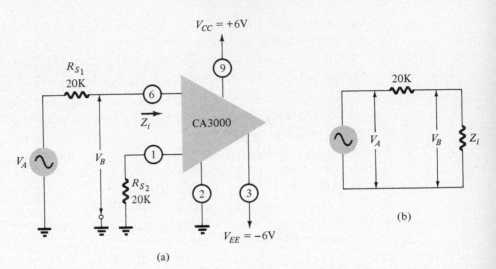

Figure 11.24. Measurement of input impedance (Z_i).

Figure 11.24 shows how the input impedance can be measured. As shown, a voltage (V_A) is applied through a resistor of fixed value (20 K), and due to the amplifier input impedance, the voltage, V_B, is less than V_A. Supply voltages are properly connected and the second input is grounded through a matching 20-K resistor. Output voltage is present although it is of no concern for the present measurements. Note that although an impedance calculation would require measurements of input voltage and current, the present method avoids having to measure current. From the equivalent circuit showing the input as an impedance Z_i we can express the voltage V_B from the voltage divider rule as

$$V_B = \frac{Z_i}{Z_i + 20 \text{ K}} V_A \tag{11.31}$$

Solving for Z_i, we get

$$Z_i = \frac{20 \text{ K} \times V_B}{V_A - V_B} = \frac{20 \text{ K}}{(V_A/V_B) - 1}$$

From the manufacturer's measurements a typical value of $Z_i = 195$ K was obtained. Since this value is only typical, it might be desired to repeat the measurement for the particular unit on hand. If, then, the measurements gave $V_A = 1$ V and $V_B = 0.90$ V, then

$$Z_i = \frac{R_x}{(V_A/V_B) - 1} = \frac{20\text{ K}}{(1/0.9) - 1} = \frac{20\text{ K}}{1.11 - 1} = \frac{20\text{ K}}{0.11} = 198\text{ K}$$

A more direct measurement can be obtained by using a variable resistance (R_x) in place of the fixed 20-K resistor. If this resistance is adjusted until $V_B = \frac{1}{2}V_A$, we note that[3]

$$Z_i = \frac{R_x}{(V_A/V_B) - 1} = \frac{R_x}{(V_A/\frac{1}{2}V_A) - 1} = R_x$$

That is, the value of R_x is then the same value as Z_i. If R_x is a calibrated resistor box, then the value of Z_i can be read directly with no additional calculation required. Thus, with the amplifier properly biased and operating, the use of a single variable resistor unit, and a voltmeter to measure V_A and V_B, the value of the input impedance is directly obtained.

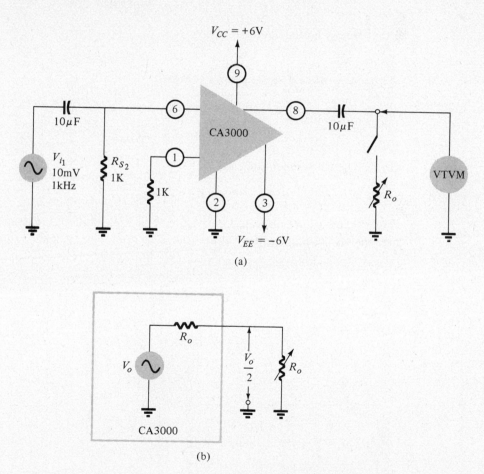

(a)

(b)

Figure 11.25. Measurement of output impedance.

[3]Value of R_x cannot be too large because offset current through R_x should still keep $V_B \cong 0$ V.

Single-Ended Output Impedance (Z_o)

The output impedance of either output terminal with respect to ground is listed at, typically, 8 K. If a measurement is to be made on a particular unit, the technique shown in Fig. 11.25 can be used. First the output voltage, unloaded, is measured. Then the switch is closed, connecting R_o as a load on the amplifier and R_o is adjusted until the output voltage is $V_o/2$. The value of R_o is then equal to the output impedance of the amplifier ($Z_o = R_o$).

Note that if CA3000s are connected in cascade (series), the loading of one stage on another is small; that is, if a source voltage of V_o and source resistance of $Z_o = 8$ K from the output of one stage is connected as input to a second stage ($Z_i = 195$ K), the resulting loaded input signal will be

$$V_i = \frac{Z_i}{Z_i + Z_o}(V_o) = \frac{195}{195 + 8}(V_o) = 0.96\,V_o$$

so that almost no loading takes place.

Common-Mode Rejection Ratio (CMRR)

The common-mode rejection ratio defined as

$$\text{CMRR} = \left| \frac{A_d}{A_c} \right|$$

may also be calculated in decibel units as

$$\text{CMRR} = 20 \log \frac{(2A_1)}{V_o/V_i}$$

where V_i and V_o are measured in the circuit of Fig. 11.26 and A_1 was previously measured as the single-ended voltage gain. The ratio of V_o/V_i is the value of A_c, the

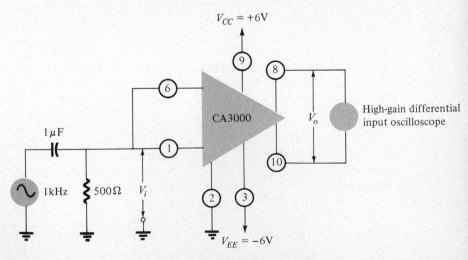

Figure 11.26. Measurement of common-mode gain (A_c).

gain of the amplifier with common-input signal (ideally zero). If the measurements are $V_i = 0.3$ V and $V_o = 0.12$ mV, and if A_1 previously was 20, the value of CMRR is calculated to be

$$\text{CMRR} = 20 \log = \frac{2A_1}{V_o/V_i} = 20 \log \frac{2(20)}{0.12 \text{ mV}/0.3 \text{ V}}$$

$$= 20 \log \frac{40}{0.4 \times 10^{-3}} = 20 \log 10^5 = 20(5) = 100 \text{ dB}$$

Notice that the value of A_c is 0.4×10^{-3}, which is small—the smaller the better.

Maximum Output Voltage Swing, V_o (p-p)

The voltage gain of the amplifier is meaningful only within the linear operating range of the amplifier. Within the linear range the output is an amplified representation of the input signal. It should be clear that the overall voltage swing at the amplifier output is limited at least by the power-supply voltages. The manufacturer lists the maximum output voltage swing as V_o(p-p) = 6.4 V. Thus, the output can vary from the quiescent operating point by a peak-to-peak swing of 6.4 V, at most.

Referring to the static characteristics, the typical quiescent operating point is +2.6 V. If the output voltage is then caused to rise by 3.2 V due to the input signal, the output will rise to +5.8 V, which is about the largest value (limited by the supply voltage). If the output also varies by 3.2 V down from the bias point, it goes to −0.6 V, which is about the value of the emitter voltage of the amplifier. So we see that the overall swing of 6.4 V (p-p) can take place but that any larger voltage swing will result in clipping of the output signal (amplitude distortion).

If we use the typical gain of $A_1 = 20$, we can determine the largest input signal (peak to peak) that can be applied without causing the output to be clipped. Since

$$V_o = A_1 V_i$$

$$V_i = \frac{V_o}{A_1} = \frac{6.4 \text{ V(p-p)}}{20} = 0.32 \text{ V(p-p)}$$

Thus, any input larger than 0.32 V or 320 mV (p-p) will result in the output clipping.

Bandwidth at −3-dB Point (BW)

The amplifier can operate from 0 Hz, or dc, to some upper frequency. When the gain has dropped by 3 dB, the frequency at that point is considered the upper frequency of the amplifier and is, in this case, equal to the bandwidth of the amplifier. Figure 11.27 shows a typical gain-bandwidth curve for the CA3000. The gain has been normalized so that the gain of 32 dB is now 0 dB. When the gain drops by 3 to 29 dB, or on the curve to −3 dB, then the upper frequency (f_u) can be read (not too easily from this graph) as 650 kHz, so that the value of bandwidth is BW = 650 kHz. The frequency is plotted on a logarithmic scale because of its very wide range.

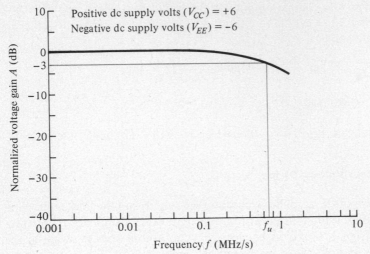

Positive dc supply volts $(V_{CC}) = +6$
Negative dc supply volts $(V_{EE}) = -6$

Figure 11.27. Gain-bandwidth curve for CA3000.

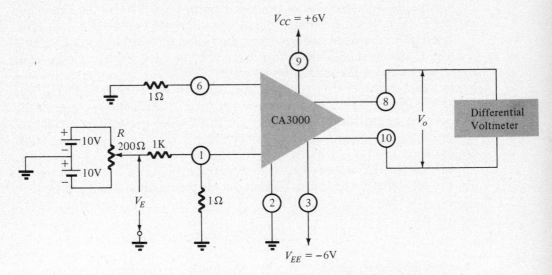

Figure 11.28. Test circuit to measure V_{IO}: (1) adjust R for $V_o(dc) = 0V$; (2) measure V_E and record V_{IO} in millivolts.

Input Offset Voltage (V$_{IO}$)

The input offset voltage (V_{IO}) provides a measure of how much the input voltage must be offset from 0 V to result in both output voltages being biased at exactly the same voltage point. Figure 11.28 shows a test circuit to apply an offset voltage until the output voltages are exactly the same. Notice that the typical value is only 1.4 mV so that reasonably matched circuits can be expected in IC form with the components being made in the same process. A voltage divider of 1 Ω and 1 K provides attenuation of 1000 so that the actual voltage applied to terminal 1 is 1000 times smaller than V_E, or is in millivolts.

11.5 BASICS OF OPERATIONAL AMPLIFIERS (OPAMP)

An operational amplifier is a very high gain differential amplifier that uses voltage feedback to provide a stabilized voltage gain. The basic amplifier used is essentially a difference amplifier having very high open-loop gain (no signal feedback condition) as well as high input impedance and low output impedance. Typical uses of the operational amplifier (OPAMP) are scale changing; analog computer operations, such as addition and integration; and a great variety of phase shift, oscillator, and instrumentation circuits.

Figure 11.29 shows an OPAMP unit having two inputs and a single output. Recall how the two inputs affect an output in a difference amplifier. Here the inputs are marked with *plus* (+) and *minus* (−) to indicate noninverting and inverting inputs, respectively. A signal applied to the *plus* input will appear in phase and amplified at the output, whereas an input applied to the *minus* (−) terminal will appear amplified but inverted at the output.

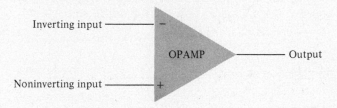

Figure 11.29. Basic OPAMP.

The basic circuit connection of an operational amplifier is shown in Fig. 11.30a. As shown the circuit operates as a scale changer or constant-gain multiplier. An input signal V_1 is applied through a resistor R_1 to the minus input terminal. The output voltage is fed back through resistor R_f to the same input terminal. The plus input terminal is connected to ground. We now wish to determine the overall gain of the circuit (V_o/V_1). To do this we must consider some more details of the OPAMP unit.

Figure 11.30b shows the OPAMP replaced by an equivalent circuit of input resistance R_i and output voltage source and resistance. An ideal OPAMP as

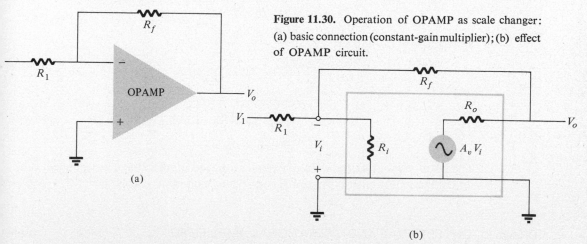

Figure 11.30. Operation of OPAMP as scale changer: (a) basic connection (constant-gain multiplier); (b) effect of OPAMP circuit.

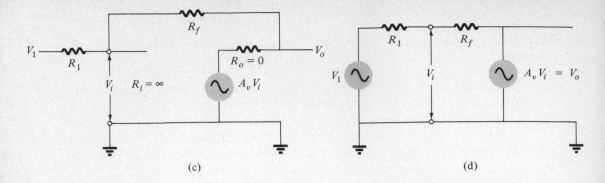

Figure 11.30. (continued): (c) ideal OPAMP; (d) ideal equivalent circuit.

shown in Fig. 11.30c has infinite resistance ($R_i = \infty$), zero output resistance ($R_o = 0$), and very high voltage gain ($A_v \gg 1$). The connection for the ideal amplifier is shown redrawn in Fig. 11.30d. Using the resulting equivalent circuit we can determine the overall gain of the circuit.

Using superposition we can solve for the voltage V_i in terms of the components due to each of the sources. For source V_1 only ($-A_v V_i$ set to zero)

$$V_{i_1} = \frac{R_f}{R_1 + R_f} V_1$$

For source $-A_v V_i$ only (V_1 set at zero)

$$V_{i_2} = \frac{R_1}{R_1 + R_f}(-A_v V_i)$$

The total voltage of V_i is then

$$V_i = V_{i_1} + V_{i_2} = \frac{R_f}{R_1 + R_f} V_1 + \frac{R_1}{R_1 + R_f}(-A_v V_i)$$

which can be solved for V_i as

$$V_i = \frac{R_f}{R_f + (1 + A_v)R_1} V_1$$

if $A_v \gg 1$ and $A_v R_1 \gg R_f$, as is usually true, then

$$V_i \cong \frac{R_f}{A_v R_1} V_1$$

Solving for V_o/V_1, we get

$$\frac{V_o}{V_1} = \frac{-A_v V_i}{V_1} = \frac{-A_v}{V_1}\left(\frac{R_f V_1}{A_v R_1}\right) = -\frac{R_f}{R_1}$$

$$\boxed{\frac{V_o}{V_1} = -\frac{R_f}{R_1}} \tag{11.32}$$

The result shows that the ratio of overall output to input voltage is dependent only on the values of resistors R_1 and R_f—provided that A_v is very large.

If $R_f = R_1$, the gain is

$$A_v = -\frac{R_1}{R_1} = -1$$

and the circuit provides a sign change with no magnitude change.

If $R_f = 2R_1$, then

$$A_v = \frac{-2R_1}{R_1} = -2$$

and the circuit provides a gain of 2 along with 180° phase inversion of the input signal.

If we select precise resistor values for R_f and R_1, we can obtain a wide range of gains, the gain being as accurate as the resistors used and only slightly affected by temperature and other circuit factors.

Virtual Ground

The output voltage is limited by the supply voltage of, typically, a few volts. Voltage gains as stated before are very high. If, for example, $V_o = -10$ V and $A_v = 10,000$, the input voltage is

$$V_i = -\frac{V_o}{A_v} = -\frac{(-10)}{10,000} = 1 \text{ mV}$$

If the circuit had an overall gain (V_o/V_1) of, say, 1, the value of V_1 would be 10 V. The value of V_i, compared to all the other voltages, is then small and may be considered 0 V. Note that although $V_i \cong 0$ V, it is not exactly 0 V since the output is the value of V_i times the gain of the amplifier ($-A_v$).

The fact that $V_i \cong 0$ V leads to the concept that at the input to the amplifier there exists a virtual short circuit or *virtual ground*. The concept of a virtual short implies that although the voltage is nearly 0 V, no current flows through the amplifier input to ground. Figure 11.31 depicts the virtual ground concept. The heavy line is used to indicate that we may consider that a short exists with $V_i \cong 0$ V, but that this is a virtual short in that no current flows through the short to ground. Current flows through resistor R_1 and through R_f as shown.

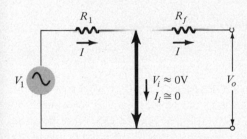

Figure 11.31. Virtual ground in an OPAMP.

If we use the virtual ground concept, we can write equations for the current I as follows:

$$I = \frac{V_1}{R_1} = -\frac{V_o}{R_f}$$

which can be solved for V_o/V_1

$$\frac{V_o}{V_1} = -\frac{R_f}{R_1}$$

The virtual ground concept, which depended on A_v being very large, allowed simple solution of overall voltage gain. It should be understood that although the circuit of Fig. 11.31 is not a physical circuit, it does allow an easy means for solving the overall circuit gain.

11.6 OPAMP CIRCUITS

Constant-Gain Multiplier

An inverting constant-multiplier circuit has already been considered but is repeated here to provide a fuller listing of basic OPAMP circuits. Figure 11.32 shows an inverting constant-gain-multiplier circuit.

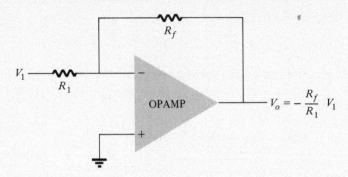

Figure 11.32. Inverting constant-gain multiplier.

EXAMPLE 11.4 The circuit of Fig. 11.32 has $R_1 = 100$ K and $R_f = 500$ K. What is the output voltage for an input of $V_1 = -2$ V?

Solution: Using Eq. (11.32) gives

$$V_0 = -\frac{R_f}{R_1}V_1 = -\frac{500\text{ K}}{100\text{ K}}(-2) = +10\text{ V}$$

Noninverting Amplifier

The connection of Fig. 11.33 shows an OPAMP circuit that works as a non-inverted constant-gain multiplier. To determine the voltage gain of the circuit we can use the equivalent virtual ground representation in Fig. 11.33b. Note that the voltage across R_1 is V_1, since $V_i \cong 0$ V. This must be equal to the voltage due to the output, V_o, through a voltage divider of R_1 and R_f so that

$$V_1 = \frac{R_1}{R_1 + R_f}V_o$$

and

$$\boxed{\frac{V_o}{V_1} = \frac{R_1 + R_f}{R_1} = 1 + \frac{R_f}{R_1}}$$
(11.33)

EXAMPLE 11.5 Calculate the output voltage of a noninverting constant-gain multiplier (as in Fig. 11.33) for values of $V_1 = 2$ V, $R_f = 500$ K, and $R_1 = 100$ K.

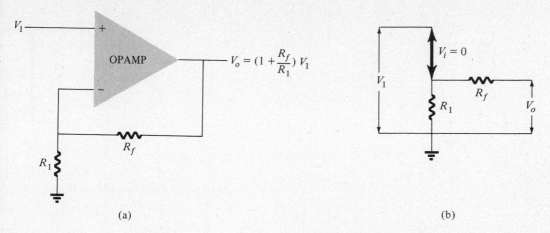

(a)   (b)

Figure 11.33. Noninverting constant-gain multiplier.

Solution: Using Eq. (11.33), we get

$$V_0 = \left(1 + \frac{R_f}{R_1}\right)V_1 = \left(1 + \frac{500\ K}{100\ K}\right)(2\ V) = 6(2) = +12\ V$$

Unity Follower

The unity follower, as in Fig. 11.34, provides a gain of 1 with no phase reversal. From the equivalent circuit with virtual ground it is clear that

$$\boxed{V_o = V_1} \tag{11.34}$$

and that the output is the same polarity and magnitude as the input. The circuit acts very much like an emitter follower except that the gain is very much closer to being exactly unity.

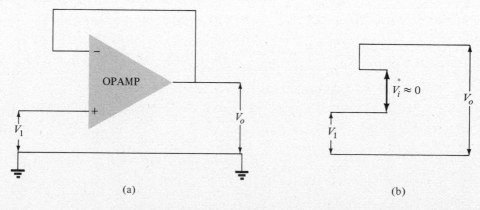

(a)   (b)

Figure 11.34. (a) Unity follower; (b) virtual ground equivalent circuit.

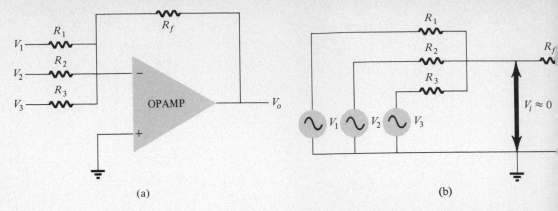

(a)   (b)

Figure 11.35. (a) Summer; (b) virtual ground equivalent circuit.

Summer

Probably the most useful of the OPAMP circuits used in analog computers is the summer circuit. Figure 11.35 shows a three-input summing circuit, which provides a means of algebraically summing (adding) three-input voltages, each multiplied by a constant-gain factor.

If the virtual equivalent circuit is used, the output voltage can be expressed in terms of inputs as

$$V_o = -\left(\frac{R_f}{R_1}V_1 + \frac{R_f}{R_2}V_2 + \frac{R_f}{R_3}V_3\right) \qquad (11.35)$$

In other words, each input adds a voltage to the output as obtained for an inverting constant-gain circuit. If more inputs are used, they add additional components to the output.

> **EXAMPLE 11.6** What is the output voltage of an OPAMP summer for the following sets of input voltages and resistors ($R_f = 1$ M in all cases)?
>
> (a) $\qquad V_1 = +1$ V, $V_2 = +2$ V, $V_3 = +3$ V,
> $\qquad\qquad R_1 = 500$ K, $R_2 = 1$ M, $R_3 = 1$ M
>
> (b) $\qquad V_1 = -2$ V, $V_2 = +3$ V, $V_3 = +1$ V,
> $\qquad\qquad R_1 = 200$ K, $R_2 = 500$ K, $R_3 = 1$ M
>
> **Solution:** Using Eq. (11.35), we get
>
> (a) $\quad V_0 = -\left[\dfrac{1000\text{ K}}{500\text{ K}}(+1) + \dfrac{1000\text{ K}}{1000\text{ K}}(+2) + \dfrac{1000\text{ K}}{1000\text{ K}}(+3)\right]$
>
> $\qquad = -[2(1) + 1(2) + 1(3)] = -7$ **V**
>
> (b) $\quad V_0 = -\left[\dfrac{1000\text{ K}}{200\text{ K}}(-2) + \dfrac{1000\text{ K}}{500\text{ K}}(+3) + \dfrac{1\text{ M}}{1\text{ M}}(+1)\right]$
>
> $\qquad = -[5(-2) + 2(+3) + 1(1)] = -[-10 + 6 + 1]$
>
> $\qquad = +3$ **V**

Integrator

So far the input and feedback components have been resistors. If the feedback component used is a capacitor, as in Fig. 11.36, the resulting circuit is an integrator.

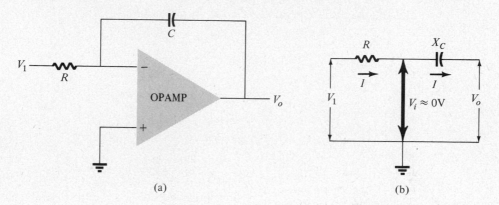

(a)   (b)

Figure 11.36. Integrator.

The virtual ground equivalent circuit shows that an expression between input and output voltages can be derived from the current I, which flows from input to output. Recall that virtual ground means that we can consider the voltage at the junction point of R and X_C to be ground (since $V_i \cong 0$ V) but that no current flows to ground at that point. The capacitive impedance can be expressed as

$$X_C = \frac{1}{j\omega C} = \frac{1}{sC}$$

where $s = j\omega$ is the Laplace notation.
Solving for V_o/V_1

$$I = \frac{V_1}{R} = -\frac{V_o}{X_C} = \frac{-1}{1/sC} = -sCV_o$$

$$\boxed{\frac{V_o}{V_1} = \frac{-1}{sCR}}$$
(11.36a)

The last expression can be rewritten in the time domain as

$$v_o(t) = -\frac{1}{RC} \int v_1(t)\, dt$$
(11.36b)

Equation (11.36b) shows that the output is the integral of the input, with an inversion and scale multiplier of $1/RC$. The ability to integrate a given signal provides the analog computer with the ability to solve differential equations and therefore allows setup of a wide variety of electrical circuit analogs of physical system operations.

As an example, consider an input step voltage showing in Fig. 11.37a. The integral of the step voltage is a ramp or linearly changing voltage. The circuit scale factor of $-1/RC$ is

$$-\frac{1}{RC} = -\frac{1}{10^6 \times 10^{-6}} = -1$$

so that
$$v_o(t) = -\int v_i(t)\, dt$$

and the output is a negative ramp as shown in Fig. 11.37b.

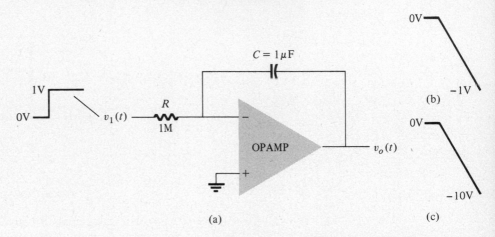

Figure 11.37. Operation of integrator with step input.

If the scale factor is changed by making $R = 100$ K, for example, then

$$-\frac{1}{RC} = -\frac{1}{10^5 \times 10^{-6}} = -10$$

and the output is
$$v_o(t) = -10 \int v_i(t)\, dt$$

which is shown in Fig. 11.37c.

More than one input may be applied to an integrator as shown in Fig. 11.38 with the resulting operation given by

$$v_o(t) = -\left[\frac{1}{R_1 C}\int v_1(t)\, dt + \frac{1}{R_2 C}\int v_2(t)\, dt + \frac{1}{R_3 C}\int v_3(t)\, dt\right] \qquad (11.37)$$

An example, showing a summing integrator as used in an analog computer, is given in Fig. 11.39. The actual circuit is shown with input resistors and feedback capacitor, whereas the analog computer representation only indicates the scale factor for each input.

Differentiator

The differentiator circuit of Fig. 11.40 is not as useful a computer circuit as the integrator because of practical problems with noise. The resulting relation for the circuit is

$$v_o(t) = -RC\frac{dv_1(t)}{dt} \qquad (11.38)$$

where the scale factor is $-RC$. Reference to any text on analog computers will show how differential equations are set up for solution using mainly summing and integrator circuits.

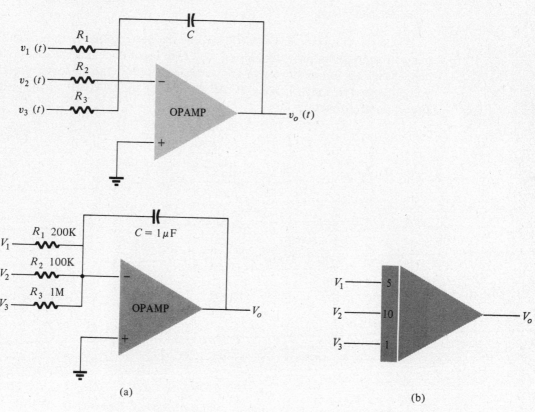

Figure 11.39. (a) OPAMP and (b) analog computer integrator circuit representation.

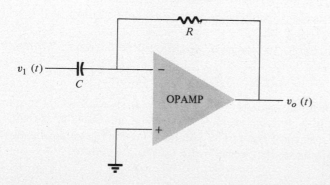

Figure 11.40. Differentiator circuit.

11.7 OPAMP APPLICATIONS

To provide some indication of how useful OPAMPs can be in other than just analog computer circuits (which is a very large OPAMP application area) a few miscellaneous applications will be considered here, with additional oscillator applications in Chapter 12.

dc Millivoltmeter

Figure 11.41 shows an OPAMP used as the basic amplifier in a dc millivoltmeter. The amplifier provides a meter with high input impedance and scale factors dependent only on resistor value and accuracy. Notice that the meter reads millivolts of signal at the circuit input. An analysis of the OPAMP circuit yields the circuit transfer function

$$\frac{I_o}{V_1} = -\frac{R_f}{R_i}\left(\frac{1}{R_s}\right)$$

$$= -\frac{100\text{ K}}{100\text{ K}} \times \frac{1}{10} = \frac{1\text{ mA}}{10\text{ mV}}$$

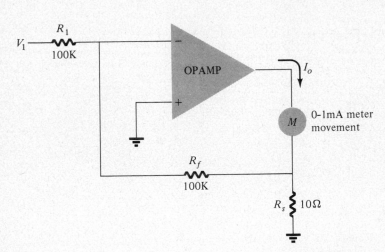

Figure 11.41. OPAMP dc millivoltmeter.

Thus, an input of 10 mV will result in a current through the meter of 1 mA. If the input is 5 mV, the current through the meter will be 0.5 mA, which is half-scale deflection.

Changing R_f to 200 K, for example, would result in a circuit scale factor of

$$\frac{I_o}{V_1} = -\frac{200\text{ K}}{100\text{ K}} \times \frac{1}{10} = \frac{1\text{ mA}}{5\text{ mV}}$$

showing that the meter now reads 5 mV, full scale. It should be kept in mind that building such a millivoltmeter requires purchasing an OPAMP circuit, a few resistors, and a meter movement. The ability to obtain a completely operating, tested OPAMP unit makes the overall meter unit easy to set up.

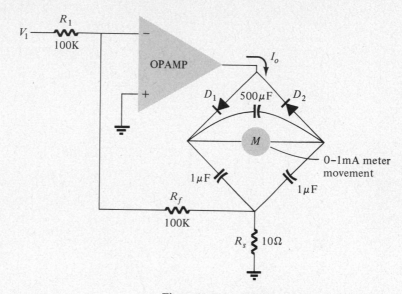

Figure 11.42. ac millivoltmeter using OPAMP.

ac Millivoltmeter

As another example, an ac millivoltmeter circuit, is shown in Fig. 11.42. The resulting circuit transfer function is

$$\frac{I_o}{V_1} = -\frac{R_f}{R_1}\left(\frac{1}{R_s}\right) = \frac{100\ \text{K}}{100\ \text{K}} \times \frac{1}{10} = \frac{1\ \text{mA}}{5\ \text{mV}}$$

which appears the same as the dc millivoltmeter, except that in this case it is for ac signals. The meter indication provides a full-scale deflection for an ac input voltage of 10 mV. An ac input signal of 10 mV will result in full-scale deflection, while an ac input of 5 mV will result in half-scale deflection, and the meter reading can be interpreted in millivolt units.

Peak Follower

A peak follower circuit accepts an ac input voltage and provides as output a voltage whose magnitude follows only the peak of the input signal. It remains at the level of the highest peak of the ac input signal and changes only if the input signal goes higher than the previous peak level to a new higher peak level.

Figure 11.43a shows an OPAMP peak follower circuit. The diode-capacitor circuit acts as a simple capacitor filter, charging the capacitor up to the peak of the input voltage. If the input voltage then decreases below this level, the diode becomes reverse biased and the capacitor remains charged. The OPAMP provides the output voltage to a load without "loading down" the capacitor. In effect, the OPAMP isolates the capacitor from the load to allow the capacitor to remain

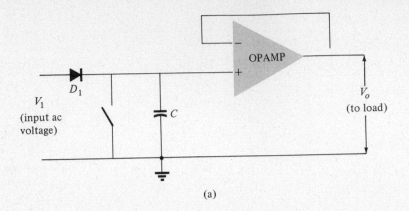

(a)

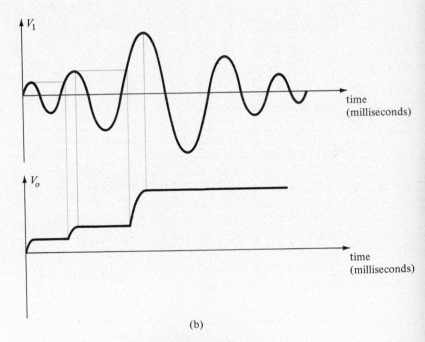

(b)

Figure 11.43. (a) Peak follower circuit; (b) typical operation.

charged. To restart a new peak follower operating cycle, the switch is closed and the capacitor discharged. Figure 11.43b shows an input ac voltage waveform and the peak follower output voltage.

Constant-Current Generator

A constant-current generator circuit provides a fixed constant current to a load, regardless of the load value. A constant-current source is shown as a possible part of the differential amplifier. An OPAMP version of a constant-current circuit is

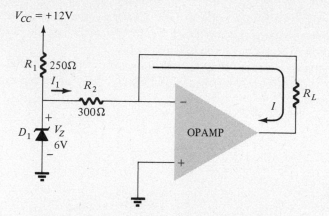

Figure 11.44. Constant-current generator.

shown in Fig. 11.44. The resistor-diode connection of R_1 and D_1 operates the Zener diode in its Zener region with the input voltage held fixed at $V_Z = 6$ V, in this case. The voltage at the minus input terminal is about 0 V so that the current I_1 is

$$I_1 = \frac{V_Z}{R_2} = \frac{6 \text{ V}}{300 \text{ } \Omega} = 20 \text{ mA}$$

We know from OPAMP theory that

$$I = I_1 = 20 \text{ mA}$$

so that the Zener and resistor R_2 fix the current through the output load resistor, R_L, for a wide range of value of R_L.

Voltage-Level Indicator

The circuit of Fig. 11.45 provides a bistable output, which is either at 0 V or $+V_{CC}$. If a reference voltage at the input is used, the circuit provides the following operation:

$$\text{if } V_1 < 1.5 \text{ V}, \ V_o = +V_{CC}$$
$$\text{if } V_1 > 1.5 \text{ V}, \ V_o = 0 \text{ V}$$

The circuit thus operates as a comparator that provides indication of whether

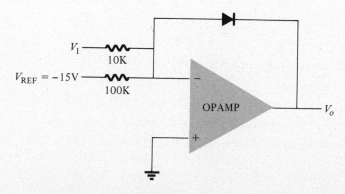

Figure 11.45. Voltage-level indicator.

an input voltage exceeds a preset voltage level. Changing the resistor values or reference voltage allows changing the voltage level. Also reversing the diode and reference supply allows the output to be 0 V and $-V_{CC}$.

These few examples provide some small indication of how versatile the OPAMP can be in applications. A number of additional oscillator circuit examples will be given in Chapter 12.

PROBLEMS

§ 11.1

1. Draw the output waveforms for the input signal and differential amplifier of Fig. 11.46.

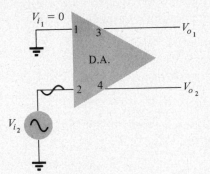

Figure 11.46. Differential amplifier and input waveform for Problem 11.1.

2. Draw the output waveform for the differential amplifier and difference input signal of Fig. 11.47.

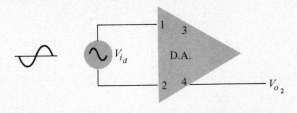

Figure 11.47. Differential amplifier and input for Problem 11.2.

§ 11.2

3. Determine the dc voltages and currents in the circuit of Fig. 11.48.

4. Calculate the ac voltage gain of the difference amplifier in Fig. 11.48.

5. Calculate the input and output resistance of the circuit in Fig. 11.48.

6. Calculate the value of the constant current for the circuit of Fig. 11.49.

7. Calculate the dc voltages and currents in the circuit of Fig. 11.49.

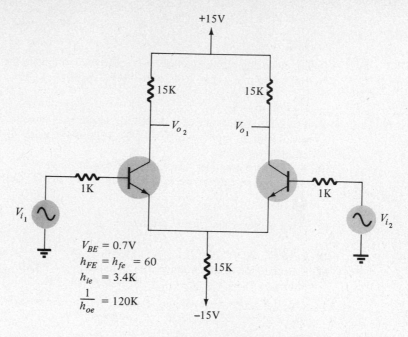

Figure 11.48. Circuit for Problems 11.3–11.5.

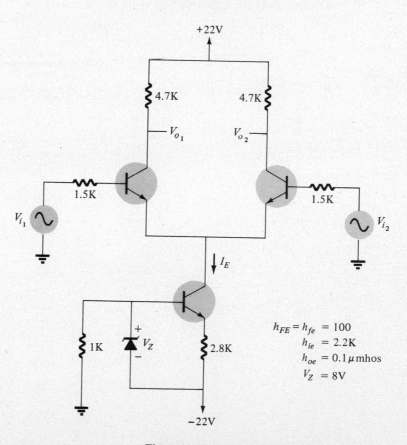

Figure 11.49. Circuit for Problems 11.6 and 11.7.

§ 11.3

8. Calculate the output voltage of a difference amplifier for inputs of $V_{i_1} = 0.5$ mV, $V_{i_2} = 0.45$ mV, $A_d = 4500$, and CMRR $= 10^4$.

§ 11.4

9. The input impedance of a difference amplifier is measured using a 25-K resistor in series with an input voltage of 5 V. What is the value of R_i if the voltage into the amplifier is 1.5 V?

10. A difference amplifier has single-ended gain of $A_1 = 120$. When determining A_c the circuit measurements are $V_i = 2$ V and $V_0 = 20$ mV. Calculate CMRR in decibels.

§ 11.5

11. What is meant by "virtual ground"?

§ 11.6

12. Calculate the output voltage of a noninverting OPAMP circuit (as in Fig. 11.33) for values of $V_1 = 4$ V, $R_f = 250$ K, and $R_1 = 50$ K.

13. Calculate the output voltage of a three-input summer (as in Fig. 11.35) for the values: $R_1 = 200$ K, $R_2 = 250$ K, $R_3 = 500$ K, $R_f = 1$ M, $V_1 = -2$ V, $V_2 = +2$ V, and $V_3 = 1$ V.

14. Determine the output voltage of the circuits of Fig. 11.50.

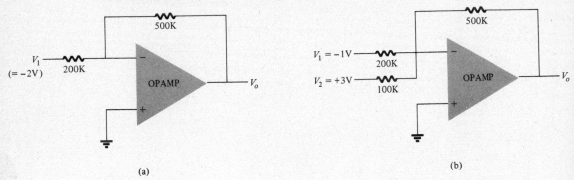

(a)

(b)

Figure 11.50. Circuits for Problem 11.14.

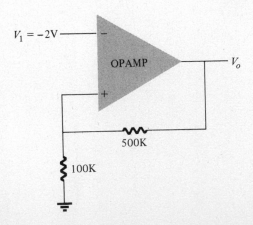

Figure 11.51. Circuit for Problem 11.15.

15. Repeat Problem 11.14 for the circuit of Fig. 11.51.

16. Draw the output waveform for the circuit and inputs of Fig. 11.52.

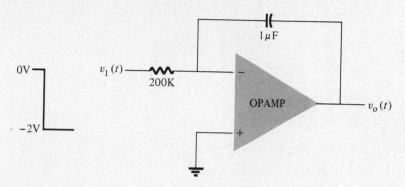

Figure 11.52. Circuit for Problem 11.16.

17. Repeat Problem 11.16 for Fig. 11.53.

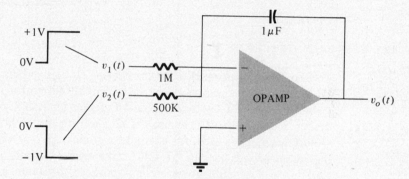

Figure 11.53. Circuit for Problem 11.17.

feedback amplifiers and oscillator circuits

12

12.1 FEEDBACK CONCEPTS

Feedback was mentioned when considering dc bias stabilization in Chapters 4 and 11. Amplifier gain was sacrificed in the circuit design for improvement in dc bias stability. We might say that a trade-off of gain for stability was made in the circuit design. Such trade-off is typical of engineering design compromises. If negative voltage feedback is used, for example, a circuit can be designed to couple some of the output voltage back to the input, reducing the overall voltage gain of the circuit. For this loss of gain, however, it is possible to obtain higher input impedance, lower output impedance, more stable amplifier gain, or higher cutoff frequency operation.

If the feedback signal is connected in order to aid or add to the input signal applied, however, *positive* feedback occurs, which could drive the circuit into operation as an oscillator.

Voltage Feedback Connection

As an example of voltage feedback the circuit of Fig. 12.1 shows a FET amplifier with negative voltage feedback. Resistor R_f and capacitor C_f (used here to block dc bias voltage) form the feedback path. Because of the amplifier inversion any signal at the output is opposite in phase to the signal at the input. The output signal fed back will then be opposite in phase or polarity and is thus a negative feedback signal. The net result of the feedback action will be to decrease the overall voltage gain to a lower amount dependent on the original gain of the amplifier without feedback and on the amount of the feedback.

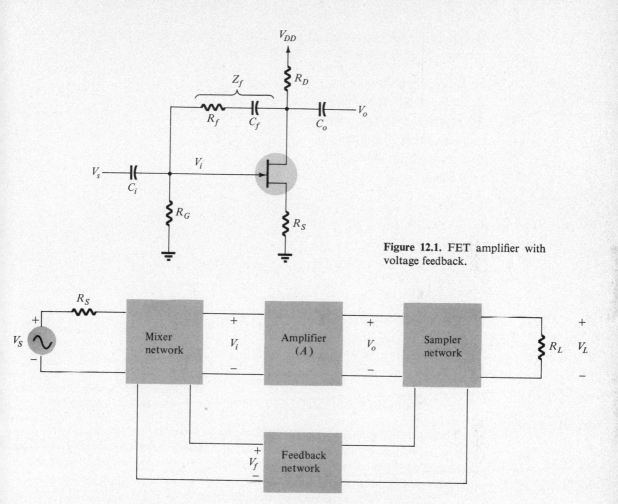

Figure 12.1. FET amplifier with voltage feedback.

Figure 12.2. Feedback amplifier, block diagram.

A general block diagram of a feedback circuit is shown in Fig. 12.2. The input signal (V_s) and feedback signal (V_f) are *mixed* or combined to form the single signal (V_i) which is then amplified by the amplifier section of the circuit. The amplifier output then goes into a sampler circuit which feeds part of the amplified signal to the load and part of the signal to the feedback network.

In the circuit of Fig. 12.1 the components C_i and R_g form the mixer network combining input and feedback signals. No sampling network is used in this circuit because the output and signal to the feedback network are the same. The feedback network is comprised of resistor R_f and capacitor C_f.

A simpler version of the feedback amplifier of Figs. 12.1 and 12.2 is that of Fig. 12.3. The mixer is shown as a circle with two inputs that are opposite in polarity as indicated by the plus and minus input signs. The basic amplifier gain is A and the gain (attenuation, normally) of the feedback network is given as β (beta). Usually, a part of the output signal is coupled back to the input in order to oppose the applied input signal V_s. In return for this gain reduction a number of improvements can be obtained, such as the following:

1. Higher input impedance.
2. Better stabilized voltage gain.
3. Improved frequency response.
4. More linear operation.
5. Lower output impedance.
6. Reduced noise.

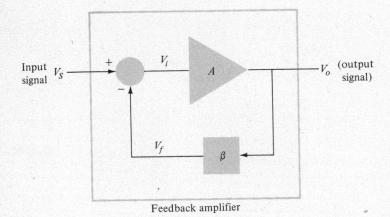

Feedback amplifier

Figure 12.3. Simpler block diagram of feedback amplifier.

These improvements all occur with a voltage-series type of feedback. In addition to the voltage feedback just discussed, there is also current feedback, for which the list of changes with and without feedback is somewhat different than those listed above. At present we wish to obtain some concept of what feedback is all about. Only the voltage-series feedback connection of Figs. 12.2 or 12.3 will be used for the present. We shall demonstrate mathematically how some of the listed improvements are obtained and shall provide a means of numerically specifying the result of using feedback. Some relations between the amplifier operation with and without feedback will now be considered.

Voltage Gain with Feedback

In the feedback circuit of Fig. 12.3 the gain without feedback is A and the feedback factor is β. It is assumed that signal transmission goes only from input (V_i) to output (V_o) for the amplifier state and only from output (V_o) to feedback input (V_f) for the feedback network.

The input voltage to the basic amplifier is the difference between signal and feedback voltage

$$V_i = V_s - V_f \qquad (12.1)$$

where the feedback voltage is a portion of the output voltage

$$V_f = \beta V_o \qquad (12.2)$$

the proportionality factor being β. The gain of the basic amplifier is simply

$$A = \frac{V_o}{V_i}$$

so that the output voltage is given by

$$V_o = AV_i \qquad (12.3)$$

With feedback employed, the overall gain of the circuitry represented by Fig. 12.3 is

$$A_f = \frac{V_o}{V_s} \qquad (12.4)$$

We can solve for this factor using Eqs. (12.1), (12.2), and (12.3) as follows:

$$\frac{V_o}{A} = V_i = V_s - V_f = V_s - \beta V_o$$

$$V_o + \beta A V_o = AV_s$$

$$\boxed{A_f = \frac{V_o}{V_s} = \frac{A}{1 + \beta A}} \qquad (12.5)$$

Thus, Eq. (12.5) shows that the gain with feedback depends on the basic amplifier gain and the amount of the feedback factor.

If, for example, the quantities β and A are $\beta = 1/10$, $A = 90$, then the gain *without* feedback is 90 and the gain *with* feedback is

$$A_f = \frac{A}{1 + \beta A} = \frac{90}{1 + (1/10)(90)} = \frac{90}{10} = 9$$

the gain being reduced by a factor of 10. This is *negative feedback*.

We can also show that the gain *with* feedback is more stable than that without feedback. A change in amplifier gain from 90 to 100 due to component value changes with temperature represents a change of 11.1%. With feedback the resulting gain is

$$A_f = \frac{100}{1 + (1/10)(100)} = 9.1$$

which represents a change of only 1.11%, an improvement by a factor of nearly 10.

If the amplifier gain and feedback values are $A = 90$, $\beta = -1/100$, the gain with feedback is then

$$A_f = \frac{90}{1 + (-1/100)(90)} = \frac{90}{1 - 0.9} = \frac{90}{0.1} = 900$$

The gain has been increased by a factor of 10. This is *positive feedback* and is the principle by which a feedback amplifier can be made into an oscillator circuit. Detailed discussion of the oscillator is deferred to Section 12.7.

Thus, as a general statement: if $|A_f| < |A|$, feedback is negative; if $|A_f| > |A|$, feedback is positive.

For a negative feedback amplifier we see that $|1 + \beta A| > 1$. Typically, the value of $|\beta A| \gg 1$ so that

$$A_f = \frac{A}{1 + \beta A} \cong \frac{A}{\beta A} = \frac{1}{\beta} \qquad (12.6)$$

In other words, the feedback gain is dependent mainly on the factor β for the case of negative feedback with $|\beta A| \gg 1$. Whereas the amplifier gain A is dependent on temperature, device parameters, and so on, and may vary considerably, the reduced gain with negative feedback can be very stable, typically depending on a resistor feedback network. Since resistors can be selected precisely and with small change in resistive value due to temperature, highly precise and stable gain with negative feedback is possible.

EXAMPLE 12.1 Calculate the gain of a negative feedback amplifier circuit having $A = 1000$ and $\beta = 1/10$.

Solution: Since $\beta A = (1/10) \cdot (1000) = 100 \gg 1$, the gain with feedback is

$$A_f \cong \frac{1}{\beta} = \frac{1}{0.1} = 10$$

In addition to the β factor setting a precise gain value, we are also interested in how stable the feedback amplifier is compared to an amplifier without feedback. Differentiating Eq. (12.5) leads to

$$\frac{dA_f}{A_f} = \frac{1}{|1 + \beta A|} \frac{dA}{A} \qquad (12.7a)$$

$$\frac{dA_f}{A_f} \cong \frac{1}{\beta A} \frac{dA}{A}, \qquad \text{for } \beta A \gg 1 \qquad (12.7b)$$

This shows that the change in gain (dA) is reduced by the factor βA when feedback is employed.

EXAMPLE 12.2 If the amplifier in Example 12.1 has a gain change of 20% due to temperature, calculate the change in gain of the feedback amplifier.

Solution: Using Eq. (12.7b), we get

$$\frac{dA_f}{A_f} \cong \frac{1}{\beta A} \frac{dA}{A} = \frac{1}{0.1(1000)}(20\%) = 0.2\%$$

The improvement is 100 times. Thus, while the amplifier gain changes from $A = 1000$ by 20%, the feedback gain changes from $A_f = 100$ by only 0.2%.

12.2 FEEDBACK CONNECTION TYPES

There are four basic ways of connecting the feedback signal. Both *voltage* and *current* can be fed back to the input either in *series* or *parallel*. Specifically, there can be

1. Voltage-series feedback (Fig. 12.4a).
2. Voltage-shunt feedback (Fig. 12.4b).
3. Current-series feedback (Fig. 12.4c).
4. Current-shunt feedback (Fig. 12.4d).

In the above listing *voltage* refers to connecting the output voltage as input to the feedback network; *current* refers to tapping off some output current through the feedback network. *Series* refers to connecting the feedback signal in series with

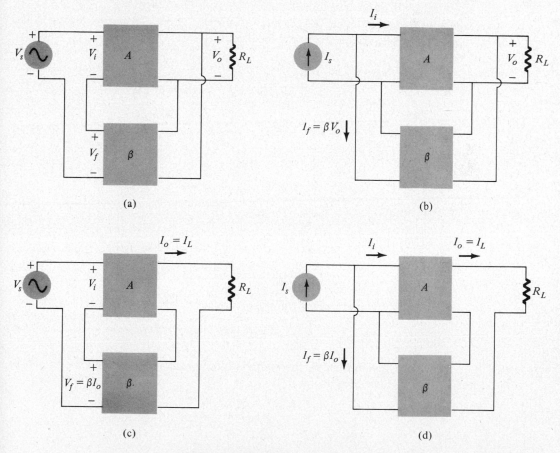

Figure 12.4. Feedback amplifier connection types: (a) voltage-series feedback; (b) voltage-shunt feedback; (c) current-series feedback; (d) current-shunt feedback.

the input signal voltage; *shunt* refers to connecting the feedback signal in shunt (parallel) with an input current source.

Series feedback connections tend to *increase* the input resistance while shunt feedback connections tend to *decrease* the input resistance. Voltage feedback tends to *decrease* the output impedance while current feedback tends to *increase* the output impedance. Typically, higher input and lower output impedances are desired for most cascade amplifiers. Both of these are provided using the voltage-series feedback connection. We shall therefore concentrate first on this amplifier connection for practical feedback circuits.

Voltage-Series Feedback Amplifier

A block diagram of a voltage-series feedback amplifier is shown in Fig. 12.5. A practical voltage source having voltage V_S and source resistance R_S is shown in *series* with the feedback signal, V_f, the resulting input signal to the amplifier stage being V_i. The output voltage of the amplifier stage (with gain A_v) is V_o, which is developed directly across load resistor R_L. A parallel *voltage* pickup of the output voltage is connected to the feedback network.

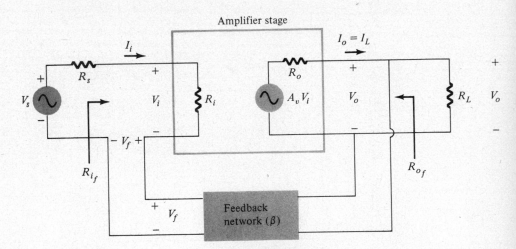

Figure 12.5. Practical voltage-series feedback amplifier.

If $V_f = 0$ (no feedback), the overall voltage gain A, including loading effects of R_s and R_i, is defined as

$$A_v = \frac{V_o}{V_s}\bigg|_{R_L=\infty \text{ (output open circuit)}}$$

where
$$V_o = A_v V_i - I_o R_o$$

$$= \underbrace{A_v \frac{R_i}{R_i + R_s}}_{A} V_s - I_o R_o$$

$$V_o = AV_s - I_oR_o \quad \text{(without feedback)}$$

(if $R_s = 0$, then $A = A_v$).

Including feedback connection ($V_f \neq 0$) the voltage gain is

$$A_f = \frac{V_o}{V_s}\bigg|_{R_L=\infty}$$

where
$$V_o = AV_i - I_oR_o$$
$$= A(V_s - V_f) - I_oR_o$$

Now,
$$A(V_s - V_f) = AV_s - AV_f = AV_s - A(\beta V_o)$$
$$= A(V_s - \beta V_o)$$

We have then

$$V_o = AV_s - A\beta V_o - I_oR_o$$
$$V_o(1 + A\beta) = AV_s - I_oR_o$$
$$V_o = \left(\frac{A}{1 + \beta A}\right)V_s - I_o\frac{R_o}{1 + \beta A}$$
$$V_o = A_fV_s - I_oR_{of}$$

with
$$\boxed{A_f = \frac{A}{1 + \beta A}} \tag{12.8}$$

$$\boxed{R_{of} = \frac{R_o}{1 + \beta A}} \tag{12.9}$$

The last two equations show that the gain without feedback is reduced by the factor $(1 + \beta A)$ with feedback connected. In addition, the output resistance is seen to be reduced from R_o (without feedback) by the factor $(1 + \beta A)$ with feedback. The larger the factor $(1 + \beta A)$, the lower the output resistance.

The input resistance with feedback is defined by

$$R_{if} \equiv \frac{V_s}{I_i} - R_s$$

which can be shown to be

$$\boxed{R_{if} = R_i(1 + \beta A)} \tag{12.10}$$

This time the factor $(1 + \beta A)$ makes the feedback input resistance larger than for the amplifier without feedback. In summary, then, a voltage-series feedback amplifier circuit can improve the operation of a nonfeedback amplifier circuit (having gain A, input resistance R_i, and output resistance R_o) as follows:

1. Stabilized voltage gain $A_f = A/(1 + \beta A)$.
2. Higher input resistance $R_{if} = R_i(1 + \beta A)$.
3. Lower output impedance $R_{of} = R_o/(1 + \beta A)$.

EXAMPLE 12.3 Calculate the gain and input and output impedance of a voltage-series feedback amplifier if the amplifier without feedback has $A = 100$, $R_i = 2$ K, $R_o = 40$ K, and the amount of feedback is $\beta = 1/10$.

Solution: For a feedback amplifier using voltage-series feedback we can use Eqs. (12.8)–(12.10)

$$A_f = \frac{A}{1 + \beta A} = \frac{100}{1 + (1/10)(100)} = \frac{100}{11} = \mathbf{9.1}$$

$$R_{if} = R_i(1 + \beta A) = 2 \text{ K}(11) = \mathbf{22 \text{ K}}$$

$$R_{of} = \frac{R_o}{1 + \beta A} = \frac{40 \text{ K}}{11} = \mathbf{3.63 \text{ K}}$$

Reduction in Frequency Distortion

Recall that Eq. (12.6) shows that for a negative feedback amplifier having $\beta A \gg 1$ the gain with feedback is $A_f \cong 1/\beta$. It follows from this that if the feedback network is purely resistive, the gain with feedback is not dependent on frequency even though the basic amplifier gain is frequency dependent. Practically, the frequency distortion arising because of varying amplifier gain with frequency is considerably reduced in a negative-voltage feedback amplifier circuit.

Reduction in Noise and Nonlinear Distortion

Signal feedback connected to oppose the input signal as in a negative feedback amplifier tends to hold down the amount of noise signal (such as power supply hum) and nonlinear distortion. The factor $(1 + \beta A)$ reduces both input noise and resulting nonlinear distortion for considerable improvement. However, it should be noted that there is a reduction in overall gain (the price required for the improvement in circuit performance). If additional stages are used to bring the overall gain up to the level without feedback, it should be noted that the extra stage(s) might introduce as much noise back into the system as that reduced by the feedback amplifier. This problem can be somewhat alleviated by readjusting the gain of the feedback amplifier circuit to obtain higher gain while also providing reduced noise signal.

Effect of Negative Feedback on Gain and Bandwidth

In Eq. (12.6) the overall gain with negative feedback is shown to be

$$A_f = \frac{A}{(1 + \beta A)} \cong \frac{A}{\beta A} = \frac{1}{\beta} \qquad \text{for } \beta A \gg 1$$

As long as $\beta A \gg 1$ the overall gain is approximately $1/\beta$. We should realize that for a practical amplifier (for single low- and high-frequency breakpoints) the open-loop gain drops off at high frequencies due to the active device and circuit capacitances. Gain may also drop off at low frequencies for capacitively coupled amplifier stages. Once the open-loop gain A drops low enough and the factor βA is no longer much larger than 1, the conclusion of Eq. (12.6) that $A_f \cong 1/\beta$ no longer holds true.

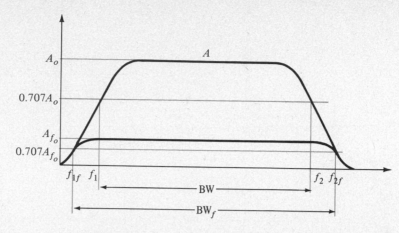

Figure 12.6. Effect of negative feedback on gain and bandwidth.

Figure 12.6 shows that the amplifier with negative feedback has more bandwidth (BW) than the amplifier without feedback. The feedback amplifier has a higher upper 3 dB frequency and smaller lower 3 dB frequency.

It is interesting to note that the use of feedback, while resulting in a lowering of voltage gain, has provided an increase in BW and in the upper 3-dB frequency, particularly. In fact, the product of gain and frequency remains the same so that the gain-bandwidth product of the basic amplifier is the same value for the feedback amplifier. However, since the feedback amplifier has lower gain, the net operation was to *trade* gain for bandwidth (we use bandwidth for the upper 3-dB frequency since typically $f_2 \gg f_1$).

12.3 PRACTICAL VOLTAGE-SERIES NEGATIVE FEEDBACK AMPLIFIER CIRCUITS

Transistor Stage

Figure 12.7a shows a transistor amplifier circuit with the output taken from the emitter terminal. Figure 12.7b shows an approximate small-signal equivalent circuit. The feedback signal is shown connected to the input in series with input voltage (V_s).

Voltage Gain

At first glance it might not appear that any feedback exists in this simple one-stage amplifier circuit. However, the ac equivalent circuit of Fig. 12.7b shows that the input voltage (V_i) is the difference of the source voltage (V_s) and feedback voltage (V_f)

$$V_i = V_s - V_f$$

The gain of the amplifier without feedback (ground side of V_s connected to emitter instead of ground) is

$$A = \frac{h_{fe}R_E}{R_s + h_{ie}} \tag{12.11}$$

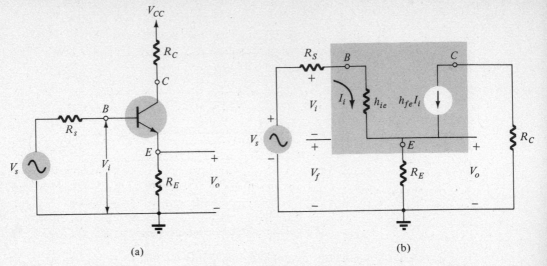

(a) (b)

Figure 12.7. Transistor amplifier with voltage-series feedback: (a) amplifier circuit; (b) equivalent ac circuit.

The feedback voltage is equal in magnitude to the output voltage so that we have $\beta = +1$. Using Eq. (12.6), we calculate the gain with feedback to be

$$A_f = \frac{A}{1 + \beta A} = \frac{h_{fe}R_E/(R_S + h_{ie})}{1 + (1)[h_{fe}R_E/(R_S + h_{ie})]}$$

$$\boxed{A_f = \frac{h_{fe}R_E}{R_S + h_{ie} + h_{fe}R_E}} \qquad (12.12)$$

Input Resistance

Without feedback we note that

$$R_i = h_{ie}$$

With feedback and neglecting $R_S(R_S = 0)$ we find, using Eq. (12.10),

$$R_{if} = R_i(1 + \beta A) = h_{ie}\left(1 + \frac{h_{fe}R_E}{h_{ie}}\right)$$

$$\boxed{R_{if} = h_{ie} + h_{fe}R_E} \qquad (12.13)$$

Output Resistance

Without feedback the equivalent circuit used has an output resistance of infinity. Had $1/h_{oe}$ been included in the transistor equivalent circuit, a practical value of around 50 K would be present. Including feedback and using Eq. (12.9), we get

$$\boxed{R_{of} = \frac{R_o}{1 + \beta A} \simeq \frac{R_S + h_{ie}}{h_{fe}}} \qquad (12.14)$$

A numerical example using the above equations should show that negative feedback in a voltage-series connection reduces the gain while increasing input resistance and lowering output resistance.

EXAMPLE 12.4   For the circuit of Fig. 12.7 and the following circuit values, calculate the voltage gain and input and output resistances, with and without feedback: $R_E = 1.5$ K, $R_S = 1$ K, $h_{ie} = 2$ K, $h_{fe} = 50$, $h_{oe} = 12.5$ μmhos.

Solution: Without feedback

$$A = \frac{h_{fe}R_E}{R_S + h_{ie}} = \frac{50(1.5 \text{ K})}{1 \text{ K} + 2 \text{ K}} = 25$$

$$R_i = h_{ie} = 2 \text{ K}$$

$$R_o = \frac{1}{h_{oe}} = 80 \text{ K}$$

With feedback

$$A_f = \frac{h_{fe}R_E}{R_S + h_{ie} + h_{fe}R_E} = \frac{50(1.5 \text{ K})}{1 \text{ K} + 2 \text{ K} + 50(1.5 \text{ K})} = \frac{75}{78} = 0.96$$

$$R_{if} = h_{ie} + h_{fe}R_E = 2 \text{ K} + 50(1.5 \text{ K}) = 77 \text{ K}$$

$$R_{of} = \frac{R_S + h_{ie}}{h_{fe}} = \frac{1 \text{ K} + 2 \text{ K}}{50} = 60 \text{ }\Omega$$

FET Stage

Figure 12.8 shows a single-stage RC-coupled FET amplifier with negative feedback. A part of the output signal (V_o) is picked off by a feedback network made up of resistors R_1 and R_2. The feedback voltage V_f is connected in series with the source signal V_s, and their difference is the input signal V_i (measured from gate to drain). Without feedback the amplifier gain is

$$A = -g_m R_L \tag{12.15}$$

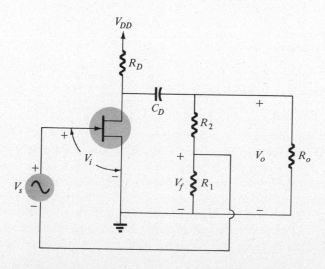

Figure 12.8. FET amplifier stage with voltage-series feedback.

where R_L is the parallel combination of resistors R_D, R_o, and a series equivalent of R_1 and R_2.

The feedback network provides a feedback factor of

$$\beta = \frac{R_1}{R_1 + R_2} \tag{12.16}$$

Using the above values of A and β in Eq. (12.6), we find the gain with negative feedback to be

$$A_f = \frac{A}{1 + \beta A} = \frac{g_m R_L}{1 + [R_1 R_L/(R_1 + R_2)]g_m} \tag{12.17a}$$

If $\beta A \gg 1$, we have

$$\boxed{A_f \cong \frac{1}{\beta} = \frac{R_1 + R_2}{R_1}} \tag{12.17b}$$

EXAMPLE 12.5 Calculate the gain without and with feedback for the FET amplifier circuit of Fig. 12.8 and the following circuit values: $R_1 = 20$ K, $R_2 = 80$ K, $R_o = 10$ K, $R_D = 10$ K, and $g_m = 4000$ μmhos.

Solution:

$$R_L \cong \frac{R_o R_D}{R_o + R_D} = \frac{10(10)}{10 + 10} = 5 \text{ K}$$

(neglecting 100 K resistance of R_1 and R_2 in series)

$$A = g_m R_L = (4000 \times 10^{-6})(5 \text{ K}) = \mathbf{20}$$

The feedback factor is

$$\beta = \frac{R_1}{R_1 + R_2} = \frac{20}{20 + 80} = 0.2$$

The gain with feedback is

$$A_f = \frac{A}{1 + \beta A} = \frac{20}{1 + 0.2(20)} = \frac{20}{5} = \mathbf{4}$$

12.4 OTHER PRACTICAL FEEDBACK CIRCUIT CONNECTIONS

Two-Stage Voltage-Series Feedback

A popular means of incorporating negative feedback to stabilize the gain of an amplifier is signal feedback in a multistage circuit. As an example, Fig. 12.9 shows two cascaded stages with voltage-series feedback. Cascaded amplifier stages with transistors Q_1 and Q_2 provide an overall gain A. A feedback network of resistors R_1 and R_2 is coupled by a capacitor (to block dc) between output and input while allowing feedback of the ac output signal. The feedback signal is taken from the collector of the second stage and is connected to the emitter (neglecting the bypassed 3.6-K resistor) of the first amplifier stage. Negative feedback results from this connection since the in-phase output signal of the second stage collector con-

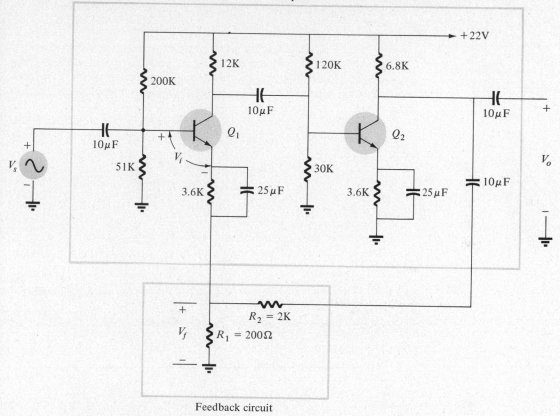

Figure 12.9. Cascaded voltage-series feedback amplifier.

nected through the feedback network to the emitter opposes the input signal between the base emitter of the first stage.

Calculations of gain, input, and output resistance with and without feedback require no new theory or equations. The techniques for cascaded stages developed in Chapter 7 and the basic equations [Eqs. (12.8)–(12.10)] for voltage-series feedback are used in the following example to show how a circuit such as in Fig. 12.9 can be analyzed.

EXAMPLE 12.6 Calculate A, R_i, and R_o for the cascaded amplifier of Fig. 12.9, omitting feedback, and then A_f, R_{if}, and R_{of} with the feedback connection considered. Use transistor parameters $h_{fe} = 65$, $h_{ie} = 1.8$ K, and $1/h_{oe} = \infty$.

Solution: *Without the feedback network:* Looking into the base of transistor Q_1, we see that the resistance is $R_{i_1} = h_{ie} = 1.8$ K. The emitter 3.6-K resistor is neglected here for ac calculations due to the bypass capacitor. Since the parallel combination of the 51-K and 200-K bias resistors is in parallel with the input (as seen by the source signal), the overall amplifier input impedance is calculated to be

$$R_i = 1.8 \text{ K} \,\|\, 51 \text{ K} \,\|\, 200 \text{ K} \cong \textbf{1.7 K}$$

The output impedance looking back into stage 2 is approximately

$$R_o = 6.8 \text{ K} \,\|\, 2.2 \text{ K} \cong \mathbf{1.7 \text{ K}}$$

where $1/h_{oe}$ is neglected as being much larger than 6.8 K, the bypassed emitter resistor is neglected, and the capacitive ac impedance is neglected—these being valid assumptions in the amplifier mid-frequency range, and feedback resistors (2-K and 0.2-K) provide an effective resistor in parallel with the output.

The gain of each stage can be obtained (including the loading of the second stage on the first) as follows: Effective load resistances of each stage are

$$R_{L_1} = 12 \text{ K} \,\|\, 120 \text{ K} \,\|\, 30 \text{ K} \,\|\, 1.8 \text{ K} \cong 1.5 \text{ K}$$

$$R_{L_2} = 6.8 \text{ K} \,\|\, 2.2 \text{ K} \cong 1.7 \text{ K}$$

(where 2.2 K is an output load resistance of the 2-K and 0.2-K resistors connected in series to ground).

The voltage gains of each stage (magnitude only) are then

$$A_{v_1} \cong - \frac{h_{fe} R_{L_1}}{h_{ie}} = \frac{65(1.5)}{1.8} \cong 54$$

$$A_{v_2} \cong - \frac{h_{fe} R_{L_2}}{h_{ie}} = \frac{65(1.7)}{1.8} \cong 61$$

The overall gain of the cascaded amplifier, neglecting feedback, is then

$$A = A_{v_1} A_{v_2} = 54(61) \cong \mathbf{3300}$$

We can calculate the feedback factor to be

$$\beta = \frac{R_1}{R_1 + R_2} = \frac{0.2}{0.2 + 2} = \frac{1}{11} = 0.091$$

With the feedback network: Using Eqs. (12.8)–(12.10)

$$R_{of} = \frac{R_o}{1 + \beta A} = \frac{1.7 \text{ K}}{1 + 0.091(3300)} = \frac{1.7 \text{ K}}{301} = \mathbf{5.67 \,\Omega}$$

$$R_{if} = R_i(1 + \beta A) = 1.7 \text{ K}(301) = \mathbf{510 \text{ K}}$$

$$A_f = \frac{A}{1 + \beta A} = \frac{3300}{301} \cong \mathbf{11}$$

Current-Series Feedback Amplifier

So far we have considered only a feedback connection that samples the output voltage and feeds a portion of that voltage back to the input in series opposition with the source signal. Another feedback technique is to sample the output current (I_o) and return a proportional voltage in series with the input. While stabilizing the amplifier gain, the current-series feedback connection increases *both* input and output resistance.

Figure 12.10 shows a simple version of an amplifier with current-series negative feedback. If the input current is negligible as in a tube or FET amplifier, then the current through resistor R is the output current I_o. The voltage developed across R is a feedback voltage connected in series with the source signal. We can consider the amplifier as providing an output current dependent on the input source voltage

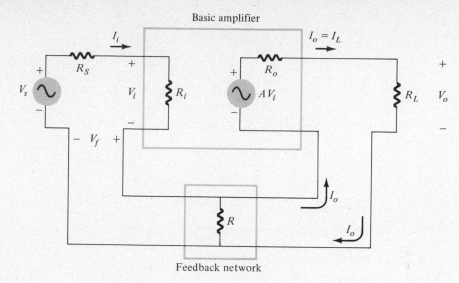

Figure 12.10. Amplifier with current-series negative feedback connection.

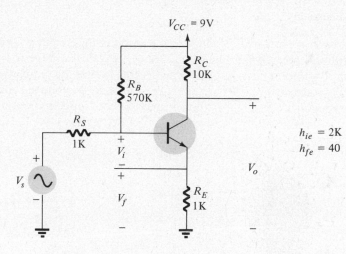

$h_{ie} = 2K$

$h_{fe} = 40$

Figure 12.11. Transistor amplifier with unbypassed emitter resistor (R_E) for current-series negative feedback.

—the amplifier then acting as a transconductance amplifier. Feedback acts to stabilize the transconductance so that the load current depends on the signal voltage and resistance R only and is stabilized in regard to any other circuit changes.

Figure 12.11 shows a single transistor amplifier stage. Since the emitter of this stage has an unbypassed emitter, it effectively has current-series feedback. The current through resistor R_E results in a feedback voltage that opposes the source signal applied so that the output voltage V_o is reduced. To remove the current-series feedback the emitter resistor must be either removed or bypassed by a capacitor (as is usually done). An example will show how the amplifier operation is affected by current-series feedback due to resistor R_E.

> **EXAMPLE 12.7** For the circuit of Fig. 12.11 calculate the gain, input resistance, and output resistance without feedback (bypassed emitter resistor, $R_E = 0$) and with feedback (R_E present), and $1/h_{oe} = \infty$.

SEC. 12.4 OTHER PRACTICAL FEEDBACK CIRCUIT CONNECTIONS   535

Solution: With R_E removed (bypassed)

$$A = \frac{-h_{fe}R_C}{h_{ie}} = \frac{-40(10\text{ K})}{2\text{ K}} = -200$$

$$R_i = h_{ie} = 2\text{ K}$$

$$R_o = R_C = 10\text{ K}$$

With R_E in the circuit

$$A_f = \frac{-h_{fe}R_C}{h_{ie} + h_{fe}R_E} = \frac{-40(10\text{ K})}{2\text{ K} + 40(1\text{ K})} = \frac{-400}{42} = -9.5$$

$$R_{if} = h_{fe}R_E + h_{ie} = 40(1\text{ K}) + 2\text{ K} = 42\text{ K}$$

$$R_{of} \cong R_C = 10\text{ K}$$

Example 12.7 shows that current-series feedback

1. Reduces amplifier gain.
2. Increases input resistance.

Voltage-Shunt Feedback

Negative feedback can be obtained by coupling a portion of the output voltage in parallel (shunt) with the input signal. Figure 12.12 shows a typical voltage-shunt feedback connection. A portion of the output voltage (which is opposite in polarity to the input voltage) is connected to the base through resistor R_f. A voltage-shunt connection stabilizes amplifier overall gain while decreasing both input and output resistances.

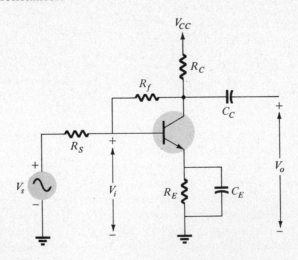

Figure 12.12. Voltage-shunt negative feedback amplifier.

Current-Shunt Feedback

A fourth feedback connection samples the output current and develops a feedback voltage in shunt with the input signal. A practical circuit version is the two-stage amplifier of Fig. 12.13. The unbypassed emitter resistor of stage 2 provides

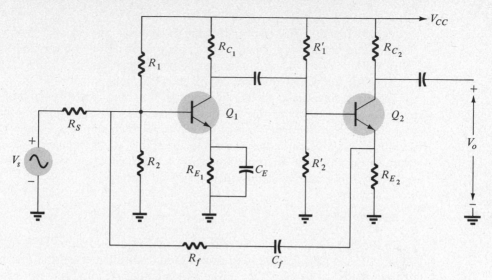

Figure 12.13. Amplifier with current-shunt negative feedback connection.

current sensing. The feedback signal is connected in shunt with the first stage input through a feedback network.

Checking the feedback signal polarity for an input to the base of stage 1, we see that the output of stage 1 is opposite in phase. The input to the base of stage 2 and the voltage across emitter R_{E_1} is then opposite in phase to the input to stage 1 so that negative feedback is achieved. A current-shunt feedback circuit typically incresses output resistance and decreases input resistance while holding the gain with feedback constant.

The operation of the four types of feedback connections is summarized in Table 12.1. All types provide stabilized but reduced gain, increased bandwidth, and decreased nonliear distortion.

TABLE 12.1 Effect of Feedback Connection Type on Input and Output Resistance

	VOLTAGE-SERIES	CURRENT-SERIES	VOLTAGE-SHUNT	CURRENT-SHUNT
R_{if}	increased	increased	decreased	decreased
R_{of}	decreased	increased	decreased	increased

12.5 FEEDBACK AMPLIFIER STABILITY— PHASE AND FREQUENCY CONSIDERATIONS

So far we have considered the operation of a feedback amplifier in which the feedback signal was *opposite* to the input signal—negative feedback. In any practical circuit this condition occurs only for some mid-frequency range of operation. We know that an amplifier gain will change with frequency, dropping off at higher frequencies from the mid-frequency value. In addition, the phase shift of an amplifier will also change with frequency so that a shift of 180° in the mid-frequency range will no longer be the situation at higher frequencies.

If, as the frequency increases, the phase shift changes from 180° then some of the feedback signal *adds* to the input signal. It is then possible for the amplifier to break into oscillations due to positive feedback. If the amplifier oscillates at some low or high frequency, it is no longer useful as an amplifier. Proper feedback-amplifier design requires that the circuit be stable at *all* frequencies, not merely those in the range of interest. Otherwise, a transient disturbance could cause a seemingly stable amplifier to suddenly start oscillating.

Nyquist Criterion

In judging the stability of a feedback amplifier, as a function of frequency, the factors of loop gain A_f, amplifier gain A, and feedback attenuation β as functions of frequency, can be used. One of the most popular techniques used to investigate stability is the Nyquist method. A Nyquist diagram is used to plot gain and phase shift as a function of frequency on a complex plane. The Nyquist plot, in effect, combines the two Bode plots of gain versus frequency and phase-shift versus frequency on a single plot. A Nyquist plot is used to quickly show whether an amplifier is stable for all frequencies and how stable the amplifier is relative to some gain or phase-shift criteria.

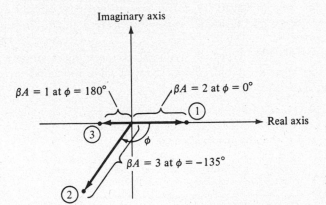

Figure 12.14. Complex plane showing typical gain-phase points.

As a start, consider the *complex plane* shown in Fig. 12.14. A few points of various gain (βA) values are shown at a few different phase-shift angles. By using the positive real axis as reference (0°) a magnitude of $\beta A = 2$ is shown at a phase shift of 0° at point 1. Additionally, a magnitude of $\beta A = 3$ at a phase shift of $-135°$ is shown at point 2 and a magnitude/phase of $\beta A = 1$ at 180° is shown at point 3. Thus, points on this plot can represent *both* gain magnitude of βA and phase shift. If the points representing gain and phase shift for an amplifier circuit are plotted at increasing frequency, then a Nyquist plot is obtained as shown by the plot in Fig. 12.15. At the origin the gain is 0 at a frequency of 0 (for RC-type coupling). At increasing frequency points f_1, f_2, and f_3 the phase shift increased as did the

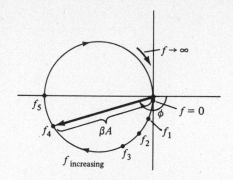

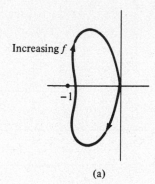

Figure 12.15. Nyquist plot.

magnitude of βA. At a representative frequency f_4 the value of A is the vector length from the origin to point f_4 and the phase shift is the angle ϕ. At a frequency f_5 the phase shift is 180°. At higher frequencies the gain is shown to decrease back to 0.

The Nyquist criteria for stability can be stated as follows:

The amplifier is unstable if the Nyquist curve plotted encloses (encircles) the −1 point, and it is stable otherwise.

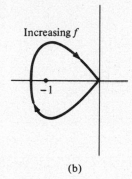

(a)

(b)

Figure 12.16. Nyquist plots showing stability conditions: (a) stable; (b) unstable.

An example of the Nyquist criteria is demonstrated by the curves in Fig. 12.16. The Nyquist plot in Fig. 12.16a is stable since it does not encircle the −1 point, whereas that shown in Fig. 12.16b is unstable since the curve does encircle the −1 point. Keep in mind that encircling the −1 point means that at a phase shift of 180° the loop gain (βA) is greater than 1; therefore, the feedback signal is in phase with the input and large enough to result in a larger input signal than that applied, with the result that oscillation occurs.

Gain and Phase Margins

From the Nyquist criterion we know that a feedback amplifier is stable if the loop gain (βA) is less than unity (0 dB) when its phase angle is 180°. We can additionally determine some margins of stability to indicate how close to instability the unit is. That is, if the gain (βA) is less than unity but, say, 0.95 in value, this would not be as relatively stable as another amplifier having, say, (βA) = 0.7 (both measured at 180°). Of course, amplifiers with loop gains 0.95 and 0.7 are both stable, but one is closer to instability, if the loop gain increases, than the other. We can define the following terms:

Gain margin (GM) is defined as the value of βA in decibels at the frequency at which the phase angle is 180°. Thus, 0 dB, equal to a value of $\beta A = 1$, is on the border of stability and any negative decibel value is stable. The more negative the

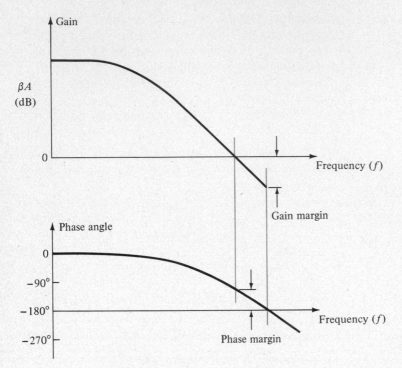

Figure 12.17. Bode plots showing gain and phase margins.

decibel gain value the more stable the feedback circuit. The GM may be evaluated in decibels from the curve of Fig. 12.17.

Phase margin (PM) is defined as the angle of 180° minus the magnitude of the angle at which the value βA is unity, 0 dB. The PM may also be evaluated directly from the curve of Fig. 12.17.

An example of these two amplifier factors is shown on the Bode plots of Fig. 12.17. Instability occurs, therefore, with a positive GM and PM greater than 180°.

12.6 OPERATION OF FEEDBACK CIRCUIT AS AN OSCILLATOR

The use of positive feedback which results in a feedback amplifier having closed-loop gain A_f greater than 1 and satisfies the phase conditions will result in operation as an oscillator circuit. An oscillator circuit then provides a constantly varying output signal. If the output signal varies sinusoidally, the circuit is referred to as a *sinusoidal oscillator*. If the output voltage rises quickly to one voltage level and later drops quickly to another voltage level, the circuit is generally referred to as a *pulse* or *square-wave oscillator*.

To understand how a feedback circuit performs as an oscillator consider the feedback circuit of Fig. 12.18. When the switch at the amplifier input is open, no oscillation occurs. Consider that we have a *fictitious* voltage at the amplifier input (V_i). This results in an output voltage $V_o = AV_i$ after the base amplifier stage and in a voltage $V_f = \beta(AV_i)$ after the feedback stage. Thus, we have a feedback voltage

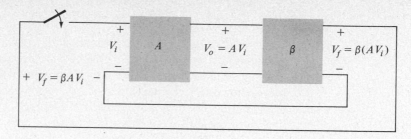

Figure 12.18. Feedback circuit used as an oscillator.

$V_f = \beta A V_i$, where βA is referred to as the *loop gain*. If the circuits of the base amplifier and feedback network provide βA of a correct magnitude and phase, V_f can be made equal to V_i. Then, when the switch is closed and fictitious voltage V_i is removed, the circuit will continue operating since the feedback voltage is sufficient to drive the amplifier and feedback circuits resulting in a proper input voltage to sustain the loop operation. The output waveform will still exist after the switch is closed if the condition

$$\beta A = 1 \tag{12.18}$$

is met. This is known as the *Barkhausen criterion* for oscillation.

In reality, no input signal is needed to start the oscillator going. Only the condition $\beta A = 1$ must be satisfied for self-sustained oscillations to result. In practice βA is made greater than 1, and the system is started oscillating by amplifying noise voltage which is always present. Saturation factors in the practical circuit provide an "average" value of βA of 1. The resulting waveforms are never exactly sinusoidal. However, the closer the value βA is to exactly 1 the more nearly sinusoidal is the waveform. Figure 12.19 shows how the noise signal results in a build-up of a steady-state oscillation condition.

Another way of seeing how the feedback circuit provides operation as an oscillator is obtained by noting the denominator in the basic feedback equation, (12.6) $A_f = A/(1 + \beta A)$. When $\beta A = -1$ or magnitude 1 at a phase angle of

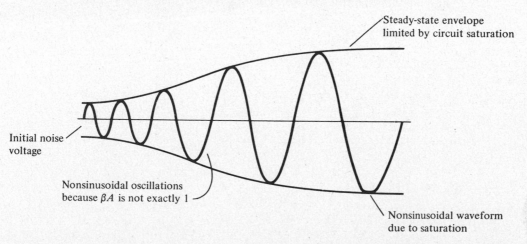

Figure 12.19. Build-up of steady-state oscillations.

$180°$, the denominator becomes 0 and the gain with feedback, A_f, becomes infinite. Thus, an infinitesimal signal (noise voltage) can provide a measurable output voltage, and the circuit acts as an oscillator even without an input signal.

The remainder of this chapter is devoted to various oscillator circuits that use a variety of components. Practical considerations are included so that workable circuits in each of the various cases are discussed.

12.7 PHASE-SHIFT OSCILLATOR

An example of an oscillator circuit that follows the basic development of a feedback circuit is the *phase-shift oscillator*. An idealized version of this circuit is shown in Fig. 12.20. Recall that the requirements for oscillation are that the loop gain, βA, is greater than unity *and* that the phase shift around the feedback network is $180°$ (providing positive feedback). In the present idealization we are considering the feedback network to be driven by a perfect source (zero source impedance) and the output of the feedback network is connected into a perfect load (infinite load impedance). The idealized case will allow development of the theory behind the operation of the phase-shift oscillator. Practical circuit versions will then be considered.

Concentrating our attention on the phase-shift network we are interested in the attenuation of the network at the frequency at which the phase shift is exactly $180°$. Using classical network analysis, we find that the results are

$$f = \frac{1}{2\pi RC\sqrt{6}} \qquad (12.19a)$$

$$\beta = \frac{1}{29} \qquad (12.19b)$$

and the phase shift is $180°$.

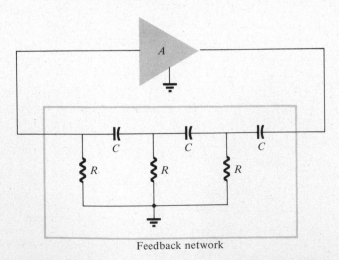

Feedback network

Figure 12.20. Idealized phase-shift oscillator.

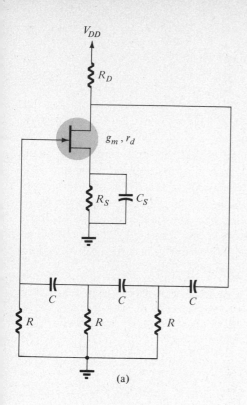

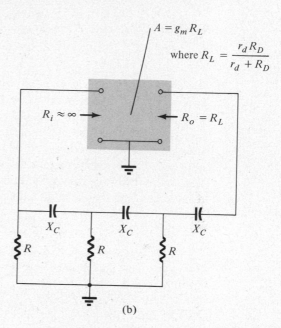

$$A = g_m R_L$$

where $R_L = \dfrac{r_d R_D}{r_d + R_D}$

$R_i \approx \infty \rightarrow$ $\leftarrow R_o = R_L$

(a)   (b)

Figure 12.21. FET phase-shift oscillator circuit.

For the loop gain βA to be greater than unity the gain of the amplifier stage must be greater than $1/\beta$ or 29

$$\boxed{A > 29} \tag{12.19c}$$

When considering the operation of the feedback network one might naively select the values of R and C to provide (at a specific frequency) 60°-phase shift per section for three sections, resulting in 180°-phase shift as desired. This, however, is not the case, since each section of the RC in the feedback network loads down the previous one. The net result that the *total* phase shift be 180° is all that is important. The frequency given by Eq. (12.19a) is that at which the *total* phase shift is 180°. If one measured the phase shift per RC section, each section would not provide the same phase shift (although the overall phase shift is 180°). If it were desired to obtain exactly 60°-phase shift for each of three stages, then emitter-follower stages would be needed after each RC section to prevent each from being loaded from the following circuit.

FET Phase-Shift Oscillator

A practical version of a phase-shift oscillator circuit is shown in Fig. 12.21a. The circuit is drawn to show clearly the amplifier and feedback network. The amplifier stage is self-biased with a capacitor bypassed source resistor R_S and a drain bias resistor R_D. The FET device parameters of interest are g_m and r_d. From FET amplifier theory the amplifier gain is calculated from

$$A = g_m R_L \qquad (12.20)$$

where R_L in this case is the parallel resistance of R_D and r_d

$$R_L = \frac{R_D r_d}{R_D + r_d} \qquad (12.21)$$

We shall assume as a very good approximation that the input impedance of the FET amplifier stage is infinite (see Fig. 12.21b). This assumption is valid as long as the oscillator operating frequency is low enough so that FET capacitive impedances can be neglected. The output impedance of the amplifier stage given by R_L should also be small compared to the impedance seen looking into the feedback network so that no attenuation due to loading occurs. In practice, these considerations are not always negligible, and the amplifier stage gain is then selected somewhat larger than the needed factor of 29 to assure oscillator action.

EXAMPLE 12.8 It is desired to design a phase-shift oscillator (as in Fig. 12.21a) using a FET having $g_m = 5000$ μmhos, $r_d = 40$ K, and feedback circuit value of $R = 10$ K. Select the value of C for oscillator operation at 1 kHz and R_D for $A > 29$ to ensure oscillator action.

Solution: Equation (12.19a) is used to solve for the capacitor value. Since $f = 1/2\pi RC\sqrt{6}$, we can solve for C

$$C = \frac{1}{2\pi R f \sqrt{6}} = \frac{1}{(6.28)(10 \times 10^3)(10^3)(2.45)}$$

$$C = 0.0065 \text{ } \mu\text{F}$$

Using Eq. (12.20), we solve for R_L to provide a gain of, say, $A = 40$ (this allows for some loading between R_L and the feedback network input impedance)

$$A = g_m R_L$$

$$R_L = \frac{A}{g_m} = \frac{40}{5000 \times 10^{-6}} = 8 \text{ K}$$

Using Eq. (12.21), we solve for R_D

$$R_L = \frac{R_D r_d}{R_D + r_d}$$

$$8 \text{ K} = \frac{R_D(40 \text{ K})}{R_D + 40 \text{ K}}$$

$$R_D = 10 \text{ K}$$

Transistor Phase-Shift Oscillator

If a transistor is used as the active element of the amplifier stage, the output of the feedback network is loaded appreciably by the relatively low input resistance (h_{ie}) of the transistor. Of course, an emitter follower input stage followed by a common-emitter amplifier stage could be used. If a single transistor stage is desired, however, the use of voltage-shunt feedback (as shown in Fig. 12.22a) is more suitable. In this connection, the feedback signal is coupled through the feedback resistor R_S in *series* with the amplifier stage input resistance (R_i).

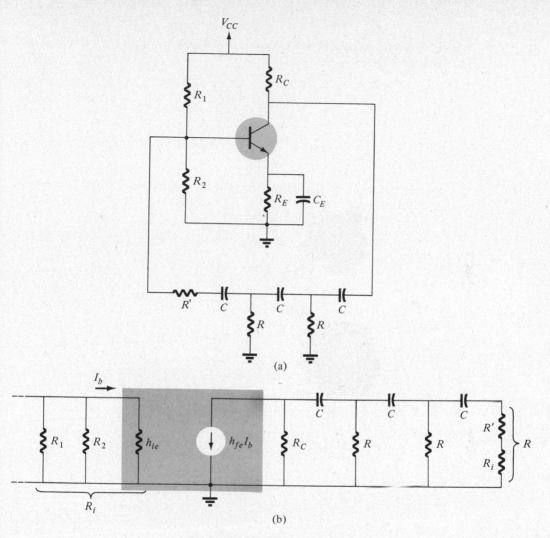

Figure 12.22. Transistor phase-shift oscillator: (a) transistor circuit; (b) ac equivalent circuit.

An ac equivalent circuit is shown in Fig. 12.22b. The figure shows that the input resistance R_i in series with feedback resistor R', is the parallel combination of resistors R_1, R_2, and h_{ie}. Also, the effective resistance for the third leg of the feedback network, the series combination of resistors R' and R_i, is made the same value as the resistance of the other two resistors of the feedback network to make calculations simpler. We shall assume that the transistor output impedance $1/h_{oe}$ is much larger than R_C.

Analysis of the ac circuit provides the following equation for the resulting oscillator frequency:

$$f = \left(\frac{1}{2\pi RC}\right)\frac{1}{\sqrt{6 + 4(R_C/R)}} \qquad (12.22)$$

SEC. 12.7 PHASE-SHIFT OSCILLATOR

545

For the loop gain to be greater than unity, the requirement on the current gain of the transistor is found to be

$$h_{fe} > 23 + 29\frac{R_C}{R} + 4\frac{R}{R_C} \qquad (12.23)$$

A practical example will demonstrate the use of the above information in designing an oscillator circuit.

EXAMPLE 12.9 Select the value of capacitor C and transistor gain h_{fe} to provide an oscillator frequency of $f = 2$ kHz. Circuit values are $h_{ie} = 2$ K, $R_1 = 20$ K, $R_2 = 80$ K, $R_C = 10$ K, and $R = 8$ K.

Solution: Using Eq. (12.22), we can determine the required value of C

$$f = \left(\frac{1}{2\pi RC}\right)\frac{1}{\sqrt{6 + 4R_C/R}}$$

$$2 \times 10^3 = \left[\frac{1}{6.28(8 \times 10^3)C}\right]\frac{1}{\sqrt{6 + 4(1.25)}}$$

$$C = \left[\frac{1}{6.28(8 \times 10^3)(2 \times 10^3)}\right]\frac{1}{3.32} = \frac{10^{-6}}{332} = 3 \times 10^{-9}$$

$$= 0.003\mu F$$

Calculating R_i as the parallel resistance of R_1, R_2, and h_{ie} gives

$$R_i = 20 \text{ K} \,\|\, 80 \text{ K} \,\|\, 2 \text{ K} \cong 1.8\text{K}$$

For

$$R' + R_i = R = 8 \text{ K}$$

$$R' = R - R_i = 8 - 1.8 = 6.2 \text{ K}$$

To determine the value of h_{fe} necessary [using Eq. (12.23)]

$$h_{fe} > 23 + 29\frac{R}{R_C} + 4\frac{R_C}{R} = 23 + 29\left(\frac{8}{10}\right) + 4\left(\frac{10}{8}\right)$$

$$= 23 + 23.2 + 5 = \textbf{51.2}$$

A transistor with $h_{fe} > 51.2$ will provide sufficient loop gain for the circuit to operate as an oscillator. Practically, a transistor with at least $h_{fe} > 60$ would be selected.

Phase-shift oscillators are suited to operating frequencies in the range of a few hertz to a few hundred kilohertz. Other oscillator configurations (typically the tuned circuits) are more suitable to frequency in the megahertz range. To adjust the frequency of a phaseshift oscillator it is necessary to operate the three feedback capacitors as a single ganged component so that all three capacitor values are kept the same. A phase-shift oscillator is operated class-A to keep distortion low while providing a sinusoidal output waveform taken from the output of the amplifier stage. It should be clear that the output signal should not directly drive any low-impedance circuit that will load down the output. This would decrease the output voltage and thereby drop the loop gain below the necessary value for oscillator action. Feeding the output to a high-impedance stage, such as a FET amplifier stage or emitter-follower transistor stage, provides negligible loading of this oscillator circuit.

12.8 THE *LC*-TUNED OSCILLATOR CIRCUIT

An oscillator circuit can be made using a transformer for the feedback network. In addition, the inductance of the transformer and a parallel capacitor can be used to tune the circuit to the desired oscillator frequency. Figure 12.23a shows a FET amplifier with positive feedback provided by the transformer and tuning by an *LC* circuit. The circuit is described as a tuned-drain, untuned-gate oscillator.

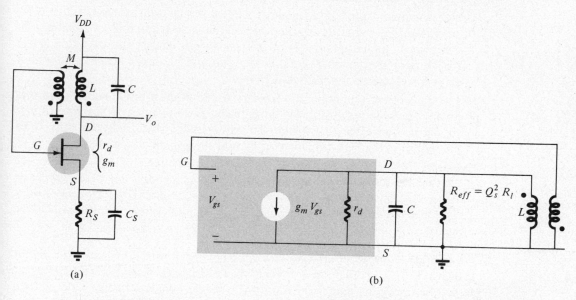

(a)

(b)

Figure 12.23. Tuned-drain oscillator circuit.

An ac equivalent circuit is shown in Fig. 12.23b. The input resistance to the gate is assumed very high and is shown as an open circuit. The FET ac equivalent circuit is shown as a current source $(g_m V_{gs})$ in parallel with an output resistance r_d. The transformer can be represented as an inductance L in the primary side and a mutual coupling factor M. The series resistance of the transformer (representing its losses) can be accounted for by an effective resistance in parallel with the primary shown as $R_{eff} = Q_s^2 R_s$ in the circuit. The factor Q_s is the series Q of the transformer, defined as

$$Q_s = \frac{\omega_o L}{R_s} \qquad (12.24)$$

Analysis of the ac equivalent circuit provides the oscillator frequency

$$\boxed{f_o = \frac{1}{2\pi\sqrt{LC}}} \qquad (12.25)$$

and FET g_m

$$g_m = \frac{1}{R}\frac{L}{M} \qquad (12.26)$$

where $$R = r_d \| Q_s^2 R_s \qquad\qquad (12.27)$$

A summary of the results provided in the above equations follows:

1. The oscillator frequency is determined by the LC-resonant ("tank") circuit. This relationship is sometimes modified slightly due to circuit nonlinearities and unaccounted for resistances, capacitances, and so on. However, it is good as a first-order approximation.
2. The minimum required FET g_m is dependent on the transformer effective resistance, the FET output resistance, the coil inductance and mutual coupling—in other words, on the parameters of the transformer chosen and the value of the FET output resistance. The value of g_m should be larger than this minimum vlaue for oscillator action to take place.
3. The transformer primary and secondary windings must be connected in proper polarity sense to result in positive feedback. For this to occur the transformer should be connected to provide 180°-phase shift, which, added to the 180°-phase shift of the FET amplifier stage, results in overall positive feedback.
4. Loading of the circuit provided by either a lower value of r_d, lower effective transformer resistance, or external loading due to a connection of the oscillator to another circuit results in the value of the resulting resistance R being lower. This then requires a larger value of g_m to provide sufficient loop gain for oscillator action.

Another example of a tuned-circuit oscillator is the tuned-collector circuit shown in Fig. 12.24. The primary side of the transformer forms a tuned-tank circuit to set the oscillator frequency. The transformer is connected to provide positive feedback and the amplifier provides sufficient gain for oscillator action to take place.

Resistors R_1, R_2, and R_E are used to dc bias the transistor. Capacitors C_E and C_2 act to bypass resistors R_E and R_2, respectively, so that they have no effect on

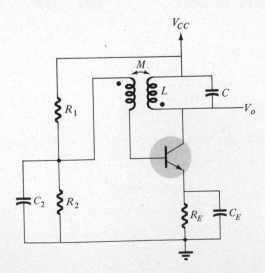

Figure 12.24. Bipolar transistor-tuned LC oscillator circuit.

the ac operation of the circuit. Notice that although the low-resistance secondary winding of the transformer provides the dc bias voltage set by R_1 and R_2 to be connected to the base, the secondary essentially provides an ac feedback voltage in shunt with the transistor base emitter since the junction point of R_1 and R_2 is at ac ground (due to bypass capacitor C_2).

> EXAMPLE 12.10   For the oscillator circuit of Fig. 12.23 and the following circuit values calculate the circuit frequency of oscillation and the minimum gain (g_m) of the FET unit: $r_d = 40$ K, $L = 4$ mH, $M = 0.1$ mH, $R_s = 50$ Ω, and $C = 0.001$ μF.
>
> **Solution:**   The resonant frequency of the oscillator circuit is calculated using Eq. (12.25)
>
> $$f_o = \frac{1}{2\pi\sqrt{LC}} = \frac{1}{6.28\sqrt{(4 \times 10^{-3})(0.001 \times 10^{-6})}}$$
>
> $$= \frac{10^6}{6.28(2)} \cong \textbf{80 kHz}$$
>
> $$\omega_o = 2\pi f_o = 6.28(80 \text{ KHz}) = 500 \times 10^3 \text{ rad/sec}$$
>
> The coil Q is then
>
> $$Q_s = \frac{\omega_o L}{R_s} = \frac{(500 \times 10^3)(4 \times 10^{-3})}{50} = 40$$
>
> The effective coil resistance is
>
> $$R_{\text{eff}} = Q_s^2 R_s = (40)^2(50) = 80 \text{ K}$$
>
> Using Eq. (12.27) to calculate R gives
>
> $$R = \frac{r_d R_{\text{eff}}}{r_d + R_{\text{eff}}} = \frac{40(80)}{40 + 80} = 26.7 \text{ K}$$
>
> The minimum value of g_m can now be calculated using Eq. (12.26)
>
> $$g_m = \frac{1}{R}\frac{L}{M} = \frac{1}{26.7 \times 10^3}\frac{4 \times 10^{-3}}{0.1 \times 10^{-3}} = 1.5 \times 10^{-3} = \textbf{1500 } \boldsymbol{\mu}\textbf{mhos}$$
>
> The FET selected should have a value of g_m greater than 1500 μmhos.

12.9 TUNED-INPUT, TUNED-OUTPUT OSCILLATOR CIRCUITS

A variety of circuits shown in Fig. 12.25 provide tuning in both the input and output sections of the circuit. Analysis of the circuit of Fig. 12.25 reveals that the following types of oscillators are obtained when the reactance elements are as designated.

OSCILLATOR TYPE	REACTANCE ELEMENTS		
	X_1	X_2	X_3
1. Colpitts oscillator	C	C	L
2. Hartley oscillator	L	L	C
3. Tuned input, tuned output	LC	LC	—

SEC. 12.9 TUNED-INPUT, TUNED-OUTPUT OSCILLATOR CIRCUITS

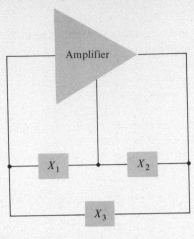

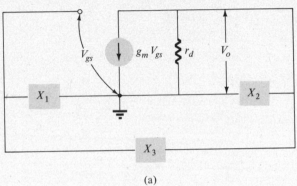

Figure 12.25. Basic configuration of resonant circuit oscillator.

(a)

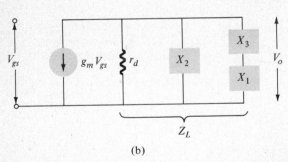

(b)

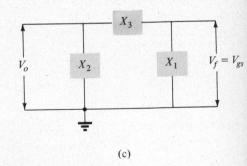

(c)

Figure 12.26. Basic resonant circuit oscillator using FET amplifier.

The circuit form of the oscillator using a FET amplifier is shown in Fig. 12.26a. For operation as an oscillator the Barkhausen criterion is

$$\beta A = 1$$

The amplifier gain A is given simply by

$$A = -g_m Z_L \qquad (12.28)$$

where Z_L is a parallel combination of impedances

$$Z_L = r_d \| X_2 \| (X_1 + X_3) \qquad (12.29)$$

The circuit is redrawn in Fig. 12.26b so that the gain network is clearly shown. Figure 12.26b shows the parallel components that form an equivalent ac impedance, Z_L, as given in Eq. (12.29).

Figure 12.26c shows the feedback section of the circuit. Notice that the output voltage (V_o) is developed across X_2 and that the resulting feedback voltage (V_f) across X_1 is the input voltage to the amplifier (V_g). The feedback factor β is given by

$$\beta = \frac{X_1}{X_1 + X_3} \tag{12.30}$$

Plugging Eq. (12.28) for A and Eq. (12.30) for β into the basic equation (12.18) and using Eq. (12.29) for Z_L provide a means of determining the necessary device gain (g_m).

$$A = \frac{-g_m r_d X_2 (X_1 + X_3)}{r_d(X_1 + X_2 + X_3) + X_2(X_1 + X_3)} \tag{12.31}$$

The oscillator frequency is obtained from

$$(X_1 + X_2 + X_3) = 0 \tag{12.32}$$

When X_1 and X_2 are capacitors and X_3 is an inductor, the circuit is called a *Colpitts oscillator*. When X_1 and X_2 are inductors and X_3 is a capacitor, the circuit is called a *Hartley oscillator*.

12.10 COLPITTS OSCILLATOR

FET Colpitts Oscillator

A practical version of a FET Colpitts oscillator is shown in Fig. 12.27a. The circuit is basically the same form as shown in Fig. 12.26 with the addition of the

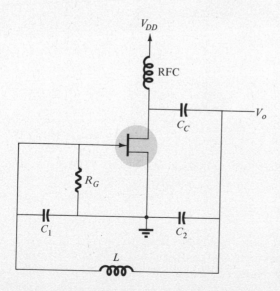

Figure 12.27. FET Colpitts oscillator.

components needed for dc bias of the FET amplifier. The oscillator frequency can be found to be

$$f_o = \frac{1}{2\pi\sqrt{LC_{eq}}}$$

(12.33)

where

$$C_{eq} = \frac{C_1 C_2}{C_1 + C_2}$$

Transistor Colpitts Oscillator

A transistor Colpitts oscillator circuit can be made as shown in Fig. 12.28. The circuit frequency of oscillation is given by Eq. (12.33).

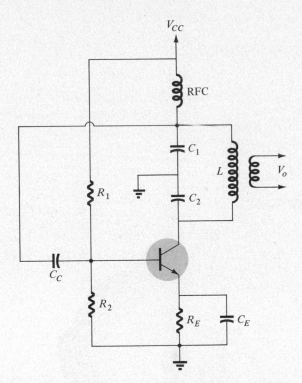

Figure 12.28. Transistor Colpitts oscillator.

12.11 HARTLEY OSCILLATOR

If the elements in the basic resonant circuit of Fig. 12.26 are X_1 and X_2 (inductors), and X_3 (capacitor), the circuit is a Hartley oscillator.

FET Oscillator

A FET Hartley oscillator circuit is shown in Fig. 12.29. The circuit is drawn so that the feedback network conforms to the form shown in the basic resonant circuit (Fig. 12.26). Note, however, that inductors L_1 and L_2 have a mutual coupl-

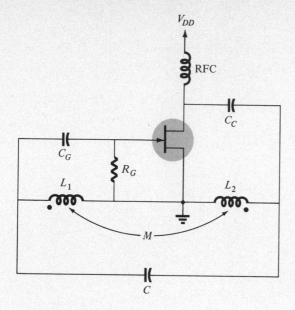

Figure 12.29. FET Hartley oscillator.

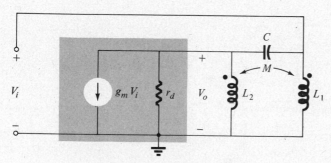

Figure 12.30. FET Hartley oscillator ac equivalent circuit.

ing, M, which must be taken into account in determining the equivalent inductance for the resonant tank circuit.

ac Circuit Analysis

We can obtain an ac equivalent circuit for Fig. 12.29 by replacing capacitors C_c and C_G by shorts, RFC coil by an open circuit, and the FET by its equivalent circuit as shown in Fig. 12.30. The tank circuit of mutually linked inductors L_1 and L_2 and capacitor C can be shown equivalent to a tank circuit of capacitor C in parallel with an equivalent inductance

$$L_{eq} = L_1 + L_2 + 2\,M \qquad (12.34)$$

The circuit frequency of oscillation is then given approximately by

$$\boxed{f_o = \frac{1}{2\pi\sqrt{L_{eq}C}}} \qquad (12.35)$$

with L_{eq} given in Eq. (12.34).

SEC. 12.11 HARTLEY OSCILLATOR

553

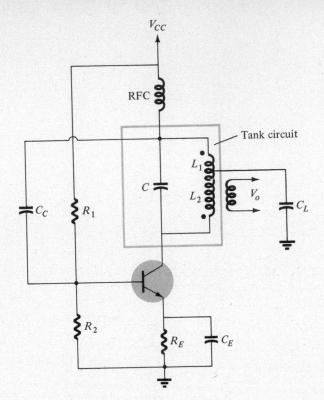

Figure 12.31. Transistor Hartley oscillator circuit.

Transistor Hartley Oscillator

Figure 12.31 shows a transistor Hartley oscillator circuit. The circuit operates at a frequency given by Eq. (12.35).

12.12 CRYSTAL OSCILLATOR

A crystal oscillator is basically a tuned-circuit oscillator using a piezo-electric crystal as a resonant tank circuit. The crystal (usually quartz) has a greater stability in holding constant at whatever frequency the crystal is originally cut to operate. Crystal oscillators are used whenever great stability is required, for example, in communication transmitters and receivers.

Characteristics of a Quartz Crystal

A quartz crystal (one of a number of crystal types) exhibits the property that when mechanical stress is applied across the faces of the crystal, a difference of potential develops across opposite faces of the crystal. This property of a crystal is called the *piezoelectric effect*. Similarly, a voltage applied across one set of faces of the crystal causes mechanical distortion in the crystal shape.

When alternating voltage is applied to a crystal, mechanical vibrations are set up—these vibrations having a natural resonant frequency dependent on the crystal. Although the crystal has electromechanical resonance, we can represent the crystal

CHAP. 12 FEEDBACK AMPLIFIERS AND OSCILLATOR CIRCUITS   554

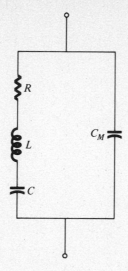

Figure 12.32. Electrical equivalent circuit of a crystal.

action by an equivalent electrical resonant circuit as shown in Fig. 12.32. The inductor L and capacitor C represent electrical equivalents of crystal mass and compliance while resistance R is an electrical equivalent of the crystal structure's internal friction. The shunt capacitance C_M represents the capacitance due to mechanical mounting of the crystal. Because the crystal losses, represented by R, are small, the equivalent crystal Q (quality factor) is high—typically 20,000. Values of Q up to almost 10^6 can be achieved by using crystals.

The crystal as represented by the equivalent electrical circuit of Fig. 12.32 can have two resonant frequencies. One resonant condition occurs when the reactances of the series RLC leg are equal (and opposite). For this condition the *series-resonant* impedance is very low (equal to R). The other resonant condition occurs at a higher frequency when the reactance of the series resonant leg equals the reactance of capacitor C_M. This is a parallel resonance or antiresonance condition of the crystal. At this frequency the crystal offers a very high impedance to the external circuit. The impedance versus frequency of the crystal is shown in Fig. 12.33. In order to use the crystal properly it must be connected in a circuit so that its low impedance in the series-resonant operating mode or high impedance in the antiresonant operating mode is selected.

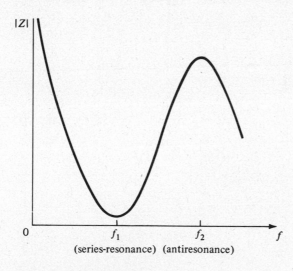

Figure 12.33. Crystal impedance vs. frequency.

Series-Resonant Circuits

To excite a crystal for operation in the series-resonant mode it may be connected as a series element in a feedback path. At the series-resonant frequency of the crystal its impedance is smallest and the amount of (positive) feedback is largest. A

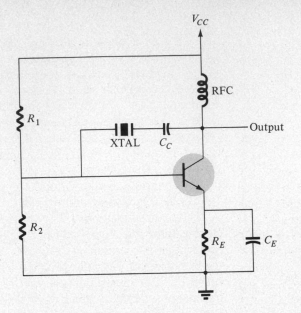

Figure 12.34. Crystal controlled oscillator using crystal in series-feedback path.

typical transistor circuit is shown in Fig. 12.34. Resistors R_1, R_2, and R_E provide a voltage-divider stabilized dc bias circuit. Capacitor C_E provides ac bypass of the emitter resistor and the RFC coil provides for dc bias while decoupling any ac signal on the power lines from affecting the output signal. The voltage feedback from collector to base is a maximum when the crystal impedance is minimum (in series-resonant mode). The coupling capacitor C_C has neglibible impedance at the circuit operating frequency but blocks any dc between collector and base.

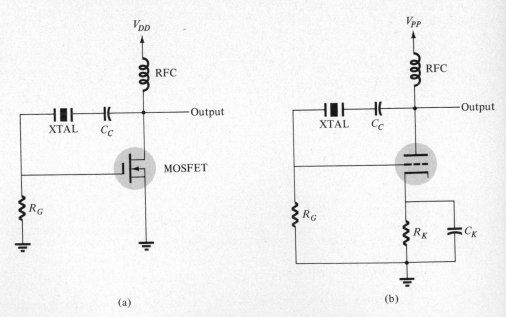

(a) (b)

Figure 12.35. FET and vacuum-tube Pierce crystal controlled oscillator circuits: (a) FET circuit: (b) vacuum-tube circuit.

The resulting circuit frequency of oscillation is set, then, by the series-resonant frequency of the crystal. Changes in supply voltage, transistor device parameters, and so on, have no effect on the circuit operating frequency which is held stabilized by the crystal. The circuit frequency stability is set by the crystal frequency stability, which is good.

The circuit shown in Fig. 12.34 is generally called a Pierce crystal-controlled oscillator. Other versions of the circuit using FETs and vacuum tubes are shown in Fig. 12.35.

Another transistor circuit is shown in Fig. 12.36. The circuit provides tuning by an *LC* tank circuit in the collector and tuning by a series-resonant excited crystal connected as feedback from a capacitive voltage divider. The *LC* circuit is adjusted near the desired operating crystal frequency, but the exact circuit frequency is set by the crystal and stabilized by the crystal.

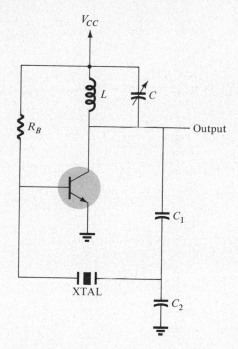

Figure 12.36. Transistor crystal oscillator.

Parallel-Resonant Circuits

Since the parallel-resonant impedance of a crystal is a maximum value, it is connected in shunt. At the parallel-resonant operating frequency a crystal appears as an inductive reactance of largest value. Figure 12.37 shows a crystal connected as the inductor element in a modified Colpitts circuit. The basic dc bias circuit should be evident. Maximum voltage is developed across the crystal at its parallel-resonant frequency. The voltage is coupled to the emitter by a capacitor voltage divider—capacitors C_1 and C_2.

A *Miller* crystal-controlled oscillator circuit is shown in Fig. 12.38. A tuned *LC* circuit in the drain section is adjusted near the crystal parallel-resonant fre-

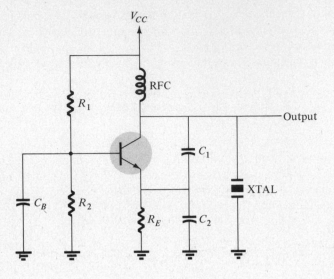

Figure 12.37. Crystal controlled oscillator operating in parallel-resonant triode.

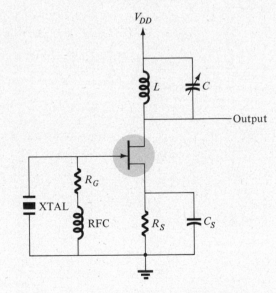

Figure 12.38. Miller crystal controlled oscillator.

quency. The maximum gate-source signal occurs at the crystal antiresonant frequency controlling the circuit operating frequency.

12.13 OPAMP OSCILLATOR CIRCUITS

Phase-Shift Oscillator

As IC circuits have become more popular they have been adapted to operate in oscillator circuits. One need buy only an OPAMP to obtain an amplifier circuit of stabilized gain setting and incorporate some means of signal feedback to pro-

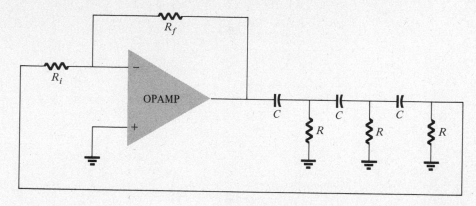

Figure 12.39. Phase-shift oscillator using OPAMP.

duce an oscillator circuit. For example, a phase-shift oscillator is shown in Fig. 12.39. The output of the OPAMP is fed to a three-stage *RC* network which provides the needed 180° of phase shift (at an attenuation factor of 1/29). If the OPAMP provides gain (set by resistors R_i and R_f) of greater than 29, a loop gain greater than unity results and the circuit acts as an oscillator [oscillator frequency is given by Eq. (12.19a)].

Colpitts Oscillator

An OPAMP Colpitts oscillator circuit is shown in Fig. 12.40. Again, the OPAMP provides the basic amplification needed while the oscillator frequency is set by an *LC* feedback network of a Colpitt configuration. The oscillator frequency is given by Eq. (12.33).

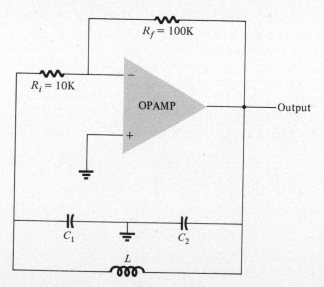

Figure 12.40. OPAMP Colpitts oscillator.

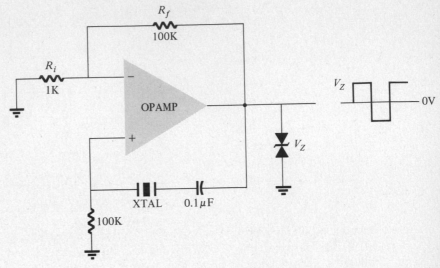

Figure 12.41. Crystal oscillator using OPAMP.

Crystal Oscillator

An OPAMP can be used in a crystal oscillator as shown in Fig. 12.41. The crystal is connected in the series-resonant path and operates at the crystal series-resonant frequency. The present circuit has a high gain so that an output square-wave signal results as shown in the figure. A pair of Zener diodes is shown at the output to provide output amplitude at exactly the Zener voltage (V_Z).

Wien Bridge Oscillator

A practical oscillator circuit uses an OPAMP and RC bridge circuit, with the oscillator frequency set by the R and C components. Figure 12.42 shows a basic version of a Wien bridge oscillator circuit. Note the basic bridge connection. Resistors R_1, R_2 and capacitors C_1, C_2 form the frequency adjustment elements, while resistors R_3 and R_4 form part of the feedback path. The OPAMP output is connected as the bridge input at points a and c. The bridge circuit output at points b and d is the input to the OPAMP.

Neglecting loading effects of the OPAMP input and output impedances, the analysis of the bridge circuit results in

$$\frac{R_3}{R_4} = \frac{R_1}{R_2} + \frac{C_2}{C_1} \tag{12.36a}$$

and

$$\omega_o = \frac{1}{\sqrt{R_1 C_1 R_2 C_2}} \tag{12.36b}$$

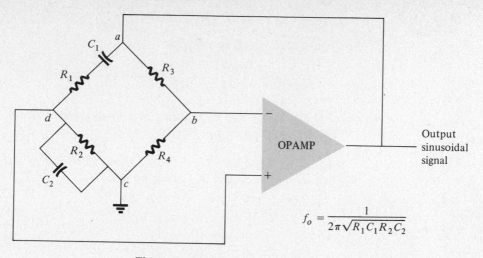

Figure 12.42. Wien bridge oscillator circuit using OPAMP amplifier.

$$f_o = \frac{1}{2\pi\sqrt{R_1 C_1 R_2 C_2}}$$

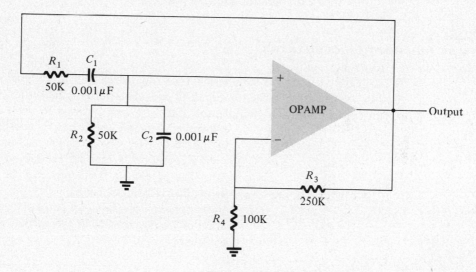

Figure 12.43. Practical Wien bridge oscillator circuit.

If, in particular, the values are $R_1 = R_2 = R$ and $C_1 = C_2 = C$, the resulting oscillator frequency is

$$\omega_o = \frac{1}{RC} \qquad (12.37a)$$

and

$$\frac{R_3}{R_4} = 2 \qquad (12.37b)$$

Thus, a ratio of R_3 to R_4 greater than 2 will provide sufficient loop gain for the circuit to oscillate at the frequency calculated using Eq. (12.37a).

A practical circuit design is shown in Fig. 12.43. Although the circuit is shown in somewhat different form from Fig. 12.42, you should compare the two to satisfy yourself that they are indeed identical in form.

SEC. 12.13 OPAMP OSCILLATOR CIRCUITS

For the circuit values given we calculate

$$f_o = \frac{1}{2\pi RC} = \frac{1}{6.28(50 \times 10^3)(0.001 \times 10^{-6})} = 3.18 \text{ kHz}$$

EXAMPLE 12.11   Design the RC elements of a Wien bridge oscillator as in Fig. 12.43 for operation at $f_o = 10$ kHz.

Solution: Using equal values of R and C we can select $R = 100$ K and calculate the required value of C using Eq. (12.37a):

$$f_o = \frac{1}{2\pi RC}$$

$$C = \frac{1}{2\pi f_o R} = \frac{1}{(6.28(10 \times 10^3)(100 \times 10^3)} = \frac{10^9}{6.28} = \textbf{159 pF}$$

We can use $R_3 = 250$ K and $R_4 = 100$ K to provide a ratio R_3/R_4 greater than 2 for oscillation to take place.

12.14 UNIJUNCTION OSCILLATOR

A particular device, the unijunction transistor (discussed in Chapter 9), can be used in a single-stage oscillator circuit to provide a pulse signal suitable for digital circuit applications. The unijunction transistor can be used in what is called a *relaxation oscillator* as shown by the basic circuit of Fig. 12.44. Resistor R_T and capacitor C_T are the timing components that set the circuit oscillating rate. The oscillating frequency may be calculated using Eq. (12.38) which includes the unijunction transistor *intrinsic stand-off ratio* η, as a factor (in addition to R_T and C_T) in the oscillator operating frequency.

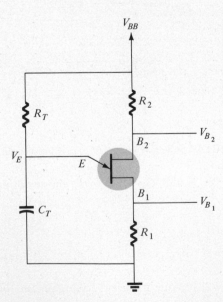

$$\boxed{f_o \cong \frac{1}{R_T C_T \ln{[1/(1 - \eta)]}}} \quad (12.38)$$

Typically, a unijunction transistor has a stand-off ratio from 0.4 to 0.6. Using a value of $\eta = 0.5$, we get

$$f_o \cong \frac{1}{R_T C_T \ln{[1/(1 - 0.5)]}} = \frac{1}{R_T C_T \ln{2}} = \frac{1.44}{R_T C_T}$$

$$\cong \frac{1.5}{R_T C_T} \quad (12.39)$$

Figure 12.44. Basic unijunction oscillator circuit.

Capacitor C_T is charged through resistor R_T toward supply voltage V_{BB}. As

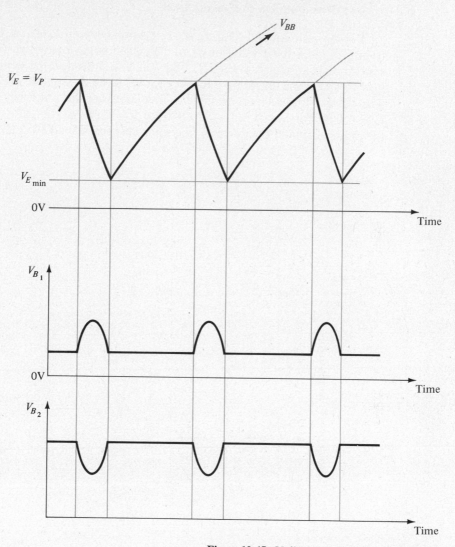

Figure 12.45. Unijunction oscillator waveforms.

long as the capacitor voltage V_E is below a stand-off voltage (V_P) set by the voltage across $B_1 - B_2$, and the transistor stand-off ratio η is given by Eq. (12.40),

$$V_P = \eta V_{B_1} V_{B_2} - V_D \qquad (12.40)$$

and the unijunction emitter lead appears as an open circuit. When the emitter voltage across capacitor C_T exceeds this value (V_P), the unijunction circuit fires, discharging the capacitor, after which a new charge cycle begins. When the unijunction fires, a voltage rise is developed across R_1 and a voltage drop is developed across R_2 as shown in Fig. 12.45. The signal at the emitter is a sawtooth voltage waveform, that at base 1 is a positive-going pulse, and that at base 2 is a negative-going pulse.

Use of Nomograph to Design Circuit

For the basic circuit of Fig. 12.44 the relation between R_T, C_T, and the oscillator frequency (at a fixed value of η) can be given as a nomograph (see Fig. 12.46 instead of as an equation [Eq. (12.38)]. The line shown on the nomograph shows how to obtain the operating frequency for values of $R_T = 10$ K and $C_T = 0.2$ μF. The straight line connecting these points intersects the frequency axis at about $f_o = 650$ Hz for $\eta = 0.56$, or $f_o = 480$ for $\eta = 0.68$. A transistor with $\eta = 0.6$, for example, would operate at a frequency between 480 and 650 Hz.

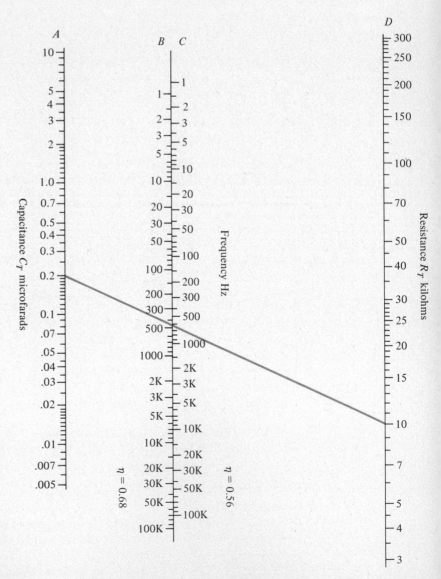

Figure 12.46. Nomograph to calculate unijunction oscillator frequency.

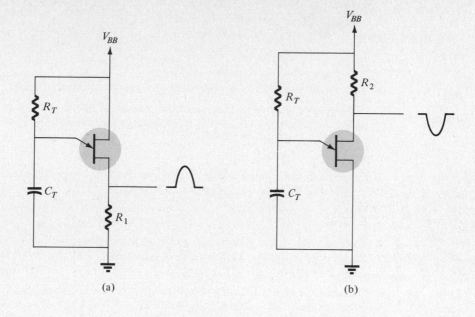

(a)

(b)

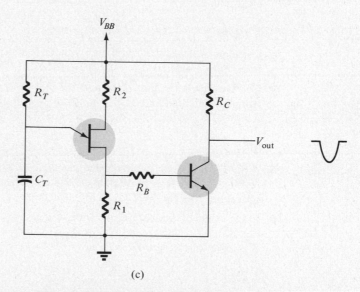

(c)

Figure 12.47. Some unijunction oscillator circuit configurations.

EXAMPLE 12.12 Using the nomograph determine the value of R_T for $C_T = 0.01$ μF and $\eta = 0.56$ for operation at $f_o = 10$ kHz.

Solution: By connecting a straight line between $C_T = 0.01$ μF and $f_o = 10$ kHz (on $\eta = 0.56$ scale) points, the intersection with the R_T axis is read as

$$R_T = 12 \text{ K}$$

[As an exercise, check the result using Eq. (12.38).]

A few circuit variations of the unijunction oscillator are provided in Fig. 12.47.

SEC. 12.14 UNIJUNCTION OSCILLATOR

PROBLEMS

§ 12.1

1. Calculate the gain of a negative feedback amplifier having $A = 2000$, $\beta = 1/10$.

2. If the gain of an amplifier changes from a value of 1000 by 10%, calculate the gain change if the amplifier is used in a feedback circuit having $\beta = 1/20$.

§ 12.2

3. Calculate the gain, input, and output impedances of a voltage-series feedback amplifier having $A = 300$, $R_i = 1.5$ K, $R_o = 50$ K, and $\beta = 1/15$.

§ 12.3

4. Calculate the voltage gain, input, and output impedance with and without feedback for a circuit as in Fig. 12.7 for circuit values $R_e = 2$ K, $R_s = 600\ \Omega$, $h_{ie} = 1.5$ K, and $h_{fe} = 100$.

5. Calculate the gain with and without feedback for a FET amplifier as in Fig. 12.8 for circuit values $R_1 = 200$ K, $R_2 = 800\ \Omega$, $R_o = 40$ K, $R_D = 8$ K, and $g_m = 5000\ \mu$mhos.

§ 12.4

6. Calculate A, R_i, and R_o with and without feedback for a circuit as in Fig. 12.19 for transistor parameters $h_{fe} = 80$, $h_{ie} = 2.2$ K.

7. For a circuit as in Fig. 12.11 and the following circuit values, calculate the circuit gain and the input and output impedances with and without feedback: $R_B = 600$ K, $R_E = 1.2$ K, $R_C = 12$ K, $h_{ie} = 2$ K, and $h_{fe} = 75$. Use $V_{CC} = 16$ V.

§ 12.7

8. A FET phase-shift oscillator having $g_m = 6000\ \mu$mhos, $r_d = 36$ K, and feedback resistor $R = 12$ K is to operate at 2.5 kHz. Select R_D and C for specified oscillator operation.

9. Select values of capacitor C and transistor gain h_{fe} to provide operation of a transistor phase-shift oscillator at 5 kHz for circuit values $R_1 = 24$ K, $R_2 = 75$ K, $R_C = 18$ K, $R = 6$ K, and $h_{ie} = 2$ K.

§ 12.8

10. Calculate the minimum gain (g_m) of the FET in the oscillator circuit of Fig. 12.23 and the circuit frequency of oscillation for circuit values $r_d = 50$ K, $L = 5$ mH, $M = 0.2$ mH, $R_s = 60\ \Omega$, and $C = 0.002\ \mu$F.

11. For a FET Colpitts oscillator as in Fig. 12.27a and the following circuit values determine the circuit oscillation frequency: $C_1 = 750$ pF, $C_2 = 2500$ pF, $L = 40\ \mu$H, $R_g = 750$ K, $L_{RFC} = 0.2$ mH, $C_C = 2000$ pF.

12. For the transistor Colpitts oscillator of Fig. 12.28 and the following circuit values calculate the oscillation frequency: $L = 100\ \mu$H, $L_{RFC} = 0.5$ mH, $C_1 = 0.005\ \mu$F, $C_2 = 0.01\ \mu$F, $C_c = 10\ \mu$F.

§ 12.11

13. Calculate the oscillator frequency for a FET Hartley oscillator as in Fig. 12.29 for the following circuit values: $C = 250$ pF, $L_1 = 1.5$ mH, $L_2 = 1.5$ mH, $M = 0.5$ mH.

14. Calculate the oscillation frequency for the transistor Hartley circuit of Fig. 12.31 and the following circuit values: $L_{RFC} = 0.5$ mH, $L_1 = 750$ μH, $L_2 = 750$ μH, $M = 150$ μH, $C = 150$ pF.

§ 12.12

15. Draw circuit diagrams of (a) a series-operated crystal oscillator and (b) a shunt-excited crystal oscillator.

§ 12.13

16. Design the RC elements of a Wien bridge oscillator circuit (as in Fig. 12.42) for operation at $f_o = 2$ kHz.

§ 12.14

17. Design a unijunction oscillator circuit for operation (a) at 1 kHz and, (b) at 150 kHz.

pulse and digital circuits

13

13.1 GENERAL

Digital circuits are built using bipolar transistors, or MOSFETs, in integrated circuits (ICs). Basically, digital circuits are simple since they are operated either fully saturated (ON) or in cutoff (OFF), usually represented by voltage levels. Our concern, then, is with the two output states of a digital circuit. These can be at either one of two preselected voltage levels: $+5$ V and 0 V, for example. Logically, these levels may be related to the binary conditions of 1 and 0. For example, $+5$ V $= 1$ and 0 V $= 0$ or -10 V $= 1$ and 0 V $= 0$ are possible relations that may be assigned to voltage levels and logic conditions. Digital circuits provide manipulation of the logical conditions representing the two different voltage levels of the circuit. A number of logic gates or circuits are covered in this chapter. A logical AND gate, for example, provides an output only when *all* inputs are present, whereas a logical OR gate provides an output if *any* one input is present. An inverter circuit provides logical inversion—a 1 output for 0 input, or vice versa. The most popular configurations for a logic gate include an AND or OR gate, followed by an inverter, an AND-gate-inverter combination called a not-AND or NAND gage, and an OR inverter called a not-OR or NOR gate.

Also important in digital circuits is a class of multivibrator circuits. These circuits have two opposite output terminals (if one output is logical-1, the other is logical-0, or vice versa). The most useful of these is the bistable multivibrator or *flip-flop*, which can remain in either stable condition of output. The flip-flop can be used as a memory device—holding the state it was placed in after the initial pulse operating the circuit has passed. It can also be used to build a binary counter or a shift-register for use in digital computers.

A second circuit, the monostable multivibrator, provides opposite voltage outputs but, as the name implies, it can remain stable in only one state. If the circuit is triggered to operate, it can go into the opposite state of output voltages but can remain in this state only for a fixed time interval. The monostable multivibrator, or ONE-SHOT, is used for pulse shaping, time delay, and other timing actions.

An astable multivibrator or CLOCK circuit has no stable operating state and provides a changing output voltage (pulse train). The circuit is therefore a square-wave oscillator providing pulses to activate various computer operations.

A circuit that appears similar to the multivibrator class is the Schmitt trigger. This circuit operates from a slowly varying input signal and switches output voltage state when the input voltage goes above a preset voltage level or below a second voltage level.

13.2 DIODE LOGIC GATES—AND, OR

A logic gate provides an output signal for a desired logical combination of input signals. An AND gate, for example, provides an output of "logical-1" only if all the inputs are present, that is, if each is at 1 (logical-1). This circuit can be compared to one using switches connected in series. Only if, as in Fig. 13.1, all switches are closed (at 1) does an output voltage appear (and drive the light indicator ON). Thus, a light ON (logical-1) appears if switch A, AND switch B, and AND switch C are closed. This can also be written as the Boolean or logical expression

$$\text{Light ON} = A \cdot B \cdot C$$

(The dot indicates AND; read as A *AND* B *AND* C.)

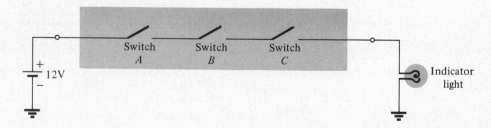

Figure 13.1. AND circuit using switches.

Diode AND Gate

A more practical form of AND gate using diodes is shown in Fig. 13.2. Using positive-voltage levels of $+V$ for 1 and 0 V as 0, we see that the circuit provides a 1 output of $+V$ only if *all* inputs are $+V$ or 1. More specifically, any input at 0 V will hold the output at about 0 V, as shown below.

1. A zero-volt input at any (one or more) input terminal(s) will cause that diode to short the output to ground. In logic form this means that 0 at any input produces 0 output.

2. A 1 input at inputs 1 and 2, but a 0 input at 3, will produce a 0 output because diode 3 shorts the output to ground.

3. Only when a 1 input is provided at inputs 1 AND 2 AND 3, none of the diodes is conducting and the output is 1.

Figure 13.3 shows a typical diode gate circuit and voltage truth table. Figure 13.4 shows the logic symbol and logic truth table for the circuit of Fig. 13.3 for positive logic operation (defined below).

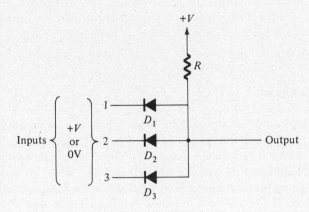

Figure 13.2. Diode AND-gate circuit.

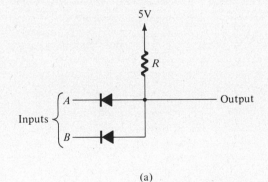

(a)

A	B	Output
0V	0V	0V
0V	+5V	0V
+5V	0V	0V
+5V	+5V	+5V

(b)

Figure 13.3. (a) Diode logic circuit; (b) voltage truth table.

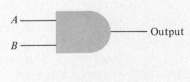

(a)

A	B	Output
0	0	0
0	1	0
1	0	0
1	1	1

(b)

Figure 13.4. (a) Diode AND-gate symbol; (b) logic truth table, for positive logic ($+5$ V $= 1$, 0 V $= 0$).

Diode OR Gate

The circuit of Fig. 13.5 shows an OR-gate circuit connection using three switches connected in parallel. If either switch *A* OR *B* OR *C* (or any combination of these) is closed (logical-1), the indicator light will be turned ON. A practical version of such a circuit uses diodes as shown in Fig. 13.6. The circuit is a positive-logic diode OR gate, and for this example it uses the same $+V$ and 0-V levels as the AND-gate circuit previously considered.

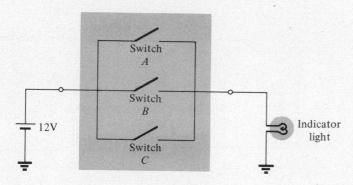

Figure 13.5. Logical OR gate using switches (Output = A + B + C).

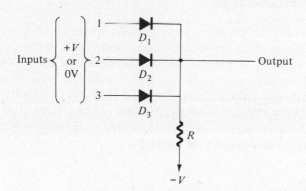

Figure 13.6. Diode OR-gate circuit.

The operation of the circuit of Fig. 13.6 is the following:

1. A 1 $(+V)$ input at any (one or more) input terminal(s) will cause that diode to conduct, placing the output at the 1 level.
2. A 0 input at inputs 1 and 2, but a 1 input at 3, will produce a 1 output because diode 3 conducts, placing the output at $+V$, and thereby holding diodes 1 and 2 in cutoff.
3. Only when a 0 input is provided at all three inputs will the output be 0.

Figures 13.7 and 13.8 show a typical OR-gate circuit and logic symbol and the respective voltage and logic truth tables.

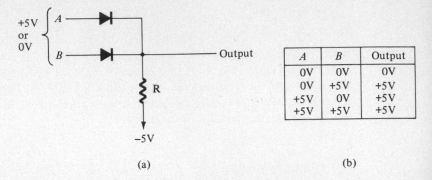

A	B	Output
0V	0V	0V
0V	+5V	+5V
+5V	0V	+5V
+5V	+5V	+5V

(a)   (b)

Figure 13.7. (a) OR-gate circuit; (b) voltage truth table.

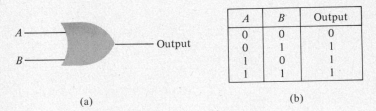

A	B	Output
0	0	0
0	1	1
1	0	1
1	1	1

(a)   (b)

Figure 13.8. (a) OR-gate logic symbol; (b) logic truth table.

Positive Logic—Negative Logic

In the two circuits just considered the voltage operation of these circuits is described by the voltage truth tables of Figs. 13.3 and 13.7. The logic operation of these two circuits, however, is dependent on the definitions of logical-1 and logical-0. Positive-logic definitions were used so far, where **positive logic** meant that the *more positive* voltage was assigned as the logical-1 state. It is possible to use other logic definitions. Using the same two circuits and voltage levels of $-V$ and 0 V provides **negative-logic** operation—with the definition of the *more negative* voltage $(-V)$ as the logical-1 level (and 0 V as logical-0).

13.3 TRANSISTOR INVERTER

A simple but important digital circuit is the inverter. Using voltage levels of 0 and +5 V, as an example, the inverter circuit will provide an output of 0 V for an input of +5 V and an output of +5 V for an input of 0 V. In logical terms the inverter, having single-input and single-output terminals, provides the opposite output—a 1 output for 0 input, or vice versa.

Figure 13.9 shows the circuit diagram of an inverter. The transistor in this circuit is operated either in saturation (ON) or in cutoff (OFF). To operate the transistor in saturation there must be sufficient base current drive corresponding to the current drawn by the collector. This requires that for a particular amount of collector current the base current must be larger than the amount specified by the transistor current gain (h_{fe} or β).

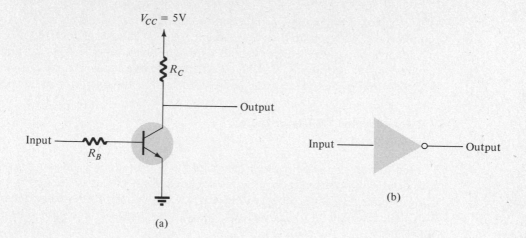

Figure 13.9. Inverter circuit and characteristics: (a) transistor inverter circuit; (b) inverter logic symbol.

13.4 LOGIC NAND AND NOR GATES

It was previously pointed out that the combination of logic gate followed by inverter is extremely popular. A loading problem exists when connecting AND or OR gates one after the other as is required in logic systems. An inverter circuit provides a buffer that minimizes loading effects. If a basic logic circuit is built using an inverter after each logic AND or OR circuit, then the overall circuit has much better characteristic in terms of loading and in terms of how fast the gate operates

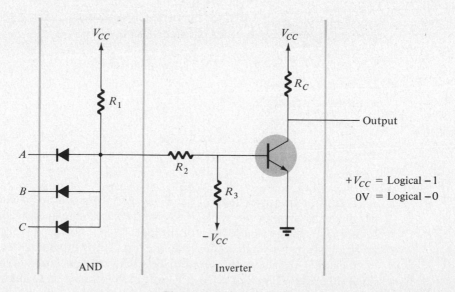

Figure 13.10. DCTL NAND-gate circuit (positive-logic).

(passes signals). Using mass production to build these logic circuits makes it more economical in the long run than to build only NAND gates or only NOR gates for a particular logic system, either type of logic gate being capable of performing all required logical operations. Having considered some rationale for the existence of NAND or NOR gates, let us examine the construction and operation of these gates.

Diode-Coupled Transistor Logic (DCTL Gate)

One version of a logic-inverter gate uses the type of diode logic gate previously considered followed by an inverter. A complete form of this combination circuit using an *npn* transistor is shown in Fig. 13.10. Considering positive-logic operation, with logic levels of $+5$ V as logical-1 and 0 V as logical-0, the gate shown is a NAND gate.

13.5 INTEGRATED-CIRCUIT (IC) LOGIC DEVICES

Digital integrated circuits (ICs) of various types find widespread use. It is economical and attractive to purchase a complete circuit of small size so that users of digital logic circuits depend on the IC units provided by the numerous manufacturers. At present there are a number of different types of logic circuits popular, each having some advantages and disadvantages. Since no one circuit type has been universally accepted, it seems reasonable to consider some of the more popular circuit types to understand their operation and their relative advantages and disadvantages. It should be clear that each type provides the same basic logical function and that other more practical factors about the overall system generally dictate which particular logic circuit type is chosen.

As a partial summary of the important factors used in selecting a circuit type we have (not necessarily in order of importance)

1. Cost.
2. Power dissipation.
3. Speed of operation.
4. Noise immunity.

Transistor-Transistor Logic (TTL) Circuit

Transistor-transistor logic (T^2L) is one circuit form of logic gate that, although possible as discrete components, is appropriate for manufacture in integrated form. Figure 13.11 shows a basic form of the logic circuit. Notice that each input is made using an emitter, a single multiple emitter transistor providing the inputs.

Figure 13.12a shows the circuit operation for output logical-0 (0 V). All inputs must either have high inputs ($+5$ V) or not be connected, so that transistor Q_1 is off. Transistor Q_2 is then driven by a base current from $+V_{CC}$ through R_1, the base

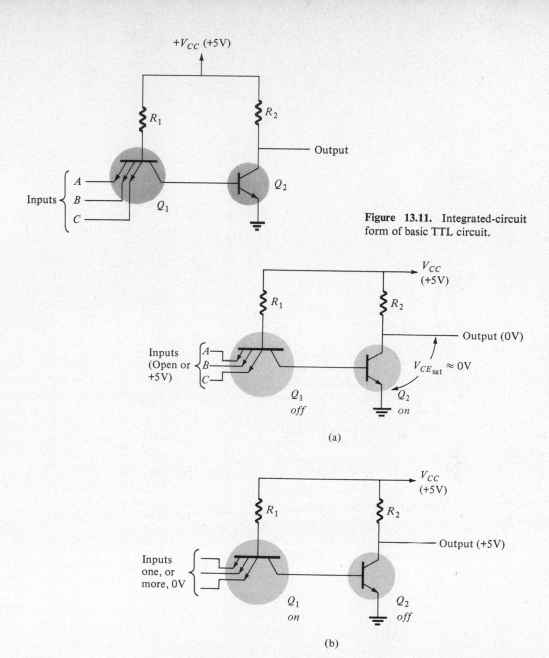

Figure 13.11. Integrated-circuit form of basic TTL circuit.

(a)

(b)

Figure 13.12. Operation of TTL (T²L) logic gate: (a) output transistor *on*; (b) output transistor *off*.

collector of Q_1 and forward-biased base emitter or Q_2. When Q_2 is on, the output voltage taken from the collector is $V_{CE_{sat}}$, approximately 0 V.

Figure 13.12b shows one (or more) input of 0 V allowing Q_1 to be biased *on*, resulting in the collector voltage of Q_1 to be near 0 V, thereby holding Q_2 *off*. When Q_2 is *off*, the collector voltage provides an output of $+5$ V. The circuit operates as a positive logic NAND gate.

SEC. 13.5 INTEGRATED-CIRCUIT (IC) LOGIC DEVICES

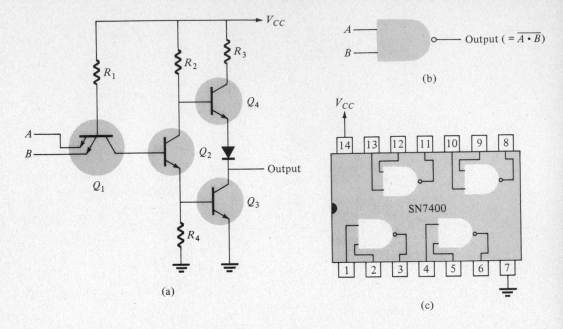

Figure 13.13. T²L logic unit: (a) circuit schematic; (b) logic symbol; (c) IC package of SN7400.

The T²L unit is now very popular because it has fast speed and good noise immunity and because of its low cost. A practical T²L logic unit is shown in Fig. 13.13a, the NAND logic symbol in Fig. 13.13b, and the package pin connections in Fig. 13.13c. Four T²L NAND gates are contained in the single SN7400 IC package, the entire unit being a quad, 2-input positive NAND gate. The basic T²L gate can switch state in typically 10 ns at power dissipation of 10 mW.

CMOS Logic Circuits

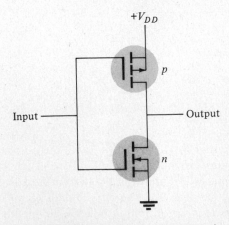

Figure 13.14. CMOS inverter circuit.

Another popular IC logic circuit is made using MOSFET devices of both *p*-channel and *n*-channel transistors, the complementary symmetry circuit being COS/MOS, COSMOS, or more simply CMOS logic. A basic CMOS inverter is shown in Fig. 13.14. A positive input voltage drives the *n*-channel FET *on* with output 0 V. An input of 0 V results in the *p*-channel FET *on* with the output a positive voltage.

A positive logic NOR gate is made as shown in Fig. 13.15. When either input is positive, an *n*-channel FET is driven *on* with output then 0 V. When both inputs are 0 V, the two *p*-channel FETs are turned *on* with output then a positive voltage. The CD4001A, for example, is an IC package of four 2-input NOR gates as shown in Fig. 13.16.

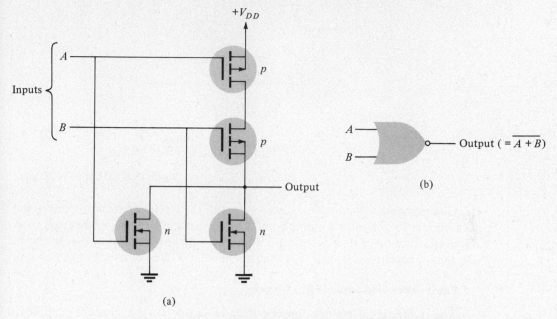

(a)

(b)

Figure 13.15. CMOS positive logic NOR gate: (a) circuit, (b) logic symbol.

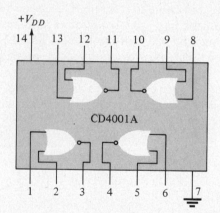

Figure 13.16. CMOS quad two-input positive NOR gate IC: CD4001A.

CMOS logic units offer very low-power dissipation at relatively slow speed. Typical propoagation delay is 100 ns, with 5-V supply at only microwatts of power dissipation, while providing noise immunity of over 2 V. A CMOS logic gate is at least five times smaller than the comparable T²L gate.

Integrated Injection Logic (I²L)

A relatively new IC logic circuit uses bipolar transistors in a circuit and physical arrangement called *integrated injection logic* (I²L) and has the better characteristics of T²L and CMOS units: The I²L unit has low-power dissipation, fast speed, and small size.

SEC. 13.5 INTEGRATED-CIRCUIT (IC) LOGIC DEVICES

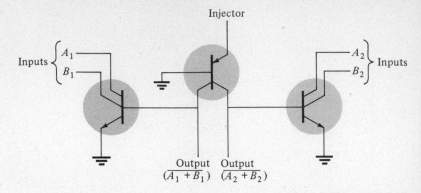

Figure 13.17. I²L dual NOR gate.

Complementary bipolar transistors are used, the *pnp* injector transistor serving as a current source, with a multiple collector *npn* transistor forming the basic gate. Figure 13.17 shows a dual gate unit. I²L gates use microwatts of power at speeds in the tens of nanoseconds, and they require the least IC area of the gates considered.

Emitter-Coupled Logic (ECL) Circuit

The technique of emitter-coupled logic (ECL) or current-mode logic (CML) circuits differs from those covered previously in one main respect. All previous circuits allowed the transistors to saturate. This means that an amount of charge is stored in the transistor base and collector regions resulting in a time delay in turning off the transistor. Current-mode logic operates the transistor in a non-saturated condition, thereby providing shorter propagation delays through the circuits.

An emitter-coupled logic circuit, such as that of Fig. 13.18, uses a transistor for each input of the circuit, which is desirable for IC manufacture. Whereas a direct-

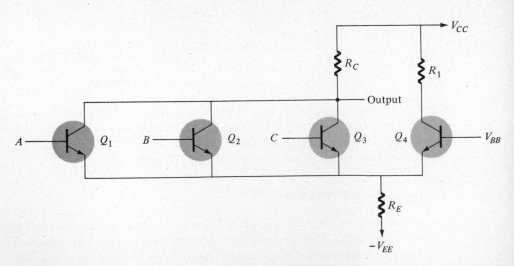

Figure 13.18. Emitter-coupled logic (ECL) circuit.

coupled logic circuit would connect the transistor emitters to ground (thereby allowing the base drive to be dependent on the input voltage), the present circuit uses an additional transistor stage providing a reference point for the common-emitter terminals. If more than the needed turn-on drive is applied, this will cause the emitter point to rise in voltage, maintaining the logic transistor in a non-saturated mode of operation.

In summary, then, the nonsaturated operation of an emitter-coupled logic (ECL) circuit provides very fast switching speeds, simultaneous OR and NOR outputs from each circuit, high fan-in and fan-out capacity, and constant noise immunity of the power supply due to the relative constant-current demand of the logic circuit whether *off* or *on*. Disadvantages are the need for three voltage supplies (which is reduced to only one using ground as one), the need for a bias driver to provide the reference supply voltage, and the higher cost of ECL circuits.

Compatibility of Integrated-Logic Circuits

Having shown that a number of different types of logic circuits are presently available and utilized, it would be helpful at this time to indicate some general factors about the various types of logic techniques. In more general terms there are three types of logic systems—current-sinking, current-sourcing, and current-mode. These types are basically incompatible. Different logic circuits of the same type can be used in the same operating system with the possibility of driving one directly from another.

Current-Sinking Logic

As indicated in Fig. 13.19a current-sink circuits draw current from the input of the following stage into the driving stage. Examples, of current-sink operating logic are T²L and CMOS logic units. If it is assumed that speed and voltage levels are set compatibly, these units could be used in the same system to drive each other.

Figure 13.19. Modes of logic operation: (a) current-sinking; (b) current-sourcing.

(a) (b)

Current-Sourcing Logic

Figure 13.19b shows that a current-sourcing circuit provides current to the input being driven by the logic circuit. It should be obvious that the opposite current-flow directions of current-sink and -source circuits make these two types completely incompatible. RTL units (not discussed previously) operate as a current-source logic.

Current-Mode Logic

Although current-mode logic circuits, such as an ECL, appear as current-sourcing operating circuits, the logic levels, switching speeds, and power supply requirements make this type incompatible for use with the other two types, as I^2L is also directly incompatible with other logic types.

13.6 BISTABLE MULTIVIBRATOR CIRCUITS

Of equal importance to logic circuits in digital circuitry is the class of multivibrator circuits. There are three basic forms of the multivibrator—bistable, monostable, and astable; the most important of these, by far, is the bistable multivibrator or flip-flop. As an indication of the applications of the multivibrator circuits, consider the following:

Bistable (FLIP-FLOP)—storage stage, counter, shift-register
Monostable (ONE-SHOT)—delay circuit, waveshaping, timing circuit
Astable (CLOCK)—timing oscillator (square-wave)

In a logic system there will typically be a large number of flip-flop stages used as counters, storage registers, shift-registers, a few one-shot circuits in special timing or pulse-shaping uses, and a limited number of CLOCK circuits (typically only one).

Characteristic of all three circuits is the availability of two outputs, where the outputs are logically inverse signals. One output is selected as the reference, this designation being indicated in a number of ways. The two outputs are sometimes marked as 0 and 1, FALSE and TRUE, or $\bar{A}$ and A, etc. The main point of the designation is to indicate that the outputs are logically opposite and to mark the output chosen as the reference output. Another means of indicating the state of the multivibrator circuit is the use of the designation of SET and RESET. When referring to the state of the circuit the definitions of SET and RESET are the following:

$$\begin{array}{ll} \text{SET:} & Q \text{ output is logical-1} \\ & \bar{Q} \text{ output is logical-0} \\ \text{RESET:} & Q \text{ output is logical-0} \\ & \bar{Q} \text{ output is logical-1} \end{array}$$

Bistable Multivibrator (Flip-Flop)

The flip-flop circuit, the most important of the multivibrator circuits, will be covered first. To provide some basic consideration of this circuit's operation a simple form of bistable circuit using two inverters is shown in Fig. 13.20 . The inverters are essentially connected in series, with two output points indicated. The two outputs are labelled Q and $\bar{Q}$, respectively. If the Q output is a logical-1, then inverter I_2 will provide $\bar{Q}$ as a logical-0. Since $\bar{Q}$ is connected as input to inverter I_1, it will cause the output of that stage to be a logical-1, as assumed. Thus, the state of logical conditions, or the voltage they represent, forms a stable situation with Q output logical-1 and $\bar{Q}$ output logical-0. If some external means is used to cause the

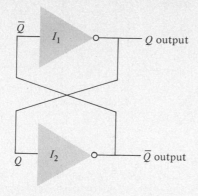

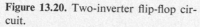

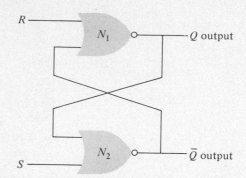

Figure 13.20. Two-inverter flip-flop circuit.

Figure 13.21. RESET and SET operation of flip-flop.

Q signal to change to logical-0, then, through inverter I_2, $\bar{Q}$ would change to logical-1. The $\bar{Q}$ input of logical-1 would then result in A being logical-0 as initially proposed. Thus, the circuit will also remain in a stable condition if the Q output is logical-0 and the $\bar{Q}$ output is logical-1. In effect, then, the circuit has two stable operating states that act as a memory of the last state it was placed into. Some external means is necessary, however, to cause the circuit to change state.

RS Flip-Flop

Figure 13.21 shows the use of NOR gates connected in series with additional inputs providing signals to cause the circuit to change state. The inputs are marked R for RESET and S for SET. Recall that a NOR gate provides logical-0 output if any of its inputs is logical-1. A logical-1 input to the S terminal will cause the output of N_2 to be logical-0. If it is assumed that no input is connected to the R terminal at this time, the inputs to N_1 are both logical-0 with output of logical-1. The result of a SET input signal then is to cause the circuit to become SET, where the SET state was previously defined as Q output = logical-1 and $\bar{Q}$ output = logical-0. Similarly, the application of only a logical-1 to the R input will cause the Q output to become logical-0 and $\bar{Q}$ output logical-1, which is the RESET state of the circuit. It should be obvious that simultaneous application of logical-1 signals to both S and R inputs is ambiguous, forcing both outputs to the logical-0 condition. This would not be an accepted operation of this circuit in which the two output signals should be always logically opposite. If the R and S inputs are both logical-0, then the circuit remains in whatever state it was last placed into.

T-Type Flip-Flop

One of the more common type of flip-flop circuits is the T-type or triggered flip-flop. This circuit is also called a *complementing* flip-flop, or *toggle* flip-flop, since its action is to change state every time an input pulse is applied to the single T-input terminal. Figure 13.22a shows a block diagram of a T-type flip-flop. The waveforms

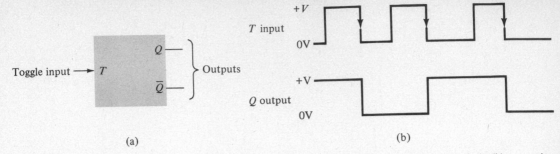

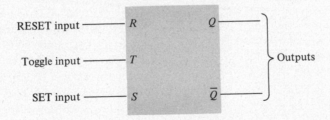

Figure 13.22. Toggle flip-flop (T-type): (a) block symbol; (b) operating waveforms.

in Fig. 13.22b show that the output changes stage, or toggles, every negative edge of the input signal. Notice that the output goes through a full cycle every two input cycles so that the output is at one-half the frequency of the input or that every two input pulses result in one full cycle of the output. The T-type flip-flop is the basic circuit used as a counter stage in binary counters.

RST Flip-Flop

A versatile multivibrator circuit combines the SET, RESET, and toggle features in a single unit called an RST flip-flop. The R input (see Fig. 13.23) provides direct RESET operation, the S input provides direct SET operation, and the T input provides the toggle or complementing operation.

Figure 13.23. RST flip-flop.

When a logical-1 input is applied to the R input, the circuit will be RESET. If the logical-1 dc voltage is maintained at the R terminal, the circuit will be held "locked" or forced in the RESET state. Even if an input signal is applied to the T input, the circuit will remain in RESET. Similarly, a logical-1 input only to the S terminal will hold the circuit locked in the SET state. Only if both R and S inputs are logical-0 can the T input signal operate the flip-flop. With R and S inputs logical-0, a CLOCK input to the T terminal will complement the stage output for every CLOCK pulse received. If, however, both R and S inputs are logical-1, the output state is not defined (usually, the two outputs will be logical-0—not a proper condition).

JK Flip-Flop

The J and K input terminals (Fig. 13.24a) are used to provide information or data inputs. When a trigger pulse is then applied, the circuit changes state corresponding to the inputs to the J and K terminals. A JK circuit is built as an integrated

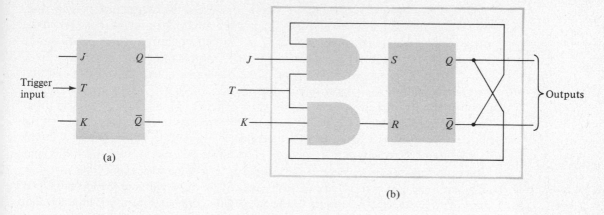

(a)

(b)

J	K	Circuit Action
0	0	Remains in same state
0	1	RESETS
1	0	SETS
1	1	Circuit toggles

(c)

Figure 13.24. Basics of *JK* flip-flop: (a) logic symbol; (b) *JK* flip-flop made using *RS* flip-flop, plus gating; (c) truth table.

circuit form using TTL or CMOS logic manufacture. It is not at all important here to consider the circuit details and in fact one only purchases a complete unit in IC form and has little to do with the details of circuit operation. It is important to be aware of the differences in using these different logic types, in knowing whether they use current-sourcing, sinking, or emitter-coupled logic, in details of speed of operation, noise margin, power supply voltages, etc. These factors, however, are descriptive of the overall circuit, and the details of what actually goes into building the actual circuit are often of little importance. For the JK flip-flop, then, we shall discuss its operation from the logic or block diagram point of view and consider some applications using the circuit. The *JK* circuit is versatile and is presently the most popular version of the flip-flop circuit. A logic symbol of a *JK* flip-flop is shown in Fig. 13.24a. The *J* and *K* terminals shown are the data input terminals receiving the information of logical-1 or logical-0. These information inputs do not, however, change the state of the flip-flop circuit, which will remain in its present state until a trigger pulse is applied. Thus, for example, if the circuit were presently RESET and the input data were such as to result in the SET condition, the circuit would still maintain the RESET condition, even with the *J* and *K* input data signals applied. Only when the trigger pulse occurs are the input data used to determine the new state of the circuit—the SET state for the present example.

Figure 13.24b shows how an *RS* flip-flop may be modified to form a *JK* flip-flop. This connection is not typically used to make *JK* flip-flops although the practical circuit version acts essentially in the manner to be described. Using an *RS* flip-flop and two AND gates provides the *JK* operation. When the trigger input is logical-0, the AND gates are disabled and the circuit will maintain its present state. When the trigger pulse occurs (becomes logical-1), the circuit operation still depends on the other inputs to the AND gates. Since each AND gate had one input fed back from the outputs of the circuit, at least one of the AND gates has a logical-1 in addition to the logical-1 of the trigger signal (when it occurs). Finally, we have the *J* and *K* inputs to each of the respective AND gates. There are four possible combinations of the *J* and *K* inputs. To consider the complete operation of the circuit each of the possible conditions is listed in the truth table of Fig. 13.24c. If both *J* and *K* inputs are logical-0, then both AND gates are disabled and the circuit remains in the same state (no change takes place). If the *J* input is logical-0 and *K* input logical-1, and, further, if the *Q* output is logical-1, then the trigger pulse going to the logical-1 level will provide all logical-1 inputs to the AND gate connected to the *R* input so that the circuit ends up in the RESET state. If the *K* input is logical-0 and the *J* input logical-1, and, further, if the $\bar{Q}$ output is logical-1, then the occurrence of the trigger pulse will cause the circuit to be SET. Finally, if both *J* and *K* inputs are logical-1, the action of the trigger pulse's becoming logical-1 is to toggle or complement the circuit. If the 0 output were logical-1 (circuit RESET), the trigger pulse occurring would result in the circuit's being SET, thereby causing it to change state. Similarly, with the *Q* output logical-1, the occurrence of a trigger pulse would result in a RESET signal, thereby changing the circuit state. Thus, the circuit would toggle (change state) from whichever condition it happened to be in on application of the trigger pulse. In this last case, with *J* and *K* inputs both logical-1, the circuit would operate as a *T* flip-flop and could be used as such. When opposite data input signals are applied as *J* and *K* inputs, the trigger pulse will shift the data into the present flip-flop state, the stage then acting as a shift-register stage. Thus, the *JK* flip-flop can be used as a shift-register stage, a toggle stage for counting operations, or generally as a control logic stage.

If positive logic is used (0 V = logical-0 and +V as logical-1, for example), the circuit triggers when the trigger pulse goes from the logical-0 to logical-1 condition —this being referred to as positive-edge triggering since the circuit is triggered at the time the voltage goes positive (from 0 to +V). When the circuit triggers on a voltage change from +V to 0 V, as does the *T*-type flip-flop of Fig. 13.22, the triggering is trailing-edge triggering. The manufacturer's information sheets should indicate the type of triggering required to operate the particular circuit so that it may be properly used.

13.7 MONOSTABLE AND ASTABLE MULTIVIBRATOR CIRCUITS AND SCHMITT TRIGGER CIRCUIT

Monostable Multivibrator (ONE-SHOT)

As a characteristic property of a multivibrator circuit the monostable provides two opposite-state output signals. As the name implies, the outputs are stable in only one of the two possible states (SET and RESET). Figure 13.25a shows a logic

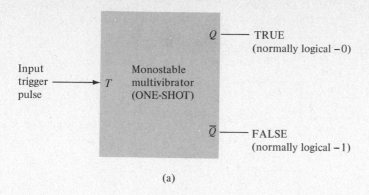

(a)

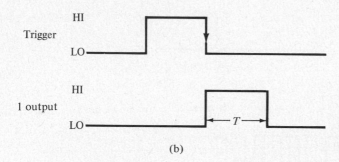

(b)

Figure 13.25. Operation of mono-stable multivibrator: (a) ONE-SHOT logic symbol; (b) input trigger and 1 output waveforms.

block symbol of a ONE-SHOT circuit in the stable RESET state (Q output = logical-0, and $\bar{Q}$ output = logical-1). The input trigger signal is a pulse that operates the circuit in an edge-triggered manner. Figure 13.25b shows a typical input trigger pulse and corresponding output waveform (assuming triggering on the trailing edge of the trigger pulse). The Q output is normally low (RESET state). When a negative-going voltage change triggers the circuit, the 1 output goes high (SET state), which is the unstable circuit state. It will remain high only for a fixed time interval, T, which is determined basically by a timing capacitor whose value may be selected externally. Thus, the output state remains in the SET state only for a preselected time T, after which the output returns to the RESET state, where it remains until another trigger pulse is applied.

In the waveform of Fig. 13.25b the output of the ONE-SHOT can be viewed as a delayed pulse whose negative-going edge occurs at some set time T after the trigger pulse. Figure 13.26 shows a number of additional actions possible using the ONE-SHOT circuit. For example, it is possible to accept a train of narrow pulses as trigger signal and provide as output a corresponding train of wider pulses as shown in Fig. 13.26a. If the pulses received are narrow, they may be widened to a pulse interval T, where T is, of course, less than the interval between trigger pulses. Any input pulse received during the timing interval T will be ignored by the usual ONE-SHOT circuit.

Figure 13.26b shows a series of wide input pulses and corresponding narrow output pulses from the ONE-SHOT circuit. Note that for this example the narrow pulse is initiated by the negative-going edge of the input trigger signal. Figure 13.26c shows how the ONE-SHOT may be used to provide only a single pulse when the

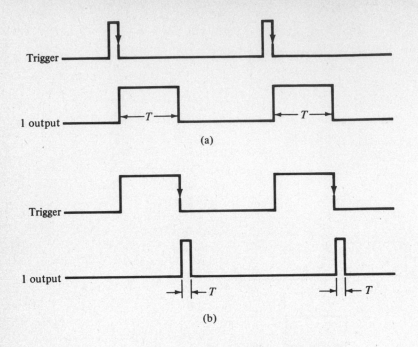

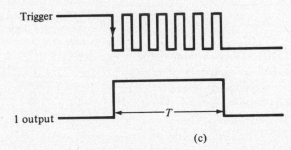

Figure 13.26. Some pulse-shaping actions of a ONE-SHOT circuit: (a) pulse shaping—widening a narrow pulse; (b) pulse shaping—narrowing a wide pulse; (c) blocking multiple pulse—providing a single output pulse.

trigger signal is a number of pulses. This is useful, for example, if the input pulses are obtained from a mechanical switch. When the switch is moved, the contact bounce will usually result in a number of transitions between low and high states. If the switch signal is used directly, one throw of the switch may result in one, two, three, or more transitions, which would result in erroneous system action. If the ONE-SHOT is used, only a single pulse is provided—even with the multiple input pulses shown—as long as the interval of contact bounce is less than the pulse interval, T, of the ONE-SHOT.

A popular TTL monostable multivibrator is the SN74121 shown in Fig. 13.27. The output provides a pulse of duration T,[1] the Q output going from its normally low state (0 V) to the high output state or SET state. Triggering can occur when either the A inputs are both grounded and the B input goes positive or when the B input is lifted high (or unconnected) and the A inputs go low. A few examples will help show how the ONE-SHOT can be used.

---

[1]For the SN74121 the pulse width of the output pulse is $T \cong 0.7 R_T C_T$ where R_T is the timing resistor, R_{ext} or R_{int} and C_T is the timing capacitor, C_{ext}.

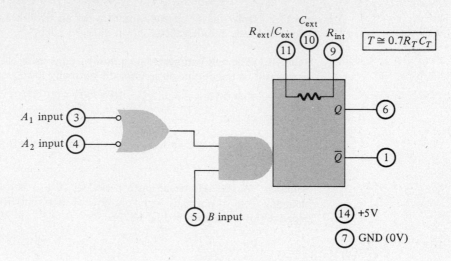

$$T \cong 0.7 R_T C_T$$

A_1 input ③

A_2 input ④

R_{ext}/C_{ext} ⑪ C_{ext} ⑩ R_{int} ⑨

Q ⑥

$\overline{Q}$ ①

⑤ B input

⑭ +5V

⑦ GND (0V)

Figure 13.27. SN74121 monostable multivibrator.

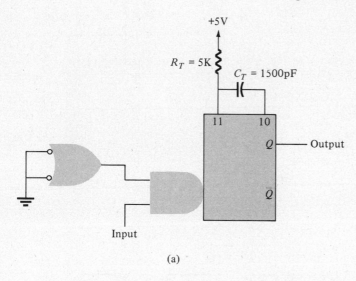

+5V

$R_T = 5K$

$C_T = 1500$pF

11   10

Q —— Output

$\overline{Q}$

Input

(a)

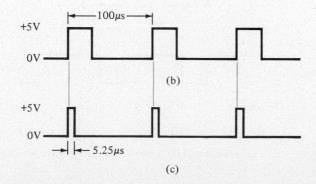

|←——— 100µs ———→|

+5V

0V

(b)

+5V

0V

|←→| ←— 5.25µs

(c)

Figure 13.28. One-shot circuit and waveforms for Example 13.1.

EXAMPLE 13.1 Draw the Q output waveform for an SN74121 (see Fig. 13.28a) triggered by an input signal as shown in Fig. 13.28b.

Solution: The unit is triggered on a positive edge using the B input, and the time width of the output pulse from the normally low Q output is

$$T = 0.7R_TC_T = 0.7(5 \times 10^3)(1500 \times 10^{-12}) = 5.25 \ \mu s$$

as shown in Fig. 13.28c.

EXAMPLE 13.2 Draw the Q output waveform for the circuit of Fig. 13.29a and input of 100 kHz clock.

Solution: A 100-kHz clock has a period of 10 μs, the resulting Q output changing from its normally high voltage level each time the input clock goes negative (low) as shown in Fig. 13.29b.

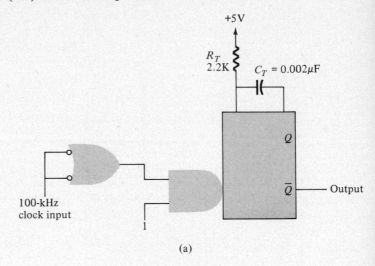

(a)

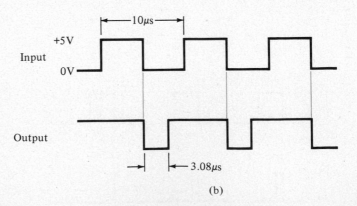

(b)

Figure 13.29. One-shot and waveforms for Example 13.2.

Astable Multivibrator (CLOCK)

A third version of multivibrator has no stable operating state—it oscillates back and forth between RESET and SET states. The circuit provides a CLOCK signal for use as a timing train of pulses to operate digital circuits. Figure 13.30 shows a circuit diagram of an astable multivibrator. Notice that both cross-coupling components are capacitors, thereby allowing no stable operating state. If the resistors and capacitors used are of equal value, the frequency of the CLOCK is

$$f = \frac{1}{2T} = \frac{1}{2(0.7RC)} = \frac{1}{1.4RC} = \frac{0.7}{RC}$$

Integrated circuit units may be used to build the CLOCK circuit. Figure 13.31 shows a few examples including the circuit parameters affecting the CLOCK frequency. A typical CLOCK output waveform is shown in Fig. 13.32.

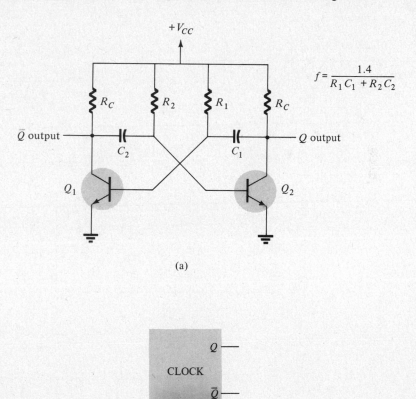

$$f = \frac{1.4}{R_1 C_1 + R_2 C_2}$$

(a)

(b)

Figure 13.30. (a) Astable multivibrator circuit; (b) logic symbol.

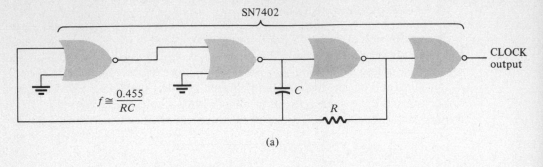

(a)

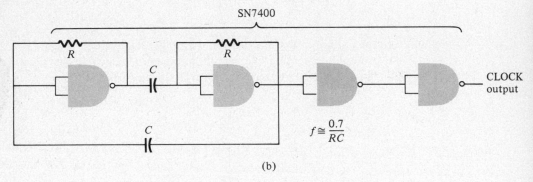

(b)

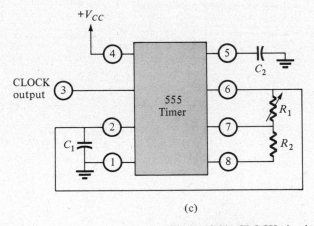

(c)

Figure 13.31. CLOCK circuits built with various IC units.

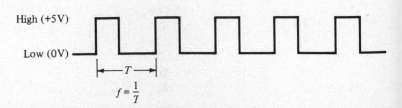

Figure 13.32. CLOCK output waveform.

Schmitt Trigger

A circuit that is somewhat like the multivibrator circuits considered is the Schmitt trigger circuit shown in Fig. 13.33. Somewhat analogous to the ONE-SHOT, the Schmitt trigger is used for waveshaping purposes. Basically, the circuit has two opposite operating states as do all the multivibrator circuits. The trigger signal, however, is not typically a pulse waveform but a slowly varying ac voltage. The Schmitt trigger is level sensitive and switches the output state at two distinct triggering levels, one called a *lower-trigger level* (LTL) and the other an *upper-trigger level* (UTL). The circuit generally operates from a slowly varying input signal, such as a sinusoidal waveform, and provides a digital output—either the logical-0 or logical-1 voltage level.

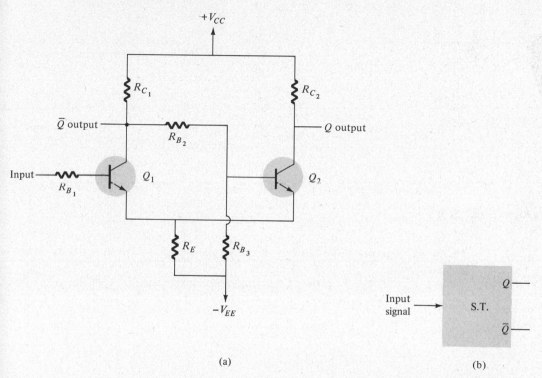

(a)   (b)

Figure 13.33. (a) Schmitt trigger circuit; (b) block symbol.

The typical waveform of Fig. 13.34 shows a sinusoidal waveform input and squared waveform output. Note that the output signal frequency is exactly that of the input signal, except that the output has a sharply shaped slope and remains at the low or high voltage level until it switches. One example of a Schmitt trigger application is converting a sinusoidal signal to one that is useful with digital circuits. Signals such as a 60-Hz line voltage or a slowly varying voltage obtained from a magnetic pickup are squared up for digital use. Another possibility is using the Schmitt trigger to provide a logic signal that indicates whenever the input goes above a threshold level (UTL).

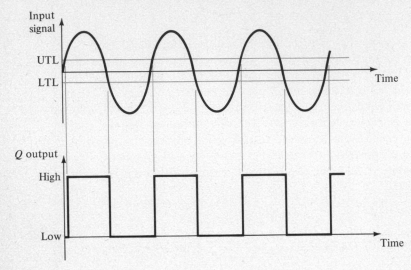

Input
signal

UTL

LTL

Time

Q output

High

Low

Time

Figure 13.34. Typical waveforms for Schmitt trigger.

PROBLEMS

§ 13.2

1. A positive-logic diode AND gate has voltage levels of 0 and $+5$ V. Draw a circuit diagram and prepare voltage and logic truth tables for a two-input gate. (Assume ideal diodes.)

2. Draw the circuit diagram of a negative-logic AND gate for voltage levels of -5 V and 0 V. Assume ideal diodes and prepare voltage and logic truth tables for a two-input gate.

3. Determine the output voltage of circuits of Fig. 13.35 (assume ideal diodes).

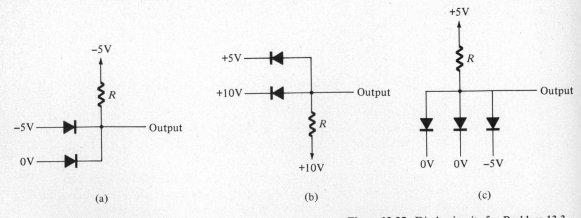

(a)

(b)

(c)

Figure 13.35. Diode circuits for Problem 13.3.

4. Draw the circuit diagram of a four-input TTL NAND gate indicating voltage levels and logic definitions.

5. State two differences between CMOS and TTL operation.

6. Draw the circuit diagram of a three-input CMOS NOR gate.

7. How does an ECL-type gate differ from TTL?

8. Draw the circuit diagram of a three-input I²L gate.

§ 13.6

9. Describe the operation of an RS flip-flop.

10. Draw the output waveform of a T flip-flop triggered by a 10-kHz clock.

11. Describe the operation of a JK flip-flop.

§ 13.7

12. Draw the circuit diagram of a ONE-SHOT circuit. Define output logic levels and show which polarity voltage change triggers the circuit.

13. Draw the output waveform of an SN74121 circuit driven by an input CLOCK signal of 10 kHz applied to B input. The timing component values are $R_T = 10$ K and $C_T = 1000$ pF.

14. What is the frequency of an astable multivibrator circuit (as in Fig. 13.31a) having timing component values of $R_T = 2.7$ K and $C_T = 750$ pF?

15. Draw the Q output voltage signal of a Schmitt trigger circuit for the input trigger signal of 5 V (rms) at 60 Hz. Circuit trigger levels are UTL $= +5$ V and LTL $= 0$ V.

regulators and miscellaneous circuit applications 14

14.1 INTRODUCTION

This chapter examines both the voltage and current regulators and introduces a few circuits of general interest. Throughout the chapter there is an application of the usually appreciated, approximate technique of semiconductor circuit analysis. In this manner, the "total" circuit behavior can more easily be understood. This chapter will also serve, in an indirect manner, as a test of the reader's comprehension of some of the important fundamental concepts introduced in earlier chapters. If the circuits described are, in general, understood with little necessity to review past discussions, the reader has every right to feel that he has reached a first, important plateau of sophistication in semiconductor circuit analysis.

REGULATORS

14.2 REGULATION DEFINED

The wide variety of circuit configurations capable of performing voltage or current regulation necessitates that only a few of the more commonly applied be considered in this text. Voltage and current regulation can best be defined through the use of the circuits of Figs. 14.1 and 14.3, respectively.

In Fig. 14.1a the no-load (open-circuit) terminal voltage of the supply is designated by V_{NL}. The corresponding load current is $I_{NL} = 0$. The full-load conditions are shown in Fig. 14.1b. The ideal situation would require that $V_L = V_{FL} = V_{NL}$ for every value of R_L between no-load and full-load conditions. In other words, the

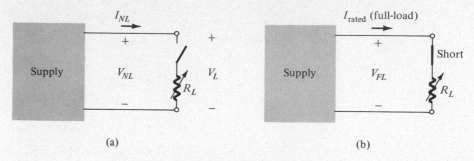

(a)   (b)

Figure 14.1. Voltage regulation: (a) no-load (*NL*) state; (b) full-load (*FL*) state.

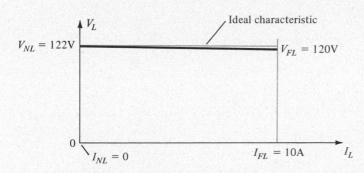

Figure 14.2. Terminal characteristics of a cumulative compound dc generator.

terminal voltage V_L would be unaffected by variations in R_L. Unfortunately, there is no supply available today, whether it be tube, semiconductor, or electromechanical (generator), that can provide a terminal voltage completely independent of the load applied (even for a specified range). However, for most applications, the level of sophistication has reached the point where the ideal response curve can be assumed. An ideal curve has been superimposed on the cumulative (flat) compound dc generator characteristics of Fig. 14.2.

Note that the load voltage V_L drops 2 V from its no-load to full-load value. The drop in terminal voltage is due to a number of changes in potential levels internal to the generator as a result of increased load.

Voltage regulation of any supply, in per cent, is defined by

$$\text{Voltage regulation (VR) (\%)} = \frac{V_{NL} - V_{FL}}{V_{FL}} \times 100\% \qquad (14.1)$$

Substituting the ideal values of $V_L = V_{FL} = V_{NL}$ into Eq. (14.1) results in VR = 0%. Obviously, a *low* voltage regulation is desirable and 0% is ideal. Semiconductor and tube supplies with voltage regulations of 0.01% or lower are common today. For the electromechanical generator of Fig. 14.2

$$\text{VR}\% = \frac{122 - 120}{120} \times 100\% \cong 1.7\%$$

The circuit configurations associated with the no-load and full-load conditions

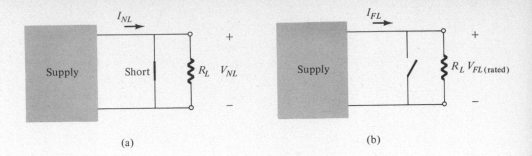

Figure 14.3. Current regulation: (a) no-load (*NL*) state; (b) full-load (*FL*) state.

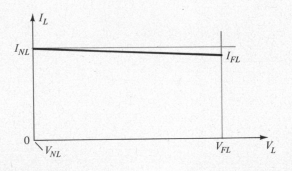

Figure 14.4. Current regulation—defining quantities.

(Courtesy Lambda Electronics Corp.)

(Courtesy Kepco, Inc.)

(a)

(Courtesy Lambda Electronics Corp.)

(b)

Figure 14.5. Regulated supplies: (a) voltage regulated; (b) current regulated.

for *current regulation* are provided in Fig. 14.3 The defining terminal characteristics are provided in Fig. 14.4.

Note for this situation that a constant I_L is desired for a variable terminal voltage. Current regulation is defined by

$$\text{Current regulation (IR) (\%)} = \frac{I_{NL} - I_{FL}}{I_{FL}} \times 100\% \qquad (14.2)$$

where I_{NL} and I_{FL} and the no-load full-load currents as defined by Fig. 14.3.

Two voltage- and one current-regulated supply appear in Fig. 14.5. The necessity for low voltage and current regulation characteristics should be obvious. Without them, the bias conditions of electronic systems, the speed of dc and ac motors, and the logic of computer circuits would all be severely affected.

14.3 ZENER AND THERMISTOR VOLTAGE REGULATORS

There are, fundamentally, two basic configurations for establishing voltage or current regulation. Each is shown in Fig. 14.6. The choice of terminology for each as indicated in the figure is obviously derived. You will find as you progress through some of the more typical circuit configurations that the more sophisticated regulators will apply the benefits to be derived from both series and shunt regulation in

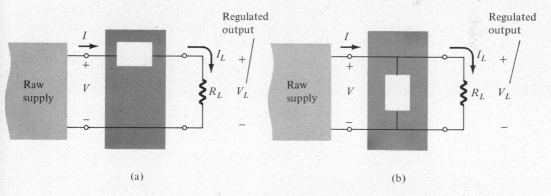

Figure 14.6. Regulators: (a) series; (b) parallel (shunt).

the same system. Before continuing, let us examine the unregulated supply of Fig. 14.7. It clearly demonstrates the need for voltage and/or current regulators. Recall from your experimental work that when your dc supply indicates 20 V, you want it to be that value for any load you apply to its terminals. At any except infinite ohms (open circuit) in Fig. 14.7, would this be the case? Obviously not. As R_L increases, the voltage across R_L also increases, and V_L does not remain fixed. The function of a voltage regulator is to maintain V_L at 20 V for any R_L from 0 to 1 K. Therefore, the voltage regulator appears between the unregulated supply and the load as shown in Fig. 14.8.

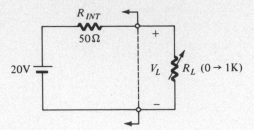

Figure 14.7. Circuit demonstrating the need for voltage and current regulators.

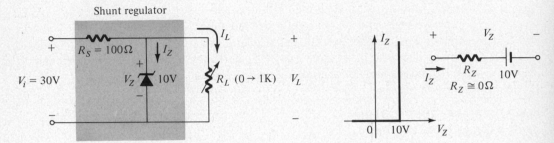

Figure 14.8. Zener diode shunt regulator.

Our interest for the moment is in the voltage-regulating device. A simple shunt voltage regulating system is shown in Fig. 14.8. As indicated, it consists simply of a Zener diode and series resistance, R_s. Proper operation requires that the Zener diode be in the *on* state. The first requirement, therefore, is to find the minimum R_L (and corresponding I_L) to ensure that this condition is established. Before Zener conduction, the Zener diode is fundamentally an open circuit and the circuit of Fig. 14.8 can be replaced by that indicated in Fig. 14.9. As indicated in the figure, the load voltage will be determined by the voltage divider rule (similar to the system of Fig. 14.7). At Zener conduction, $V_L = V_Z = 10$ V. If this data is used, it is possible to find the minimum value of R_L that the Zener will tolerate in maintaining constant V_L. Applying the voltage divider rule to the circuit of Fig. 14.10a results in

$$V_L = \frac{R_L V_i}{R_L + R_s}$$

Substituting values results in

$$10 = \frac{R_L \times 30}{R_L + 0.1 \text{ K}}$$

$$10R_L + 1 \text{ K} = 30R_L$$

$$20R_L = 1 \text{ K}$$

and

$$R_L = 50 \ \Omega$$

The *minimum* load R_L for this supply is, therefore, 50 Ω, corresponding to a *maximum* load current of

$$I_{\max} = \frac{10 \text{ V}}{0.05 \text{ K}} = 200 \text{ mA}$$

$$V_{L \text{ (before firing)}} = \frac{R_L V_i}{R_L + 100\Omega}$$

Figure 14.9. Regulator of Fig. 14.8 before firing.

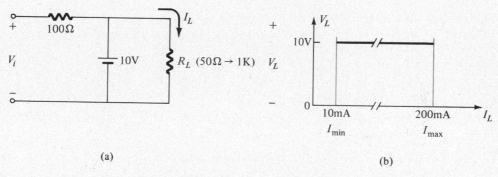

(a)

(b)

Figure 14.10. Zener diode shunt regulator: (a) after firing; (b) regulated output.

The maximum current is indicated on the plot of Fig. 14.10b. For any value of R_L from 50 Ω to 1 K the Zener diode will be in the *on* state. An application of the voltage divider rule to the circuit of Fig. 14.9 for any value of R_L less than 50 Ω will result in $V_L < V_Z$ and the diode will remain in the *off* state. At $R_L = 1$ K, $I_L = (10/1 \text{ K}) = 10$ mA (indicated in Fig. 14.10b as I_{min}), and

$$I_{R_s} = \frac{30 - 10}{0.1 \text{ K}} = 200 \text{ mA}$$

with

$$I_Z = I_{R_s} - I_L = 190 \text{ mA}$$

Our approximation, $R_Z \cong 0\ \Omega$, has resulted in the ideal characteristics of Fig. 14.10b between 10 and 200 mA. For this regulator the per cent regulation would be determined between these two points of normal operation. The effect of R_Z on the regulation can be determined easily if we find the Thévenin equivalent circuit for the portion of the network of Fig. 14.11 at the left of points 1 and 2.

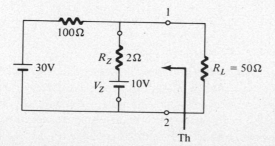

Figure 14.11. Determining the effect of R_z on the output of the Zener shunt regulator.

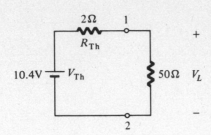

$$V_{Th} = 10 + \frac{2(30-10)}{102} = 10 + \frac{40}{102} \cong 10.4 \text{ V}$$

$$R_{Th} = 100 \| 2\,\Omega \cong 2\,\Omega$$

Substitute the Thévenin equivalent circuit (Fig. 14.12) and

$$V_L = \frac{50(10.4)}{52} = 10 \text{ V}$$

Figure 14.12. Thévenin equivalent circuit for the circuit of Fig. 14.11.

R_z, therefore, has a negligible effect at minimum R_L (maximum I_L). For $R_L = 1$ K (minimum I_L).

$$V_L = \frac{1\text{K}(10.4)}{1\text{K} + 2\,\Omega} \cong 10.4 \text{ V} > 10 \text{ V}$$

(obtained above for $R_L \cong 0\,\Omega$) and

$$VR\% = \frac{V_{1\text{K}} - V_{50\Omega}}{V_{50\Omega}} \times 100\% = \frac{10.4 - 10.0}{10.0} \times 100\%$$

$$= \frac{0.4}{10.4} \times 100\% = 4\%$$

More than one regulated output can be supplied at the same time from the same unregulated input using the circuit configuration of Fig. 14.13.

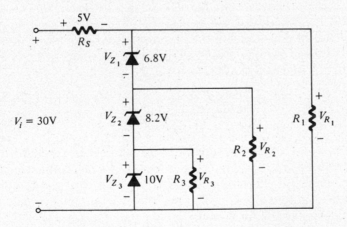

Figure 14.13. Zener regulator having three reference potentials.

It should be clear that

$$V_{R_1} = V_{Z_1} + V_{Z_2} + V_{Z_3} = 25 \text{ V}$$
$$V_{R_2} = V_{Z_2} + V_{Z_3} = 18.2 \text{ V}$$

and

$$V_{R_3} = V_{Z_3} = 10 \text{ V}$$

A shunt regulator employing a thermistor appears in Fig. 14.14. Any tendency for V_L to decrease due to a change in load will result in a decrease in current through the thermistor. The temperature of the thermistor element will thereby decrease, resulting in an increase in its resistance. The resulting resistance $R_T = R_L \| (R_{Th} + 100\,\Omega)$ will increase somewhat and the load voltage $V_L = (R_T V_i / R_T + R_1)$ will

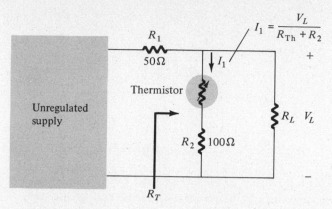

Figure 14.14. Thermistor shunt regulator.

tend to increase, offsetting the initial drop in V_L. The symbol $\uparrow$ is assigned to an increasing quantity and $\downarrow$ to a decreasing quantity in the following summary of the voltage-regulating action of this system (read from left to right):

$$V_L\downarrow, I_{\text{Th}}\downarrow, R_{\text{Th}}\uparrow, R_T\uparrow, V_L\uparrow$$
$$\underbrace{\hspace{2.5cm}}_{\text{balance}}$$

An increasing V_L will have the opposite effect on each element and quantity of the above summary.

14.4 TRANSISTOR VOLTAGE REGULATORS

The characteristics of a voltage regulator can be markedly improved by using active devices such as the transistor. The simplest of the transistor-*series*-type voltage regulator appears in Fig. 14.15a. In this configuration the transistor behaves like a simple variable resistor whose resistance is determined by the operating conditions. The basic operation of the regulator is best described using the circuit of Fig. 14.15b, in which the transistor has been replaced by a variable resistor, R_T. For

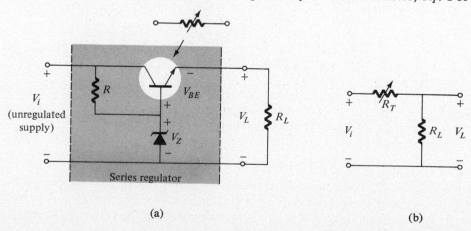

(a)

(b)

Figure 14.15. Transistor series voltage regulator.

variations in R_L, if V_L is to remain constant, the ratio of R_L to R_T must remain fixed. Applying the voltage divider rule results in

$$V_L = \overbrace{\frac{R_L}{R_L + R_T}}^{\text{[fixed for constant } V_L(V_i = \text{constant})]}} V_i$$

For
$$\frac{R_L}{R_T} = k_1 \quad \text{or} \quad R_L = k_1 R_T$$

$$\frac{R_L}{R_L + R_T} = \frac{k_1 R_T}{k_1 R_T + R_T} = \frac{k_1}{k_1 + 1} = k \text{ (constant as required)}$$

In summary, for a decreasing or increasing load (R_L), R_T must change in the same manner and at the same rate to maintain the same voltage division.

You will recall from Section 14.2 that the voltage regulation is determined by noting the variations in terminal voltage versus the load current demand. For this circuit an increasing current demand associated with a *decreasing* R_L will result in a tendency on the part of V_L to decrease in magnitude also. If, however, we apply Kirchhoff's voltage law around the output loop

$$V_{BE} = \overbrace{V_Z}^{\text{(fixed)}} - V_L$$

A decrease in V_L (since V_Z is fixed in magnitude) will result in an increase in V_{BE}. This effect will, in turn, increase the level of conduction of the transistor, resulting in *decrease* in its terminal (collector-to-emitter) resistance. This is, as described in the previous paragraphs of this section, the effect desired to maintain V_L at a fixed level.

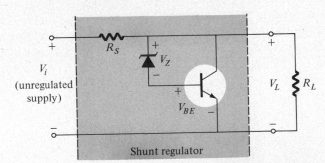

Figure 14.16. Transistor shunt voltage regulator.

A voltage regulator employing a transistor in the shunt configuration is provided in Fig. 14.16. Any tendency on the part of V_L to increase or decrease in magnitude will have the corresponding effect on V_{BE} since

$$V_{BE} = V_L - \overbrace{V_Z}^{\text{(fixed)}}$$

For decreasing V_L, the result is a decrease in the current through the resistor R_S since the conduction level of the transistor has dropped ($V_{BE\downarrow}$). The reduced drop in potential across R_S will offset any tendency on the part of V_L to decrease in magnitude. In sequential logic

$$V_L\downarrow, V_{BE}\downarrow, I_B\downarrow, I_C\downarrow, I_{R_s}\downarrow, V_{R_s}\downarrow, V_L\uparrow$$
$$\underbrace{\qquad\qquad\qquad\qquad}_{\text{balance}}$$

A similar discussion can be applied to increasing values of V_L.

A series voltage regulator employing a second transistor for control purposes can be found in Fig. 14.17. The base-to-emitter potential (V_{BE_2}) of the control transistor Q_2 is determined by the difference between V_1 and the reference voltage V_Z. The voltage level V_1 is sensitive to changes in the terminal voltage V_L. Any tendency on the part of V_L to increase will result in an increase in V_1 and therefore in V_{BE_2} since $V_{BE_2} = V_1 - V_Z$. The difference in potential is amplified by the control transistor and carried to the variable series resistive element Q_1. An increase in V_{BE_2}, corresponding to an increase in I_{B_2} and I_{C_2} will result in a decreasing I_{B_1} (assuming I_{R_3} to be relatively constant or decreasing only slightly). The net result is a decrease in the conductivity of Q_1 corresponding to an increase in its terminal resistance and a stabilization of V_L. In sequential logic

$$V_L\uparrow, V_1\uparrow, V_{BE_2}\uparrow, I_{C_2}\uparrow, I_{B_1}\downarrow, R_{(Q_1)}\uparrow, V_L\downarrow$$
$$\underbrace{\qquad\qquad\qquad\qquad}_{\text{balance}}$$

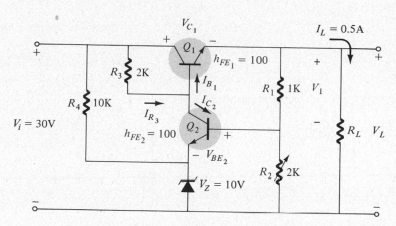

Figure 14.17. A series voltage regulator employing two transistors.

Again, a similar discussion can be applied to decreasing values of V_L.

EXAMPLE 14.1 We shall now calculate various currents and voltages of the circuit of Fig. 14.17 for the input shown. Approximations, as introduced in previous chapters, will be the working tools of this analysis. Those of primary importance include $I_C \cong h_{FE}I_B$, $V_{BE} \cong 0$ V, and $I_C \cong I_E$.

Solution:

$$V_{R_4} = V_i - V_Z = 30 - 10 = \textbf{20 V}$$

and

$$I_{R_4} = \frac{20}{10\text{ K}} = \textbf{2 mA}$$

$$V_{R_2} \cong V_Z = \textbf{10 V} \qquad \text{since } V_{BE_2} \cong 0\text{ V}$$

and

$$I_{R_2} = \frac{10}{2\text{ K}} = \textbf{5 mA}$$

Assuming $\qquad I_{B_2} \ll I_{R_1}, \quad I_{R_2}$

then $\qquad I_{R_1} = I_{R_2} = \textbf{5 mA}$

and $\qquad V_L = 5 \text{ mA} \times 3 \text{ K} = \textbf{15 V}$

with $\qquad V_{R_3} = V_i - V_L(V_{BE_1} \cong 0 \text{ V})$

$\qquad\qquad\qquad = 30 - 15 = \textbf{15 V}$

and $\qquad I_{R_3} = \dfrac{15}{2 \text{ K}} = \textbf{7.5 mA}$

Similarly, $\qquad V_{C_1} = V_i - V_L = 30 - 15 = \textbf{15 V}$

$\qquad\qquad\qquad I_{E_1} \cong h_{FE} I_{B_1} = 100 I_{B_1}$

and $\qquad I_{B_1} = \dfrac{I_{E_1}}{100} = \dfrac{(500 + 5)}{100} = \textbf{5.05 mA}$

$\qquad\qquad\qquad I_{C_2} = I_{R_3} - I_{B_1}$

$\qquad\qquad\qquad\qquad = 7.5 - 5.05$

$\qquad\qquad\qquad\qquad = \textbf{2.45 mA}$

$\qquad\qquad\qquad I_{B_2} \cong \dfrac{I_{C_2}}{100} = \dfrac{2.45}{100} = \textbf{24.5 μA}$

(Certainly, $I_{B_2} \ll I_{R_2}, I_{R_2}$ as employed above is an excellent approximation.)
and, finally,

$$I_Z = I_{R_4} + I_{C_2} = 2 + 2.45 = \textbf{4.45 mA}$$

The fact that $I_{B_2} \ll I_{R_1}, I_{R_2}$ permits the use of the circuit of Fig. 14.18 to derive a rather useful equation for the circuit of Fig. 14.17. Applying the voltage divider rule results in

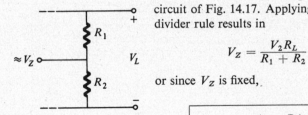

$$V_Z = \frac{V_2 R_L}{R_1 + R_2}$$

or since V_Z is fixed,

Figure 14.18. Circuit employed in the derivation of Eq. (14.3).

$$\boxed{V_L = V_Z\left(1 + \frac{R_1}{R_2}\right)} \qquad (14.3)$$

For the above case

$$V_L = 10(1 + \tfrac{1}{2}) = 15 \text{ V}$$

You will find in Fig. 14.17 that R_2 is a variable resistor. Variations in this resistance will control V_L as determined by Eq. (14.1). The maximum voltage available is obviously 30 V (for $V_i = 30$ V) since at this point $V_{C_1} = 0$ V (saturation). The minimum is 10 V attainable with either $R_1 = 0$ or $R_2 = \infty$.

14.5 COMPLETE POWER SUPPLY (VOLTAGE REGULATED)

A power supply employing a voltage regulator similar to the one described in Fig. 14.17 appears in Fig. 14.19. A *Darlington* circuit has replaced the single series transistor of Fig. 14.7 in an effort to increase the sensitivity of the regulator to changes

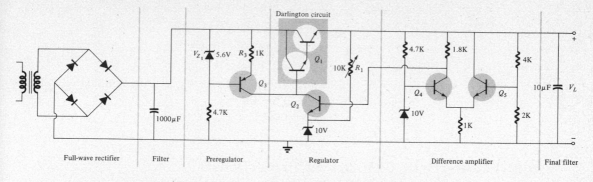

Figure 14.19. Complete voltage-regulated power supply.

in V_L. In the circuit of Fig. 14.17 changes in I_{C_2} will be reflected in changes in I_{R_3} causing a reduction in the sensitivity of I_{B_1} to changes in V_L. To minimize this undesirable effect R_3 should be as large as possible while still permitting the necessary R_3 for proper circuit behavior. This is most efficiency achieved by employing a current source in place of R_3. The current source has, ideally, infinite terminal resistance along with the capability to supply the necessary current. This portion of a power supply as indicated in Fig. 14.19 is sometimes referred to as a *preregulator*. For the circuit of Fig. 14.19

$$I_{\text{current source}} = I_{C_3} \cong \frac{V_{Z_1}}{R}$$

To improve further the sensitivity of the regulator to changes in V_L a *difference amplifier* has been introduced, the output of which is fed to the control transistor. The unregulated input is a full-wave rectified signal to be passed through a capacitive filter. The 10-μF capacitor at the output is to reduce the possibility of oscillations and further filter the supply voltage. The supply voltage V_L can be varied by changing R_1 while still maintaining regulation.

14.6 CURRENT REGULATOR

The analysis of current regulators will be limited to a brief discussion of the circuit of Fig. 14.20. Recall from the introductory discussion that a current regulator is designed to maintain a fixed current through a load for variations in terminal volt-

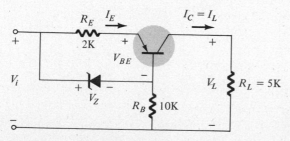

Figure 14.20. Series current regulator.

age. A decrease in $I_L = I_C$ due to a drop in V_L would result in a decrease in $I_E \cong I_C$ and, in turn, a drop in V_{R_E}. The base-to-emitter potential is

$$V_{BE} = \overbrace{V_Z}^{\text{(fixed)}} - V_{R_E}$$

A decrease in V_{R_E} will result in an increase in V_{B_E} and the conductivity of the transistor, maintaining I_L at a fixed level.

MISCELLANEOUS CIRCUIT APPLICATIONS

14.7 CAPACITIVE-DISCHARGE IGNITION SYSTEM

The capacitive-discharge (CD) ignition system has become increasingly popular in recent years due to the resulting improvement in engine performance. Tests have indicated that there is not only an increase in available power, but also better mileage and extended spark plug life. This system will rechannel the heavy firing currents that otherwise would pass through and cause rapid wear of the distributor points. In the CD system the heavy currents are carried by a semiconductor device (SCR) whose operation is simply controlled by the distributor points.

The essential components of a capacitive-discharge ignition system appear in Fig. 14.21a. The dc-to-dc inverter uses an oscillator chopper to convert the 12-V dc to an ac pulse (Fig. 14.21b) so that transformer action can be employed to step up the voltage. Recall that only time-varying primary inputs can be transformed at a higher or lower level to the secondary of a transformer. The transformer output is then filtered to produce the indicated 200-V dc level. When the distributor points are closed, the SCR will be in the open-circuit state and the capacitor (C_1) will charge to 200 V at a rapid rate since the charging time constant ($\cong R_p \times C$) is very

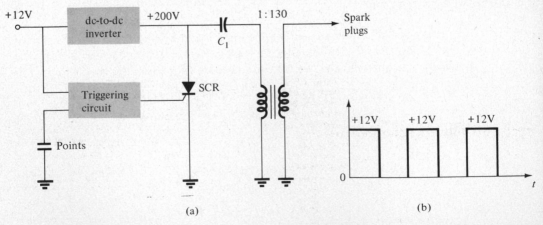

(a)　　　　　　　　　　　　　　(b)

Figure 14.21. Capacitive-discharge ignition system.

small. R_p is simply the dc resistance of the primary coil of the transformer. At the instant the points are open, the trigger circuit will turn the SCR on (short-circuit state) and the capacitor will discharge across the primary coil. The indicated turns ratio will result in a high secondary voltage, which will appear directly across the spark plugs. For a change (ΔV_p) in primary voltage of 200 V the change in secondary voltage (ΔV_s) as determined by the turns ratio will be

$$\frac{\Delta V_p}{\Delta V_s} = \frac{N_p}{N_s} \Longrightarrow \Delta V_s = \frac{N_s}{N_p} \Delta V_p = 130(200) = 26 \text{ kV}$$

14.8 COLOR ORGAN

The present interest in the unique has resulted in a rather interesting device called the *color organ*. It is an electronic package that will produce a visual display of colored lights, at various intensities, that will correspond directly with the intensity of various tones of the audio signal. One workable design appears in Fig. 14.22.

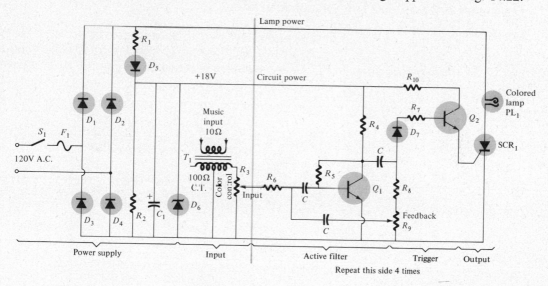

Figure 14.22. Color organ.

R_1 – 1500 ohm, 10 W res. ±10%
R_2 – 820 ohm, 1 W res. ±10%
R_3 – 50 ohm pot "Color Control"
C_1 – 15 μF, 35 V elec. capacitor
D_1, D_2 – 1N3495 rectifier (Motorola)
D_3, D_4 – 1N3495 R rectifier (Motorola)
D_5, D_7 – 1N40001 diode
D_6 – 18 V zener diode (Motorola 1N4746)
T_1 – Interstage trans. 100 ohms c.t./10 ohms c.t.
(Stancor TA-2 or equiv.)

S_1 – 10A toggle sw.
F_1 – 10A, 120V fuse
The following parts are for a single channel.
Four channels are required.
R_4 – 3300 ohm, ½ W res. ±10%
R_5 – 1 megohm, ½ W res. ±10%
R_6 – 4700 ohm, ½ W res. ±10%
R_7 – 10,000 ohm, ½ W res. ±10%
R_8 – 2700 ohm, ½ W res. ±10%
R_9 – 2000 ohm pot (Mallory "Trim-Pot" MTC-1)
R_{10} – 560 ohm, ½ W res. ±10%

SCR$_1$ – 7.4 A SCR (General Electric C-20-B)
PL$_1$ – 120V incandescent bulb–in color (20 to 450W total per channel)
Q_1, Q_2 – 2N3391 transistors
C – 0.1 μF, 50 V capacitor (for l.f. green channel)
C – 0.047 μF, 50 V capacitor (for medium-l.f. blue channel)
C – 0.022 μF, 50 V capacitor (for medium-h.f. red channel)
C – 0.01 μF, 50 V capacitor (for h.f. yellow channel)

Fundamentally, the circuit package has an active filter associated with each color, with the result that the circuit to the right of the dashed line must be repeated for each color. Each filter is designed to pass only those frequencies associated with a particular color. The greater the intensity of a particular frequency the brighter will be that particular color. The notation in Fig. 14.22 clearly describes the function of each section of the system. For this system the circuit at the right of the dashed line is repeated 4 times with the indicated value of C for each color.

The power supply is nothing more than a full-wave bridge-rectifier with an RC filter and output reference Zener diode of 18 V. The color control resistor R_1 is simply controlling the strength of the signal appearing at the base of the active filter. A signal of the proper strength and frequency will activate the trigger circuit (simple transistor amplifier) to turn on the SCR. Once the SCR is in its short-circuit state, the colored lamp will light. The brightness of the bulb is controlled by the strength of the "on" state of the SCR.

14.9 LIGHT DIMMER (MOTOR SPEED CONTROL)

The circuit of Fig. 14.23 can be employed as a light dimmer for lamps up to 400 W or as a light dc motor speed control (up to 2 A). The bridge rectifier will establish the signal appearing in Fig. 14.24a at point a of the circuit of Fig. 14.23. This is, as

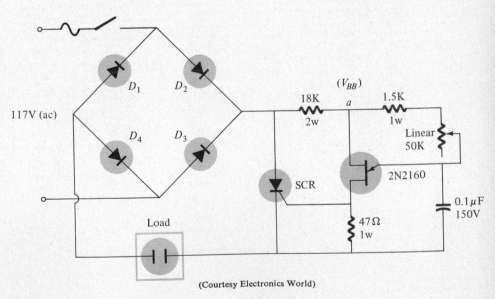

(Courtesy Electronics World)

Figure 14.23. Light dimmer (motor speed control).

indicated, V_{BB} for the unijunction transistor as discussed in Chapter 9. It differs to the extent that V_{BB} is a variable quantity here, whereas it was a fixed dc level in Chapter 9. The voltage V_E at the emitter of the unijunction transistor will increase toward the instantaneous value of V_{BB} (as shown in Fig. 14.24b) until the firing potential (V_P) is reached. At this point, the unijunction will enter the conduction state and the discharge of C through R_2 (47 Ω) will result in the waveform of Fig. 14.24c. The pulse appearing across R_2 will trigger the SCR into the conduction state and the output across the load will appear as shown in Fig. 14.24d. Note that the peak value of the load voltage is greater than the peak value of V_{BB} due to the elimination of the voltage divider action of the 18-K resistor. The delay time and, consequently, the voltage appearing across the load can be varied by the linear 50-K potentiometer. An increase in t_d will, of course, result in a reduction in the effective voltage appearing across the load, reducing the brightness of the bulb or the speed of a motor if employed as a control device.

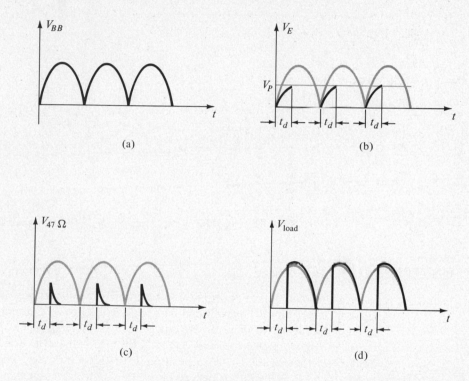

(a)   (b)   (c)   (d)

(Courtesy General Electric Co.)

Figure 14.24. Light dimmer control and load waveforms.

14.10 UNIJUNCTION CODE PRACTICE OSCILLATOR

A code practice oscillator of relatively simple construction and operation appears in Fig. 14.25. The 50-K volume control is affecting nothing more than the V_{BB} potential of the unijunction transistor. The higher its level the greater the strength of the resulting output signal. The 25-K tone control and capacitor C are controlling the frequency and thereby the tone of the output signal. You will recall from Chapter 9 that the frequency of the output signal of a unijunction oscillator is directly dependent on the magnitude of R_1, R_2, and C. So long as the key is depressed, a pulse having a definite frequency and strength will appear across the phone or speaker with a tone determined by the component values.

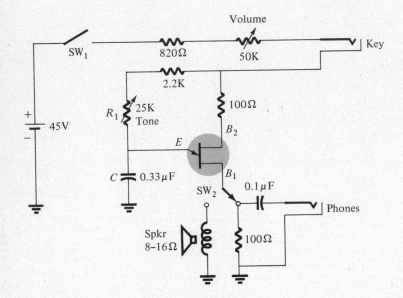

Figure 14.25. Unijunction transistor code practice oscillator.

14.11 TRANSISTOR TESTER

The circuit for and actual photographs of a transistor tester appear in Fig. 14.26a. The basic operation of the tester can best be described by examining the resultant circuit configuration for determining one of the parameters, such as h_{FE}, the static value of the short-circuit forward current gain of a transistor. The table of Fig. 14.26b direct that switch S_1 be in position 2 for this parameter. This means that the contacts SIA, SIB, and SIC are all at position 2. If we assume that the transistor is *npn*, the switch S_2 will be in the position indicated in Fig. 14.26a. The resulting circuit appears in Fig. 14.27a. It is redrawn in Fig. 14.27b for ease of further analysis. The enclosed area of Fig. 14.27b is the current-limiting, calibrating, fine adjustment for the meter movement. It is obvious from Fig. 14.27b that the movement will indicate the magnitude of I_C.

If we assume $V_{BE} = 0.6$ V, then

$$I_B = \frac{6 - 0.6}{271.2\,\mathrm{K}} = \frac{5.4}{271.2\,\mathrm{K}} \cong 20\ \mu\mathrm{A}$$

as indicated on the chart. It should prove somewhat interesting to trace through the resulting circuits for the remaining measurements using this particular transistor

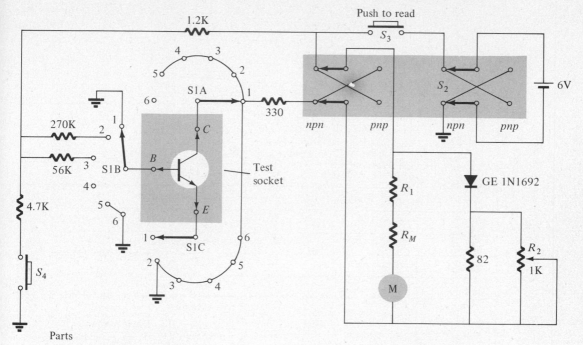

Parts

S_1 — 3 pole 6 position non-shorting
    selector switch
S_2 — 4 pole 2 position switch

S_3 — S_4 normally open push switches
M — $100\,\mu A$ full scale meter
R_M — Meters internal resistance

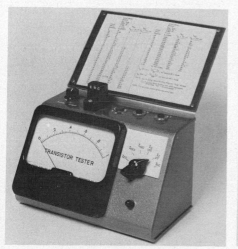

(Courtesy General Electric Co.)

(a)

Figure 14.26.(a) Transistor tester.

SEC. 14.11   TRANSISTOR TESTER

611

<table>
<tr><th>To test</th><th>When</th><th>Adjust Selector switch S_1 to position</th><th colspan="2">Result</th></tr>
<tr><td>I_{CO}</td><td>$V_{CE} = 6V$</td><td>1</td><td colspan="2">Read meter direct</td></tr>
<tr><td>I_C</td><td>$I_B = 20\mu A$</td><td>2</td><td colspan="2">Read meter direct</td></tr>
<tr><td>I_C</td><td>$I_B = 100\mu A$</td><td>3</td><td colspan="2">Read meter direct</td></tr>
<tr><td>I_{CEO}</td><td>$V_{CE} = 6V$</td><td>4</td><td colspan="2">Read meter direct</td></tr>
<tr><td>I_{CES}</td><td>$V_{CE} = 6V$</td><td>5</td><td colspan="2">Read meter direct</td></tr>
<tr><td>I_{EO}</td><td>$V_{EO} = 6V$</td><td>6</td><td colspan="2">Read meter direct</td></tr>
<tr><td>h_{FE}</td><td>$I_B = 20\mu A$</td><td>2</td><td colspan="2">Calculate: $h_{FE} = \frac{I_C}{I_B} = \frac{\text{meter reading}}{20\mu A}$</td></tr>
<tr><td>h_{FE}</td><td>$I_B = 100\mu A$</td><td>3</td><td colspan="2">Calculate: $h_{FE} = \frac{I_C}{I_B} = \frac{\text{meter reading}}{100\mu A}$</td></tr>
<tr><td>h_{fe}</td><td>$I_B = 20\mu A$</td><td>2</td><td>Calculate: $h_{fe} = \frac{I_{C_1} - I_{C_2}}{4 \times 10^{-6}}$</td><td rowspan="2">Where: I_{C_1} = meter reading
I_{C_2} = meter reading with S_4 closed</td></tr>
<tr><td>h_{fe}</td><td>$I_B = 100\mu A$</td><td>3</td><td>Calculate: $h_{fe} = \frac{I_{C_1} - I_{C_2}}{20 \times 10^{-6}}$</td></tr>
<tr><td>6V battery</td><td>——</td><td>4</td><td colspan="2">With 150Ω resistor connected to C–E of test socket, full-scale meter deflection will result when S_3 is pressed.</td></tr>
</table>

(Courtesy General Electric Co.)

(b)

Figure 14.26.(b) (continued).

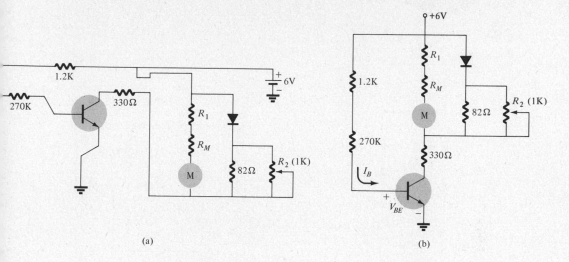

Figure 14.27. (a) Transistor tester circuit for determining h_{be}; (b) circuit redrawn.

tester. In addition, it would be time well spent to build the circuit and check its operation. A transistor tester is certainly a useful instrument to have available.

14.12 HIGH-IMPEDANCE FET VOLTMETER

The schematic for a high-impedance FET voltmeter appears in Fig. 14.28. The ultimate in voltmeter design requires that the input impedance be infinite for each scale so that the response of a circuit is not altered when the meter is introduced. The high input impedance of the FET amplifier is employed toward this end in the FET voltmeter. The wiper arm denoted a in Fig. 14.28 is set on the appropriate voltage scale. The voltage appearing at the gate of the FET is then determined by a simple voltage divider relationship if we assume the input impedance of the FET to be essentially infinite (open-circuit) ohms; that is, if we define $R_{A-B} = 2\,\text{M} + 10\,\text{M} + 8\,\text{M} + \text{M} + 800\,\text{K} + 100\,\text{K} + 80\,\text{K} + 10\,\text{K} + 10\,\text{K} = 22\,\text{M}$. Then for the 1 V scale

$$V_{\text{gate}} = \frac{(R_{A-B} - 12\,\text{M})V_{A-B}}{R_{A-B}} = \frac{10\,\text{M}\,V_{A-B}}{22\,\text{M}} \cong 0.45\,V_{A-B}$$

In addition, for the above approximation $Z_{\text{input(FET)}} = \infty\,\Omega$ the input impedance of the voltmeter is simply determined by the series resistance between A and B independent of scale employed. Therefore,

$$Z_{i\text{(meter)}} = 22\,\text{M}$$

A moment's investigation should reveal that the FET and movement are part of a bridge circuit. This in itself should give some hint as to its mode of operation. If there is no difference in potential across $A–B$, the various calibrating resistors are

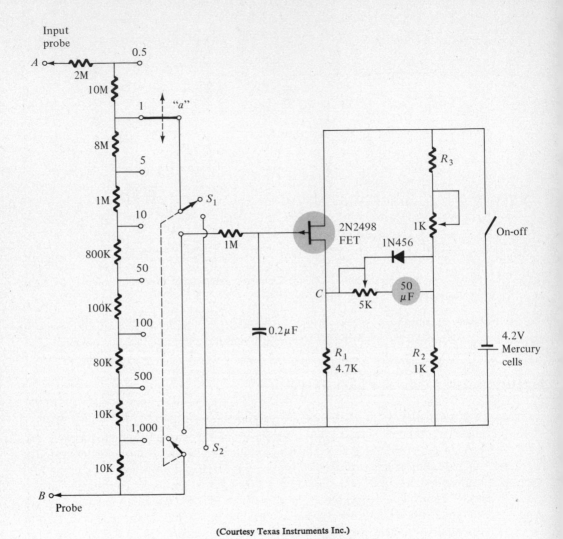

(Courtesy Texas Instruments Inc.)

Figure 14.28. High-impedance FET voltmeter.

adjusted to indicate zero deflection. Subsequently, when *A* and *B* are placed across a difference in potential levels, the resulting drain-to-source voltage of the FET will result in an unbalanced condition and a deflection of the movement. If properly calibrated, the movement will indicate the magnitude, in volts, of the voltage across points *A–B*. The 1-M resistor and 0.02-μF capacitor will filter out any stray ac voltages that appear at the input to the voltmeter. Switches S_1 and S_2 will allow the meter to read upscale even if the polarity of the potential level at *A* with respect to *B* is reversed. The vertical dashed lines indicate that they are ganged together so that they will both move to the same relative positions.

14.13 UNIJUNCTION HOME SIGNAL SYSTEM

The system of Fig. 14.29 will reveal, by the tone of the audio signal, the location at which entrance is desired. The capacitor C will charge toward a potential determined by the voltage divider rule; that is,

$$V = \frac{33\,K(30)}{33\,K + R} \text{ (determined by button depressed)}$$

The charging time constant ($\tau = RC$) will be determined by the botton depressed. The smaller the time constant the sooner the unijunction transistor will enter the conduction state and permit a pulse to reach the remote speakers. Naturally, therefore, the smaller the time constant the higher the pulse frequency and the higher the pitch of the resulting signal.

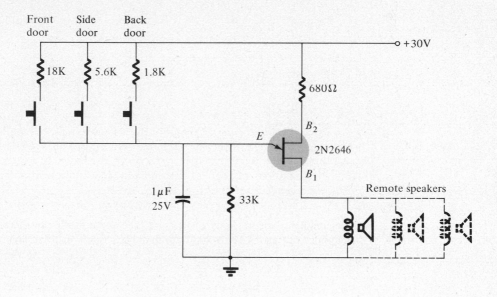

(Courtesy General Electric Co.)

Figure 14.29. Unijunction home signal system.

PROBLEMS

§ 14.2

1. Determine the per cent voltage regulation of a dc supply providing 100 V under no-load conditions and 95 V at full-load.

2. Determine the voltage regulation of the network of Fig. 14.30 if R_L is limited to a minimum value of 1 K before exceeding the current limits of the supply.

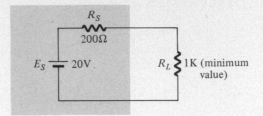

Figure 14.30

3. Determine the current regulation of the network of Fig. 14.31 if R_L is limited to a minimum value of 1 K before exceeding the voltage limits of the supply.

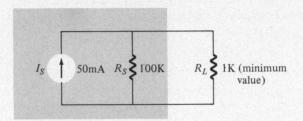

Figure 14.31

§ 14.3

4. (a) For the Zener diode shunt regulator of Fig. 14.8 determine the minimum value of R_L to ensure that the Zener diode is in the "on" state if $V_i = 60$ V, $V_Z = 12.5$ V, and $R_s = 200\ \Omega$.
   (b) For the conditions of part (a) determine the maximum load current I_L.
   (c) Find the minimum value of I_L if $R_{L(max)}$ is 5 K.
   (d) If R_Z is 1.5 Ω, determine the voltage regulation for the range indicated above.

5. A Zener regulator uses a diode with $V_Z = 22$ V and a maximum dissipation of 2 W. If the applied voltage is 300 V, find the minimum value of the resistance R_S.

§ 14.4

6. For the network of Fig. 14.15, if $R_L = 1$ K, $R = 5$ K, $V_Z = 10$ V, $V_{BE} = 0.7$ V, and $V_i = 20$ V determine the following:
   (a) The voltage V_L and the current I_L.
   (b) The collector current I_C.
   (c) The current through R.
   (d) The supply current.

7. For the network of Fig. 14.16, if $R_L = 4$ K, $R_S = 2$ K, $V_Z = 10$ V, $V_{BE} = 0.7$ V, and $V_i = 20$ V determine the following:
   (a) The voltage V_L.
   (b) The current I_L.
   (c) The supply current through R_S.
   (d) The Zener current if $\beta = 50$.

8. Determine the value of R_2 in the network of Fig. 14.17 to establish a load voltage of 20 V.

§ 14.5

9. (a) Describe the complete behavior of the voltage-regulated supply of Fig. 14.19.
   (b) Describe the resulting action to maintain a fixed V_L if V_L begins to drop in level.

§ 14.6

10. For the network of Fig. 14.20 calculate V_L, I_C, I_E, and the supply current if $V_{BE} = 0.7$ V, $V_Z = 10$ V, and $V_i = 20$ V.

cathode ray oscilloscope

15

15.1 GENERAL

One of the basic functions of electronic circuits is the generation and manipulation of electronic waveshapes. These electronic signals may represent audio information, computer data, television pictures, timing information (as used in radar work), and so on. The common meters used in electrical work—the dc or ac voltmeter—(VOM or VTVM) measure either dc, peak, or rms. These measurements are correct only for nondistorted sinusoidal signals or they measure true rms for a particular signal with no indication of how the signal varies with time. Obviously, when signal processing is being done, these overall measurements are essentially meaningless. What is necessary is to " see" what is going on in the circuit, hopefully in the small fractions of time it takes for the signal waveshape to change. The cathode ray oscilloscope (CRO) provides just this type of operation—visual presentation of the signal waveshape, allowing the technician or engineer to look at different points in the electronic circuit to see those changes taking place. In addition, the CRO may be calibrated and used to measure both voltage and time variations so that information is available on how much voltage is present, how much voltage changed, and how long it took to make the change (or a portion of the change).

Consider a radar circuit in which a pulse originates in an electronic circuit and is radiated by an antenna toward a distant object. At some later time a reflected pulse is received. The time it took to reach the object and return is measured on a CRO to provide an indication of distance. It is necessary when developing and setting up such circuits, to be able to display the pulse sent out—its amplitude, the sharpness of the pulse, etc.—and then visually compare it with the return signal after reception and processing of the signal. The return signal can be checked on the CRO for sharpness and voltage amplitude. The time difference between the pulse transmitted and that received may be read on the calibrated time scale of the CRO.

A second example is the use of a CRO to view signals through out the circuit of a television (TV) receiver (or transmitter). The video signals (information signals for TV) contain complex waveforms, which provide black-white intensity, synchronization information for the picture, and audio information. To properly test and adjust these circuits requires comparing their operating waveforms with expected waveforms. Such adjustments could not be made without the use of the CRO. (It is interesting to note that due to the cathode ray tube (CRT), the TV can be its own test scope, showing where the circuit is operating well or poorly.)

Without the scope, electronic work would virtually be impossible—it would be like groping in a dark room. Unless one is able to "see" the circuit waveforms there is no way to correct errors, understand mistakes in design, or make adjustments. As the voltmeter, ammeter, and power meter are basic tools of the power engineer or electrician, the CRO is the basic tool of the electronic engineer and technician.

There is a wide range of CROs available, some suited for general work in many areas of electronics, others for work only in a specific area. A CRO may be designed

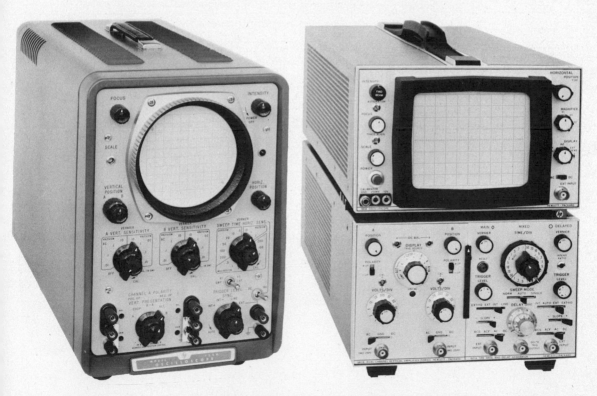

Figure 15.1. Typical cathode ray oscilloscopes (CROs): HP 122A and HP 180A scopes.

to operate from 100 Hz up to 500 kHz, or from dc up to 50 MHz; it may allow viewing signals to within a time span of 1 microsecond, or down to a few nano-seconds (10^{-9} sec); it may provide one waveform or a number of waveforms simultaneously on the face of the CRO. When a number of waveforms are shown simultaneously, voltages at different places in the circuit can be viewed at the same time. All these CRO features provide flexibility and enable one to use a CRO that is well suited for the job at hand. Another CRO feature is its ability to hold the display for either a very short duration of time or for a long duration. In fact, CROs, called *storage scopes*, provide storage of a display for many hours so that an original signal (which appeared long before) may still be analyzed or compared with another signal at a later time. Figure 15.1 shows two representative scopes.

15.2 CATHODE RAY TUBE— THEORY AND CONSTRUCTION

The cathode ray tube (CRT) is the "heart" of the CRO. It provides the visual display that makes the instrument so useful. The tube contains the following four basic parts:

1. An *electron gun* to produce a stream of electrons:
2. *Focusing and accelerating* elements to produce a well-defined *beam* of electrons.
3. *Horizontal and vertical deflecting plates* to control the path of the beam:
4. An *evacuated glass envelope* with a *phosphorescent screen* that glows visibly when struck by the electron beam.

Figure 15.2 shows an overall view of a complete electrostatic deflection CRT. We shall first briefly consider its operation. A *cathode* (K) containing an oxide

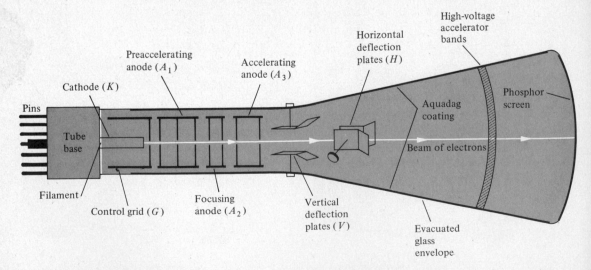

Figure 15.2. Cathode ray tube: basic construction.

coating is heated indirectly by a filament resulting in the release of electrons from the cathode surface. A control grid (G) provides for control of the number of electrons passing on into the tube. The electrons are then focused into a tight beam and accelerated to higher velocity by the focusing and accelerating anodes. The parts discussed so far comprise the *electron gun* of the CRT.

The high-velocity, well-defined electron beam then passes through two sets of deflection plates. The first set of plates is oriented to deflect the electron beam *vertically*, up or down. The *direction* of the vertical deflection is determined by the voltage *polarity* applied to the deflecting plates. The amount of deflection is set by the applied voltage *magnitude*. The beam is also deflected horizontally (to the left or right) by a voltage applied to the horizontal deflecting plates. The deflected beam is further accelerated by very high voltages applied to the tube and it finally strikes a *phosphorescent* material on the inside face of the tube. The phosphor glows when struck by the energetic electrons—the visible glow seen at the front of the tube by the person using the scope.

The CRT is a self-contained unit with leads brought out through a base to pins, as in any electron tube. Various types of CRTs are manufactured for a variety of applications. The CRT of Fig. 15.2 is basic and allows discussion of the essential elements of the device. We can now consider each part of the tube in more detail and then the external CRO controls, which provide the useful operation of the tube.

Electron Gun

The cathode (Fig. 15.3a) is cylindrical, made of nickel, and capped on the end with an oxide coating. The oxide coating, typically made of oxides of barium and strontium, is applied to the cap of the cathode facing the screen direction. A filament, made of a tungsten or tungsten alloy, indirectly heats the cathode when a current is caused to flow through the filament. The filament acts like the heating elements of a device such as a toaster or rotisserie. Electrons will be liberated from the oxide coating on the cathode surface by the heating effect of the filament. These

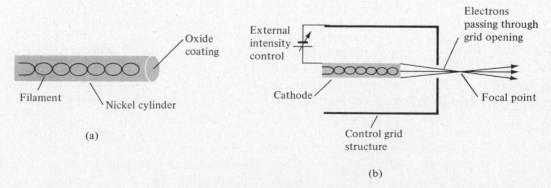

Figure 15.3. Components of the CRT electron gun: (a) cathode (K) and filament; (b) control grid (G).

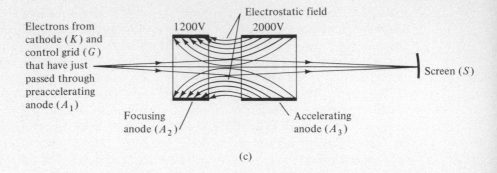

Electrons from cathode (K) and control grid (G) that have just passed through preaccelerating anode (A_1)

Electrostatic field

1200V   2000V

Screen (S)

Focusing anode (A_2)

Accelerating anode (A_3)

(c)

Figure 15.3 (continued): (c) focusing (A_1) and accelerating anodes (A_2).

liberated electrons will travel in the general direction of the screen but at various angles and with various velocities.

Focusing and Accelerating Elements

To provide some focusing of the electrons a control grid (Fig. 15.3b) with a small opening in the direction of the screen is placed after the cathode. In addition, a biasing voltage is applied to the control grid to control the flow of electrons through the small opening of the grid structure. If the grid voltage is made negative (with respect to the cathode), there will be a reduction in the number of negatively charged electrons passing through the grid opening. With a large enough negative voltage all the electrons leaving the cathode due to the heating effect of the filament will be prevented from passing through the grid aperture. The grid therefore permits adjustment of the number of electrons generated by the electron gun with the ability to completely cut off the electron flow to the screen, if desired.

Those electrons that pass through the control grid structure are given an initial acceleration toward the screen of the tube by a preaccelerator anode (A_1) having a positive potential (100–200 V) with respect to the cathode. The focusing anode (A_2) and accelerating anode (A_3) operate to focus the electrons into a tight beam and also to further accelerate or speed up the electrons coming from the cathode. Fig. 15.3c shows the electric field set up by typical voltage applied to anodes A_2 and A_3 and their effect on the electron flow. The force on the electrons due to the electric field set up by these anodes will result in a well-shaped beam having a focal point at the screen of the tube. The person using the scope can control the focusing of the beam through an external FOCUS control.

The difference in potential between anodes A_2 and A_3 will set up an electrostatic field as shown in Fig. 15.3c. The effect of these electric field force lines will be to focus the electron beam as shown. The deflection will be such as to push those electrons diverging from the center back in toward the center of the beam. Once past the focusing anodes, the electrons move past the deflecting plates to the screen in a tight beam.

Deflecting Plates
(Vertical and Horizontal)

Figure 15.4 shows the electron beam passing through a pair of plates. If the voltage of the upper plate is more positive than the voltage of the lower plate, the electron beam will be attracted upward. Reversing the plate polarity would cause the beam to be deflected downward. The voltage applied externally (of the CRT) to the deflection plates shown results in the signal's being deflected in a vertical manner; hence, the designation *vertical deflection plates*. This voltage may be a steady (dc) voltage or a varying one. As a point of interest, the home TV tube is similar to that discussed so far except for a magnetic deflection system.

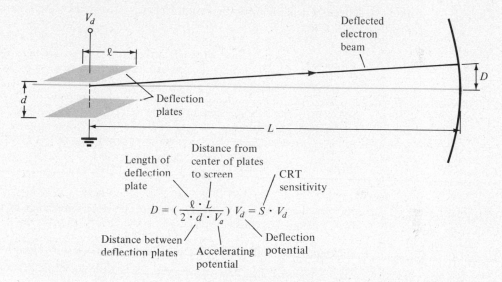

Figure 15.4. Tube factors affecting beam deflection.

An expression relating the factors in calculating the amount of deflection is (see Fig. 15.4)

$$D = \frac{lLV_d}{2dV_a} \tag{15.1}$$

where the distance terms are shown in Fig. 15.4 and the voltages are V_d, the voltage applied to the deflection plates and V_a, the accelerating voltage of the tube (typically thousands of volts). Since all the values except the deflecting voltage are usually fixed for a tube, they may be lumped together as one term defined as the electrostatic deflection sensitivity

$$S \equiv \frac{D}{V_d} = \frac{lL}{2dV_a} \tag{15.2}$$

The sensitivity has units of meters per volt, centimeters per volt, or mililimeters per volt, depending on the unit of measurement used. It indicates the amount of deflection of the electron beam at the screen per volt of signal applied to the deflection plates.

The *deflection factor* (G) of a CRT is

$$G \equiv \frac{V_d}{D} = \frac{1}{S}$$

(15.3)

Note that G is the reciprocal of the sensitivity and indicates how many volts of deflecting voltage must be applied for each meter of deflection at the screen. Either the deflection sensitivity or deflection factor is provided by the manufacturer for each type of CRT.

EXAMPLE 15.1 A manufacturer of a CRT rates his tube as having a deflection factor of 40 V/in. Calculate the amount of deflection seen on the screen for deflection voltages of 20 and 120 V.

Solution: Using the relation for the amount of deflection in Eqs. (15.2) and (15.3) gives the solution

(a) $D = SV_d = (V_d/G) = [20 \text{ V}/(40 \text{ V/in})] = \frac{1}{2} \text{ in.}$

(b) $D = [120 \text{ V}/(40 \text{ V/in})] = 3 \text{ in.}$

EXAMPLE 15.2 If a deflection of 2.5 in. is obtained from a deflection voltage of 50 V on the plates of the CRT, calculate the deflection factor of the tube. How much will the beam be deflected for a voltage of 75 V for the same tube?

Solution: The deflection factor of the tube is obtained by using Eq. (15.3):

$$G = V_d/D = 50/2.5 = 20 \text{ V/in.}$$

Again using Eq. 15.3, we get

$$D = V_d/G = 75/20 = 3.75 \text{ in.}$$

EXAMPLE 15.3 The sensitivity of a CRT is given as 0.25 mm/V. What voltage must be applied to the deflecting plates to obtain a deflection of 2 in.?

Solution: It is first necessary to convert the deflection amount of 2 in. into millimeters. The conversion factor of 25.4 mm/in. is used to obtain

$$D = 2 \text{ in.} \times \frac{25.4 \text{ mm}}{1 \text{ in.}} = 50.8 \text{ mm}$$

Using Eq. (15.2), we calculate the deflection voltage

$$V_d = \frac{D}{S} = \frac{50.8 \text{ mm}}{0.25 \text{ mm/V}} = 203.2 \text{ V}$$

The deflection plates are placed so that the vertical deflection plates are farther from the screen than the horizontal deflection plates. The reason for this is indicated in the above discussion of deflection sensitivity. The vertical plates are used to display the voltage to be viewed or measured. Since the further the deflection plates are from the screen the greater the sensitivity, the usual practice is to obtain the greater sensitivity for the vertical deflection plates.

Electron Acceleration and Phosphor Screen Action

After the beam is acted upon by the deflection plates, it passes down the tube in a straight path (although at some deflection angle due to the deflecting voltages). To ensure that sufficiently energetic electrons strike the phosphor screen the electron beam may be further accelerated by intensifier bands as indicated in Fig. 15.5.

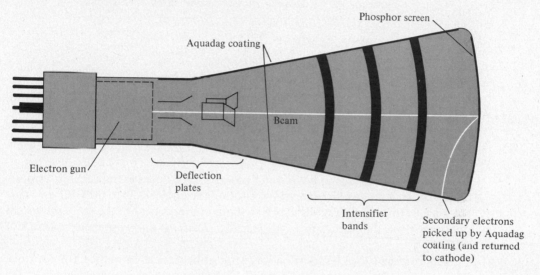

Figure 15.5. Action of intensifier bands, phosphor screen, and Aquadag coating.

The high-energy electron beam strikes the phosphor material, causing it to glow. Because of the high energy of the striking electrons a secondary emission of electrons from the phosphor screen will occur. These electrons would build up a layer of negative charge, which would deteriorate the tube operation. A layer of material called *Aquadag* is coated along the side of the tube, however, and the electrons emitted by secondary action are picked up by the coating and are returned to the cathode.

If the glow seen were to be present only as long as the beam strikes the screen, the light given off would be termed fluorescent (as the lights in a home or plant). The phosphor screen, however, will continue to glow even after the beam is turned off. The length of time the glow continues can vary from a few milliseconds to a few seconds depending on the type of phosphor material used. CRO tubes used to observe very-high-frequency signals have short *persistence* (short amount of glow time after beam is removed). Other scopes used for observing only very-low-frequency signals have persistence times of a few seconds. There is also a special type of CRT used in memory-type scopes that has an effective persistence in the range of hours.

Table 15.1 shows a number of the phosphor types as well as their persistence and color of glow. The color is determined by the phosphor material used. The tube types listed by the manufacturer include the phosphor type. For example, a 5BP4 has a phosphor screen type $P4$. CRO manufacturers also refer to the phosphor, indicating, for example, that their scope can be purchased with either $P31$, $P2$, $P7$, or $P11$ phosphors, depending on the buyer's choice.

TABLE 15.1 CRT Phosphors

PHOSPHOR NO.	COLOR OF PHOSPHORESCENT GLOW	PERSISTENCE
$P1$	green	medium
$P2$	green	long
$P4$	white	medium
$P5$	blue	very short
$P7$	greenish-yellow	very long
$P11$	blue	short
$P31$	green	medium-short

Operation and Controls Affecting CRT

As important as it is to understand how the CRT operates, it is equally important to understand how the external controls of the CRO relate to the CRT and how the various control operations are carried out in the scope. For the present we shall concentrate on the operation of the *intensity*, *focusing*, and *positioning* controls of the scope.

Figure 15.6a shows a typical biasing arrangement for a few of the CRO controls. The external scope dials are shown in Fig. 15.6b to provide a clearer picture of the operation of the scope. A voltage divider network containing resistors and potentiometers provides for the positioning, focusing, and intensity control of the CRT. A high-voltage (HV) power supply provides the plus (+)-to-minus (−) voltage indicated in Fig. 15.6a. A potentiometer, allowing adjustment of the difference in voltage between cathode (K) and grid ($G1$), provides the control of beam intensity. The voltage applied to the grid is more negative than that connected to the cathode. By adjusting the grid voltage to be more negative the number of electrons passing through the grid is reduced, providing a less intense electron beam. The adjustment of intensity can be made sufficiently negative (from grid to cathode) to enable complete cutoff of the beam. Thus, the user is provided with the ability to regulate the intensity of the screen glow.

The voltage at focusing anode A_2 is connected to another potentiometer, allowing adjustment and control of the focusing of the electron beam. If the adjustment is used, the focal point of the beam can be set directly at the screen so that a clear spot results. Since variation of the intensity will have some effect on the focusing, adjustment of one usually requires adjustment of the other.

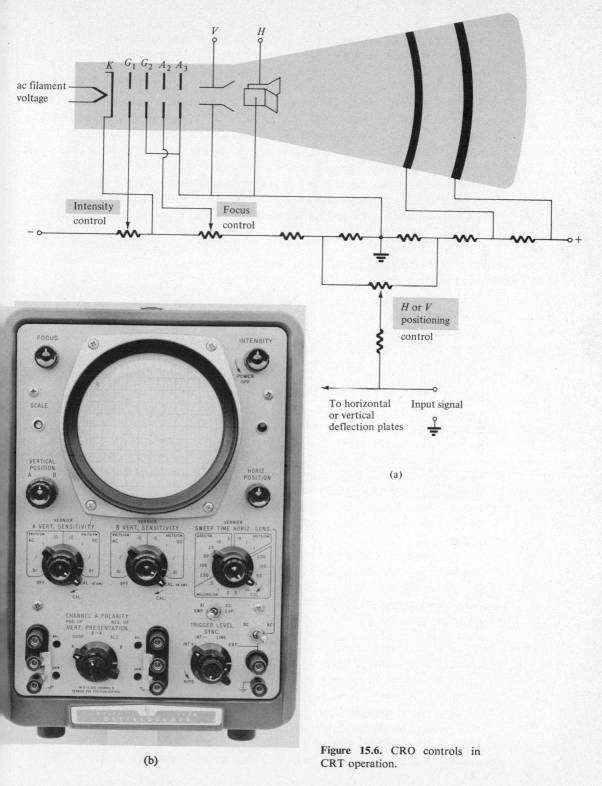

Figure 15.6. CRO controls in CRT operation.

SEC. 15.2 CATHODE RAY TUBE—THEORY AND CONSTRUCTION

The very high voltages at the right side of the voltage divider network are connected to the intensifier bands to provide high acceleration of the electron beam. In general, these intensifier band voltages are fixed.

The result of applying voltages to the vertical (or horizontal) plates is to deflect the beam vertically (or horizontally) on the face of the tube. In addition, it is usually essential to be able to reposition the picture on the screen; this is provided by external horizontal and vertical positioning controls. This action is shown simply by the addition of an adjustable dc voltage to the input signal (to either the vertical or horizontal plates). The addition of the positioning dc voltage to that applied to the deflection plates moves the electron beam additionally to the right or left (or up or down) so that the overall picture position can be adjusted on the screen. Separate control is provided for the vertical and horizontal sections of the tube.

15.3 CATHODE RAY OSCILLOSCOPE (CRO) —OPERATION AND CONTROLS

It is necessary that the electron beam be horizontally deflected across the face of the tube to view signal waveforms. In the usual repetitive mode of operation the beam must be returned back across the tube to allow the repeated horizontal sweep of the beam. The beam is deflected horizontally from the left side of the CRT face to the right side during normal operation and then the beam is "blanked" (turned off) during the return sweep of the beam. Additionally, the signal to be viewed must be applied to these electronic circuits, which drive the beam vertically, while it is being swept horizontally across the CRT face.

Typical tube phosphors used have some persistence even to signals of very short time duration. A repetitive tracing of the viewed waveform will then result in a steady continuous display. This requires synchronizing the sweeping of the beam (*sync*) with the signal being viewed. If the signal is properly synced, the display will be stationary and appear as if the same signal is being viewed all the time. In the absence of sync the picture will appear to drift or move horizontally across the screen.

The basic parts of a CRO are shown in the block diagram of Fig. 15.7. We shall first consider the operation of the CRO for this simplified general picture. To obtain a noticeable beam deflection of a fraction of an inch to a few inches the usual voltage applied to the deflection plates is on the order of tens to hundreds of volts. Since the signals measured using the CRO may be only a few volts, or even as little as a few microvolts, amplifier circuits are needed to increase the input signal to the voltage levels required to operate the tube. Thus, there are amplifier sections as part of the CRO circuitry for both the vertical and horizontal deflection of the beam. It may also happen that the signal to be viewed is too large (with the amplifier gain) to be viewed within the span of the tube face. When this happens, an attenuator is needed to cut down or reduce the signal amplitude to the amount that results in a visible display within screen dimensional limits.

A synchronizing signal must be available in order to synchronize the horizontal sweep of the electron beam with the signal to be viewed. As shown in Fig. 15.7,

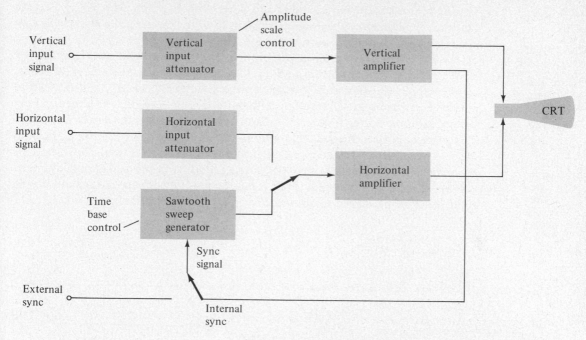

Figure 15.7. Cathode ray oscilloscope, general block diagram.

the sync signal applied to the horizontal sweep generator can be obtained from the vertical channel of the CRO as an internal sync signal, or an external sync signal may be connected. Note that the input to the horizontal amplifier is obtained from either the sweep generator or from a horizontal input attenuator. The normal scope operation uses the sweep input to the horizontal channel. A *different* external horizontal input signal is typically used to obtain a Lissajous pattern for phase or frequency measurement (see Section 15.7).

15.4 CRO—DEFLECTION AND SWEEP OPERATION

We shall now consider the operation of the CRO for various inputs to the vertical and horizontal deflection plates. Although the actual CRO inputs are not *directly* connected to the deflection plates, but are coupled through attenuator networks and amplifiers, we shall still refer to such signals (applied to the vertical and horizontal inputs or vertical and horizontal channels of the CRO) as the deflection signals. Obviously, the positioning controls and the setting of the attenuators for each channel of the scope will affect the resulting signal actually applied to the deflection plates.

With 0 V connected to the vertical input terminals the electron beam may be positioned to the vertical center of the screen. If 0 V are also applied to the horizontal input, the beam is then at the center of the CRT face and remains a stationary dot on the face of the CRT. Note that the vertical and horizontal positioning controls allow movement of this dot anywhere on the face of the CRT so that the zero input voltages cause the beam to strike the center of the screen *only* if the positioning controls are also zeroed. Unless otherwise noted in the following discussion, assume that the beam is properly centered and that any deflection off center is due to the input signals to the vertical and horizontal channels of the scope.

Zero input signals result in a dot on the screen (Fig. 15.8a). Any dc voltage applied to either input will result in shifting the dot on the screen with the resulting picture still only a dot. Figure 15.8b shows the resulting display due to a negative dc voltage applied at the vertical input and a positive dc voltage at the horizontal input. The negative voltage to the vertical plates deflects the beam downward by an amount proportional to the voltage magnitude and the voltage on the horizontal plates deflects the beam to the right, the resulting dot appearing in the lower right sector of the screen as shown. The position of the resultant dot can be con-

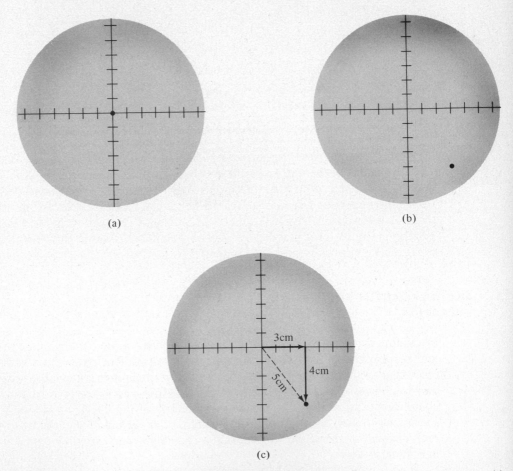

(a)   (b)

(c)

Figure 15.8. Dot on CRT screen due to stationary electron beam: (a) centered dot due to stationary electron beam; (b) off-center stationary dot; (c) stationary dot showing deflection components.

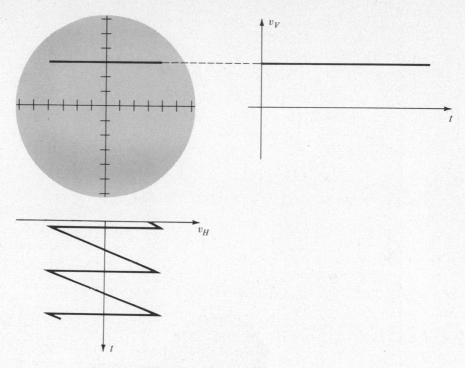

Figure 15.9. Scope display for dc vertical signal and linear horizontal sweep signal.

sidered the vector sum of the two deflection voltages. A positive horizontal deflection of 3 cm and a negative vertical deflection of 4 cm, for example, would result in the beam spot as shown in Fig. 15.8c, at a distance of 5 cm from the screen center.

To view a signal on the CRT face (note that only a dot was visible even though a steady dc voltage was applied) it is necessary to deflect the beam across the CRT with a horizontal sweep signal so that the variations of the vertical signal can be observed. Figure 15.9 shows the resulting straight line display for the positive dc voltage applied to the vertical input using a linear (sawtooth) sweep signal on the horizontal channel. With the electron beam held at a constant vertical distance, the horizontal voltage, going from negative to zero to positive voltage, causes the beam to move from the left side of the tube to the center and over to the right side. The resulting display is a straight line above the vertical center and a dc voltage now properly appears as a steady voltage line. The sweep signal is indicated to be a continuous waveform and not just a single sweep. This is a necessary if a long-term display is to be obtained. A single sweep across the tube would quickly fade out. By repeating the sweep, the display is generated over and over again and if enough sweeps are generated per second, the display always appears to be present. If the sweep rate is slowed down (the time scale controls of the scope allow this), the actual travel of the beam across the face of the tube can be observed.

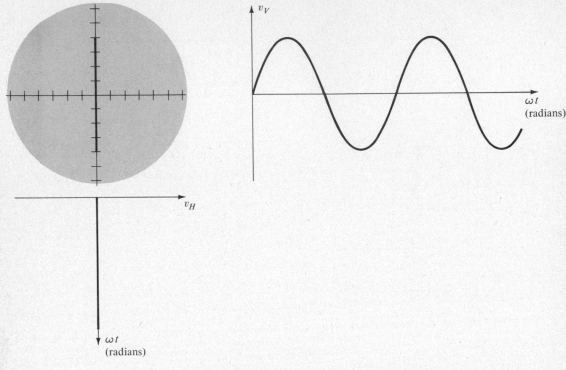

Figure 15.10. Resulting scope display for sinusoidal vertical input and no horizontal input.

Applying only a sinusoidal signal to the vertical input (no horizontal signal) results in a vertical straight line as shown in Fig. 15.10. If the sweep speed (frequency of the sinusoidal signal in this case) is reduced, it will be possible to see the electron beam moving up and down along the straight line path.

To view a sinusoidal signal it is necessary to use a linear sweep signal on the horizontal plates. Then, if the linear sweep is applied to the horizontal plates, the signal viewed on the CRO is that applied to the vertical input. Figure 15.11 shows the resulting CRO display due to a horizontal linear sweep input and a sinusoidal input to the vertical channel. For 1 cycle of the input signal to appear as shown it is necessary that the signal and linear sweep frequencies be synchronized. If there is any difference, the display will appear to move (not be synchronized), unless the sweep frequency is some multiple of the sinusoidal frequency. If, for example, the time for 1 cycle of the sinusoidal signal is 5 msec ($f = 200$ Hz) and that for 1 cycle of the linear sweep is 10 msec ($f = 100$ Hz), the vertical input signal will go through 2 cycles before the sweep has moved the beam across the face of the CRT and the display appears as 2 cycles of the input signal (Fig. 15.11b). Lowering the sweep frequency allows more cycles of the sinusoidal signal to be displayed, whereas increasing the sweep frequency results in less of the sinusoidal vertical input to be displayed, thereby appearing as a *magnification* of a part of the input signal.

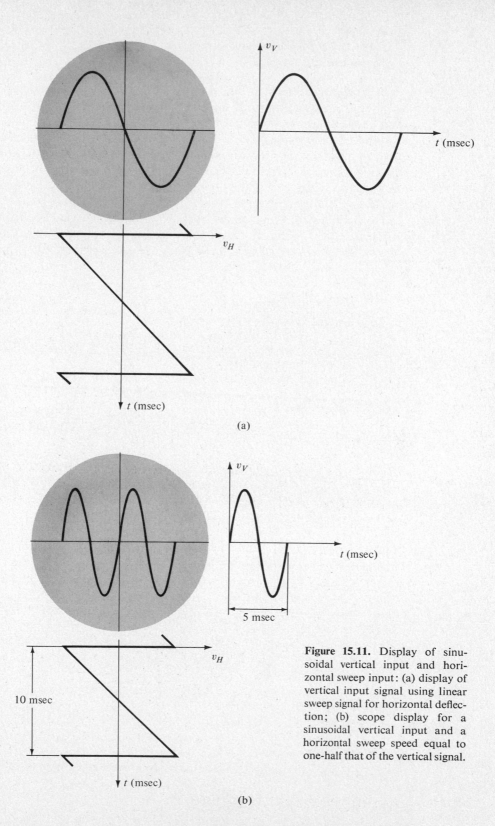

Figure 15.11. Display of sinusoidal vertical input and horizontal sweep input: (a) display of vertical input signal using linear sweep signal for horizontal deflection; (b) scope display for a sinusoidal vertical input and a horizontal sweep speed equal to one-half that of the vertical signal.

(a)

(b)

Use of Linear Sawtooth Sweep to
Display Vertical Input

A signal applied only to the vertical input will cause the electron beam to be deflected only up and down. If the vertical input is a dc voltage, the result will be a dot on the screen displaced from the screen center. If a varying voltage is applied to the vertical input, the beam will move up and down resulting in either the dot moving up and down or the appearance of a line, depending on how fast the input signal repeats a cycle and the persistence of the screen phosphor. Similar action results for an input to only the horizontal channel with resulting displacement to the right or left or a horizontal line.

To obtain a display that shows the form of the input signal applied to the vertical channel it is necessary to apply a linear sawtooth sweep signal to the horizontal channel as well. This is the normal operation of the CRO. The sawtooth

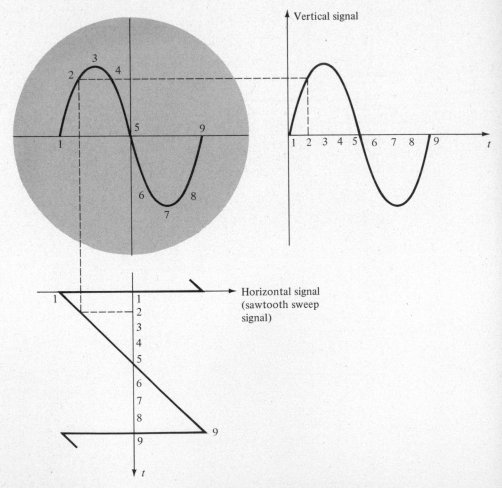

Figure 15.12. Use of linear sweep signal to view input signal.

sweep signal is provided as part of the scope circuitry with adjustment of the sweep rate provided as an external control. To understand the operation of the CRO we must consider how the electron beam is deflected as a result of signals applied to both the vertical and horizontal input at the same time.

To understand how the two deflecting voltages result in the CRT display we shall determine the path the beam takes by observing where it is at a few points during a single sweep of the electron beam (see Fig. 15.12). The voltage applied to the horizontal deflection plates is the sawtooth voltage. For convenience, the horizontal signal time axis is shown with increasing time in the downward vertical direction. The amount of the sweep voltage is shown horizontally with 0 V at the center of the screen display, positive voltage to the right of center, and negative voltage to the left of center. If the beam deflection is considered due to the sweep signal alone, the voltage starting off as negative deflects the beam to the left side of the screen. As the sweep voltage gets less negative the beam will move toward the center. The sweep voltage will pass through zero and go to some positive value at which time the beam will be deflected to the right side of the screen. (Note that the vertical deflection is not being considered for the moment.) When the sweep voltage reaches the largest positive voltage shown (point 9), it very quickly drops back to a negative value and the beam moves back to the starting point on the left of the screen.

The action of the sweep voltage, then, is to move the electron beam across the screen (from left to right) at a constant rate. When we now consider the additional deflection that occurs because of the vertical input signal, we shall obtain the actual display shown in Fig. 15.12. It should be clearly understood that the vertical signal applied alone (without horizontal sweep) results in a straight vertical line on the screen. The linear (horizontal) sawtooth sweep voltage must also be applied to cause the beam to move across the CRT so that a display of the input signal is obtained. Another example of the operation of vertical and horizontal inputs and resulting screen display is the pulse-type signal shown in Fig. 15.13.

The two examples shown in Figs. 15.12 and 15.13 had the same frequencies; that is, the time for a complete cycle of the vertical signal and for a cycle of the sawtooth sweep signal were the same. When this is true, the display is a single cycle of the input signal. The horizontal sweep speed, however, is adjustable and need not be exactly the same as the input signal to be viewed. Figure 15.14 shows the resulting display if the sweep speed is faster than the input signal. In the example shown the time for 1 cycle of the input signal is 4 msec, whereas the time for 1 cycle of the sawtooth signal is only 1 msec. In this case the beam is moved across the tube in 1 msec during which time the vertical input is deflected from zero up to the maximum positive point as shown in the figure. Only a part of the input signal is shown in this case. If the time base of the sawtooth sweep is adjusted even shorter, the amount of the signal displayed will be an even shorter portion. This is an effective *magnification* of the signal, since the full screen width can now display a smaller portion of the input signal. For pulse-type computer signals such magnification is extremely important. It may also be necessary to view more than one full cycle of the input signal. In this case the sweep speed is made slower so that it takes longer for the sweep beam to be deflected once across the screen, thereby allowing

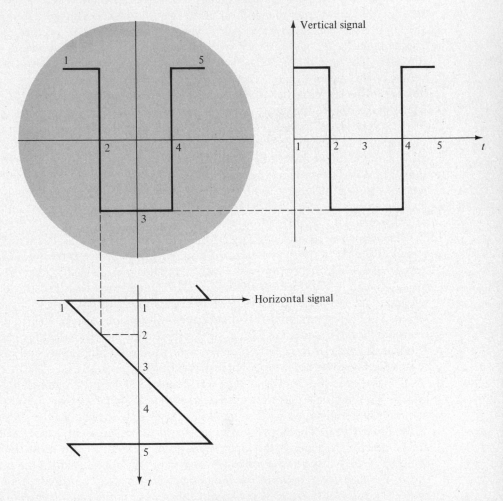

Figure 15.13. Use of the linear sweep for a pulse-type wave form.

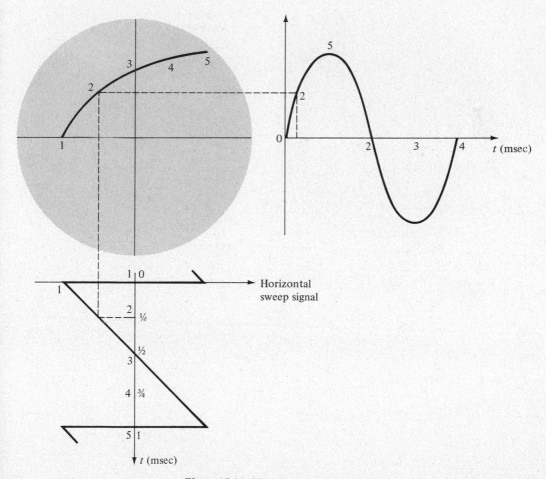

Figure 15.14. Time base at higher frequency than input signal—magnification of displayed signal.

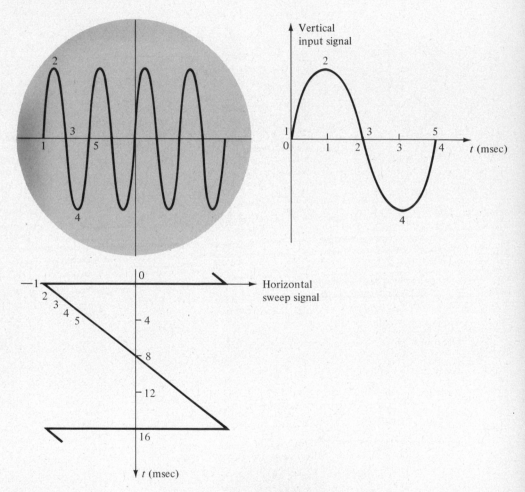

Figure 15.15. Time base slower than input signal—many cycles viewed in one sweep.

a number of cycles of the input signal to be displayed. Figure 15.15 shows the resulting display when the sweep signal takes 16 msec and the input signal only 4 msec for one full cycle. In this case the horizontal sweep speed is 4 times slower and the display shown in 4 cycles of the input signal.

> EXAMPLE 15.4   A sinusoidal waveform at a frequency of 4 kHz is applied to the vertical input of a scope. (a) The horizontal sweep speed is set so that a full cycle takes 0.5 msec. Show the resulting display for one sweep of the beam. (b) Repeat part (a) for a sweep frequency of 8 kHz.
>
> **Solution:**   (a) At a frequency of 4 kHz the time for one full cycle of the input signal is

$$T = \frac{1}{f} = \frac{1}{4 \times 10^3} = 0.25 \times 10^{-3} \text{ sec} = \textbf{0.25 msec}$$

> During the sweep time of 0.50 msec the input signal will go through 2 cycles. (b) At a sweep speed of 2 kHz the time for 1 cycle of sweep is

$$T = \frac{1}{f} = \frac{1}{8} \times 10^3 = \textbf{0.125 msec}$$

> During this amount of time the input signal will only go through 1 half-cycle so that the display will be only 1 half-cycle of the input signal.

15.5 SYNCHRONIZATION AND TRIGGERING

Synchronization

The CRO display can be adjusted by setting the sweep speed to display either a number of cycles, 1 cycle, or part of a cycle. This is a very valuable feature of the CRO and helps make it the useful instrument it is. However, in discussing the sweep of the beam for a single cycle, we have only considered the case in which the input signal and sweep signal frequencies are related. More generally the horizontal sweep frequency setting is not the same, or even proportional to the frequency of the input signal. When this occurs, the display is not synchronized and either appears to drift or is not recognizable.

Figure 15.16 shows the resulting display for a number of cycles of the sweep signal. Each time the horizontal sawtooth voltage goes through a linear sweep cycle (from maximum negative to zero to maximum positive voltage), the electron beam is caused to move once horizontally across the face of the tube. The sawtooth voltage then drops very quickly back to the negative starting voltage and the electron beam is suddenly caused to move back to the left side of the screen. In most CROs the electron beam is *blanked* during this *retrace* of the beam so that no line is shown on the screen. After the very short retrace time the beam begins another sweep across the tube. If the input voltage is not the same each time a new sweep begins, the same display will not be seen each time. To see a steady display it is necessary that the input signal repeat exactly the same pattern for each sweep of the beam. In Fig. 15.17 the sweep signal frequency is too low and the CRO display will have an apparent "drift" to the left. Actually, a different display is

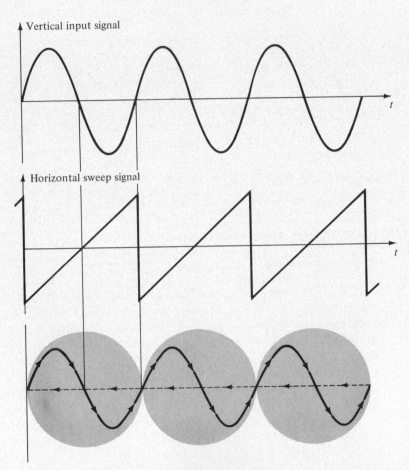

Figure 15.16. Steady scope display—input and sweep signals synchronized.

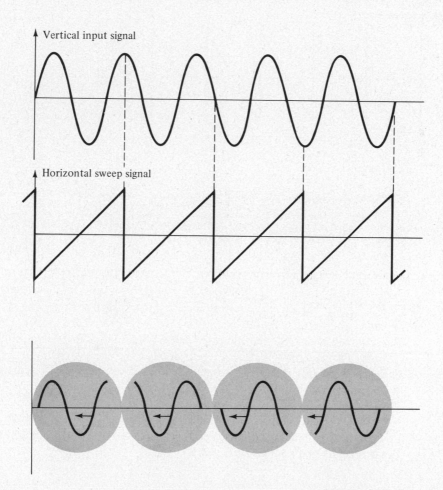

Vertical input signal

Horizontal sweep signal

Figure 15.17. Sweep frequency too *low*—apparent drift to left.

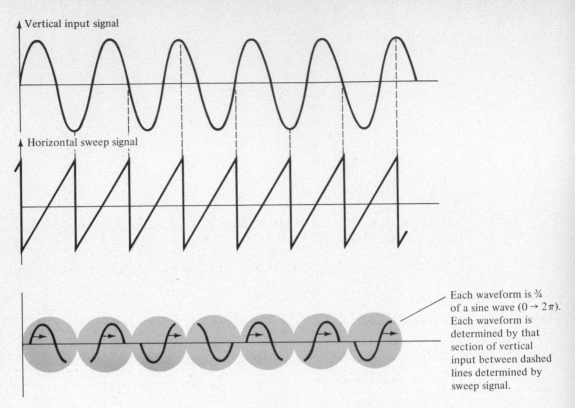

Vertical input signal

Horizontal sweep signal

Each waveform is ¾ of a sine wave ($0 \to 2\pi$). Each waveform is determined by that section of vertical input between dashed lines determined by sweep signal.

Figure 15.18. Sweep frequency too *high*—apparent drift to right.

formed each sweep of the beam as shown in Fig. 15.17. When viewed on the CROs such a display continuously drifts to the left. Observe carefully that each sweep of the beam starts at a different point in the cycle of the input signal and that a different display is formed. Adjustment of the sweep speed to a faster sweep time synchronizes or brings the display to a standstill—assuming that neither the sweep generator frequency nor the input signal frequency changes.

Figure 15.18 shows the result of setting the sweep frequency too high. Less than 1 cycle of the input signal is viewed by each sweep of the beam and the drift in this case appears to be toward the right.

Triggering

The usual method of synchronizing the input signal uses a portion of the input signal to *trigger* the sweep generator so that the rate of the sweep signal is locked or synchronized to the input signal. This is easily done in most CROs using the INTERNAL sync. Figure 15.19 shows that portion of the control panel of a CRO indicating the trigger and sync inputs and controls. We shall refer to these during the following discussion.

When INTERNAL sync is used, a portion of the vertical input signal is taken from some point in the vertical amplifier circuit and fed as the trigger input signal to the synchronizing circuit section of the horizontal sweep generator. When triggered sweep is used, the start of a horizontal linear sweep voltage does not begin immediately after the end of the retrace time (as previously considered) but only when the triggering signal occurs. Thus, the sweep occurs not at a repetitive rate set by the cycle time of the sawtooth signal but by the cycle time of the *triggering* signal. Using the same triggering signal as that viewed on the vertical input achieves the desired synchronization without any necessary sweep speed adjustment. Figure 15.20 shows the operation of a few cycles of the sweep signal and the triggering of the sweep generator. Since the trigger signal shown in Fig. 15.20 occurs at the beginning of each sinusoidal cycle, the display starts only when the input

Figure 15.19. Scope sync and trigger controls.

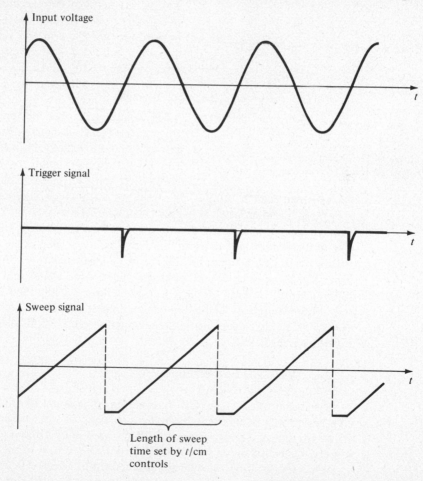

Figure 15.20. Triggered sweep.

voltage, at the beginning of a positive-going cycle, triggers the sweep generator, and a steady picture is thereby obtained.

The exact triggering point at which the sweep begins can be adjusted using front panel controls (see Fig. 15.19). Setting the trigger control to *INT+*, for example, means that the triggering signal to the sweep circuits will be obtained when the vertical input to the CRO has positive slope (voltage getting more positive with time). When the position marked *INT−* is used, the sweep will be started during the time the input signal is going more negative (having negative slope). Figure 15.21 shows the display of the same sinusoidal signal for INT+ and INT− trigger settings, respectively. In addition to these two control settings, the triggering *level* may also be adjusted. Setting the level to zero results in the display's starting when the input signal level crosses 0 V. This level can be adjusted so that the sweep is triggered to start at *any* point of voltage during either the positive slope or negative slope part of the cycle. Using both trigger level and trigger slope on plus (+) or minus (−) allows a wide range of trigger time adjustment for synchronization of a given waveform.

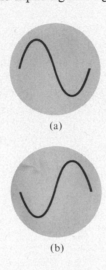

(a)

(b)

Figure 15.21. Use of internal sync modes: (a) trigger set to INT+ ; (b) trigger set to INT−.

The controls shown in Fig. 15.19 also allow two other sync modes of operation. The *LINE* sync provides triggering at the line frequency rate (60 Hz) for measurements of signals derived from the main power line. An equally important sync mode is the *EXT* (external) mode of synchronization. A completely separate signal from that applied to the vertical input can be applied to the input terminals marked EXT. This external input signal is then used to trigger the CRO. In many applications a signal taken from a point in the circuit being tested is used as the external sync signal. Thus, any measurements made using the CRO at any other point in the same circuit will be synced by a signal having the same frequency rate. More important, these other signals may be out of phase with the sync signal and this relative phase difference will be both displayed and *measurable* using the CRO. The use of the scope both to display and allow measurement of phase difference (or more generally of time displacement) between signals is important and will be covered in detail in Section 15.5.

Basic Synchronization and Triggering Circuit

Synchronization of the CRO deflection is obtained by triggering the start of a new sweep cycle using the signal to be displayed (or some other signal derived from it) as the trigger signal. Figure 15.22 shows a simple block diagram of how the trigger signal and sweep signal are related. Either a part of the vertical input signal to be viewed (INT) or a separate external input signal (EXT) is connected as the trigger input signal. The *coupling* to the trigger circuit (ac or dc), the *slope* of the

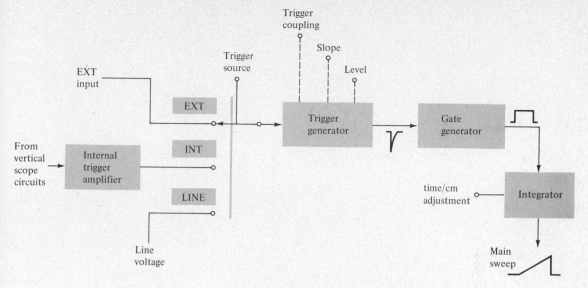

Figure 15.22. Block diagram showing trigger operation of scope.

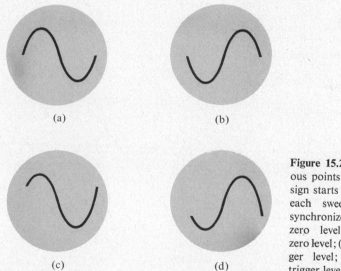

(a)

(b)

(c)

(d)

Figure 15.23. Triggering at various points of signal level (Note: sign starts at same point in cycle each sweep and is therefore synchronized): (a) positive-going zero level; (b) negative-going zero level; (c) positive-voltage trigger level; (d) negative-voltage trigger level.

signal selected (+ or −), and the *level* at which the trigger signal is set to operate are all adjustable using the CRO external controls (see Fig. 15.23). The trigger signal so determined is then used to trigger the gategenerator circuit. This causes the gate circuit to begin a timing operation providing an output pulse, which begins when the trigger pulse is received and ends after the amount of time for a single sweep of the beam, which is determined by the external time base settings (e.g., 1 msec, 10 μsec, etc.). A sweep voltage used to drive the horizontal deflection plates is derived from the pulse gate generator signal using an integrator circuit.

Figure 15.24 shows some detail of a simple sweep circuit to provide an indication of how a linear sweep may be obtained. The RC circuit (Fig. 15.24a) provides

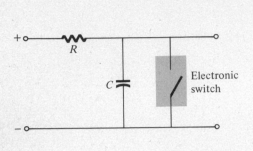

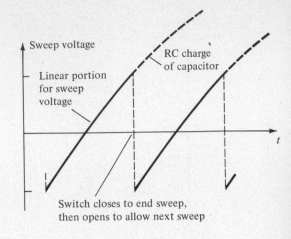

Sweep voltage

RC charge
of capacitor

Linear portion
for sweep
voltage

t

Switch closes to end sweep,
then opens to allow next sweep

(a)

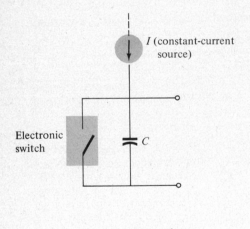

I (constant-current
source)

Electronic
switch

C

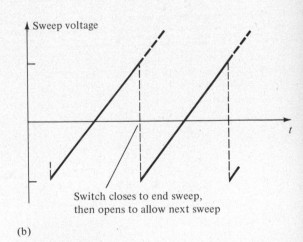

Sweep voltage

t

Switch closes to end sweep,
then opens to allow next sweep

(b)

Figure 15.24. Linear sweep voltage circuits: (a) part of an RC exponential charge used as a linear sweep; (b) use of constant-current charging of a capacitor to obtain a linear sweep voltage.

the integrator operation. As shown, the sweep voltage is obtained using a part of the charge cycle of the capacitor. Although the capacitor charges exponentially at a time rate set by the values of R and C, using only the lower part of the charging voltage waveform provides a fairly linear sweep voltage. Even better linearity is obtained charging a capacitor using a constant-current source, as shown in Fig. 15.24b. Thus, we see that the sweep voltage required to drive the horizontal deflection plates is obtained using some kind of integrator circuit that is triggered by a pulse derived from the input vertical signal to be viewed. This provides the necessary synchronization required to obtain a steady display on the CRT screen.

Multitrace Operation

A useful CRT display is sometimes obtained by using a type of CRO that will provide display of more than one waveform at the same time. This can be done using a multielement CRT gun (more than one electron gun and more than one beam). A CRO using a dual-beam CRT is called a *dual-beam* CRO. It can also be obtained using a single-electron gun and some external electronic switching to obtain the necessary multidisplay on the screen; the CRO is then referred to as *dual-trace*. Two dual-trace features are the ALTERNATE and the CHOPPED modes of display. When these features of a CRO are used, it is possible to connect and "simultaneously" display two separate signals via the CRO. Actually, they are not both displayed at the same time since there is only a single beam. It is possible, however, to switch fast enough so that the illusion of two images appearing simultaneously on the screen is produced.

When the ALTERNATE mode of electronic switching is used, the two input signals to be viewed, connected to channels *A* and *B* of the scope (to differentiate between the two vertical input signals), are connected to the vertical channel of the scope for alternate cycles of the sweep. Thus, on 1 cycle of the sweep the input to channel *A* is connected to the vertical section of the scope and drives the vertical deflection plates. After 1 cycle of sweep is completed, the electronic switching

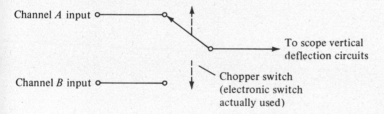

Figure 15.25. Dual-trace obtained using electronic chopping switch.

circuits (see Fig. 15.25) connect the channel *B* input to the vertical section of the scope and on the next sweep of the electron beam the B-channel signal is displayed on the screen. If (as is usually the case) the sweep times are fast enough, the alternate sweep of the beam will trace out a second display *before* the first display has disappeared. Thus, two displays will appear to be present at the same time. Figure 15.26a shows the resulting display for a sine wave and square wave applied as inputs to the CRO when using the ALTERNATE mode of presentation.

When the signals to be viewed are of low frequency, a CHOPPED mode of display is used. The CHOPPED mode of electronic switching switches the vertical amplifier input signal from the input connected at channel *A* to the signal at channel *B* and back again, repetitively—many times—for a single sweep of the beam. Thus, the display obtained is actually comprised of small pieces of each signal with enough of these small pieces to provide the illusion of two steady display signals. Figure 15.26b shows the resulting display of two signals using the CHOPPED mode of electronic switching.

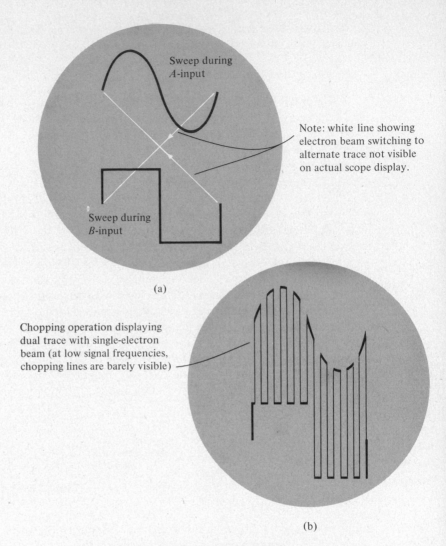

Note: white line showing electron beam switching to alternate trace not visible on actual scope display.

Sweep during *A*-input

Sweep during *B*-input

(a)

Chopping operation displaying dual trace with single-electron beam (at low signal frequencies, chopping lines are barely visible)

(b)

Figure 15.26. ALTERNATE and CHOPPED mode displays for dual-trace operation: (a) ALTERNATE mode for dual-trace using single electron beam; (b) CHOPPED mode for dual-trace using single electron beam.

An important point to keep in mind in regard to the two modes of display is the triggering operation for each mode. In CHOPPED operation a single trigger signal starts each sweep, which then provides two displays on the screen. Whatever phase displacement exists between the two signals will be displayed *exactly*, since the two inputs are being shown, in time, as they occur. The ALTERNATE mode of display, however, may provide some problem. For example, the INTERNAL mode of trigger is used with ALTERNATE sweep and, say, the trigger setting is the zero voltage level with positive slope. When the channel *A* signal crosses 0 V, with positive slope, a trigger signal starts the sweep. The ALTERNATE sweep will be when the channel *B* signal crosses 0 V with positive slope—*whenever* that occurs. The display will thus be that of two in-phase signals and the true phase displacement between the signals is *not* shown. To get around this difficulty, EXTERNAL sync may be used so that each sweep is started by the same input signal and any phase offset from *that* reference signal will be observed. Using one of the inputs as the EXTERNAL sync signal reference provides a display of the two signals in proper phase relation. Some newer multitrace CROs provide additional triggering for ALTERNATE mode display using the same input from, say, channel B as the trigger for initiating the sweep of both input signals. The ALTERNATE B-trigger mode, for example, uses the B input as the trigger for *each* sweep (for both channel-A and channel-B display). Any phase shift displayed between the two signals will then be properly shown.

15.6 MEASUREMENTS USING CALIBRATED CRO SCALES

The CRT face has two display axes—vertical and horizontal. In the normal operation of the CRO the input signal to be observed is applied to the vertical input (either single-channel or dual-channel) and the horizontal sweep is obtained using internal sweep circuitry (see Fig. 15.27). If the vertical amplifiers and attenuators are calibrated (or set to the calibrated settings), then the *amplitude* of the vertical (input) signal (or any part of it) can be accurately *measured*.

If the calibrated horizontal sweep scales of the CRO are used, the amount of time for 1 cycle of the input signal (the *period*) may be measured and used to cal-

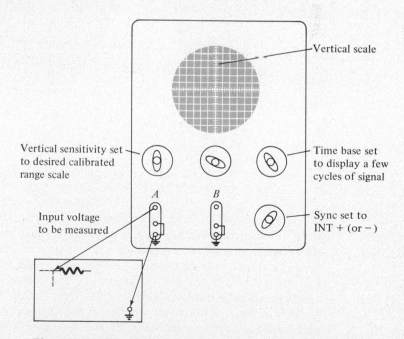

Figure 15.27. Connections to measure voltage amplitude using the scope.

culate the signal frequency. In addition, the amount of time between two sinusoidal signals crossing 0 V can be read and used to calculate the phase shift between the two signals. It is also possible to use the horizontal scales to measure the amount of time that separates the two signals being observed. Measurements of this type provide important information in pulse and digital circuitry.

Amplitude Measurements

The vertical scale of the scope is generally calibrated in units of volts per centimeter (V/cm). Figure 15.28a shows the typical CRT screen of a CRO with the vertical scale marked off in centimeters (cm). (The centimeter is a full box as indicated and the scale is sometimes referred to as volts per box.) Each centimeter or box is further subdivided into five parts so that each minor division mark represents 02. cm.

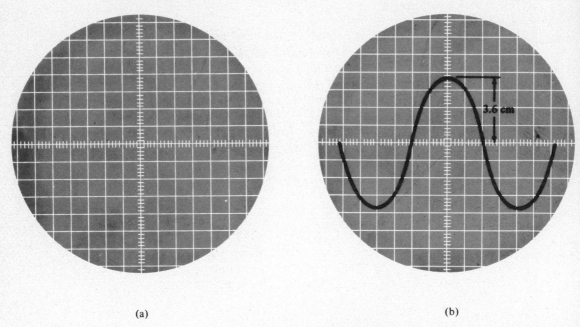

(a)   (b)

Figure 15.28. Scope scale for amplitude measurement: (a) scope scales; (b) measurement of the peak amplitude of a sinusoidal waveform.

As an example of voltage amplitude measurement using the scope calibrated scale, observe the waveform in Fig. 15.28b. The peak amplitude of the sinusoidal waveform shown can be read off the CRT screen as 3.6 cm. If the scale selected by the vertical sensitivity dial setting is 1 V/cm, this reading would represent 3.6 cm × 1 V/cm = 3.6 V. If the vertical sensitivity dial setting were 0.1 V/cm, then the voltage amplitude would be 3.6 cm × 0.1 V/cm = 0.36 V.

The vertical and horizontal *position* of the displayed waveform may be adjusted without affecting the amplitude value. *It is important, however, that the vernier part of the sensitivity dial be set to the calibrated (CAL) position.* In the CRO shown in

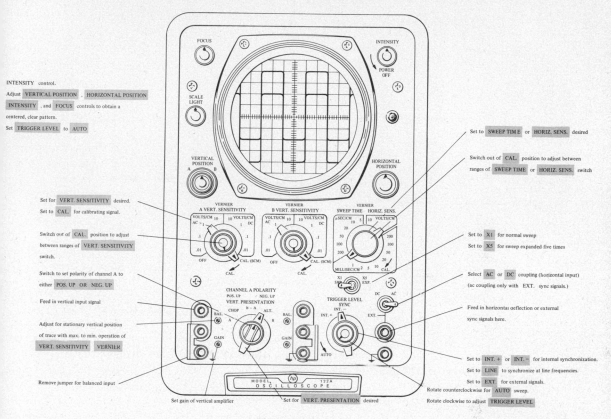

Figure 15.29. HP 122A scope showing front panel controls.

Fig. 15.29 the calibrated position of the vertical sensitivity dial is the fully *clockwise* position of the dial. For the measurement considered above, the peak value of the waveform was measured with respect to the center line of the CRT screen. For this to correspond to the center of the waveform the vertical position of the beam would have had to be centered previously. When centering is *not* desired, the measurement of peak-to-peak amplitude is a more reasonable and accurate choice. The pulse-type waveform shown in Fig. 15.30a has a peak-to-peak amplitude of 4.8 cm so that a dial setting of 1 V/cm would indicate a voltage amplitude (peak-to-peak) of 4.8 V/cm $\times$ 1 V/cm = 4.8 V.

Figure 15.30b shows another pulse-type waveform. To find the amplitude (voltage) difference between the peaks of the two pulses shown the measurement is

$$1\text{st peak} = 3.5 \text{ cm}$$

$$2\text{nd peak} = 2.4 \text{ cm}$$

$$\text{pulse amplitude difference} = 3.5 - 2.4 = 1.1 \text{ cm}$$

If the scale setting were 10 V/cm, this would correspond to a voltage difference of 1.1 cm $\times$ 10 V/cm = 11 V.

SEC. 15.6 MEASUREMENTS USING CALIBRATED CRO SCALES

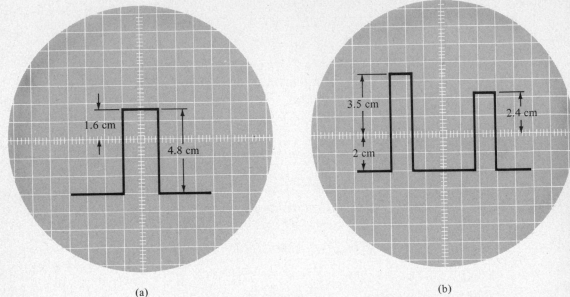

(a)   (b)

Figure 15.30. Measurement of pulse-type waveform amplitudes.

EXAMPLE 15.5 A sinusoidal waveform is observed on a scope as having a peak-to-peak amplitude of 6.4 cm. If the CRO vertical sensitivity setting is 5 V/cm, calculate the peak-to-peak and the rms values of the voltage.

Solution: The peak-to-peak amplitude is

$$6.4 \text{ cm} \times 5 \text{ V/cm} = 32 \text{ V} = V_{p-p}$$

$$V_p = \frac{V_{p-p}}{2} = \frac{32 \text{ V}}{2} = 16 \text{ V}$$

The rms value of voltage is

$$V_{\text{rms}} = 0.707 V_p = 0.707(16) = \mathbf{11.3 \text{ V}}$$

EXAMPLE 15.6 How many centimeters (peak-to-peak) of a sinusoidal waveform should be observed corresponding to an rms voltage of 120 V for a scale setting of 50 V/cm?

Solution: Converting the rms voltage to peak-to-peak, we get

$$2 \times (120 \times 1.414) = 339.4 \text{ V} \cong 340 \text{ V, peak-to-peak}$$

The number of centimeters corresponding to this would be

$$\frac{1 \text{ cm}}{50 \text{ V}} \times 340 \text{ V} = \mathbf{6.8 \text{ cm}}$$

EXAMPLE 15.7 A pulse-type waveform is measured as having a peak amplitude of 15 V. If the vertical dial setting was 10 V/cm, how many centimeters of signal amplitude were observed (peak-to-peak)?

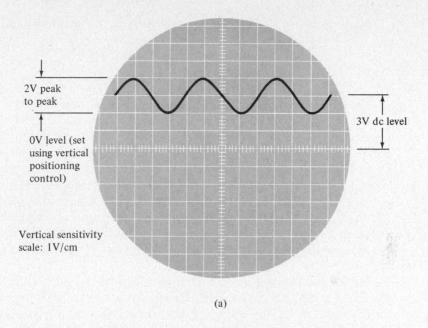

2V peak
to peak

0V level (set
using vertical
positioning
control)

3V dc level

Vertical sensitivity
scale: 1V/cm

(a)

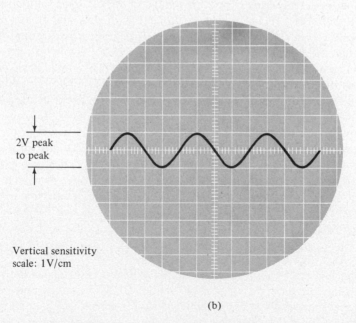

2V peak
to peak

Vertical sensitivity
scale: 1V/cm

(b)

Figure 15.31. Use of dc and ac input modes: (a) dc input mode; (b) ac input mode.

Solution: The number of centimeters for a peak reading of 15 V is

$$15 \text{ V} \times \frac{1 \text{ cm}}{10 \text{ V}} = 1.5 \text{ cm}$$

The peak-to-peak amplitude is 2(1.5 cm) = **3 cm**.

The vertical sensitivity dial shown in Fig. 15.29 indicates separate positions for ac and dc readings. (On many other quality CROs there is one set of sensitivity scales with a separate switch to change from ac to dc operation.) The difference between these two modes of measurement is simple but important in using the scope properly. The *dc input* results in the displayed waveform showing the dc level of the signal being measured. If, for example, the signal to be measured is a 2-V, peak-to-peak sinusoidal voltage riding on a 3-V dc level as shown in Fig. 15.31a (assuming the scope position controls were previously centered), the display indicates the presence of dc. We are then able to measure not only the ac variation of the signal, but also the exact dc levels at all parts of the signal, as shown in Fig. 15.31a.

When only the ac variation is of interest, then the ac scale setting may be used. When the ac input setting is used, the same input signal shown in Fig. 15.31a is displayed in Fig. 15.31b. Notice that the dc level has been removed and only the ac variation is shown. Essentially, the difference between the two is that the ac input scale position couples the signal through a capacitor to eliminate the dc level of the input signal and provide only the ac variation for measurement.

A practical example of the advantage of the ac over the dc setting is in the measurement of a signal having, say, a dc level of 80 V and an ac variation around this level of only 4 V. Displaying this, using the dc scale setting of 20 V/cm, for example, results in the scope display of Fig. 15.32a. Notice how small the ac variation is compared to the large dc level of 80 V. (If a smaller scale setting, of, say, 2 V/cm were used, the ac signal would not be seen at all since a deflection for 80 V would be well off the face of the screen.) Using the ac input scale setting of, say, 1 V/cm provides the display of Fig. 15.32b in which only the ac part of the signal is shown and is explanded to a reasonable viewing and measuring amplitude.

Although the ac scales are good for observing the ac part of a signal having a dc level, the dc scale setting is still important when dc levels must be measured. One additional feature is often found in a good scope—a zero voltage or GROUND position, which connects the input of the vertical amplifier to 0 V without requiring the input signal connection to be removed.

Using Calibrated Sweep for Time Measurements

The horizontal sweep signal can be adjusted in calibrated steps from a few seconds to microseconds of time per centimeter. If the sweep time selector were set at 1 msec/cm, each box (or centimeter) on the screen would correspond to a time of 1 msec. If a pulse signal, such as the one in Fig. 15.33, were observed to have a pulse width of 2.6 cm at a scale setting of 50 μsec/cm, the pulse width could be calculated

$$2.6 \text{ cm} \times 50 \ \mu\text{sec/cm} = 130 \ \mu\text{sec}.$$

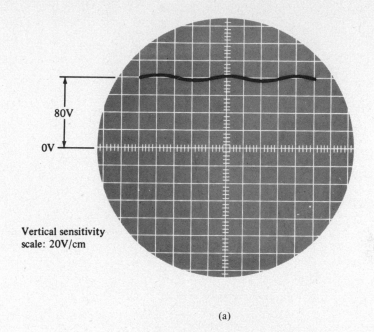

80V

0V

Vertical sensitivity
scale: 20V/cm

(a)

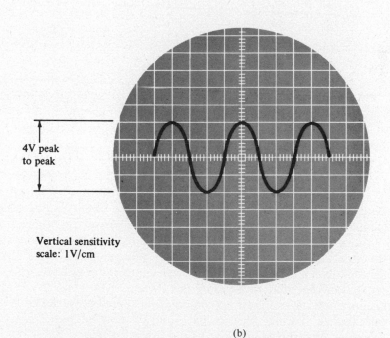

4V peak
to peak

Vertical sensitivity
scale: 1V/cm

(b)

Figure 15.32. Using ac mode to view only signal variation: (a) small ac variation around large dc level shown using dc mode for input; (b) observing *only* ac variation of input signal using ac mode for input.

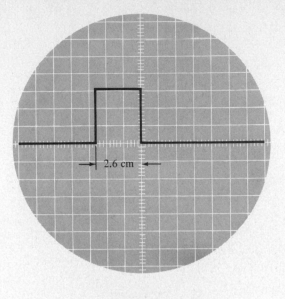

Figure 15.33. Pulse waveform for time measurement.

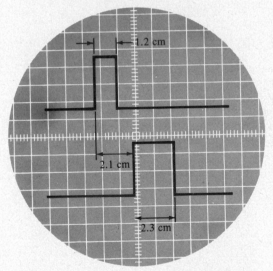

Figure 15.34. Pulse waveforms showing time measurement using calibrated sweep scales.

Thus, the CRO allows display of waveforms of all shapes and permits measurements of time so that all aspects of the signal observed can be measured accurately.

With the dual-trace feature of the CRO two different waveforms can be observed simultaneously and any time differences between parts of the respective signals may be observed and measured. For example, the waveforms of Fig. 15.34 show two pulses viewed at once (using, say, external sync to preserve their proper time displacement). If we use the calibrated horizontal sweep scales, we can measure, for example, the time for each pulse and the time between the two pulses. From the waveforms shown the pulse widths are calculated as

$$1.2 \text{ cm} \times 20 \ \mu\text{sec/cm} = 24 \ \mu\text{sec}$$
$$2.3 \text{ cm} \times 20 \ \mu\text{sec/cm} = 46 \ \mu\text{sec}$$

The time from the start of the first pulse until the start of the second is calculated as

$$2.1 \text{ cm} \times 20 \ \mu\text{sec/cm} = 42 \ \mu\text{sec}$$

Using the scope in this way provides time period or interval information about the two waveforms and their relation to each other.

> **EXAMPLE 15.8** Two pulse-type signals are observed on a CRO using a sweep scale setting of 5 msec/cm. If the pulse widths are measured as 3.5 and 4.2 cm, respectively, and the distance between the start of each pulse is 2.8 cm, calculate the time measurements for the pulse widths and delay between pulses.
>
> **Solution:** Pulse width 1 = 3.5 cm × 5 msec/cm = 16.5 msec
> Pulse width 2 = 4.2 cm × 5 msec/cm = 21 msec
> Delay time = 2.8 cm × 5 msec/cm = 14 msec

> **EXAMPLE 15.9** Two pulses delayed by 15 μsec are observed on an oscilloscope using a time base setting of 10 μsec/cm. Both pulses have pulse widths of 2 μsec. Calculate the readings in centimeters on the scope for the pulse width and delay time.
>
> **Solution:** Pulse delay measurement = 15 μsec/10 μsec/cm = 1.5 cm
> Pulse width measurement = 2 μsec/10 μsec/cm = 0.2 cm

Frequency Measurements Using Calibrated Scope Scales

It is also possible to use the calibrated time scales of the CRO to calculate the *frequency* of the observed signals. This requires using the calibrated horizontal sweep scale to measure the time for 1 cycle of the observed signal and then calculating the signal frequency using the relation

$$f = 1/T \tag{15.4}$$

where f is the signal frequency and T is the period or the time for one full cycle of the signal.

> **EXAMPLE 15.10** A square-wave signal is observed on the scope to have 1 cycle measured as 8 cm at a scale setting of 20 μsec/cm. Calculate the signal frequency.
>
> **Solution:** Calculating, first, the period for 1 cycle of the observed signal:
>
> $$T = 8 \text{ cm} \times 20 \ \mu\text{sec/cm} = 160 \ \mu\text{sec}$$
>
> The frequency is then calculated to be
>
> $$f = \frac{1}{T} = \frac{1}{160} \ \mu\text{sec} = \frac{1}{160} \times 10 = \textbf{6.7 kHz}$$

> **EXAMPLE 15.11** A sinusoidal signal is observed on the scope to repeat 1 cycle in 4.8 cm. If the scale setting was 50 μsec/cm, calculate the frequency of the sinusoidal signal.

Solution:

$$T = 4.8 \text{ cm} \times 50 \ \mu\text{sec/cm} = 240 \ \mu\text{sec}$$

$$f = \frac{1}{T} = \frac{1}{240} \ \mu\text{sec} = \textbf{4.16 kHz}$$

EXAMPLE 15.12   A 500-kHz signal is observed on the CRO. How many centimeters should be observed for one full cycle of the signal if the sweep setting is 1 μsec/cm?

Solution: Calculating first the time for 1 cycle of the signal, we get

$$T = \frac{1}{f} = \frac{1}{500} \text{kHz} = 2 \ \mu\text{sec}$$

Calculating the number of centimeters for a full cycle, we get

$$\text{No. of cm} = 2 \ \mu\text{sec} \times \frac{1 \text{ cm}}{1 \ \mu\text{sec}} = 2 \text{ cm}$$

Phase-Shift Measurements Using Calibrated Scope Scales

The calibrated time scales can also be used to calculate phase shift between two sinusoidal signals (of the same frequency, of course). If a dual-trace or dual-beam CRO is used to display the two sinusoidal signals simultaneously so that one signal is used for the EXT sync input, the two waveforms will appear in proper time perspective and the CRO can be used to measure the amount of time between the start of 1 cycle of each of the waveforms. This amount of time can then be used to

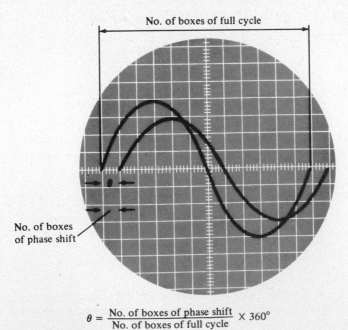

$$\theta = \frac{\text{No. of boxes of phase shift}}{\text{No. of boxes of full cycle}} \times 360°$$

Figure 15.35. Phase shift measurement using horizontal scope scale.

calculate the phase angle between the two signals. Figure 15.35 shows two sinusoidal signals having a phase shift of theta (θ) degrees. We can measure distance in centimeters on the CRO scale and use these readings to obtain θ as follows. The value of the phase angle is related to the degrees in one full cycle of the sinusoidal signal. We can set up a simple relation between these values by equating the number of centimeters or boxes for one full cycle to 360° and the number of centimeters or boxes for the phase shift to the desired phase angle in degrees. The relation is

$$\frac{\text{No. of boxes for one full cycle}}{360°} = \frac{\text{No. of boxes of phase shift}}{\theta} \tag{15.5}$$

where θ is the phase angle (phase shift) in degrees. Using Eq. (15.5), we can calculate the phase angle from

$$\boxed{\theta = \frac{\text{No. of boxes of phase shift}}{\text{No. of boxes of full cycle}} \times 360°} \tag{15.6}$$

Note that the calculation does not involve the actual calibrated time base setting and, in fact, the observed waveform can be varied using the horizontal amplifier vernier adjustment to obtain as many boxes for one full cycle as desired. This will not affect the actual phase shift calculated since the proportionality between the phase shift and one full cycle is preserved at any gain or scale setting used. Adjusting the time base so that, say, 12 boxes (or centimeters) correspond to one full cycle would rescale the measurement so that each box is 30° of phase angle (360°/12). Then, reading 2 boxes of phase shift would quickly convert to 60° phase shift by multiplying the boxes of phase shift by the 30°/box scale factor. If the reading obtained were 1.5 boxes, the phase shift would be 1.5 boxes × 30°/box = 45° phase shift. The relation for calculating phase angle is then

$$\boxed{\theta = \text{scale factor} \times \text{phase distance measured} \atop \text{(in boxes or centimeters)}} \tag{15.7}$$

EXAMPLE 15.13   In measuring phase shift between two sinusoidal signals on a CRO the scale setting is adjusted so that one full cycle is 8 boxes. Calculate the scale factor for this adjustment and the amount of phase shift for a reading of 0.75 boxes.

Solution: Setting one full cycle to 8 boxes results in a scale factor of 360°/8 boxes = 45°/box.

$$\theta = \text{scale factor} \times \text{phase distance measured}$$

$$= \frac{45°}{\text{box}} \times 0.75 \text{ boxes} = \mathbf{33.85°}$$

EXAMPLE 15.14   One full cycle is set to 9 boxes. The phase displacement is measured as 0.4 boxes. Calculate (a) the scale factor and phase shift in degrees and (b) the number of boxes to observe for a phase shift of 60°.

SEC. 15.6 MEASUREMENTS USING CALIBRATED CRO SCALES

659

Solution: (a) The scale factor set is

$$\text{scale factor} = \frac{360°}{9 \text{ boxes}} = \mathbf{\frac{40°}{box}}$$

$$\text{phase shift} = 0.4 \text{ boxes} \times \frac{40°}{box} = \mathbf{16°}$$

(b) No. of boxes for 60° phase shift $= \dfrac{60°}{40°/box} = \mathbf{1.5 \text{ boxes}}$

15.7 USE OF LISSAJOUS FIGURES FOR PHASE AND FREQUENCY MEASUREMENTS

Another method for measuring either phase shift between two sinusoidal signals or the frequency of an unknown sinusoidal signal is the use of Lissajous figures. The technique can be applied to a single-channel CRO and does not require the fine calibration scales previously considered. Basically, the two signals under study (to determine the phase shift between the two signals) are connected as vertical and horizontal inputs to the CRO. The usual (internal) horizontal sweep signal is *not* used at this time. A pattern or Lissajous figure is developed on the CRT and is used to determine the amount of the phase shift or frequency of the unknown signal.

Lissajous figure techniques for measurement are more popular for low-quality CROs and for single-trace scopes (where two inputs cannot be compared at one time). Although not as popular a measurement technique as those considered in Section 15.6, the use of Lissajous figures is still interesting and sometimes helpful.

Use of Lissajous Figures to Calculate Phase Shift

Lissajous figures are obtained on the scope by applying the two sinusoidal inputs to be compared to the vertical and horizontal channels of the oscilloscope. The value of the phase shift is then calculated from measured values taken from the resulting Lissajous pattern.

Figure 15.36 shows a circuit arrangement with two different sinusoidal signals connected to the vertical and horizontal inputs of the CRO, respectively. The resulting pattern on the CRT face is a straight line, a circle, or an ellipse. To understand the relationship between the resulting figure and the applied inputs we shall investigate a variety of signals having different phase angles and determine the resulting Lissajous pattern.

The procedure for measuring phase difference between two sinusoidal signals is simply to apply the two signals to the vertical and horizontal inputs and then make two measurements from the resulting display. The usual horizontal linear sweep is

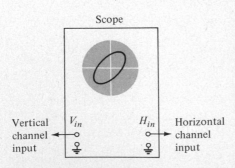

Figure 15.36. Signal input connection for Lissajous figure measurement of phase angle.

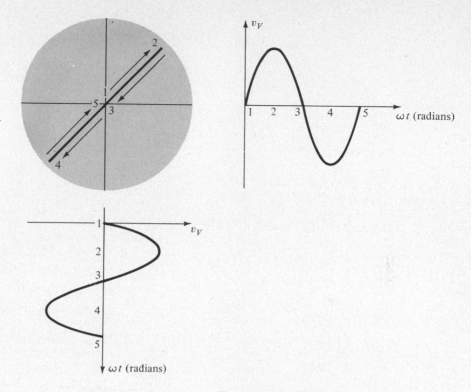

Figure 15.37. Lissajous figure for 0° phase shift.

not used at all for this procedure. Note that the two signals must be the same frequency (otherwise the parameter phase angle is meaningless). Figure 15.37 shows the resulting display when the *same* signal is applied to *both* channels (0° phase shift). The overall display is that of a straight line. In the example the line is at 45° slope for equal-amplitude signals. If the signals were in phase but not of equal amplitude, the line would have slope other than 45°, but the important factor here, the fact that it is a straight line, indicates that the phase shift is zero.

To see more clearly how the straight line is developed on the CRT screen let us break the sinusoidal cycle into, say, quarter-cycles and follow the resulting beam deflection due to the two sinusoidal signals applied to the vertical and horizontal inputs. Figure 15.37 shows the two input signals and the resulting display. If the signal frequency is very low (a few hertz), the beam movement can be observed on the tube. If the signal frequency is high enough, the beam will travel up and back so fast that the display will appear as a solid line. The appearance of a straight line when measuring phase angle between two sinusoidal signals can be directly interpreted as no phase shift (0° phase angle between signals).

Figure 15.38 shows the resulting waveform for two inputs having a phase shift between 0° and 90°. The resulting pattern is an ellipse (at 45° if the two amplitudes are the same). The angle at which the ellipse is generated is of no importance for the phase angle calculation. Noting that the vertical signal amplitude at time 1 is $V = V_m \sin \theta$, we can calculate the angle θ from

$$\theta = \sin^{-1}(V/V_m) \tag{15.8}$$

SEC. 15.7 USE OF LISSAJOUS FIGURES FOR MEASUREMENTS

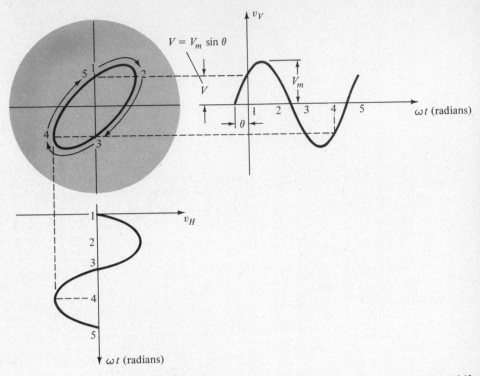

Figure 15.38. Lissajous figures for θ between 0 and 90° phase shift.

The values of V and V_m can be easily obtained from the ellipse by measuring the distance (amplitude) of the signal from the center line to where it crosses the center vertical axis, and V_m as the distance from the vertical center line to the top of the ellipse. Using these values in the above relation, we can calculate the phase angle θ.

Since the measurements of V and V_m will be used as a ratio, the actual size or values are not important—only their ratio. This being the case, the actual scale settings of the input signals are unimportant and the CRO adjustments may be used to get an ellipse on the CRT face of about maximum possible size for greatest accuracy. Once the deflection controls are properly centered by adjusting the beam to the center of the tube with no input signals, the two measurements marked A and B in Fig. 15.39 are read and the angle θ is calculated from

$$\theta = \sin^{-1}\left(\frac{A}{B}\right) \tag{15.9}$$

If, for example, the distance B is set to 10 boxes (for whatever scale setting and fine adjustments are necessary), the number of boxes of A can be read and the ratio of A/B obtained. The actual voltages are not important, only their ratio as obtained by the number of scale divisions or boxes on the CRT face.

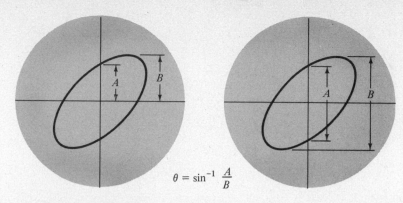

$$\theta = \sin^{-1}\frac{A}{B}$$

Figure 15.39. Calculation of phase shift from Lissajous figure for θ between 0 and 90°.

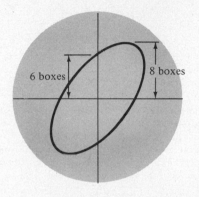

6 boxes   8 boxes

Figure 15.40. Lissajous figure for Example 15.15.

EXAMPLE 15.15 Calculate the phase shift θ for the Lissajous figure in Fig. 15.40.

Solution:

$$\theta = \sin^{-1}\frac{A}{B} = \sin^{-1}\frac{6}{8} = \sin^{-1} 0.75 = \mathbf{49°}$$

If the two signals are out of phase by exactly 90°, the resulting waveform is a circle. This result also follows from the above calculation, since for a circle the measured values A and B are equal and the value of θ calculated is $\theta = \sin^{-1}(1) = 90°$. The relation for calculating θ would also show that for the straight line the measured value of $A = 0$ gives $0/B = 0$, and $\theta = \sin^{-1}(0) = 0°$. Thus, in summary, the values A and B can be measured as indicated above and used to calculate the phase angle within the range of 0 and 90°.

For phase angles of 90–180° the ellipse has a negative slope, as in Fig. 15.41, and the angle calculated by the above method must be subtracted from 180° to obtain the phase shift. Phase angles above 180° result in Lissajous figures such as those below 180°, and they cannot be directly distinguished. One technique for determining if the measured angle is less or more than 180° is to add an extra

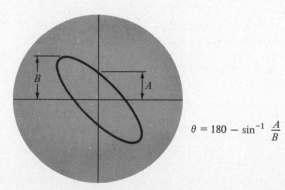

$$\theta = 180 - \sin^{-1}\frac{A}{B}$$

Figure 15.41. Calculation of phase shift from Lissajous figure for θ from 90 to 180°.

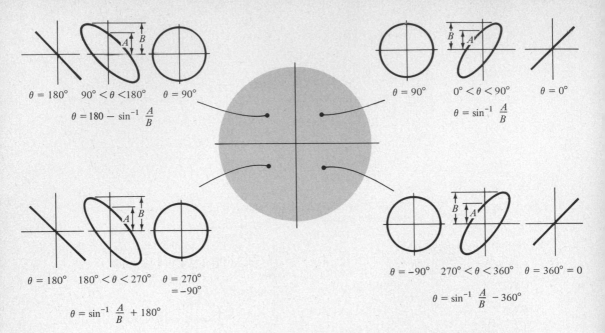

Figure 15.42. Lissajous phase angle calculation in all quadrants.
Note: Additional test adding phase angle shift to signal to be measured is necessary to determine whether observed figure is for upper two quadrants or lower two quadrants. If added phase shift causes positive-sloped ellipse to become larger or negative-sloped ellipse smaller, upper quadrants are indicated; otherwise, lower quadrants.

(slight) phase shift to the signal being measured. If the phase angle measured increases, the angle was less than 180°. If it decreases, the angle was greater than 180°, and the correct angle is then calculated by adding 180° for the angle computed with negative-sloped ellipse. A comprehensive summary, which clearly shows the required methods to compute the phase angle, is shown in Fig. 15.42.

EXAMPLE 15.16   Calculate the phase angle (between 0° and 180°) for the following Lissajous figures shown in Fig. 15.43.

Solution:
(a) $\theta = \sin^{-1}(A/B) = \sin^{-1}(3/5) = \mathbf{37°}$
(b) $\theta = \sin^{-1}(A/B) = \sin^{-1}(6/6) = \mathbf{90°}$ (could have been determined by inspection—circle indicates phase shift of 90°)
(c) $\theta = 180° - \sin^{-1}(A/B) = 180° - \sin^{-1}(4/5) = 180° - 53° = \mathbf{127°}$
(d) by inspection, $\theta = \mathbf{180°}$

Use of Lissajous Figures for Frequency Measurements

If a well-calibrated CRO time base is not available, a signal generator can be used to measure the frequency of an unknown sinusoidal signal. Figure 15.44a shows the equipment setup to perform the measurement. The unknown signal is

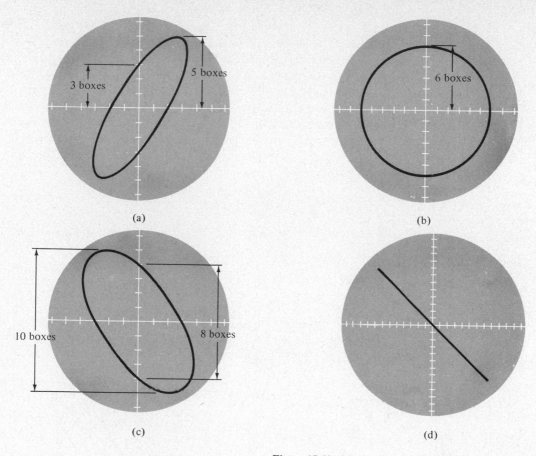

(a)

(b)

(c)

(d)

Figure 15.43. Lissajous figures for Example 15.16.

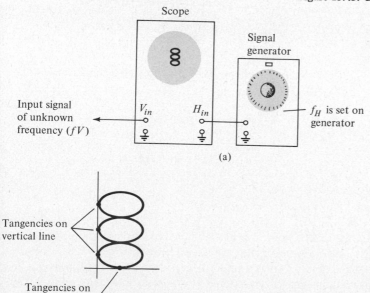

Scope

Signal
generator

Input signal
of unknown
frequency (fV)

V_{in} H_{in}

f_H is set on
generator

(a)

Tangencies on
vertical line

Tangencies on
horizontal line

(b)

Figure 15.44. Frequency calculation using Lissajous figure: (a)
equipment setup for frequency
measurement; (b) Lissajous figure for frequency calculation.

connected to the vertical channel (it could have been fed to the horizontal as well) and the calibrated signal source input is fed to the horizontal channel. The frequency of the signal generator is adjusted until a steady Lissajous pattern is obtained. A sample figure resulting from this procedure appears in Fig. 15.44b. The Lissajous pattern can become very interesting and involved to analyze. However, for the frequency measurement, all that is needed is the number of tangencies (points at the edge of the arcs) along a vertical and horizontal line at the side of the figure as shown in Fig. 15.44b. The frequency relation between horizontal and vertical inputs is given by

$$\frac{f_H}{f_V} = \frac{\text{No. of tangencies vertical}}{\text{No. of tangencies horizontal}} \qquad (15.10)$$

from which we obtain the relation used to calculate the unknown input signal:

$$\boxed{f_{\text{unknown}} = f_V = \frac{\text{No. of tangencies horizontal}}{\text{No. of tangencies vertical}} \times f_H} \qquad (15.11)$$

EXAMPLE 15.17 Calculate the frequency of the unknown signal applied to the vertical input for the Lissajous figures shown in Fig. 15.45. The frequency indicated on the figure is that of the known horizontal input.

Figure 15.45. Lissajous figure for Example 15.17.

Solution:

(a) $f_u = \frac{2}{1}(1000) = 2000$ Hz

(b) $f_u = \frac{4}{1}(600 \text{ kHz}) = 2.4$ MHz

(c) $f_u = \frac{5}{3}(50 \text{ kHz}) = 83.3$ kHz

15.8 SPECIAL CRO FEATURES

The CRO is becoming increasingly more sophisticated and specialized in use. Whereas CROs were originally general in range and usage, modern CROs can be geared specifically to the field or area of interest providing those measurements of importance to a particular area of electronics. One important feature of many

Figure 15.46. Plug-in scope attachments.

CROs is the use of *plug-ins*. Rather than manufacture a single integral unit with certain features of interest to a particular area or range of operation, the CRO is manufactured with only the power supply and deflection circuitry as integral to the unit. Figure 15.46 shows a few typical plug-in units and their respective scope main frames (the main frame is the basic CRO body containing power supplies, CRT, and CRT deflection circuitry). Plug-in units are available for both the vertical and horizontal sections of the CRO. These plug-in units may be selected to have features such as the following: single input ranging from 0.005 V/cm to 20 V/cm scale sensitivity; dual-trace capability with two vertical inputs and selection of channel *A* only, channel *B* only, CHOPPED, and ALTERNATE modes of display; differential input for two separate signals; dc-50-MHz frequency range; and dc input from 100 μV to 20 V. Time base plug-ins might provide single time base from 1 μsec to 1 sec/cm selection; two time bases allowing delayed and mixed sweep modes of operation; time base with sampling operation allowing viewing down to a few nanoseconds, etc.

A useful CRO feature uses two time bases to provide a selection of a small part of the signal viewed allowing expanded presentation of only that selected part of the signal. Figure 15.47 shows a digital-type signal containing a number of separate

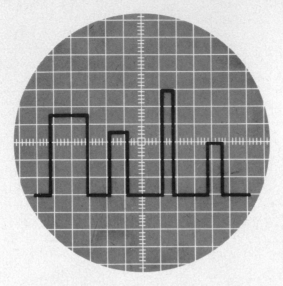

Figure 15.47. Digital-type pulse viewed on scope.

pulses. A delayed type of sweep presentation would be necessary if it were desired to view, say, the third pulse in Fig. 15.47. If the usual single time base were expanded by changing the sweep generator rate so that a shorter period of time were viewed on the screen, this would only magnify the signal shown in Fig. 15.47 from the left (start of the sweep) out; that is, the display would be expanded for the part of the signal starting at the beginning of the sweep, but the part of the signal over to the right would be out of the screen area for this more detailed sweep setting. The delay sweep feature to be discussed allows selecting a part of the displayed presentation and then changing the display to show only a selected portion at whatever magnification is desired.

The main time base is referred to as the A time base and is the horizontal sweep, which provides the picture as shown in Fig. 15.47. An additional time base sweep generator is also provided (called B time base) when delayed sweep operation is available. A basic description of this delayed sweep operation is shown in Fig. 15.48. The pulse-type signal shown in Fig. 15.48a indicates a selected part of that signal by the more intensified display. This part of the signal is then shown in Fig. 15.48b in a more detailed (more magnified) presentation. The use of a delayed sweep time base allowed this selection. The detailed presentation in the block diagram of Fig. 15.48c shows how the circuitry is connected to accomplish this operation. Note in Fig. 15.48c that the main and delayed sweep circuitry are two approximately identical units. Their connection in the overall CRO operation is what differentiates them. If the front panel control is set to main sweep, only the main sweep generator is activated and the input trigger signal is connected to the main sweep trigger circuit, resulting in the main sweep waveform being used as the horizontal deflection signal. When delayed sweep operation is set up, an additional sweep generator is activated—this is the delayed sweep circuit. For operational ease, the CRO is set up so that the sweep of the delayed generator can be added to that of the main generator to provide the intensified display shown in

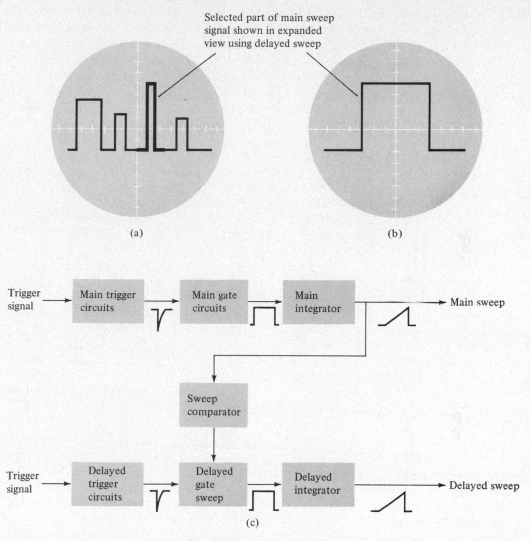

Figure 15.48. Operation of delayed sweep: (a) main sweep presentation; (b) delayed sweep presentation; (c) delayed sweep operation—block diagram.

Fig. 15.48a. In effect, then, a potentiometer adjustment is provided on the front panel controls to set the sweep comparator of Fig. 15.48c so that at some selected time in the main sweep the delayed sweep circuitry will be activated (triggered). The amount of time that the delayed sweep is generated is independently set by the delayed sweep time base controls on the scope front panel. This time setting is always less than that of the main sweep. Thus, an intensified presentation is obtained with some adjustment for when the intensified part of the display begins and for how long it lasts.

The CRO mode of operation can now be changed from main time base to delayed time base operation. When this is done, the displayed signal is then only the portion previously shown as the intensified scetion. What has happened is that the main sweep no longer drives the horizontal deflection circuitry—but it still is used to trigger the delayed sweep as before. Thus, the delayed sweep does not start until the delayed time interval previously seen on the screen as the start of the intensified display. When this time in the signal operation occurs, the delayed sweep now drives the horizontal deflection circuitry and a display (at the delayed generator sweep rate) is provided. This display, as shown in Fig. 15.48b, shows only the previously intensified part of the signal and shows it at the expanded display setting of the delayed sweep generator.

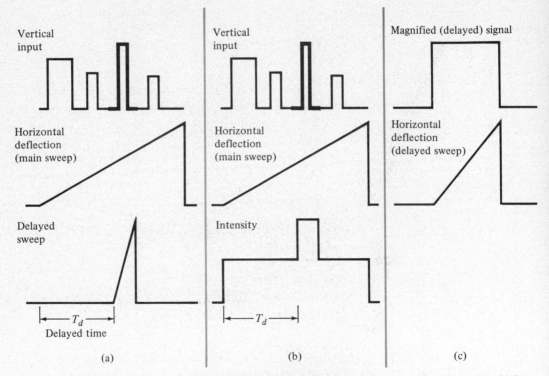

Figure 15.49. Main and delay sweep displays: (a) main sweep; (b) intensified main sweep; (c) delayed sweep.

Figure 15.49 shows the sweep waveforms for the main and delayed circuitry to show the resulting operation as described. As shown in Fig. 15.49a, the amount of the vertical input signal displayed is set by the main sweep horizontal time base. The delayed sweep time base is adjusted for some faster sweep rate providing less sweep time for a cycle, the start of the delayed sweep being set at some amount of time after the start of the main sweep. With the controls set to main sweep, the 3 pulses shown in Fig. 15.49a are displayed. Figure 15.49b shows the intensified main sweep and the intensity signal, which controls the signal intensity seen on the scope screen. (This intensity signal drives the control grid to vary the number of electrons

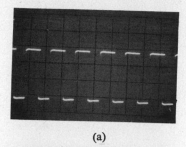

(a)

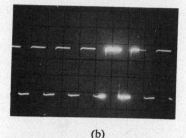

(b)

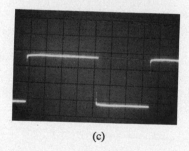

(c)

Figure 15.50. Actual main and delayed sweep scope displays: (a) main sweep; (b) intensified main sweep; (c) delayed sweep.

in the electron beam.) The intensity is kept at a normally lower level and set to a higher level during the delay time interval. The display seen then still shows the 3 pulses selected by the main sweep and additionally the intensified part of the sweep as set by the delayed sweep circuitry.

Finally, when the controls are switched to delayed sweep operation, the horizontal sweep is taken from the delayed sweep generator and the previously intensified presentation is now the complete screen display. It must be kept in mind that even when only the delayed sweep is operating the screen display, the signal seen is tied to that originally displayed by the main sweep and the delayed time is also based on the start of the main sweep.

To summarize, the use of the delayed sweep additional to the main sweep allows selection of a part of any displayed signal with complete and flexible control in displaying only the selected part of the signal at whatever shorter time display interval desired. Figure 15.50 shows actual CRO displays of main sweep, main sweep intensified, and, finally, delayed sweep.

Another special feature found in CROs having the delayed sweep operation is *mixed sweep*. This is nothing more than a mixing of the main sweep and delayed sweep signals at one time on the screen. The adjustment of the delayed trigger time sets the point at which the sweep changes from the main sweep rate to the delayed sweep rate as shown in Fig. 15.51.

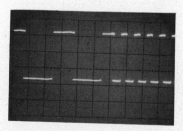

Figure 15.51. Actual mixed sweep display.

PROBLEMS

§ 15.2

1. A scope CRT has a rated deflection factor of 50 V/in. How much deflection is obtained on the CRT screen for plate deflection voltages of (a) $+40$ V, (b) -75 V?

2. What is the vertical deflection sensitivity of a CRT that has a screen deflection of 10 mm when a voltage of 50 V is applied to the vertical deflection plates?

3. A plate deflection voltage of 80 V results in a screen deflection of 2 cm. What is the tube deflection factor and how much deflection would result if 120 V were applied to the deflection plates?

4. A CRT having 0.3 mm/V sensitivity is used. How much deflection, in inches, results from a plate deflection voltage of 150 V?

5. A CRT with accelerating potential of 10,000 V has a deflection sensitivity of 0.45 mm/V. If an applied voltage causes a deflection of 6 cm, how much deflection would result with an accelerating potential of 15,000 V? What is the amount of the applied voltage?

§ 15.4

6. A 1-kHz sinusoidal signal is fed to the vertical input of an oscilloscope. Draw the scope presentation for the following horizontal time base sweep frequencies (assume sweep triggered on positive-going slope at 0-V level): (a) 1 kHz, (b) 2kHz, (c) 500 Hz.

7. (a) A 50-kHz square wave is fed into the vertical input of a CRO. If the horizontal sweep speed is set to 2 μsec/cm, draw the CRO display for a field of 10 cm on the CRT.
   (b) Repeat for a sweep speed of 4 μsec/cm.
   (c) Repeat for a sweep speed of 1 μsec/cm.

8. What is the CRO horizontal sweep frequency if 4 cycles of a 10-kHz signal are viewed?

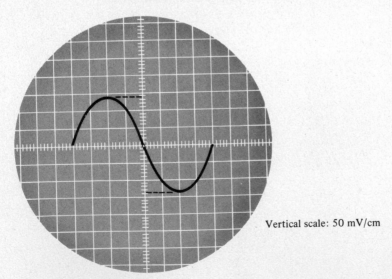

Vertical scale: 50 mV/cm

Figure 15.52. Waveform for Problem 15.11.

9. A sinusoidal signal is observed on a CRO as having a peak amplitude of 4.1 cm. If the scope vertical gain setting is 0.5 V/cm, calculate the peak and rms values of the input voltage.

10. Draw the CRO display for a 4-V rms sinusoidal waveform for a vertical scale of 2 V/cm. Indicate vertical axis and scale markings clerly.

11. What is the peak-to-peak and rms voltage for the sinusoidal waveform of Fig. 15.52?

12. A square-wave signal measured on a CRO has a peak amplitude of 650 mV. If the CRO scale setting was 200 mV/cm, how many centimeters of signal amplitude were observed (peak-to-peak)?

13. A pulse-type signal is observed on a CRO to have a width of 6.4 cm. Calculate the pulse width time for the following sweep scale setting: (a) 5 msec/cm, (b) 100 μsec/cm, (c) 2 μsec/cm.

14. Two pulse-type signals are observed on a CRO at a scale setting of 20 μsec/cm. If the pulse widths are measured as 1.8 and 3.2 cm, respectively, and both start at the same time, calculate the time width of each pulse and the time delay between the end of the pulses.

15. A sinusoidal signal observed on a CRO repeats 1 cycle in 6.3 cm. If the scale setting was 5 μsec/cm, calculate the signal frequency.

16. Calculate the signal frequency of a square-wave signal having a width for 1 half-cycle of 10.5 cm at a scale setting of 10 μsec/cm.

17. A 400-Hz signal is observed on a CRO. How many centimeters should be observed on a CRO? How many centimeters should be observed for 3 cycles of the signal if the scale setting is 1 msec/cm?

18. For the CRO display of Fig. 15.53 calculate the following: (a) peak-to-peak voltage (V_{p-p}) and V_{rms}, (b) time for one complete cycle (T), (c) frequency of waveform signal (f).

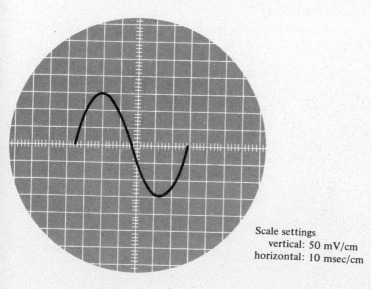

Scale settings
vertical: 50 mV/cm
horizontal: 10 msec/cm

Figure 15.53. Waveform for Problem 15.18.

19. For the CRO display of Fig. 15.54 calculate the following: (a) V_{p-p}, (b) time for 2 cycles (T), (c) pulse repetition rate (f).

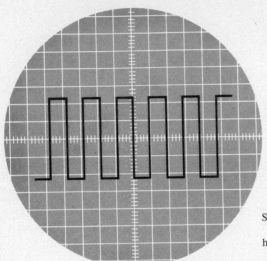

Scale settings –
vertical: $100\,\mu V/cm$
horizontal: $50\,\mu sec/cm$

Figure 15.54. Waveform for Problem 15.19.

20. The CRO scale is adjusted so that a sinusoidal signal takes 6 cm for one full cycle. Calculate the scale factor for this adjustment and the amount of phase shift for a reading of 1.5 cm.

21. (a) A full cycle is set to 8 cm. The phase displacement is measured as 0.5 cm. (a) Calculate the scale factor and phase shift in degrees. (b) Calculate the number of centimeters observed for a phase shift of 40°.

§ 15.7

22. Draw the Lissajous figures for the following phase angles: (a) $\theta = 180°$, (b) $\theta = -90°$, (c) $\theta = 45°$, (d) $\theta = 135°$

23. Calculate the phase angle for the Lissajous figures in Fig. 15.55. (Assume that phase shift is less than 180°).

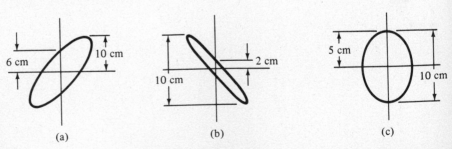

(a) (b) (c)

Figure 15.55. Lissajous figures for Problem 15.23.

24. To measure the frequency of an unknown sinusoidal signal a calibrated signal at 10 kHz is connected to the vertical input. Calculate the unknown frequency (f_u) for the Lissajous figures in Fig. 15.56.

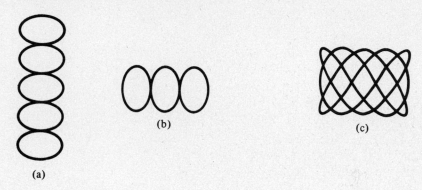

(b)

(c)

(a)

Figure 15.56. Lissajous waveforms for Problem 15.24.

hybrid parameters—conversion equations (exact and approximate)

A

A.1 EXACT

Common-Emitter Configuration

$$h_{ie} = \frac{h_{ib}}{(1 + h_{fb})(1 - h_{rb}) + h_{ob}h_{ib}} = h_{ic}$$

$$h_{re} = \frac{h_{ib}h_{ob} - h_{rb}(1 + h_{fb})}{(1 + h_{fb})(1 - h_{rb}) + h_{ob}h_{ib}} = 1 - h_{rc}$$

$$h_{fe} = \frac{-h_{fb}(1 - h_{rb}) - h_{ob}h_{ib}}{(1 + h_{fb})(1 - h_{rb}) + h_{ob}h_{ib}} = -(1 + h_{fc})$$

$$h_{oe} = \frac{h_{ob}}{(1 + h_{fb})(1 - h_{rb}) + h_{ob}h_{ib}} = h_{oc}$$

Common-Base Configuration

$$h_{ib} = \frac{h_{ie}}{(1 + h_{fe})(1 - h_{re}) + h_{ie}h_{oe}} = \frac{h_{ic}}{h_{ic}h_{oc} - h_{fc}h_{rc}}$$

$$h_{rb} = \frac{h_{ie}h_{oe} - h_{re}(1 + h_{fe})}{(1 + h_{fe})(1 - h_{re}) + h_{ie}h_{oe}} = \frac{h_{fc}(1 - h_{rc}) + h_{ic}h_{oc}}{h_{ic}h_{oc} - h_{fc}h_{rc}}$$

$$h_{fb} = \frac{-h_{fe}(1 - h_{re}) - h_{ie}h_{oe}}{(1 + h_{fe})(1 - h_{re}) + h_{ie}h_{oe}} = \frac{h_{rc}(1 + h_{fc}) - h_{ic}h_{oc}}{h_{ic}h_{oc} - h_{fc}h_{rc}}$$

$$h_{ob} = \frac{h_{oe}}{(1 + h_{fe})(1 - h_{re}) + h_{ie}h_{oe}} = \frac{h_{oc}}{h_{ic}h_{oc} - h_{fc}h_{rc}}$$

Common-Collector Configuration

$$h_{ic} = \frac{h_{ib}}{(1 + h_{fb})(1 - h_{rb}) + h_{ob}h_{ib}} = h_{ie}$$

$$h_{rc} = \frac{1 + h_{fb}}{(1 + h_{fb})(1 - h_{rb}) + h_{ob}h_{ib}} = 1 - h_{re}$$

$$h_{fc} = \frac{h_{rb} - 1}{(1 + h_{fb})(1 - h_{rb}) + h_{ob}h_{ib}} = -(1 + h_{fe})$$

$$h_{oc} = \frac{h_{ob}}{(1 + h_{fb})(1 - h_{rb}) + h_{ob}h_{ib}} = h_{oe}$$

A.2 APPROXIMATE

Common-Emitter Configuration

$$h_{ie} \cong \frac{h_{ib}}{1 + h_{fb}}$$

$$h_{re} \cong \frac{h_{ib}h_{ob}}{1 + h_{fb}} - h_{rb}$$

$$h_{fe} \cong \frac{-h_{fb}}{1 + h_{fb}}$$

$$h_{oe} \cong \frac{h_{ob}}{1 + h_{fb}}$$

Common-Base Configuration

$$h_{ib} \cong \frac{h_{ie}}{1 + h_{fe}} \cong \frac{-h_{ic}}{h_{fc}}$$

$$h_{rb} \cong \frac{h_{ie}h_{oe}}{1 + h_{fe}} - h_{re} \cong h_{rc} - 1 - \frac{h_{ic}h_{oc}}{h_{fc}}$$

$$h_{fb} \cong \frac{-h_{fe}}{1 + h_{fe}} \cong \frac{-(1 + h_{fc})}{h_{fc}}$$

$$h_{ob} \cong \frac{h_{oe}}{1 + h_{fe}} \cong \frac{-h_{oc}}{h_{fc}}$$

Common-Collector Configuration

$$h_{ic} \cong \frac{h_{ib}}{1 + h_{fb}}$$

$$h_{rc} \cong 1$$

$$h_{fc} \cong \frac{-1}{1 + h_{fb}}$$

$$h_{oc} \cong \frac{h_{ob}}{1 + h_{fb}}$$

ripple factor and voltage calculations

B.1 RIPPLE FACTOR OF RECTIFIER

The ripple factor of a voltage is defined by

$$r \equiv \frac{\text{rms value of ac component of signal}}{\text{average value of signal}}$$

which can be expressed as

$$r = \frac{V_r\,(\text{rms})}{V_{dc}}$$

Since the ac voltage component of a signal containing a dc level is

$$v_{ac} = v - V_{dc}$$

the rms value of the ac component is

$$V_r\,(\text{rms}) = \left[\frac{1}{2\pi} \int_0^{2\pi} v_{ac}^2\, d\theta\right]^{1/2} = \left[\frac{1}{2\pi} \int_0^{2\pi} (v - V_{dc})^2\, d\theta\right]^{1/2}$$

$$= \left[\frac{1}{2\pi} \int_0^{2\pi} (v^2 - 2vV_{dc} + V_{dc}^2)\, d\theta\right]^{1/2}$$

$$= [V^2\,(\text{rms}) - 2V_{dc}^2 + V_{dc}^2]^{1/2} = [V^2\,(\text{rms}) - V_{dc}^2]^{1/2}$$

where $V(\text{rms})$ is the rms value of the total voltage.

For the half-wave rectified signal

$$V_r\,(\text{rms}) = [V^2\,(\text{rms}) - V_{dc}^2]^{1/2}$$

$$= \left[\left(\frac{V_m}{2}\right)^2 - \left(\frac{V_m}{\pi}\right)^2\right]^{1/2}$$

$$= V_m\left[\left(\frac{1}{2}\right)^2 - \left(\frac{1}{\pi}\right)^2\right]^{1/2}$$

$$\boxed{V_r \text{ (rms)} = 0.385\, V_m, \qquad \text{half-wave}} \tag{B.1}$$

For the full-wave rectified signal

$$V_r \text{ (rms)} = [V^2 \text{ (rms)} - V_{\text{dc}}^2]^{1/2}$$

$$= \left[\left(\frac{V_m}{\sqrt{2}} \right)^2 - \left(\frac{2V_m}{\pi} \right)^2 \right]^{1/2}$$

$$= V_m \left[\frac{1}{2} - \frac{4}{\pi^2} \right]^{1/2}$$

$$\boxed{V_r \text{ (rms)} = 0.305\, V_m, \qquad \text{full-wave}} \tag{B.2}$$

B.2 RIPPLE VOLTAGE OF CAPACITOR FILTER

Assuming a triangular ripple waveform approximation as shown in Fig. B.1, we can write (see Fig. B.2)

$$V_{\text{dc}} = V_m - \frac{V_r(\text{p-p})}{2} \tag{B.3}$$

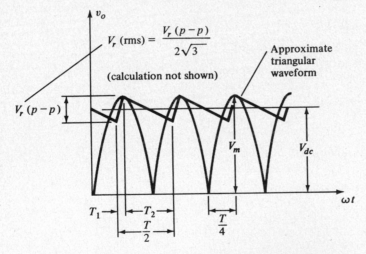

$$V_r \text{ (rms)} = \frac{V_r\,(p-p)}{2\sqrt{3}}$$

(calculation not shown)

Approximate triangular waveform

Figure B.1. Approximate triangular ripple voltage for capacitor filter.

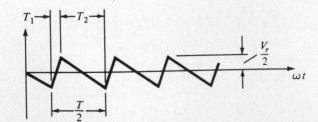

Figure B.2. Ripple voltage.

During capacitor-discharge the voltage change across C is

$$V_r(\text{p-p}) = \frac{I_{dc}T_2}{C} \qquad \text{(B.4)}$$

From the triangular waveform in Fig. B.1

$$V_r(\text{rms}) = \frac{V_r(\text{p-p})}{2\sqrt{3}} \qquad \text{(B.5)}$$

(obtained by calculations, not shown).

Using the waveform details of Fig. B.1 results in

$$\frac{V_r(\text{p-p})}{T_1} = \frac{V_m}{T/4}$$

$$T_1 = \frac{V_r(\text{p-p})(T/4)}{V_m}$$

Also,

$$T_2 = \frac{T}{2} - T_1 = \frac{T}{2} - \frac{V_r(\text{p-p})(T/4)}{V_m} = \frac{2TV_m - V_r(\text{p-p})T}{4V_m}$$

$$T_2 = \frac{2V_m - V_r(\text{p-p})}{V_m}\frac{T}{4} \qquad \text{(B.6)}$$

Since Eq. (B.3) can be written as

$$V_{dc} = \frac{2V_m - V_r(\text{p-p})}{2}$$

we can combine the last equation with B.6

$$T_2 = \frac{V_{dc}}{V_m}\frac{T}{2}$$

which, inserted into Eq. (B.4), gives

$$V_r(\text{p-p}) = \frac{I_{dc}}{C}\left(\frac{V_{dc}}{V_m}\frac{T}{2}\right)$$

$$T = \frac{1}{f}$$

$$V_r(\text{p-p}) = \frac{I_{dc}}{2fC}\frac{V_{dc}}{V_m} \qquad \text{(B.7)}$$

Combining Eqs. (B.5) and (B.7), we solve for $V_r(\text{rms})$

$$\boxed{V_r(\text{rms}) = \frac{V_r(\text{p-p})}{2\sqrt{3}} = \frac{I_{dc}}{4\sqrt{3}fC}\frac{V_{dc}}{V_m}} \qquad \text{(B.8)}$$

B.3 RELATION OF V_{dc} AND V_m TO RIPPLE, r

The dc voltage developed across a filter capacitor from a transformer providing a peak voltage, V_m, can be related to the ripple as follows:

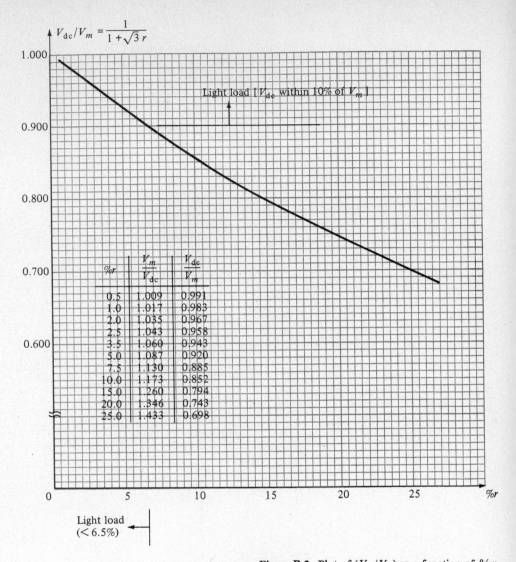

$$V_{dc}/V_m = \frac{1}{1+\sqrt{3}\,r}$$

Light load [V_{dc} within 10% of V_m]

$\%r$	$\dfrac{V_m}{V_{dc}}$	$\dfrac{V_{dc}}{V_m}$
0.5	1.009	0.991
1.0	1.017	0.983
2.0	1.035	0.967
2.5	1.043	0.958
3.5	1.060	0.943
5.0	1.087	0.920
7.5	1.130	0.885
10.0	1.173	0.852
15.0	1.260	0.794
20.0	1.346	0.743
25.0	1.433	0.698

Light load
(< 6.5%)

Figure B.3. Plot of (V_{dc}/V_m) as a function of $\% \, r$.

$$r = \frac{V_r\,(\mathrm{rms})}{V_{dc}} = \frac{\dfrac{V_r(\mathrm{p\text{-}p})}{2\sqrt{3}}}{V_{dc}}$$

$$V_{dc} = \frac{V_r(\mathrm{p\text{-}p})}{\dfrac{2\sqrt{3}}{r}} = \frac{V_r(\mathrm{p\text{-}p})/2}{\sqrt{3}\,r} = \frac{V_r(p)}{\sqrt{3}\,r} = \frac{V_m - V_{dc}}{\sqrt{3}\,r}$$

$$V_m - V_{dc} = \sqrt{3}\,r V_{dc}$$

$$V_m = (1 + \sqrt{3}\,r)V_{dc}$$

$$\boxed{\frac{V_m}{V_{dc}} = 1 + \sqrt{3}\,r} \tag{B.9}$$

The relation of Eq. (B.9) applies to both half- and full-wave rectifier-capacitor filter circuits and is plotted in Fig. B.3. As example, at a ripple of 5% the dc voltage is $V_{dc} = 0.92 V_m$, or within 10% of the peak voltage, where as at 20% ripple the dc voltage drops to only $0.74 V_m$ which is more than 25% less than the peak value. Note that V_{dc} is within 10% of V_m for ripple less than 6.5%. This amount of ripple represents the borderline of the light-load condition.

B.4 RELATION OF V_r (RMS) AND V_m TO RIPPLE, r

We can also obtain a relation between V_r (rms) V_m, and the amount of ripple for both half-wave and full-wave rectifier-capacitor filter circuits as follows:

$$\frac{V_r(\text{p-p})}{2} = V_m - V_{dc}$$

$$\frac{V_r(\text{p-p})/2}{V_m} = \frac{V_m - V_{dc}}{V_m} = 1 - \frac{V_{dc}}{V_m}$$

$$\frac{\sqrt{3} V_r (\text{rms})}{V_m} = 1 - \frac{V_{dc}}{V_m}$$

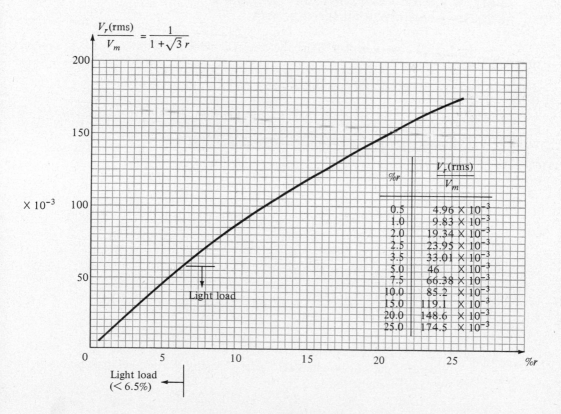

$$\frac{V_r(\text{rms})}{V_m} = \frac{1}{1 + \sqrt{3}\, r}$$

%r	$\frac{V_r(\text{rms})}{V_m}$
0.5	4.96×10^{-3}
1.0	9.83×10^{-3}
2.0	19.34×10^{-3}
2.5	23.95×10^{-3}
3.5	33.01×10^{-3}
5.0	46×10^{-3}
7.5	66.38×10^{-3}
10.0	85.2×10^{-3}
15.0	119.1×10^{-3}
20.0	148.6×10^{-3}
25.0	174.5×10^{-3}

Light load

Light load (< 6.5%)

Figure B.4. Plot of V_r (rms)/V_m as a function of % r.

APP. B RIPPLE FACTOR AND VOLTAGE CALCULATIONS

683

Using Eq. (B.9), we get

$$\frac{\sqrt{3}\,V_r\,(\text{rms})}{V_m} = 1 - \frac{1}{1 + \sqrt{3}\,r}$$

$$\frac{V_r\,(\text{rms})}{V_m} = \frac{1}{\sqrt{3}}\left(1 - \frac{1}{1 + \sqrt{3}\,r}\right) = \frac{1}{\sqrt{3}}\left(\frac{1 + \sqrt{3}\,r - 1}{1 + \sqrt{3}\,r}\right)$$

$$\boxed{\frac{V_r\,(\text{rms})}{V_m} = \frac{r}{1 + \sqrt{3}\,r}} \tag{B.10}$$

Equation (B.10) is plotted in Fig. B.4.

Since V_{dc} is within 10% of V_m for ripple $\leq 6.5\%$,

$$\frac{V_r\,(\text{rms})}{V_m} \simeq \frac{V_r\,(\text{rms})}{V_{\text{dc}}} = r \qquad \text{light load}$$

and we can use $V_r\,(\text{rms})/V_m = r$ for ripple $\leq 6.5\%$.

B.5 RELATION BETWEEN CONDUCTION ANGLE, % RIPPLE, AND $I_{\text{peak}}/I_{\text{dc}}$ FOR RECTIFIER—CAPACITOR FILTER CIRCUITS.

In Fig. B.1 we can determine the angle at which the diode starts to conduct, θ, as follows: Since

$$v = V_m \sin\theta = V_m - V_r(\text{p-p}) \quad \text{at} \quad \theta = \theta_1$$

$$\theta_1 = \sin^{-1}\left[1 - \frac{V_r(\text{p-p})}{V_m}\right]$$

Using Eq. (B.10) and $V_r\,(\text{rms}) = V_r(\text{p-p})/2\sqrt{3}$ gives

$$\frac{V_r(\text{p-p})}{V_m} = \frac{2\sqrt{3}\,V_r\,(\text{rms})}{V_m}$$

so that

$$1 - \frac{V_r(\text{p-p})}{V_m} = 1 - \frac{2\sqrt{3}\,V_r\,(\text{rms})}{V_m} = 1 - 2\sqrt{3}\left(\frac{r}{1 + \sqrt{3}\,r}\right)$$

$$= \frac{1 - \sqrt{3}\,r}{1 + \sqrt{3}\,r}$$

and

$$\boxed{\theta_1 = \sin^{-1}\frac{1 - \sqrt{3}\,r}{1 + \sqrt{3}\,r}} \tag{B.11}$$

where θ_1 is the angle at which conduction starts.

When the current becomes zero after charging the parallel impedances R_L and C, we can determine that

$$\theta_2 = \pi - \tan^{-1}\omega R_L C$$

An expression for $\omega R_L C$ can be obtained as follows:

$$r = \frac{V_r\,(\text{rms})}{V_{dc}} = \frac{\dfrac{I_{dc}}{4\sqrt{3}\,fC} \cdot \dfrac{V_{dc}}{V_m}}{V_{dc}} = \frac{V_{dc}/R_L}{4\sqrt{3}\,fC} \cdot \frac{1}{V_m}$$

$$= \frac{V_{dc}/V_m}{4\sqrt{3}\,fCR_L} = \frac{2\pi\left(\dfrac{1}{1+\sqrt{3}\,r}\right)}{4\sqrt{3}\,\omega CR_L}$$

so that

$$\omega R_L C = \frac{2\pi}{4\sqrt{3}\,(1+\sqrt{3}\,r)r} = \frac{1.814}{r(1+\sqrt{3}\,r)}$$

Thus, conduction stops at an angle

$$\boxed{\theta_2 = \pi - \tan^{-1}\frac{1.814}{(1+\sqrt{3}\,r)r}} \tag{B.12}$$

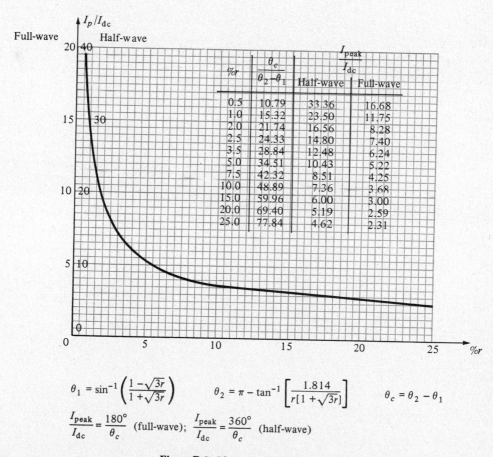

%r	$\dfrac{\theta_c}{\theta_2-\theta_1}$	$\dfrac{I_{peak}}{I_{dc}}$ Half-wave	Full-wave
0.5	10.79	33.36	16.68
1.0	15.32	23.50	11.75
2.0	21.74	16.56	8.28
2.5	24.33	14.80	7.40
3.5	28.84	12.48	6.24
5.0	34.51	10.43	5.22
7.5	42.32	8.51	4.25
10.0	48.89	7.36	3.68
15.0	59.96	6.00	3.00
20.0	69.40	5.19	2.59
25.0	77.84	4.62	2.31

$$\theta_1 = \sin^{-1}\left(\frac{1-\sqrt{3r}}{1+\sqrt{3r}}\right) \qquad \theta_2 = \pi - \tan^{-1}\left[\frac{1.814}{r[1+\sqrt{3r}]}\right] \qquad \theta_c = \theta_2 - \theta_1$$

$$\frac{I_{peak}}{I_{dc}} = \frac{180°}{\theta_c}\ (\text{full-wave}); \quad \frac{I_{peak}}{I_{dc}} = \frac{360°}{\theta_c}\ (\text{half-wave})$$

Figure B.5. Plot of I_p/I_{dc} versus % r, half and full-wave operation.

APP. B RIPPLE FACTOR AND VOLTAGE CALCULATIONS

From Eq. (2.10b) we can write

$$\frac{I_{\text{peak}}}{I_{\text{dc}}} = \frac{I_p}{I_{\text{dc}}} = \frac{T}{T_1} = \frac{180°}{\theta} \qquad \text{full-wave} \tag{B.13a}$$

$$= \frac{360°}{\theta} \qquad \text{full-wave}$$

A plot of I_p/I_{dc} as a function of ripple is provided in Fig. B.5 for both half- and full-wave operation.

charts and tables

TABLE C.1

Greek Alphabet and Common Designations

NAME	CAPITAL	LOWER CASE	USED TO DESIGNATE
alpha	A	α	Angles, area, coefficients
beta	B	β	Angles, flux density, coefficients
gamma	Γ	γ	Conductivity, specific gravity
delta	Δ	δ	Variation, density
epsilon	E	ϵ	Base of natural logarithms
zeta	Z	ζ	Impedance, coefficients, coordinates
eta	H	η	Hysteresis coefficient, efficiency
theta	Θ	θ	Temperature, phase angle
iota	I	ι	
kappa	K	κ	Dielectric constant, susceptibility
lambda	Λ	λ	Wave length
mu	M	μ	Micro, amplification factor, permeability
nu	N	ν	Reluctivity
xi	Ξ	ξ	
omicron	O	o	
pi	Π	π	Ratio of circumference to diameter = 3.1416
rho	P	ρ	Resistivity
sigma	Σ	σ	Sign of summation
tau	T	τ	Time constant, time phase displacement
upsilon	Υ	υ	
phi	Φ	ϕ	Magnetic flux, angles
chi	X	χ	
psi	Ψ	ψ	Dielectric flux, phase difference
omega	Ω	ω	Capital: ohms; lower case: angular velocity

Formulae: $\log ab = \log a + \log b$

$\log \dfrac{a}{b} = \log a - \log b$

TABLE C.2

$\log a^n = n \log a$

Common Logarithms

no.	0	1	2	3	4	5	6	7	8	9
0	....	0000	3010	4771	6021	6990	7782	8451	9031	9542
1	0000	0414	0792	1139	1461	1761	2041	2304	2553	2788
2	3010	3222	3424	3617	3802	3979	4150	4314	4472	4624
3	4771	4914	5051	5185	5315	5441	5563	5682	5798	5911
4	6021	6128	6232	6335	6435	6532	6628	6721	6812	6902
5	6990	7076	7160	7243	7324	7404	7482	7559	7634	7709
6	7782	7853	7924	7993	8062	8129	8195	8261	8325	8388
7	8451	8513	8573	8633	8692	8751	8808	8865	8921	8976
8	9031	9085	9138	9191	9243	9294	9345	9395	9445	9494
9	9542	9590	9638	9685	9731	9777	9823	9868	9912	9956
10	0000	0043	0086	0128	0170	0212	0253	0294	0334	0374
11	0414	0453	0492	0531	0569	0607	0645	0682	0719	0755
12	0792	0828	0864	0899	0934	0969	1004	1038	1072	1106
13	1139	1173	1206	1239	1271	1303	1335	1367	1399	1430
14	1461	1492	1523	1553	1584	1614	1644	1673	1703	1732
15	1761	1790	1818	1847	1875	1903	1931	1959	1987	2014
16	2041	2068	2095	2122	2148	2175	2201	2227	2253	2279
17	2304	2330	2355	2380	2405	2430	2455	2480	2504	2529
18	2553	2577	2601	2625	2648	2672	2695	2718	2742	2765
19	2788	2810	2833	2856	2878	2900	2923	2945	2967	2989
20	3010	3032	3054	3075	3096	3118	3139	3160	3181	3201
21	3222	3243	3263	3284	3304	3324	3345	3365	3385	3404
22	3424	3444	3464	3483	3502	3522	3541	3560	3579	3598
23	3617	3636	3655	3674	3692	3711	3729	3747	3766	3784
24	3802	3820	3838	3856	3874	3892	3909	3927	3945	3962
25	3979	3997	4014	4031	4048	4065	4082	4099	4116	4133
26	4150	4166	4183	4200	4216	4232	4249	4265	4281	4298
27	4314	4330	4346	4362	4378	4393	4409	4425	4440	4456
28	4472	4487	4502	4518	4533	4548	4564	4579	4594	4609
29	4624	4639	4654	4669	4683	4698	4713	4728	4742	4757
30	4771	4786	4800	4814	4829	4843	4857	4871	4886	4900
31	4914	4928	4942	4955	4969	4983	4997	5011	5024	5038
32	5051	5065	5079	5092	5105	5119	5132	5145	5159	5172
33	5185	5198	5211	5224	5237	5250	5263	5276	5289	5302
34	5315	5328	5340	5353	5366	5378	5391	5403	5416	5428
35	5441	5453	5465	5478	5490	5502	5514	5527	5539	5551
36	5563	5575	5587	5599	5611	5623	5635	5647	5658	5670
37	5682	5694	5705	5717	5729	5740	5752	5763	5775	5786
38	5798	5809	5821	5832	5843	5855	5866	5877	5888	5899
39	5911	5922	5933	5944	5955	5966	5977	5988	5999	6010
40	6021	6031	6042	6053	6064	6075	6085	6096	6107	6117
41	6128	6138	6149	6160	6170	6180	6191	6201	6212	6222
42	6232	6243	6253	6263	6274	6284	6294	6304	6314	6325
43	6335	6345	6355	6365	6375	6385	6395	6405	6415	6425
44	6435	6444	6454	6464	6474	6494	6493	6503	6513	6522
45	6532	6542	6551	6561	6571	6580	6590	6599	6609	6618
46	6628	6637	6646	6656	6665	6675	6684	6693	6702	6712
47	6721	6730	6739	6749	6758	6767	6776	6785	6794	6803
48	6812	6821	6830	6839	6848	6857	6866	6875	6884	6893
49	6902	6911	6920	6928	6937	6946	6955	6964	6972	6981
50	6990	6998	7007	7016	7024	7033	7042	7050	7059	7067
no.	0	1	2	3	4	5	6	7	8	9

TABLE C.2

Common Logarithms (continued)

no.	0	1	2	3	4	5	6	7	8	9
50	6990	6998	7007	7016	7024	7033	7042	7050	7059	7067
51	7076	7084	7093	7101	7110	7118	7126	7135	7143	7152
52	7160	7168	7177	7185	7193	7202	7210	7218	7226	7235
53	7243	7251	7259	7267	7275	7284	7292	7300	7308	7316
54	7324	7332	7340	7348	7356	7364	7372	7380	7388	7396
55	7404	7412	7419	7427	7435	7443	7451	7459	7466	7474
56	7482	7490	7497	7505	7513	7520	7528	7536	7543	7551
57	7559	7566	7574	7582	7589	7597	7604	7612	7619	7627
58	7634	7642	7649	7657	7664	7672	7679	7686	7694	7701
59	7709	7716	7723	7731	7738	7745	7752	7760	7767	7774
60	7782	7789	7796	7803	7810	7818	7825	7832	7839	7846
61	7853	7860	7868	7875	7882	7889	7895	7903	7910	7917
62	7924	7931	7938	7945	7952	7959	7966	7973	7980	7987
63	7993	8000	8007	8014	8021	8028	8035	8041	8048	8055
64	8062	8069	8075	8082	8089	8096	8102	8109	8116	8122
65	8129	8136	8142	8149	8156	8162	8169	8176	8182	8189
66	8195	8202	8209	8215	8222	8228	8235	8241	8248	8254
67	8261	8267	8274	8280	8287	8293	8299	8306	8312	8319
68	8325	8331	8338	8344	8351	8357	8363	8370	8376	8382
69	8388	8395	8401	8407	8414	8420	8426	8432	8439	8445
70	8451	8457	8463	8470	8476	8482	8488	8494	8500	8506
71	8513	8519	8525	8531	8537	8543	8549	8555	8561	8567
72	8573	8579	8585	8591	8597	8603	8609	8615	8621	8627
73	8633	8639	8645	8651	8657	8663	8669	8675	8681	8686
74	8692	8698	8704	8710	8716	8722	8727	8733	8739	8745
75	8751	8756	8762	8768	8774	8779	8785	8791	8797	8802
76	8808	8814	8820	8825	8831	8837	8842	8848	8854	8859
77	8865	8871	8876	8882	8887	8893	8899	8904	8910	8915
78	8921	8927	8932	8938	8943	8949	8954	8960	8965	8971
79	8976	8982	8987	8993	8998	9004	9009	9015	9020	9025
80	9031	9036	9042	9047	9053	9058	9063	9069	9074	9079
81	9085	9090	9096	9101	9106	9112	9117	9122	9128	9133
82	9138	9143	9149	9154	9159	9165	9170	9175	9180	9186
83	9191	9196	9201	9206	9212	9217	9222	9227	9232	9238
84	9243	9248	9253	9258	9263	9269	9274	9279	9284	9289
85	9294	9299	9304	9309	9315	9320	9235	9330	9335	9340
86	9345	9350	9355	9360	9365	9370	9375	9380	9385	9390
87	9395	9400	9405	9410	9415	9420	9425	9430	9435	9440
88	9445	9450	9455	9460	9465	9469	9474	9479	9484	9489
89	9494	9499	9504	9509	9513	9518	9523	9528	9533	9538
90	9542	9547	9552	9557	9562	9566	9571	9576	9581	9586
91	9590	9595	9600	9605	9609	9614	9619	9624	9628	9633
92	9638	9643	9647	9652	9657	9661	9666	9671	9675	9680
93	9685	9689	9694	9699	9703	9708	9713	9717	9722	9727
94	9731	9736	9741	9745	9750	9754	9759	9763	9768	9773
95	9777	9782	9786	9791	9795	9800	9805	9809	9814	9818
96	9823	9827	9832	9836	9841	9845	9850	9854	9859	9863
97	9868	9872	9877	9881	9886	9890	9894	9899	9903	9908
98	9912	9917	9921	9926	9930	9934	9939	9943	9948	9952
99	9956	9961	9965	9969	9974	9978	9983	9987	9991	9996
100	0000	0004	0009	0013	0017	0022	0026	0030	0035	0039
no.	0	1	2	3	4	5	6	7	8	9

answers to selected odd-numbered problems

CHAPTER 1

11. 13.73 mA. **13.** 12 μA. **15.** 200 mW; 10 μW; $\cong$ 6.14 mA.
17. Vacuum-tube: $V_{PK} = 18$ V, $I_{PK} = 3$ mA, $P = 54$ mW; Semiconductor: $v_d \cong 1$ V, $i_d \cong 18$ mA, $P = 18$ mW. **19.** Vacuum-tube: $R_{dc} = 6$ K; Semiconductor: $R_{dc} = 55$ Ω. **21.** Eq. 1.9: $r'_d = 1.3$ Ω; Eq. 1.8: $r_d \cong 2$ Ω. **23.** Vacuum-tube: $r_{av} = 2.2$ K; Semiconductor: $r_{av} = 15$ Ω. **25.** Values of problem 23. **27.** v_o (peak) $= 39.3$ V. **29.** $v_d(p - p) = 25.75$ mV. **41.** (a) R_L(min) $= 244.44$ Ω; (b) I_L(max) $= 90$ mA; (c) I_L(min) $= 4.4$ mA; (d) —; (e) $I_Z = 68$ mA, $I_s = 90$ mA; (f) $P_Z = 1.496$ W. **45.** (a) Difference $= (34 - 8)$ pF $= 26$ pF; (b) (-8 V): $\Delta C/\Delta V_r = (18 - 10)/(10 - 6) = 2$ pF/V; (-2 V): $\Delta C/\Delta V_r = (51 - 33)/(3 - 1) = 9$ pF/V. **47.** (a) $I_{pk_Q} = 2.7$ μA, $V_{pk_Q} \cong 97$ V; (b) $I_{pk_Q} \cong 1$ μA, $V_{pk_Q} \cong 164$ V.

CHAPTER 2

1. 28.6 V. **3.** 566 V, PIV. **5.** $V_{dc} = 80$ V; PIV $= 251.6$ V. **7.** $V_{dc} = 57.2$ V. **9.** PIV $= 251.6$ V. **13.** (a) Transformer $V_m = 314$ V, diode PIV $= 314$ V, turns ratio $(N_2 : N_1) = 1.85 : 1$; (b) transformer $V_m = 314$ V, diode PIV $= 314$ V, turns ratio $(N_2 : N_1) = 0.93 : 1$; (c) transformer $V_m = 157$ V, diode PIV $= 314$ V, turns ratio $(N_2 : N_1) = 0.93 : 1$. **15.** 0.143. **17.** 38.4 V. **19.** 0.067 ($= 6.7\%$). **21.** 0.0718 ($= 7.18\%$). **23.** 37.5 μF. **25.** 2.3%. **27.** 1.6 V. **29.** 72.7 mA. **31.** 32%. **33.** $Z_l = 6$ K, $X_{C_2} \| R_L = 50$ Ω. **35.** 0.885 V. **37.** $C_2 = 22.5$ μF, $L = 2.78$ H.

CHAPTER 3

9. $I_C = 7.92$ mA.   11. $A_v = 25$.   13. (a) $I_C \cong 5$ mA; (b) $I_C \cong I_E \cong 4$ mA;
(c) $V_{EB} \cong 725$ mV.   17. (a) $\beta = 118$; (b) $\alpha = 0.992$; (c) $I_{CEO} \cong 350$ μA;
(d) $I_{CBO} \cong 2.966$ μA.   27. $\Delta I_p \cong (5.2 - 5)$ mA $= 0.2$ mA.

CHAPTER 4

1. $I_E = 1.81$ mA, $I_C = 1.78$ mA, $V_{CB} = -2.1$ V.   3. $I_B = 55.3$ μA, $I_C =$
2.5 mA, $V_{CE} = 3.75$ V.   5. $I_B = 194.5$ μA, $I_C = 10.7$ mA, $V_{CE} = 4.6$ V.
7. $V_B = 1.86$ V, $V_E = 1.16$ V, $I_E = 1.55$ mA, $V_C = 13.5$ V, $V_{CE} = 12.34$ V.
9. $I_B = 23.25$ μA, $I_C = 1.4$ mA, $V_{CE} = 3$ V.   11. $I_B = 45$ μA, $I_C = 1.8$ mA,
$V_{CE} = 9.05$ V.   13. 10.7 V.   15. (b) $V_{CE_Q} = 10$ V, $I_{C_Q} = 2.5$ mA; (c) V_{CE_Q}
$= 1$ V, $I_{C_Q} = 2.4$ mA; (d) $V_{CE_Q} = 7.2$ V, $I_{C_Q} = 1.8$ mA.   17. $V_{PK_Q} \cong 100$ V,
$I_{P_Q} \cong 5.5$ mA.   19. $V_{PK_Q} \cong 285$ V, $I_{P_Q} \cong 7$ mA.   21. $R_C = 6$ K, $R_B \cong$
772 K.   23. $R_E = 300$ Ω, $R_C = 700$ Ω, $R_B = 657.5$ K.   25. 10.5 V.
27. 3.6 mA.

CHAPTER 5

1. (a) $h_{11} = 2$ Ω, $h_{12} = \frac{2}{3}$, $h_{21} = -\frac{2}{3}$, $h_{22} = \frac{4}{9}$.   3. $h_{fe} = 110$ vs. 100, $h_{oe} =$
40 μA/V vs. 33 μA/V.   5. Greatest change $= h_{re}$, least $= h_{fe}$.   7. (a) 55.56;
(b) -560; (c) $\cong 397.78$ Ω; (d) 3.9 K; (e) 3.11×10^4.   9. (a) 79.78; (b)
-276; (c) 1.126 K; (d) 3.87 K; (e) 22×10^3.   11. (a) 60 vs. 55.56; (b)
-571.43 vs. -560; (c) 420 vs. 397.78 Ω; (d) 4 K vs. 3.9 K; (e) 3.43×10^4 vs.
3.11×10^4.   13. (a) 85 vs. 79.78; (b) -278.57 vs. -276; (c) 1.19 K vs.
1.129 K; (d) 3.9 K vs. 3.87 K; (e) 23.7×10^3 vs. 22×10^3.   15. (a) -23.4;
(b) $Z_i \cong 2.35$ K, $Z_o \cong 1.1$ K; (c) -16.4.   17. $Z_i \cong 122.45$ K, $Z_o \cong 5.6$ K,
$A_v = -4.67$.   19. $Z_i = 62.86$ K, $Z_o = 18.8$ Ω, $A_v \cong 1$, $A_i \cong 19$.   21. $r_e \cong$
7 Ω, $\beta_{r_e} = h_{ie} = 0.42$ K, same results as #7.   23. $r_e \cong 15.04$ Ω, $\beta_{r_e} = h_{ie} =$
3 K, same results as #17.   25. $r_e \cong 19.03$ Ω, $\beta_{r_e} = h_{ie} = 1.9$ K, same results
as #19.   27. $A_v = 14.55$, $A_i = 0.25$, $Z_i = 52.8$ K, $Z_o \cong 12$ K.   29. $A_v =$
R_C/R_{E_1}, $Z_i \cong R_{B_1} \| R_{B_2} \| (h_{ie} + h_{fe}R_{E_1})$, $Z_o \cong R_C$.   31. $A_v = -3.09$, $Z_i \cong$
49.5 K, $A_i \cong -22.5$.   33. $r_p \cong 11.36$ K, $g_m \cong 2.15 \times 10^{-3}$, $\mu \cong 22.5$.   35.
$A_v = 0.91$, $Z_o = 609.76$ Ω.

CHAPTER 6

1. $I_{D_Q} = 0.2$ mA, $V_{DS_Q} = 11.5$ V.   3. $V_{GG} = V_{GS} \cong -0.8$ V.   5. $V_{GS_Q} =$
-2.3 V, $I_{D_Q} = 1.9$ mA.   7. $m \cong 0.42$, $V_{GS_Q} = -1.6$ V, $I_{D_Q} = 1.3$ mA.   9.
$g_m = 1,500$ μmhos.   11. A_v -3.3.   13. $V_o = -370$ mV.   15. (a) 177 K;
(b) 3.5K.   17. (a) 13.3 K; (b) 6.86 K.   19. $R_D = 4.4$ K; $R_S = 1.5$ K; $R_G =$
1M   21. $R_D = 25$ K, $R_{S1} = 250$ Ω, $R_{S2} = 1$ K, $R_G = 1$M.

CHAPTER 7

1. (a) $|A_{i_T}| = 80$, $|A_{v_T}| = 160$; (b) $|A_v| = 12.65$, $|A_i| = 8.94$. **3.** $Z_i = 1.09$ K, $Z_o \cong 3.3$ K, $A_{v_T} = 9220.4$, $A_{i_T} = 4568.29$, $A_{p_T} = 42.1 \times 10^6$. **5.** $Z_i = 13.85$ K, $Z_o \cong 24.51$ Ω, $A_{v_T} \cong -2$, $A_{i_T} \cong -27.7$. **7.** $Z_i = 1.19$ K, $Z_o \cong 10$ K, $A_{v_T} = 16.64 \times 10^3$, $A_{i_T} = 1.98 \times 10^3$. $A_{p_T} = 32.95 \times 10^6$. **9.** $A_{v_1} = -11.278$, $A_{v_2} = -2.318$, $A_{v_T} = 26.142$. **11.** $a = 20$. **13.** $V_{E_2} = 11$ V, $V_{B_2} = V_{C_1} = 11.7$ V, $V_{B_1} = 2.416$ V; $r_{e_2} = 2.6$ Ω, $r_{e_1} = 18.18$ Ω; ac gain unaffected. **15.** (a) $r_{e_1} = 5.28$ Ω, $r_{e_2} = 6.13$ Ω; (b) $A_{VT} = 473.08$, $V_o = 4.73$ V; (c) $Z_i = 0.239$ K, $Z_o \cong 2.5$ K. **17.** $A_i = 1783.38$, $Z_i = 0.89$ M, $Z_o \cong 2.2$ K, $A_v = 4.4$. **19.** $A_i = 2716$, $Z_i = 0.44$ M, $Z_o \cong 2.2$ K, $A_v = 13.5$. **21.** (a) 13 dB; (b) 13 dB; (c) $\cong 7$ dB. **23.** 67.96 dB. **25.** (a) $f_{L_S} = 85.61$ Hz, $f_{L_C} = 35.78$ Hz, $f_{L_E} = 129.9$ Hz; (b) -103.8; (c) $f_{H_i} = 1.738$ MHz, $f_{H_o} = 8.52$ MHz. **27.** (a) $f_{L_S} = 85.61$ Hz, $f_{L_C} = 35.78$ Hz, $f_{L_E} = 6.5 \times 10^3$ Hz; (b) -103.8; (c) $f_{H_i} = 1.738$ MHz, $f_{H_o} = 8.52$ MHz. **29.** $f_2' = 1.1$ MHz. **31.** (a) $f_{L_S} = 199$ Hz, $f_{L_C} = 5.475$ Hz, $f_{L_E} = 7.077$ Hz; (b) first stage: $f_{H_i} = 0.726$ MHz, $f_{H_o} = 0.651$ MHz, second stage: $f_{H_o} = 11.1$ MHz; (c) 468.77. **33.** $f_1 = 1.58$ Hz, $f_2 = 88.9$ kHz. **35.** (a) $f_1 = 159.23$ Hz, $f_S = 7.494$ Hz, $f_2 = 0.833$ MHz; (b) 31.8 nF; (c) $\cong 57.8$ mV.

CHAPTER 8

1. 2.5 K. **3.** 44.7 :1. **5.** 37%. **7.** (a) 42.3%; (b) 64.2%. **15.** (a) 56.25 W; (b) 84.25 W; (c) %η = 66.8%; (d) $P_t = 14$ W. **17.** 25 W. **19.** 4.375 W.

CHAPTER 9

5. (a) yes; (b) no; (c) no; (d) 6 V, 800 mA—excellent, 4 V, 1.6 A-no. **11.** (a) $\cong 0.7$ mW/cm²; (b) $((0.8 - 0.15)/0.8) \times 100\% = 81.25\%$. **15.** 1.53 MΩ > $R_1 > 4.875$ K. **19.** $I_B = 27$ μA, $I_C = 1.08$ mA.

CHAPTER 11

3. $V_E = 0.7$ V, $I_E = 0.955$ mA, $I_{E_1} = I_{E_2} = 0.48$ mA, $V_{C_1} = 7.8$ V. **5.** $R_i = 3.6$ K, $R_o = 13.3$ K. **7.** 1.3 mA, 15.9 V **9.** 10.7 K. **13.** 0 V. **15.** -12 V.

CHAPTER 12

1. 10. **3.** $A_f = 14.3$, $R_{if} = 31.4$ K, $R_{of} = 2.38$ K. **5.** $A = 33.4$, $A_f \cong 1$. **7.** Without feedback; $A = 450$, $R_i = 2$ K, $R_o = 12$ K; with feedback: $A_f = 9.8$, $R_{if} = 92$ K, $R_{of} = 12$ K. **9.** $C = 1250$ pF, $h_{fe} = 45$. **11.** $f_o = 1.05$ MHz. **13.** $f_o = 159$ kHz.

CHAPTER 13

3. (a) 0 V; (b) $+5$ V; (c) -5 V.

CHAPTER 14

1. 5.26%. **3.** 1.01%. **5.** 3.06 K. **7.** (a) 9.3 V; (b) 2.33 mA; (c) 4.65 mA; (d) 46.4 μA.

CHAPTER 15

1. (a) $D = 0.6$ in.; (b) $D = -1.5$ in. **3.** $G = 40$ V/cm, $D = 3$ cm. **5.** $D_2 = 4$ cm, $V_d = 133.3$ V. **9.** $V_p = 2.05$ V, $V_{rms} = 1.45$ V. **11.** $V_{p-p} = 270$ mV, $V_{rms} = 95.6$ mV. **13.** (a) $T_{pw} = 32$ msec; (b) $T_{pw} = 6.4$ msec; (c) $T_{pw} = 12.8$ μsec. **15.** $f = 31.8$ kHz. **17.** 7.5 cm. **19.** (a) $V_{p-p} = 460$ μV; (b) 190 μsec; (c) $f = 10.5$ kHz. **21.** (a) Scale factor $= 45°$/cm, phase shift $= 22.5°$; (b) 0.89 cm. **23.** (a) $36.7°$; (b) $23.6°$; (c) $90°$.

index

S

T